GLOBAL URBANISM

Global Urbanism is an experimental examination of how urban scholars and activists make sense of, and act upon, the foundational relationship between the 'global' and the 'urban'.

What does it mean to say that we live in a global-urban moment, and what are its implications? Refusing all-encompassing answers, the book grounds this question, exploring the plurality of understandings, definitions, and ways of researching global urbanism through the lenses of varied contributors from different parts of the world. The contributors explore what global urbanism means to them, in their context, from the ground and the struggles upon which they are working and living. The book argues for an incremental, fragile and in-the-making emancipatory urban thinking. The contributions provide the resources to help make sense of what global urbanism is in its varieties, what's at stake in it, how to research it, and what needs to change for more progressive urban futures. It provides a heterodox set of approaches and theorisations to probe and provoke rather than aiming to draw a line under a complex, changing and profoundly contested set of global-urban processes.

Global Urbanism is primarily intended for scholars and graduate students in geography, sociology, planning, anthropology and the field of urban studies, for whom it will provide an invaluable and up-to-date guide to current thinking across the range of disciplines and practices which converge in the study of urbanism.

Michele Lancione is Professor of Geography at the Polytechnic of Turin, Italy, and Visiting Professor of Urban Studies at the University of Sheffield, United Kingdom. He is a member of the Common Front for Housing Rights (Bucharest), co-founder and editor of the *Radical Housing Journal* and corresponding editor at *IJURR*. His work focuses on radical forms of inhabitation and housing struggles (through a five-year European Research Council project) and the politics of life at the margins in the contemporary urban.

Colin McFarlane is Professor of Geography at Durham University, United Kingdom. His current work is on the politics and experience of urban densities (through a European Research Council project), the relationship between urban waste and life in the city, thinking the city through the idea of the fragment and the potentials for urban equalities (through a Global Challenges Research Fund project led by University College London).

GLOBAL URBANISM

Knowledge, Power and the City

Edited by Michele Lancione and Colin McFarlane

Routledge
Taylor & Francis Group
LONDON AND NEW YORK

First published 2021
by Routledge
2 Park Square, Milton Park, Abingdon, Oxon OX14 4RN

and by Routledge
52 Vanderbilt Avenue, New York, NY 10017

Routledge is an imprint of the Taylor & Francis Group, an informa business

British Library Cataloguing-in-Publication Data
A catalogue record for this book is available from the British Library

Library of Congress Cataloging-in-Publication Data
Names: Lancione, Michele, editor. | McFarlane, Colin, 1979- editor.
Title: Global urbanism : knowledge, power and the city / edited by Michele Lancione and Colin McFarlane.
Description: Abingdon, Oxon ; New York, NY : Routledge, 2021. | Includes bibliographical references and index.
Identifiers: LCCN 2020053290 (print) | LCCN 2020053291 (ebook) | ISBN 9780367200961 (hardback) | ISBN 9780367745349 (paperback) | ISBN 9780429259593 (ebook)
Subjects: LCSH: Urbanization. | Sociology, Urban. | Cities and towns--Growth.
Classification: LCC HT361 .G568 2021 (print) | LCC HT361 (ebook) | DDC 307.76--dc23
LC record available at https://lccn.loc.gov/2020053290
LC ebook record available at https://lccn.loc.gov/2020053291

ISBN: 978-0-367-20096-1 (hbk)
ISBN: 978-0-367-74534-9 (pbk)
ISBN: 978-0-429-25959-3 (ebk)

DOI: 10.4324/9780429259593

Typeset in Bembo
by SPi Global, India

CONTENTS

Contributors *ix*
Acknowledgements *xviii*

PART I
Introduction **1**

1 Navigating the global urban 3
Michele Lancione and Colin McFarlane

PART II
Rethinking global urbanisms **15**

2 Thinking urban grammars: An interview with Ash Amin 17

3 Decentering global urbanism: An interview with Ananya Roy 25

4 Hinterlands of the Capitalocene 34
Neil Brenner and Nikos Katsikis

5 Making space for queer desire in global urbanism 49
Gavin Brown and Dhiren Borisa

6 Seeing like an Italian city: Questioning global urbanism from an "in-between space" in Turin 56
Francesca Governa

7 Theorising from where? Reflections on de-centring global (southern) urbanism 62
Hyun Bang Shin

8 Globalizing postsocialist urbanism 71
Liviu Chelcea, Slavomíra Ferenčuhová and Gruia Bădescu

9 Beyond the noosphere? Northern England's 'left behind' urbanism 80
John Flint and Ryan Powell

10 Footnote urbanism: The missing East in (not so) global urbanism 88
Martin Müller

11 Comparative urbanism and global urban studies: Theorising the urban 96
Jennifer Robinson

PART III
Everyday global urbanisms 105

12 Global urbanism inside/out: Thinking through Jakarta 107
Helga Leitner and Eric Sheppard

13 Tiwa's morning 116
Grace Adeniyi-Ogunyankin and Linda Peake

14 "Out there, over the hills, on the other side of the tracks": A horizon of the global urban 124
AbdouMaliq Simone

15 Constructing the South-East Asian ascent: Global vertical urbanisms of brick and sand 131
William Jamieson, Katherine Brickell, Nithya Natarajan and Laurie Parsons

16 Nairobi city, streets and stories: Young lives stay in place while going global through digital stages 140
Tatiana Thieme

17 Rethinking global urbanism from a 'fripe' marketplace in Tunis 148
Katharina Grüneisl

18 Liminal spaces and resistance in Mexico City: Towards an everyday global urbanism 156
Alicia Lindón

19 Death and the city: Necrological notes from Kinshasa 166
Filip De Boeck

20 Pathways toward a dialectical urbanism: Thinking with the contingencies of crisis, care and capitalism 176
Suraya Scheba

21 Global self-urbanism: Self-organisation amidst regulatory crisis and uneven urban citizenship 183
Francesco Chiodelli and Margherita Grazioli

PART IV
Governing global urbanisms **191**

22 Unlocking political potentialities 193
Edgar Pieterse

23 Climate changed urbanism? 203
Harriet Bulkeley, Laura Tozer and Emma Lecavalier

24 The global urban condition and politics of thermal metabolics: The chilling prospect of killer heat 211
Simon Marvin

25 On the deployment of scientific knowledge for a new urbanism of the anthropocene 219
Vanesa Castán Broto

26 Global cities and the bioeconomy of health innovation 227
Donald McNeill

27 Hacking the urban code: Notes on durational imagination in city-making 235
Swati Chattopadhyay

28 Global urbanism: Urban governance innovation in/for a world of cities 243
Pauline McGuirk

29 Corridor urbanism 251
Jonathan Silver

30 Beyond-the-network urbanism: Everyday infrastructures in states of mutation 259
Yaffa Truelove

31 Still construction and already ruin 267
Mariana Cavalcanti

32 The migration of spaces: Monumental urbanism beyond materiality 276
Morten Nielsen

33 Land as situated spatio-histories: A dialogue with global urbanism 284
Wing-Shing Tang and Solomon Benjamin

PART V
Contesting global urbanism **291**

34 Women organising, advocacy and Indian cities in between informal dwelling and informal economies: An interview SEWA's Renana Jhabvala 293

35 From a Neapolitan perspective, reaching out beyond prevailing cultural models: An interview Emma Ferulano 298

36 Urban struggles and theorising from Eastern European cities: A collective interview with Ana Vilenica, Ioana Florea, Veda Popovici and Zsuzsi Pósfai 306

37 Planning, community spaces and youth urban futures: From Accra, in conversation with Victoria Okoye and Yussif Larry Aminu 317

38 A counter-dominant global urbanism? Experiments from Lebanon 325
Mona Harb

39 Living in the city beyond housing: Urbanism of the commons
Belen Desmaison 333

Index *343*

CONTRIBUTORS

Grace Adeniyi-Ogunyankin is Assistant Professor in the Departments of Geography and Planning and Gender Studies at Queen's University. She is a feminist scholar who is interested in African urban futures and Black futurity. She does urban ethnography work concerning issues of gender, spirituality, spatial exclusion and financial inclusion. She also has a research focus on popular culture that explores issues of subjectivity and belonging, as well as the use of Afrofuturism and Afropolitan Imagineering in geographic projects that address the politics of difference.

Ash Amin is Professor of Geography at the University of Cambridge. His recent books include *Land of Strangers* (2013, Polity Press), *Arts of the Political* (with Nigel Thrift, Duke University Press, 2013), *Releasing the Commons* (edited with Philip Howell, Routledge, 2016), *Seeing Like a City* (with Nigel Thrift, Polity Press, 2016) and *European Union and Disunion* (edited with Philip Lewis, British Academy, 2017). His current work is on ecologies of mental health in Shanghai and Delhi, on grammars of the urban (with Michele Lancione) and on post-identity politics.

Yussif Larry Aminu is founder and executive director of Spread-Out Initiative NGO, based in Nima (Accra, Ghana). Spread-Out Initiative uses creative arts, technology and entrepreneurship to support children and youth from diverse backgrounds in Nima to realise their potential and attain their dreams of a better future. Larry Aminu, a Pan-African and philanthropist, believes in preserving local history and culture through community art. He has worked with international researchers and university professors, as well as assisted international student and volunteer programmes in Nima-Mamobi and within West Africa.

Gruia Bădescu is an Alexander von Humboldt research fellow at the University of Konstanz. His research examines the relationship between urban interventions and political ruptures, particularly in postsocialist and post-war contexts. His work has focused on Central and Eastern Europe, as well as its dialogues with other regions, specifically South America and the Middle East. He is currently finalising a book on urban reconstruction after war and political violence.

Hyun Bang Shin is Professor of Geography and Urban Studies at the London School of Economics and Political Science and directs the LSE Saw Swee Hock Southeast Asia Centre. His research centres on the critical analysis of the political economy of speculative urbanisation, gentrification and displacement; urban spectacles; and urbanism, with particular attention to Asian cities. He is a trustee of the Urban Studies Foundation and editor of the *International Journal of Urban and Regional Research.*

Solomon Benjamin is Associate Professor at the Humanities and Social Science Department of the Indian Institute of Technology Madras, India. Two interconnected themes explore metro and non-metro contexts: non-teleological situated histories of land tenures creating and entangling varied realms – from bureaucratic realms in ideas of sacredness shaping spatialities of territorial transformation. Second, researching smartphone re-engineering connecting China's Shenzhen to various 'China Bazaars' in Indian cities and towns helps explore the dialectics of economy whose societal embedding operate transnationally.

Filip De Boeck is Professor of Anthropology at the University of Leuven, writer, filmmaker and curator. He lives and works between Brussels and the Democratic Republic of Congo (DRC). Over the past 30 years, he has conducted extensive field research in both rural and urban communities in the DRC. Co-authored with photographer Sammy Baloji, his most recent book is *Suturing the City. Living Together in Congo's Urban Worlds* (London: Autograph ABP, 2016). De Boeck has also curated several international exhibitions and is the author of a long documentary film about Kinshasa.

Dhiren Borisa is Assistant Professor at Jindal Global Law School, India, and an Honorary Visiting Fellow at School of Geography, Geology and the Environment, University of Leicester, United Kingdom. An urban social geographer by training with a PhD from Jawaharlal Nehru University, New Delhi, their research pushes for decolonising and de-brahmanising knowledges through intersectional frameworks from the Global South. Their research interests lie at the intersections of caste and queerness and how they shape cartographies of everyday survival in Indian cities.

Neil Brenner is Lucy Flower Professor of Urban Sociology at the University of Chicago. His writing and teaching focus on the theoretical, conceptual and methodological dimensions of urban questions and on the challenges of reinventing our approach to urbanisation in relation to the crises, contradictions and struggles of our time. Brenner's most recent books are *New Urban Spaces: Urban Theory and the Scale Question* (Oxford, 2019) and *Critique of Urbanization: Selected Essays* (Bauwelt Fundamente, 2016).

Katherine Brickell is Professor of Human Geography at Royal Holloway, University of London. Her feminist-oriented research cross-cuts social, political, development and legal geography, with a focus on the domestic sphere as a precarious space of contemporary life. Katherine's scholarship aims to reaffirm and reprioritise the home as a political entity foundational to the concerns of human geography. Her monograph *Home SOS: Gender, Violence and Survival in Crisis Ordinary Cambodia* (Wiley 2020) brings domestic violence and forced eviction into twin view to achieve this.

Gavin Brown is Professor of Political Geography and Sexualities in the School of Geography, Geology and the Environment, University of Leicester, United Kingdom. He has wide-ranging research interests spanning critical geopolitics, protest and solidarity; diverse economies; and geographies of sexualities. He is currently preoccupied with thinking about lesbian, gay, bisexual, transgender generations and exploring how new generation HIV treatment and prevention technologies are reshaping gay men's lives. He tweets as @lestageog.

Harriet Bulkeley holds joint appointments as Professor in the Department of Geography, Durham University, and at the Copernicus Institute of Sustainable Development, Utrecht University. Her research focuses on environmental governance and the politics of climate change, energy and sustainable cities. She currently coordinates the H2020 NATURVATION project examining the role of urban innovation with nature-based solutions for sustainable development and is a co-investigator on the H2020 REINVENT project examining the political and financial challenges of decarbonisation.

Vanesa Castán Broto is Professor of Climate Urbanism at the Urban Institute at the University of Sheffield. Her main research interests are the politics of climate change in urban areas, urban energy transitions and sustainability innovation. Her latest books are *Urban Energy Landscapes* (2019, Cambridge University Press) and *Urban Sustainability and Justice* (w/ Linda Westman, 2019, ZED Books).

Mariana Cavalcanti is Associate Professor in the Sociology Department of the Institute of Social and Political Studies at the State University of Rio de Janeiro (IESP-UERJ). Her research interests centre on the anthropology of urbanism and city-making, with a particular focus on issues of housing, infrastructure and public policy. She co-edited Occupy All Streets: Olympic urbanism and contested futures in Rio de Janeiro (Terreform, 2016) and co-directed the documentary film Favela Fabril (2012, 49').

Liviu Chelcea is Professor at the Department of Sociology, University of Bucharest, Romania. He earned both a PhD and a master of arts degree in Anthropology from the University of Michigan, Ann Arbor. His research focuses on housing, inequalities and infrastructures, with a current focus on water and water infrastructures in New York City.

Swati Chattopadhyay is Professor in the Department of History of Art and Architecture at the University of California, Santa Barbara. She is the author of *Representing Calcutta: Modernity, Nationalism, and the Colonial Uncanny* (2005), *Unlearning the City: Infrastructure in a New Optical Field* (2012) and the co-editor (with Jeremy White) of *City Halls and Civic Materialism: Towards a Global History of Urban Public Space* (2014) and *Routledge Companion to Critical Approaches to Contemporary Architecture* (2019) and a founding editor of *PLATFORM* http://www.platform-space.net/.

Francesco Chiodelli is Associate Professor of Urban and Legal Geography at the University of Turin, Italy. Previously. He directed the international PhD programme in Urban Studies at the Gran Sasso Science Institute (L'Aquila). He is currently working on different manifestations of illegality and informality with reference to housing, urban planning and governance. He also investigated the spatial dimension of the Israeli-Palestinian conflict over Jerusalem. His papers appeared in journals such as *Urban Studies*, *IJURR*, *Political Geography*, *Urban Geography*, *Progress in Planning* and *Planning Theory*.

Belen Desmaison is an architect and urbanist. She is a Lecturer at the Department of Architecture of the Pontifical Catholic University of Peru and holds a master's in building and urban design in development from University College London. She is currently a PhD candidate in human geography at Durham University specialising in climate refugees, human resettlement and the co-production of socially and environmentally just habitats.

Slavomíra Ferenčuhová is a researcher at the Institute of Sociology of the Czech Academy of Sciences. Her specialisation is in research on cities in Central and Eastern Europe with a focus on (post)socialist urban planning. Her research interests include the history of knowledge production in interdisciplinary urban studies and municipal responses to climate change in Central Europe. At the time of working on the chapter, she was a visiting fellow at the Imre Kertész Kolleg Jena.

Emma Ferulano is an activist, social worker and street educator involved in pedagogical workshops, social investigations, intercultural works in public and informal spaces of Naples and suburbs. She is co-founder of chi rom e…chi no association, Arrevuoto Teatro e Pedagogia, La Kumpania Social Enterprise, AntropicA, engaged in processes of empowerment side by side with young people, women, Italian, Roma, foreign communities and for the dissemination of audio-visual works with an anthropological approach. She writes for local and national cultural magazines.

John Flint is Professor of Urban Studies in the Department of Urban Studies and Planning and Director of Research in the Faculty of Social Sciences, University of Sheffield, United Kingdom.

Ioana Florea is part of the Common Front for Housing Rights (Bucharest), the national activist platform The Block for Housing (Romania), the European Action Coalition for the Right to Housing and the City, the new regional network Essential Autonomous Struggles Transnational and the editorial team of the *Political Art Gazette*. She is a postdoctoral researcher at the University of Gothenburg, Department of Sociology and Work Science, working on housing policies, urban transformations and social movements in Bucharest's context.

Francesca Governa, PhD, is an urban geographer and professor of economic and political geography at the Interuniversity Department of Regional and Urban Studies and Planning – DIST, Politecnico di Torino. She is part of the China Room and deputy director of FULL – Future Urban Legacy Lab at Politecnico di Torino. She has carried on fieldwork in European, North African and Chinese cities working on local development processes, urban margins and spatial justice; regional urbanisation; and contemporary Urban China.

Margherita Grazioli is a postdoctoral research fellow in the Social Sciences Area of the Gran Sasso Science Institute (L'Aquila, Italy). Dwelling upon her commitment as an activist in the Housing Rights Movements in Rome, she has conducted ethnographic research about squatting and housing policies in the context of Rome while interacting with the theoretical debates about informality, neoliberal urbanisation, commons and right to the city.

Katharina Grüneisl is a PhD candidate in geography at Durham University and an affiliated researcher at the Institut de Recherche sur le Maghreb Contemporain. Her dissertation explores the genealogies and present-day mechanisms of second-hand clothes circulation in Tunis, analysing how an ostensibly marginal realm of the economy (re)configures spatial, social and political relations in the city. Her research interests thus lie at the intersection of the 'urban' and 'economic', with a focus on rethinking valuation and market-making as productive urban processes at the micro-level.

Mona Harb is Professor of Urban Studies and Politics at the American University of Beirut where she is also co-founder and research lead at the Beirut Urban Lab. Her ongoing research investigates matters of governance in contexts of limited statehood and displacement, configurations of public life in contested cities and intersections of urban activism and oppositional politics.

William Jamieson is a PhD candidate in geography at Royal Holloway, University of London. His work is concerned with the integration of political geography and literary theory through critical creative writing methods to enhance our understanding of how space is 'read' and 'written' by capital. His project concerns the dynamics of land reclamation in Singapore and sand extraction across Southeast Asia. His fiction pamphlet, *Thirst for Sand*, was published by Goldsmiths Press in 2019.

Renana Jhabvala is an economist best known for her 40-year-long association with the Self-Employed Women's Association, a trade union of women in the informal economy, and for her writings on women in the informal economy. In 1990, she received the Padmashree, an honour from the government of India, and from 2012 to 2017, she served as the chancellor of Gandhigram Rural University. She was a member of the UN Secretary General's High-Level Panel on Women's Economic Empowerment and was honoured with the FICCI Lifetime Achievement Award.

Nikos Katsikis is an urbanist working at the intersection of urbanisation theory, design and geospatial analysis. He is Postdoctoral Researcher at the University of Luxembourg and research affiliate at Urban Theory Lab – Chicago, where he collaborates with Neil Brenner on research projects investigating the use of geospatial data in the study of urbanisation and on the hinterlands of the Capitalocene. He holds a doctor of design from the Harvard GSD.

Michele Lancione is Professor of Urban Studies at the Urban Institute in Sheffield. He is a member of the Common Front for Housing Rights (Bucharest), co-recipient of two Antipode Awards, co-founder and editor of the open-source *Radical Housing Journal* and corresponding editor for Europe at *IJURR*. His current work focuses on radical forms of inhabitation and housing struggles (through a five-year European Research Council project) and the politics of life at the margins in the contemporary urban.

Emma Lecavalier is a doctoral candidate and Vanier Scholar in the Department of Political Science at the University of Toronto. Her research focuses on the global governance of climate change in and through cities. She is a Fellow at the Global Policy Institute at Durham University and a Research Assistant with the Environmental Governance Lab at the Munk School (University of Toronto).

Helga Leitner is Professor at the University of California Los Angeles, Geography. Her research has been located in and about cities in different parts of the globe. Current research projects examine urban transformations, their impact on people's livelihoods, urban resilience and grassroots movements. Her recent books include *Urban Studies Inside/Out: Theory, Method, Practice* (co-edited with J. Peck and E. Sheppard) and *Everyday Equalities: Making Multicultures in Settler Colonial Cities* (co-authored with R. Fincher, K. Iveson and V. Preston).

Alicia Lindón is Professor-Researcher in the Department of Sociology at Universidad Autónoma Metropolitana-Iztapalapa, Mexico City; researcher at the National Council of Science and Technology (level III); and editor in chief of *Iztapalapa. Revista de Ciencias Sociales y Humanidades.* She is the director of the Editorial Board of Social Sciences and Humanities at the same university. Her research focuses on the sociogeography of everyday life, biographical approach and urban ways of life.

Simon Marvin is Director of the Urban Institute at the University of Sheffield. He is interested in the interrelationships between technological change and the urban condition, and his work has focused on the analysis of infrastructural liberalisation, low carbon transitions, digital technologies, living labs, urban operating systems and AI and robotics. Recently, he has focused on the role of the socio-technical systems in the production of micro-climatically controlled volumetric environments, including both indoor enclosures and outdoor thermally controlled spaces.

Colin McFarlane is Professor of Geography at Durham University, United Kingdom. His current work is on the politics and experience of urban densities (through a European Research Council project), the relationship between urban waste and life in the city, thinking the city through the idea of the fragment, and the potentials for urban equalities (through a Global Challenges Research Fund project led by University College London). His recent book is *Fragments of the City: Making and Remaking Urban Worlds* (2021, University of California Press).

Pauline McGuirk is Senior Professor in Urban Geography and Director of the Australian Centre for Culture, Environment, Society and Space at the University of Wollongong, New South Wales. Her research focuses on critical studies of urban governance, its changing geographies, practices and politics, with a current focus on urban governance innovations. She is a Fellow of the Academy of Social Sciences Australia and former editor of *Progress in Human Geography*.

Donald McNeill is Professor of Urbanism, School of Architecture, Design and Planning at the University of Sydney. He is interested in the political economy of the built environment and particularly the intersection of finance, governance, and expertise in (1) urban megaprojects and central business districts and (2) the innovation economy, including innovation districts, bioeconomies and start-up ecosystems. His books include *Global Cities and Urban Theory* (Sage, 2019) and *The Global Architect: Firms, Fame and Urban Form* (Routledge, 2009).

Martin Müller is Professor of Human Geography at the University of Lausanne, Switzerland. He is an urban and political geographer working on and with the Global Easts: a concept through which he attempts to think global theory beyond North and South. He is interested in global-urban types, such as mega-events and cultural flagships, and how these are mobilised in cities outside the West in a quest to turn economic into cultural capital. His recent publications include 'In Search of the Global East: Thinking between North and South' (Geopolitics). www.martinmuller.net @DrMartinMuller

Nithya Natarajan is Lecturer in International Development at King's College, London. Her work focuses on South India and Cambodia and explores agrarian change, rural-urban livelihoods, labour precarity, gender and debt. She completed an ESRC-funded PhD at SOAS, University of London. Nithya's current GCRF-funded research explores gender, climate, credit and nutrition in Cambodia and South India. She has published widely through academic outputs and is committed to an activist ethics of research through wider media, policy briefs and legal activism.

Morten Nielsen has a PhD in social anthropology (University of Copenhagen) and is currently a research professor at the National Museum of Denmark and head of the interdisciplinary research project 'Middle Class Urbanism: An Interdisciplinary Study of the Physical Reordering of Urban Sub-Saharan Africa'. Based on his fieldwork in Mozambique, Scotland, and United States, he has published on issues such as urban citizenship, time and temporality, comedy, human creativity, urban aesthetics, materiality, infrastructure and political cosmologies.

Victoria Okoye is a PhD candidate in urban studies and planning at the University of Sheffield. She has worked on community and youth projects centring on residents' understandings of space to inform design and planning. She collaborated with Yussif Larry Aminu and Spread-Out Initiative NGO for her PhD research, using creative methods to explore Nima youths' embodied experiences and spatial practices in community space. In her research, Victoria makes connections between African urban contexts and Black geographies globally.

Laurie Parsons is Lecturer in Human Geography at Royal Holloway, University of London. His work examines the contested politics of climate change on socio-economic inequalities, patterns of work and mobilities. Strongly committed to policy engagement, Laurie has conducted large-scale projects examining inequalities in Cambodia's economic development for Transparency International, Plan International, Save the Children, CARE International, ActionAid, UN Women and UNDP, amongst others. His first book, *Going Nowhere Fast*, was published in 2020 by Oxford University Press.

Linda Peake is Professor in the Faculty of Environmental and Urban Change and Director of the City Institute at York University, Toronto, Canada. She is currently PI on the SSHRC Partnership Grant, Urbanization, Gender and the Global South: A Transformative Knowledge Network (GenUrb); a Trustee of the Urban Studies Foundation; and sits on the editorial board of the Urban Geography and the International Advisory Board of *Signs: Journal of Women in Culture and Society*.

Edgar Pieterse is the founding director of the African Centre for Cities at the University of Cape Town. His research and teaching explore urban development politics, everyday culture, publics, radical social economies, responsive design and adaptive governance systems. He publishes different kinds of text and curates exhibitions, as well as difficult conversations about pressing urban problems.

Veda Popovici is a political worker based in Bucharest with an interest in decolonial thought, intersectional feminism, anti-fascism and material possibilities of creating the commons. Her political work has developed through radical collectives, such as Macaz, the Alternative Library, Dysnomia and the Gazette of Political Art. Dedicated to radical housing action, she is part of the Common Front for Housing Rights (Bucharest) and the national federation for radical housing justice the Block for Housing. Since 2019, she has been the facilitator of the European Action Coalition for the Right to Housing and the City.

Zsuzsi Pósfai (based in Budapest) is a founding member of Periféria Policy and Research Center, which is a professional collective working for spatial and housing justice. Her research focuses on the financialisaton of housing in Hungary and Eastern Europe, on the debt of households and possibilities for alternative forms of housing finance. She is also active in promoting rental housing cooperatives – both locally and regionally (for the latter, see moba.coop).

Ryan Powell is Reader in Urban Studies in the Department of Urban Studies and Planning, University of Sheffield, United Kingdom.

Jennifer Robinson is Professor of Human Geography at University College London. Her book, *Ordinary Cities* (Routledge, 2006) developed a postcolonial critique of urban studies. She is finalising a new book on comparative urbanism, proposing methodological foundations for a more global-urban studies. Earlier empirical research explored the history of apartheid cities and the politics of post-apartheid city-visioning. Current empirical projects focus on the politics of large-scale urban developments (London, Johannesburg, Shanghai) and the transnational circuits shaping African urbanisation (Accra, Dar es Salaam, Lilongwe).

Ananya Roy is Professor of Urban Planning, Social Welfare, and Geography and the Meyer and Renee Luskin Chair in Inequality and Democracy at the University of California, Los Angeles (UCLA). She is inaugural Director of the Institute on Inequality and Democracy at UCLA, a research centre that advances scholarship concerned with displacement and dispossession in Los Angeles and elsewhere in the world.

Suraya Scheba is Lecturer in the Environmental & Geographical Sciences Department at the University of Cape Town. She is on the editorial board of *Urban Forum* and a member of the Situated Urban Political Ecologies collective. Her research interests are located in both water and housing struggles, and she is currently leading a comparative project exploring land occupations and evictions in Cape Town, São Paulo, and Bogota.

Eric Sheppard is Alexander von Humboldt Chair at UCLA. His scholarship embraces economic and urban geography and socio-spatial theory, currently focusing on uneven geographies of capitalist globalisation, more-than-capitalist practices, environmental justice, southern urban theory and urban social movements, applying this to understanding urban land transformations in Jakarta. His recent books include *Urban Studies Inside-Out: Theory, Method, Practice* (co-edited with Helga Leitner and Jamie Peck), *Spatial Histories of Radical Geography* (co-edited with Trevor Barnes) and *Limits to Globalization.*

Wing-Shing Tang was formerly Professor in the Department of Geography, Hong Kong Baptist University. His research interest is to comprehend the nature of cities and urban development, especially Chinese cities (including Hong Kong). This is achieved by interrogating Foucault, Lefebvre, Harvey and others by subjecting Western discourses on space to urban experiences in other parts of the world. His research outcomes include a modest attempt to challenge Lefebvre's thesis of complete urbanisation and his argument of urbanising Gramsci by the experience of Hong Kong and China. The other is a critique of urban China research by coining the double concepts of random conceptual indigenisation and random conceptual appropriation. Currently, he is interested in exploring the applicability of Chinese philosophies to unentangle the ever-lasting debate in urban studies about town-country contradictions. In doing so, he has also initiated urban comparative projects between urban China and urban South Asia.

Jonathan Silver is a Senior Research Fellow at the Urban Institute, University of Sheffield. His work has focused on the deployment, operations and experiences of infrastructure across urban space. Developing a comparative focus across global North and Souths, infrastructural sectors and various scales of the urban this research has sought to understand and politicise the inequalities that emerge through these systems and across the urbanisation process. His fieldwork experiences have included cities in Ghana, Uganda, South Africa, the United Kingdom and United States, amongst others.

AbdouMaliq Simone is Senior Professorial Fellow, the Urban Institute, University of Sheffield and Honorary Professor at the ACC, University of Cape Town. His work centres on trajectories of urban change in the Global South, with emphasis on popular economy, urban peripheries and socio-cultural formations. His most recent publication is *Improvised Lives: Rhythms of Endurance in an Urban South* (Polity, 2019).

Ana Vilenica is a member of the Roof, an anti-eviction organisation from Serbia; the Radical Housing Journal collective; editorial collective for Central and East Europe of the *Interface: A Journal for and about Social Movements*; and the EAST-Essential Autonomous Struggles Transnational. Her research interests are housing struggles, housing and migration, issues related to social reproduction and housing, culture-led urban regeneration, developer-led art and artwashing.

Tatiana Thieme is Associate Professor in Human Geography at University College London. Tatiana's research projects have been funded by the Economic and Social Research Council, a Cambridge Humanities Grant, and the British Academy. Broadly, her research and writing focus on 'hustle economies' amongst youth cut off from mainstream institutional support, urban environmental politics and grassroots development. While she has worked in cities across the Global North and Global South, her primary field site remains in Nairobi, Kenya.

Laura Tozer is Assistant Professor at the University of Toronto in the Department of Physical and Environmental Sciences. Her interdisciplinary research focuses on urban environmental governance and examines the reconfiguration of urban systems to put cities on more sustainable footing. Her current focus is on the policies and politics driving transitions to achieve zero-carbon, resilient and equitable cities.

Yaffa Truelove is Assistant Professor of Geography and International Affairs at the University of Colorado at Boulder. Her research examines urban and feminist political ecologies of water, southern urbanism and the everyday infrastructures and governance of Indian cities.

ACKNOWLEDGEMENTS

This is a book about global urbanism, and there are a great many to thank far and wide who have helped make it happen. First and foremost, to our contributors, we could not have asked for a more talented, inspiring and organised group. You have made editing this text a pleasure. We have loved engaging with your ideas, and we're delighted to have done our part in bringing your excellent work together here. Thank you all. Second, we thank Andrew Mould and Egle Zigaite at Routledge. From Andrew's initial excitement about the idea through our discussions on what the final book might look like to Egle's excellent support in bringing the text home, you have been great to work with.

We would like to thank a few others whose friendship and scholarship have been with us from the start of this project. They have played a role, knowingly or not, in shaping this book: Ash Amin, Simon Marvin, Jenny Robinson, Edgar Pieterse, Ananya Roy and AbdouMaliq Simone. Finally, Colin extends a heartfelt thanks to Rachael for her amazing and enduring support and Keir and Arran for always asking the most interesting questions about cities. Michele wishes to thank his FCDL and RHJ comrades, his beautiful sister Silvia with her daughters Bianca and Laura, and Leo for sharing every bit of it. We dedicate this book to the struggle for just and reparative cities and to all those fighting the uneven makings of our global-urban time.

PART I
Introduction

1

NAVIGATING THE GLOBAL URBAN

Michele Lancione and Colin McFarlane

Situated urban thinking

The twenty-first century, we keep hearing, will be urban. More and more of us will live in cities, and cities play increasingly central roles in the global economy, environmental change, social inequality, and political transformation. Cities are pivotal for the Sustainable Development Goals, from water and sanitation for all to affordable energy and safe, resilient places to live. The *New Urban Agenda*, adopted in 2016, positions cities as central to tackling climate change and poverty, as well as to developing prosperity. Beyond institutional politics, it is largely from cities that grassroots groups contest violent extractions, racialized policing, undemocratic regimes, and uneven modes of production and consumption.

Urbanists, both within and beyond the academy, are asking whether, in an increasingly global 'urban age', we might need new epistemologies and politics of cities that take us beyond the remit of twentieth-century and typically Western normative approaches. Recent contributions in this sense include efforts of thinking urbanism and urbanization in genuinely global terms, questioning the role of cities in the economic, political, cultural, and environmental processes shaping global change but also include broader, and more explicitly radical, attempts at reworking epistemologies of knowledge production around, and beyond, how we come to know and to speak about what we call the 'urban' (Sheppard et al., 2013; Brenner and Schmid, 2015; Robinson, 2015; de Sousa Santos, 2016; Amin and Thrift, 2017; Oswin 2020; Roy 2020; Amin and Lancione, forthcoming).

New inquiries and experiments abound, seeking to understand what the growing prominence of cities and urban ways of life mean for theory, practice, and transformation. In this context, if it has become taken for granted to say that we live in a time of global urbanism, it's for good reason. It is almost impossible to disassociate the urban from the global. When we look across the economic, the ecological, the cultural, and the political realms of the city, so many of the questions we ask, agendas we debate, and futures we plan or hope for are shaped by and depend on connections and conditions that we can consider 'global' on the basis of their interdependence and constitutive interconnectedness. This does not mean their uneven structuration is leveled out but rather that unjust outcomes are intimately related to their translocal relationalities. At the same time, the arena of the counter-political – in urban thinking, activism and scholarship – has

DOI: 10.4324/9780429259593-2

become defined, in part, by an effort to name and explain conditions of global urbanism and to set out agendas and localized radical praxis for more equal, radically different and ecologically progressive forms of global urban life.

Our aim here is to step back from the attempt at defining these processes univocally, common to debates ranging across multiple fields, to ask a seemingly simple question: What does it mean to say we live in a global-urban moment, and what are its implications? Several questions follow on. How does this moment differently unfold globally, and what are its consequences? If we seek more socially and ecologically just urban futures, what are the questions and agendas we need to tackle, and according to which kind of position – epistemic and geographical – is the answer to this question made to count? What can we learn by examining *how* different kinds of global urbanisms are lived, felt, and enacted from situated practices? How should we understand, theorize, and research the sheer situatedness of urbanism without policing, and silencing, in the name of theory? That is, how could we think cities with, yet always also beyond, global-urban processes and conceptual framings?

As editors, we began this project with three motivations. First, that the centrality of global urbanism, differently understood to be sure, in urban scholarly epistemologies demands a rigorous and extensive examination of what urbanists mean by it and why it matters, remaining conscious of the incremental, shifting, and contextual nature of much of these knowledges.

Second, we explicitly set out not to define global urbanism in a particular way. Our starting point was that the sheer plurality of understandings, definitions, and ways of researching global urbanism was the only possible terrain for understanding and knowledge production, with all its contradictions, uncertainties, and provocations (Haraway, 1988). In this we are inspired by scholars who have been working in the feminist and queer urban studies tradition, who have long advocated for forms of situated and plural ways of thinking the city, often against the normative and centralizing canons of the field (Massey, 1994; Oswin, 2016). Following these lines of thinking, we did not impose any particular definition of global urbanism on our contributors but instead asked them to write openly about what global urbanism means to them, in their context, from the ground and struggles upon which they are working and living.

Rather than seek a consensus, we see the diversity of interpretations and agendas here as both the necessary consequence of the different nature of urbanism and urbanization, and the experimental imperative of critical urban thought. That doesn't mean there aren't shared views or resonances – we will come to those, and drawing out common conditions and perspectives matters – but instead that plurality is both inevitable and, moreover, useful as a basis for learning across differences and forming new ways of seeing. Our hope is that readers will find across these contributions, from different parts of the world, resources that will help them to make more sense of what global urbanism is, what's at stake in it, how to research it, and what needs to change for more progressive urban futures.

Third, drawing together a diverse range of interpretations of global urbanism demanded pulling together voices from across the planet. The story of global urbanism that emerges in the book – and we outline our take on that story next – is written by contributors located all over the world, writing about different kinds of cities and urban conditions, and they reflect very different career stages and urban specialisms, disciplines, and fields. Some of them don't have the privilege of working within secure academic positions and have to fight to find the time for writing; others do and struggle to rewire their epistemic privilege to work with, and within, alternative city-makings. Ours is a modest attempt at letting the thinking that is potentially transformational in form, and resolutely ordinary and mundane in its engagements, to come through and populate an academic text beyond the urge for universal explanations and reductionism. Thinking cities with and beyond global urbanism is an orientation to the multiplicity of urban inhabitation(s), their uneven (un)makings, and their uncanny propositions on the politics of urban life (Simone, 2018; Lancione, 2019, McFarlane, 2021).

Global urbanisms

Understanding global urbanism demands plural, even contradictory, positions. As the critique from a set of distinct positions has taught us – including, for instance, postcolonial, feminist, critical race studies, and queer debates – the ways in which the urban, city, or urbanization are conceived and take place demand at the very least some modesty, provisionality, and openness to different perspectives and processes across different parts of the world (e.g. Morland and Willox, 2004; McKittrick, 2006; Robinson, 2015; Roy, 2015; Buckley and Strauss, 2016). We agree with Natalie Oswin's (2016: 4) argument "to keep thinking rather than settling on an epistemology that aims to 'pin down' that which is bound to always elude us, the truth of the urban".

Along these lines, we have found inspiration in a short manifesto on global urbanism published in 2013. There, Sheppard, Leitner, and Mariganti set out a view of global urbanism not as a coherent body of theory or practice but as a loose bundle of ideas and practices that travel across the world. The promise of this disparate work lies in the clash between the global and the urban as a spark of inquiry and creative possibility. Yet much of this work, they point out, focuses on megacities and is derived from Anglo-North American theory, preoccupations, and assumptions, which undermines the possibility of a genuinely global production of critical urban thinking. The task of what they called 'provincializing global urbanism' demands both speaking back to and contesting that work, while supporting and drawing from a genuinely global loci of knowledge production, forming new conceptual anchors, and de/reconstructing imaginaries of what the city, urban life, and urbanization consist of.

Sheppard et al. go on, interestingly, to invoke a particular version of 'planet' to describe the project of global urbanism they have in mind: one that is modest, provisional, revisable, multiple, and committed to making space for alternative voices and views. They approach urbanism and urbanization through the lens of *planetarity*, read through Gayatri Spivak's argument in her (2005) *Death of a Discipline*. They argue that "planetarity" captures the planet as – here they quote Spivak – "a species of alterity, belonging to another system … we inhabit it, on loan" (Sheppard et al., 2013: 72): "If we imagine ourselves as planetary subjects rather than global agents, planetary creatures rather than global entities, alterity remains underived from us; it is not our dialectical negation, it contains us as much as it flings us away" (ibid. 73).

There are two implications in this move. The first is that it seeks to shift our view from an effort to discern global urbanism toward an interrogation and expanding dialogue around global urbanism*s*. This shift to the plural was a point of departure for us in conceiving this volume and has remained with us throughout the project. Second, it matters too that Sheppard et al. use other spatial imaginaries of the global – here planet, but we might also think of world or earthly (for instance, see the 'worlding' of Roy and Ong, 2011) – as provocations of one another. While we have anchored this book loosely around 'global urbanisms', we are not seeking to advocate that term over another, whether 'urban world' or something else (indeed, 'urban world' is part of the focus of a forthcoming book by one of us, McFarlane 2021).

We have sought to take seriously the distinct situated knowledges that emerge in and through different global contexts and debates. There is no single epistemology of the urban that we can fashion in response to the encounter of the urban and the global. Instead, the time is right to explore an 'epistemology of epistemologies' or a 'grammar of grammars' (Amin and Lancione, forthcoming), and that is what this book stages. After all, the urban condition – or for that matter the city or urbanization – does not, as Ash Amin (2013: 206) has put it, operate like a "mechanical entity such as a clock" that can be made "transparent in all its workings" (and see Simone, 2014). Any single epistemology will at best only ever illuminate quite particular kinds of logics, systems, networks, and entities. Even then, epistemologies are themselves sometimes short-lived, given that cities are hugely divergent and changing. Patterns of urbanization, forms of urban life, and conceptions of what a city is are not only often distinct in Kampala, Indore, Manila, London, and Caracas; they are themselves subject to change (McFarlane, 2011; Davidson and Iveson, 2015; Schindler, 2017).

Crucially, being situated in the multiplicity of urban knowledges is not a simple exercise in theory-making but the only possible ground for liberatory urban struggles. Urban activists working with, and through, intersectional praxis know this well. The book does not aim to offer guidance in this sense, but through its structuring and refusal of normative prescriptions, it invites the reader into a dialogic, nonlinear, and translational thinking within and across urban difference. This approach is, in part, aligned to a nonessentialist reading of difference (Jazeel 2019), but it is also nothing more (or less) than our attempt at using the tools at our disposal to move the conversation forward – without assuming there is a univocal sense of direction toward a particular destination and knowing that it is not for us to declare whatever the direction might be.

The politics of the writings enclosed in the volume speak for themselves, ranging from the grounds of anti-racist struggles, critical of uneven and various forms capitalism axiomatic in our cities, fighting for manifold ecologies of life, and more. If there is *a* theoretical statement beyond our efforts in compiling this book, it is that *thinking* (and writing) *cities with and beyond global urbanism* means to stay closer to the action, from where generalization emerges in the unfolding of urban life in which entrenched structures can be felt, thought through and responded to (Lancione and McFarlane, 2016).

Multiple urban political grounds

Thinking Global Urbanism, then, is a heterodox set of voices that takes on global urbanism, probing and provoking rather than seeking to draw a line under a complex, changing, and profoundly important set of issues. In order to do so, we invited 46 leading urban thinkers from across the world to write short commentaries and reflections, rather than full chapters, and conducted interviews with 10 prominent scholars, activists, and practitioners.

As a result, the book covers a lot of urban ground. For some contributors, global urbanism is a vital part of the political, economic, cultural, and aesthetic power of neoliberalizations. This is, for instance, the case for Desmaison and her argument for an urbanism of the commons or Chattopadhyay's contribution, where we are offered an account of the importance of understanding the different temporalities, and durational imaginations at play in global capitalisms, which provide a window into the structural changes they bring to the fore in our cities.

Others expand on these fundamentally global questions in which the city and the global are no longer thinkable or actionable separately. Here a key domain is the climate emergency. We hear from Bulkeley, Tozer, and Lecavalier about the political work of climate politics in the city, which they argue so often fails to account for the urban basis of climatic transformations, with consequences for how we then think about both environmental change and urbanism. We hear too, this time from Marvin, about the making of new urban climates, including the politics of standards and future urbanisms wrapped up in experiments in different parts of the world.

Meanwhile, Castán Broto examines the politics of 'scientific knowledge' in the making of 'climate urbanism', and in particular the sidelining of key questions of inequality, with implications for dominant versions of global urbanism in circulation.

Others outline directions to alternative global urbanisms from the context of everyday practices and often marginalized spaces. From Adeniyi-Ogunyankin and Peake we hear about how one woman living on the economic margins of Ibadan, Nigeria, negotiates a particular day and how global urbanism makes its presence felt. Lindón focuses our attention on the gendered experience of moving through Mexico City, while Brown and Borisa examine the space made and claimed for queer desire in different cities and the alternative global urbanisms we might write from those contexts, and Thieme considers how 'digital auto-portraiture' in Nairobi reflects both attachments to place and cosmopolitan youth street cultures. Leitner and Sheppard confront global urbanism from the agentic space of alterity in marginalized kampungs – postcolonial informal neighbourhoods – in Jakarta, while Harb provides an account of the ways in which urbanity becomes political through everyday counter-dominant collective experimentations in Lebanese cities. Jamieson, Brickell, Natarajan, and Parsons provide insight into global urbanism through the labor exploitations and city aspirations caught up in city-building – through what they call the 'fevirish' use of sand and bricks – in Southeast Asia.

The collection ranges across all kinds of everyday contexts and situations in which we gain insight into how the urban and global are configured in process, opening up some possibilities for life and living while closing off others. With Grüneisl, for example, we move to a 'fripe' marketplace in Tunis and the complex cartographies of global urbanism emerging from the opening of clothes and other goods. Looking at the competing rationales at play when these bales – containing imported fripe materials – are opened in front of potential customers, Grüneisl shows how, in everyday life, contrasting forms of global urbanism are enacted: sometimes adhering to, sometimes enacting a challenge to the dominant imagery of the 'global city'. For Scheba, the unequal transformations of contemporary Cape Town can be usefully apprehended not through 'situated' or 'global' urbanism but through a dialectical urban approach connecting different contexts, moments, struggles, and forms of racialized urban capitalism. If global urbanism here is networked into the making of everyday livelihoods and transactions in the ordinary city, it is also a question of the politics of life itself. De Boeck, writing on Kinshasa, positions 'thanatopolitics' – the necropolitical governance of life and death – as part of "global urbanism's nervous condition".

Key to these latter contributions is the investigation of everyday urban life and living as the site within which global urbanism is not only territorialized but also unmade to sustain otherwise neglected and sometimes violently controlled modes of life and urban inhabitation. Simone's contribution expands on this, focusing on how urban residents in 'extended' urban souths navigate their everyday lives fueled by a tension that points 'out there' but without a clear sense of direction. This is a modality, as he stresses, of "continous attainment and improvement", of keeping going, and which contrasts with a modern urban project of clear promises and attainments, and it needs to be understood in its constitutive (and contested) relationship to blackness and anti-blackness.

As we've already intimated, a number of authors have focused their attention on the arena of policy, including global, national, and local policy making, to understand how cities are (re)made through shifting networks of economic and political interests. McGuirk grounds her contribution in the context of multiple global urbanisms, examining the politics of urban governance innovations and the constellations of actants populating that space, including its lack of grounded accountability. McNeill investigates the health industry in its constitutive relationship between 'health' (understood as an urban infrastructure) and powerful private global actors, which enact a 'bioeconomy' not adequately investigated or understood. Echoing other contributions, Pieterse prompts us to reflect on what can be done when confronted with the often-disarming power of global-urban policy and its intertwined economies. What can be done and by whom in order to change the content and tempo of urban policy from within? Exemplifying the makings of a possible rhizomatic politics through five caricatures, or characters for change, Pieterse offers a proposition for system change on the metropolitan scale.

For other contributors, a vital route to discerning the shape and politics of contemporary global urbanism lies beyond the city itself. Brenner and Katsikis set out an agenda for situating and researching the 'hinterland' fundamentally in our explanations, imaginaries, and political work on cities and urbanization. Silver focuses our attention on 'corridor urbanism' and the vast infrastructural networks that shift ecological and social worlds in the making of new forms of urban (dis)connection, such as One Belt One Road. Truelove invites us to look *beyond* in order to reapproach the *within*: to attend to the established networks making up the infrastructures of our cities and from there to identify the uncertain and ongoing socio-materials transformations making up much of the infrastructural modalities of many cities across the globe. In at least two other contributions, while staying closer to the urban, we are invited to think about global city-making processes from the site of distinct urbanisms. This is true in Cavalcanti's poetic chapter on the genealogy of the "ruin" in Brazilian cities and also in Nielsen's conceptualization of 'global urbanism' as a modulation of monumental architecture.

What, then, do the contributors have to say about 'global urbanism' as a category? With the exception of Chiodelli and Grazioli, who advocate for a 'unitary conceptualization' around practices they define as 'self-urbanism', most contributors argue for the multiplication of theories and definitions. For example, grounded in a situated understanding of occupancy of land, and embracing the *tongbian* philosophy, Tang and Benjamin question the usefulness of 'global urbanism' as a singular explanatory logic. Governa poses a similarly territorialized question – and proposition for multiple global urbanism(s) – from the ground of a Southern European city, while Flint and Powell further question global-urban imaginations from the contexts of 'left behind' cities of postindustrial North England.

Always maintaining a critical edge toward the flattening (and Westernizing) attitudes of mainstream urban theory, Müller questions the erasure of what the 'Global East' from the supposedly 'global' of urbanism, while Chelcea, Ferenčuhová, and Bădescu develop a 'comparative research agenda' from the vantage point offered by postsocialist cities. Bang Shin offers a nuanced account of the different 'planetary', 'global', and 'wordling' attempts at theorizing global urbanism, focussing on the global East and recentering on the question and uneven politics of knowledge production. In a similar vein, Robinson reflects on the relationship between the comparative urban agenda and global-urban studies, positioning the former as a specific kind of "global" urban *practice* that needs to be grounded in specific solidarities and institutional networks.

These thematic concerns and politics are also reflected, and amplified, in the six interviews scattered across the volume. We selected the interviews less by logics of thematic or geographic location and more by our hope to capture a diversity of urban knowledge and praxis that have inspired us in our thinking about the encounters of urban and global. We open with Ash Amin, a scholar who has modeled a version of experimental urban thinking that has so often transported us beyond the canonization of theory in mainstream urban research. Amin provoked us to think carefully around the politics of naming, of using particular lexicons (such as 'global urbanism') to make sense of today's urban world(s) and set out a focus on the intersections between conceptualization, the emergent city, and effecting change in policy and practice.

Ananya Roy, another scholar who has done much for rethinking the 'worlding' of urbanism and its unjust, often violent, regimes, spoke about the deep connection between theory and practice. For Roy, it means little to declare a certain kind of critical position (postcolonial, feminist, anti-racist, and more) without embodying its politics in the making of scholarship in and beyond the academy. This has consequences for individual and collective thinking and even enacting global urbanism. Both Amin and Roy's interviews are provocations not only on how to think global urbanism but also on how urban scholars might work for more progressive global urbanisms.

The interviews, beyond their distinct value, collectively represent a further attempt at theorizing cities *while* searching for a new set of vocabularies and modalities, to bridge action and theory across geographies. And so we hear from Yussif Larry Aminu (the artistic director of Spread-Out Initiative NGO, a project for the urban youth in Nima, Accra, Ghana) and Victoria Okoye (an urban scholar who has worked with Spread-Out Initiative NGO through a decolonial and participatory methodology) discussing the future of African cities from and for their youngest urbanites. We are invited to get closer to the bordering-making of Italian southern urban peripheries and their struggles, with long-standing Neapolitan activist Emma Ferulano. We are brought to the complex history of women-led activism crisscrossing informal economies and marginalised neighbourhoods in urban India through the powerful voice of the trade union organizer and Padma Shri holder Renana Jhabvala. And finally, we are invited to a nuanced, enriching, critical understanding of what the 'urban global' means from the specific urban histories of Central Eastern Europe as heard from the theorizing and the direct experiences of urban struggles of the activists and scholars Ana Vilenica, Ioana Florea, Veda Popovici, and Zsuzsi Pósfai.

Navigating the global urban

Yet, confronted with this panoply of voices and understandings, a final question remains: Is it still possible to retrieve key themes across these different groundings and probings while staying close to the idea of an epistemology of epistemologies for 'a global sense' in urban studies? We think it is if one goes back to Doreen Massey's (1994, 2005) invitation to understand the global through the sense it makes as it is differently produced, lived, and felt. From that position, as the contributions to this book so clearly show, populated by all sorts of mechanics and struggles, finding 'key themes' is not only possible but necessary. These are related to what is made political, what is made of concern, from those situated geographies. Once those concerns – and not the theory used to elucidate them or the structural paradigm evoked to explain them – are centered and kept centered in theory-making, then one can critically trace connections between uneven processes and liberatory politics across urban geographies. In those connections, 'global urbanism' does not emerge as a rubric to, once again, explain how things are – although that is part of its situated work to be sure – but to illustrate how connections remain anchored to specific groundings, inform struggles of inhabitation, shape everyday responses to climate change, unfold through and against neoliberal deprivations, and so on.

The city, in this sense, comes into view as what it has always been: a ground of politics across geography – of struggle and possibilities – not simply because processes are, relationally speaking, everywhere, but because their unfolding points to resonating modes of organizing, being, theorizing, and proposing of alternative urban futures. This resonance – a ritornello crisscrossing a thousand soft subversions, as Guattari may have described it (2009) – is embodied through divergent actuations and conflicts. Yet multiple heterogeneous processes *share* the ground of the urban political across the globe: the place where contestation and social organizing becomes not only the product but also the producer of the urban political (Vasudevan, 2015).

If there are thematic concerns emerging from thinking cities through and with their global makings, connections, ways of knowing and being, and contestations, those are to be found in the situated ways in which the *urban* is made *political*. In other words, in the ways in which the particular densities of urban inhabitation – its cultural layering, societal structuring, and criss-crossing economic circulations – are not only expressions of all kinds of political projects but also lived and felt contestations that, in their shared urban groundings, can be made to count across geographies. This position demands taking seriously the quintessential capacity of the urban to be shaped translocally and continuously (re)activated from the ground of context, and to affirm, through the urban as a dense panoply of matters and mattering, a life beyond extraction, a life beyond repression, a life beyond climate extermination.

In a Lefebvrian sense, the urban is, therefore, not only the arena where the 'right to the city' is enacted and sought for but also a particularly fertile *substratum* allowing for the assemblage of that right to come to the fore – sometimes tentatively, at other times inadequately, and never automatically. One vital task for urban theory here is to orient itself *to enable* the surfacing of the multiple ways in which the urban is made political in order to, ultimately, do its part in sustaining its most progressive expressions and instantiations as defined from the ground of action. The chapters in this book point to a number of ways through which, across, and within the relational geographies of global urbanism, the urban is made political. It seems to us that, taken together, these chapters, for all their distinctions, are suggesting a basic analytic to retrieve such a making. One that would require attention to at least three points: *geography*, *the everyday*, and *resonance*.

First, the importance of *geography* here might be self-evident, but it is worth spelling it out especially when writing – as we are as editors – as white men working in one of the centers of epistemic privilege (UK academia). By geography we mean three things. First, the importance of historicizing certain spatial processes carefully as a fundamental practice *for* situatedness. If we are to avoid flattening the ways in which 'the political' emerges out of specific city-makings, then contextualization is not only valuable but also crucial. Unprompted by us, many authors and interviewees in this book have made that point clear as the only possible starting point: the prime grounding from which to interpret, speak to, and contest global urbanisms.

At the same time, by geography we are signaling the attention paid to the makings of new spatial formations. So second, to investigate how the city is made political *through* global urbanism, authors have focused on the rise of new spatial experiments (see Marvin's urban climates or McGuirk's governmentalities) and new functional parameters (such as those highlighted by Brenner and Katsikis or Silver). What is 'new' here is not only the arrangement of things but the ways in which a particular kind of production of urban form and function is managed, adapted, and articulated to sustain particular *ideas* of what urbanity is and should be in our global-urban age. In these experiments and large-scale trans-urban developments, there is an immanent *new* urbanity coming to the fore, which is not only about the here and the now but also about the prefiguration of how capitalist circulations will (re)create extractions at the intersection of planning, climate governance, commercial infrastructures, and more.

By the same token, there is a form of urban prefiguration in the third inclination of the 'geographical', offered by our authors and, even more evidently so, by our interviewees. This has to do with an attentiveness to the process of critical and radical spaces in the making: of staying with those counter-relationalities that exceed and contest dominant processes attached to different global urbanisms and which disclose other kinds of geographical imaginations and practices (and here we think again of Doreen Massey's powerful writings). Collectively, the book is both an interrogation of the geographies of global urbanism and an opening out to contested and alternative geographies, from – to highlight just a few – the high-level political wrangling's of global discourses on climate change and cities, or the neglected and marginalized urbanisms and urban knowledges of Southeast Asia or Europe, or ways of knowing and inhabiting land that exceed logics of capital accumulation, or queer spaces in different urban contexts, or the making of social infrastructures and the urban commons in Peru, to counter-geographies of urban space in Lebanon.

Second, there is *everyday life*. Not all chapters have the urban mundane as their focus. Yet most authors are pointing at its relevance in thinking both about how urban life unfolds, and the (re) production of unequal urban experiences across the globe. Pointing at the 'everyday' has become almost a mantra of a certain kind of ethnographic-infused urban writing, but the way in which the everyday becomes a way for the urban political to emerge cannot be revealed from careful observation. It is not about ethnographically enumerating the minutiae of daily encounters, which often become no more than academically validated fetishization. The everyday needs instead to be seen as a privileged gateway, more than an object of inquiry: it is in following how subject, context, and politics are articulated through mundane practices that the orientation the urban political takes can be seen.

This is what many authors in this book have taken care in exploring: the unfolding of life from apparently insignificant details – such as treads, monuments, inscriptions but also lived and felt engagement with policies and norms – in their bringing to the fore constitutive intersections with structures of governance, as well as in identifying potential points for tensioned fractures to break through. We see this in cases ranging from the informal markets of Tunis; or the negotiation of urban mobility in Mexico City; or the entanglements of bodies, subjects, and affects in the making of everyday urban infrastructures; to the effort to track different kinds of autonomous counter-urbanisms (see, for instance, the interviews with activists in Accra, Naples, and Central Eastern Europeans cities too). In some of these accounts, the focus on material and/or social infrastructure is especially important, including an attentiveness to the agency of the nonhuman to produce a specific kind of grid for global urbanism – one that is largely needed by corporate power to extract value in a ubiquitous fashion but also one that is increasingly used to build solidarities and shared actions. A critical take on global urbanism cannot rescind the analytical richness offered by these kinds of infrastructural makings, and here there is much work to be done.

Third is *resonance*. The forms of inhabitation and politicization of global urbanism assembled here express a version that William Connolly (2011: 173), writing in a different context, calls a 'resonance machine', where 'resonance' captures "rhythms of inter-infusion", the amplifying interactions within and between different processes and contexts. Resonance is a name for holding on to echoes and recurrences marked by their own historical and contextual character, reverberating urban problematics, situations, and struggles. Caldeira (2017) has written about the 'juxtaposition of singularities', while in this volume, Robinson writes about the revisable form of theory-building that emerges from thinking comparatively about the urban. The different interventions brought together in this book – access points to the ways in which the urban's entanglements with the global is differently made political – are ways of getting closer to what Simone has called *rhythms of endurance* (Simone, 2018), reverberations from which to draw suggestions about possible connections, both in the real and actual praxis of struggle, and about liberatory nondogmatic ways of thinking and writing about cities today. Taken together, they form a set of resources that can inform orientations to urban change – for example, by forcing what Pieterse calls in his chapter the 'cracks' that allow us to move from contingent diagnostics to meaningful propositions.

Thinking with and through global urbanism means envisaging the 'urban question' as situated in a plurality of modalities of knowledge and action to retrieve the transversal reverberation (of practice, of aspiration, of willfulness) that *might* cut across these efforts. Such rhythms might be found across specific struggles, such as for the right to housing or infrastructure, or in the effort to map new domains of urban – and nonurban – space and living, or in the effort to make space for other ways of being in the city, including alternative modes of living and imagining that variously contest, evade, and exceed more dominant neoliberal global urbanisms and their attendant policy apparatuses and economic rationalities. In retrieving and listening to these evocations and tonalities, even amid their moments of dissonance and divergence, the concern of global urbanism ceases to be one focused on identifying the 'right' explanatory frame and becomes one of how to explore intersecting yet also autonomous un/remakings of cityness: a never complete and open project, conflicting according to whom and how it is carried out. Theory is finding the right orientation to learn how to hear that complexity of becomings as we study, inhabit, contest, or support.

If there is *a* common disposition across the chapters of this book, it is in the attempt at inhabiting the space of global urbanism uncomfortably from the ground of situated theorizing that is always already caught up in the conscious work of unfolding cross-local conversations and contestations. That attempt is also methodological, and it cannot be otherwise: as many chapters and interviews show, it is not just about the ways one comes to make sense of global urbanism, and its politics, but also about what one does with it and how. In this sense, the book is a staging ground, methodologically and epistemologically, of an experiment in understanding global urbanism that is at once situated and provincial in its claims and dialogue and plural in the voices that shape it. If, as de Sousa Santos (2016: 8) contends, "the emancipatory transformations in the world may follow grammars and scripts other than those developed by Western-centric critical theory", when it comes to our increasingly urban world, *that* emancipation needs to pass through the ways in which we think and make sense of and activate the 'urban political' too. This oeuvre has a lot to do with fighting the silencing sometimes operated by normative thinking and enabling an attentiveness – one could say a form of 'radical care' (Hobart and Kneese, 2020) – toward *how* the urban is made political. Reverberances and possible theoretical, methodological, and practical new common ground for action might emerge from the effort of staying close to the geographical, the everyday, and their resonances. We see a lot of this effort at play in the conceptual, methodological, and empirical diversity in the book, as well as in the different forms of the urban political that are identified and set out. Ultimately, what the book argues for is an incremental, fragile, and in the making emancipatory urban thinking for the many. These pages are just one modest attempt at moving, heterogeneously and nondogmatically together, in this sense. *Avanti!*

References

Amin, A. (2013) 'The urban condition: A challenge to social science'. *Public Culture*, 25(2), 201–208.

Amin, A. and Lancione, M. eds. (Forthcoming) *Grammars of the Urban Ground*. Duke University Press, Durham, NC.

Amin, A. and Thrift, N. (2017) *Seeing Like a City*. Polity Press, Cambridge.

Brenner, N. and Schmid, C. (2015) 'Towards a new epistemology of the urban?' *CITY*, 19(2–3), 151–182

Buckley, M. and Strauss, K. (2016) 'With, against and beyond Lefebvre: Planetary urbanization and epistemic plurality'. *Environment and Planning D: Society and Space*, 34(4), 617–636.

Caldeira, T.P. (2017) 'Peripheral urbanization: Autoconstruction, transversal logics, and politics in cities of the global south'. *Environment and Planning D: Society and Space*, 35(1), 3–20.

Connolly, W.E. (2011) *A World of Becoming*. Duke University Press, Durham and London.

Davidson, M. and Iveson, K. (2015) 'Beyond city limits: A conceptual and political defense of 'the city' as an anchoring concept for critical urban theory'. *City*, 19, 5

de Sousa Santos, B. (2016) *Epistemologies of the South: Justice Against Epistemicide.* Routledge, London.
Guattari, F. (2009) *Soft Subversion.* Semiotext(e), Los Angeles.
Haraway, D. J. (1988) 'Situated knowledges: The science question in feminism and the privilege of partial perspective'. *Feminist Studies*, 14(3), 575–599.
Jazeel, T. (2019) 'Singularity. A manifesto for incomparable geographies: Singularity'. *Singapore Journal of Tropical Geography*, 40(1), 5–21.
Kawehipuaakahaopulani Hobart, H. J. and Kneese, T. (2020) 'Radical care. Survival strategies for uncertain times'. *Social Text*, 38(142), 1–16.
Lancione, M. (2019) 'Radical housing: On the politics of dwelling as difference'. *International Journal of Housing Policy*, 1–17. doi:10.1080/19491247.2019.1611121.
Lancione, M. and McFarlane, C. (2016) 'Life at the urban margins: Sanitation infra-making and the potential of experimental comparison'. *Environment and Planning A*, 48(12), 2402–2421.
Massey, D. (1994) *A Global Sense of Place.* University of Minnesota Press, Minneapolis.
Massey, D. (2005) *For Space.* Sage, London.
McFarlane, C. (2011) *Learning the City: Knowledge and Translocal Assemblage.* Wiley Blackwell, Oxford.
McFarlane, C. (2021) *Fragments of the City: Making and Remaking Urban Worlds.* University of California Press, Los Angeles.
McKittrick, K. (2006) *Demonic Grounds: Black Women and the Cartographies of Struggle.* Minneapolis: University of Minnesota Press.
Morland, I and Willox, D. (2004) *Queer Theory.* London: Palgrave.
Oswin, N. (2016) 'Planetary urbanization: A view from outside'. *Environment and Planning D: Society and Space.* doi:10.1177/0263775816675963.
Oswin, N. (2020) 'An other geography'. *Dialogues in Human Geography*, 10(1), 9–18.
Robinson, J. (2015) 'Comparative urbanism: New geographies and cultures of theorizing the urban'. *International Journal of Urban and Regional Research.* doi:10.1111/1468-2427.12273
Roy, A. (2015) 'What is urban about critical urban theory?'. *Urban Geography*, 37(6), 810-823.
Sheppard, E., Leitner, H., and Maringanti, A. (2013) 'Provinicalizing global urbanism: A mainfesto'. *Urban Geography*, 34, 893–900.
Roy, A. and Ong, A. eds. (2011) *Worlding Cities: Asian Experiments and the Art of Being Global.* Wiley-Blackwell, London; New York.
Roy, A. (2020) '"The shadow of her wings": Respectability politics and the self-narration of geography'. *Dialogues in Human Geography*, 10(1), 19–22.
Simone, A. (2018) *Improvised Lives: Rhythms of Endurance in an Urban South.* Polity Press, Cambridge.
Vasudevan, A. (2015) *Metropolitan Preoccupations: The Spatial Politics of Squatting in Berlin.* Wiley-Blackwell, London.
Schindler, S. (2017) 'Towards a paradigm of Southern urbanism'. *City*, doi:10.1080/13604813.2016.1263494
Simone, A. (2014) *Jakarta, Drawing the City Near.* University of Minnesota Press, Minneapolis, MN.
Spivak, G.S. (2005) *Death of a Discipline.* Columbia University Press, New York.

PART II

Rethinking global urbanisms

2

THINKING URBAN GRAMMARS

An interview with Ash Amin

Why talk about 'global urbanism'? What is lost or gained through using that specific term? How might research uncovering and mapping urbanisms across the world connect to policy, practice, and the making of publics? What might a focus on urbanism as emergent offer our understanding of the city and the urban today? What is the place of provisional concepts in urban thought, and how might we think about the rigour of conceptual work?

These were the key questions we explored in our interview with Ash Amin, presented here. Ash is one of the most distinguished and influential social scientists and urbanists in the world. Across decades of insightful and powerful writing, he has elucidated the urban condition, opened up new directions of conceptual possibility, and defined new questions for how we think and intervene in the urban social. This has included work on ontologies of urbanism, the politics of urban multiculture, the changing form of the urban political, the underpinnings of urban endurance, and the possibilities of renewed urban imaginations. His recent books include *Seeing Like a City* (co-written with Nigel Thrift, 2017) and *Releasing the Commons* (co-edited with Phil Howell, 2016). He is based at the University of Cambridge, where he is professor of geography and fellow of Christ's College. The interview took place on 8 July 2019.

The interview was wide-ranging in scope. We opened with 'global urbanism' and moved from there into a series of related themes, including influencing policy, generating publics, finding a language, making concepts operational, working with emergence and provisional understandings, and more. These different themes are not, however, digressions but instead ways of thinking and doing global urbanism, even if they aren't always explicitly stated as such in the interview.

As we opened the interview, Ash began with what he called an "agnostic question": Do we need a term like 'global urbanism' at all? What can we expect a term like that to achieve for urban thought? "I think we're going to struggle", he said, "to find a word that maintains the plurality and heterogeneity of approach and thinking". This intellectual risk, he went on, is that a term like 'global urbanism' can become a label, and labels carry a 'micropolitics' that might not be up to the job of inciting inquiry to find ways of approaching the sheer diversity of "the city in the world and the world in the city". He began instead not with global urbanism but with the term 'urban world'.

DOI: 10.4324/9780429259593-4

ASH AMIN: I think it [the term 'urban world'] does two things. One is that it signals the world, the earth that we're in. A second is that it implies a worldly ontology in which all things are part of a much greater entity, at which point the intuitive distinctions between local, national, and global, disappear. I would want to work more with terms like "worlding", "worldly", "world".

COLIN MCFARLANE: In *Death of a Discipline*, Gayatri Spivak (2005) distinguishes between the term "global" and the term "planetarity". While "global" is a striated depiction inevitably caught up with a human-centredness, "planeterity" was much more about recognising that humans are just one actor, albeit often a powerfully destructive actor, that cannot adequately comprehend the plurality and scale of the planet.

ASH: Yes, and the making of the social occurs *through* the interrelationships – the indivisibilities between the human, the material, the infrastructural, the technological, the eco-habitat. Terms like "world", "worlding", "earthly", perhaps "planetary", probably have more purchase. So, for me, if "global urbanism" is to mean anything, it has to be very much about how cities and other urban formations are what they are because of the very intersectional nature of all things living and dead, which take on particular meanings and particular intensities of combinations. Perhaps "global urbanism" captures some of that, but to me, it doesn't quite capture these rhizomatic associations.

MICHELE LANCIONE: I agree with that. The way we're working towards "global urbanism" is as a placeholder for different ways of starting to get a grip on this kind of complexity. The question here is, how can we convey a sense that global urbanism is not yet another concept to explain conditions but is instead a placeholder aimed at opening out views on the complexity of what the city has become or maybe has always been?

COLIN: I think your idea, Michele, of the placeholder is precisely the way we have been thinking about it as we have gone about this volume. Global urbanism is not a position we're trying to advocate but instead as a route to getting to – to paraphrase what you've said before, Ash – the urban as a kind of clearing- and sorting house for the world and as a space of combination and transformation that is always spatially stretched in different ways.

So for us, this kind of project was important as part of that larger task, and I think you described that really well.

But I want to ask you about the place of the city in all this. But what kind of entity is the city here? Particularly at a moment when urbanisation is increasingly a story of expansion into peripheries of different sorts, and new geographies are being formed and anticipated. And how do we intervene in that context of changing urbanisation? It seems to me that to a large extent, that depends on your conception of the kind of entity a city now is.

ASH: I think the tension here is between the processual and the ascribed. Or the processual and the inherited. So, cities will expand in rhizomatic ways. That throws up a challenge to the presumptions and instruments of law, of planning, of state, less so the corporate – which kind of goes with the rhizomatic. If urban process exceeds the juridical boundaries of a city – of planning authorities, municipal authorities, and so on – that raises a problem for the latter. The spilling over is treated variously as non-urban, or peri-urban, or extra-urban, or informal, or neglected, and so on. The challenge that we face now analytically and politically has to do with the squaring of the circle between urban ontology, which is now extraordinarily trans-territorial and relational in its constitution, and the categories of government and governance, which are entirely territorially fixed, scalar, and Russian doll-like.

At the same time, this ontology also includes a lot within the local that is largely ignored or 'minoritised', for example, forms of urbanism such as neighbourhoods made and managed by poor communities, producing its own rules of habitation. The question then becomes, how do you frame all that life that is exuberant outside the ascribed territorial? That's *the* big challenge, I think.

On policy, practice, and publics

MICHELE: You've said that we're required to place what you call "urbanicity" at the heart of national, rural, and international analysis, and so on. I think some of your recent work is trying do that explicitly. I know that part of the urgency for you is to bring some of these debates down to earth in a sense. Trying to bring some of this complexity language, in ways that may allow for the practitioners and the policy makers to engage with it. But the difficulty here is enormous because it requires, really, a shift in paradigm in what we consider to be the city. As soon as you start to say the name "city", people imagine a kind of Weberian city. So how do you do that work as an intellectual? How do you see yourself doing that work of translating some of the complexity languages you have contribute to, into something that can be appropriated and used?

ASH: What's interesting is that certain quantitatively minded people who use complexity theory get quite a lot of airing. I know many people in the world of science who end up as key government advisors to helping them to apply the rules of complexity science on how for example, vast traffic flows could be governed or how the dangers of an accident occurring at any given moment could be avoided. Frank Kelly, for example, who's the ex-master of my college, Christ's College, is a complexity mathematician and was chief advisor to the UK Ministry of Transport for a really long time. They are listened to because they're able to make tangible what multiple interactions or emergence means and show how emergence can turn into emergency, for instance. They're able to use a combination of data, complex modelling, simulations, scenario building – a series of techniques of revealing and forecasting all that which occurs as a result of the clash between things. Or consider the "smart data" capture of the vast crowds of things and information and people and bodies that compose the city. People operating in these ways are able to offer scenarios which are the combination of data and data-based speculation.

And decision makers are very keen to hear from them because they're acutely conscious that the tried and tested ways of doing things don't work. You talk to government officials at local or national level – I think they will readily tell you that the available tools of intervention are constantly betrayed. So, I think that for us in the 'softer' social sciences, the challenge is to produce convincing arguments and evidence on the basis of the sorts of data that *we* gather, which is often case specific and partial. This is quite a challenge: to make the general argument from such an evidence base.

COLIN: Some of this, surely, is about the art of combining micro-urban stories with the language of liberal governance. It strikes me that a lot of people who are influential at the policy level they speak the language of the state, including ways of talking about and presenting data. That's a challenge for someone who, for example, might do deep ethnography. How do we make the necessary shifts there? There's a set of communicative skills at play but also questions of content, styles of interactions, and so on. And it really matters where you are and who you're talking to. Sometimes the biggest challenge is about building any kind of working relationship with the state, which can take years in some cases.

ASH: But the world of science in the last decade or so has made *huge* steps towards making what it does credible and apprehensible. We in the social sciences and humanities are often looked upon more quizzically, particularly if we're trying to launch new ideas and grammars of the city, reveal the hidden and that which is not numerically capturable. There is a question of finding a language with traction, as there is a question of the policy world wanting to work interactively and experimentally with us. Look at the effort that goes into teaching science in schools, where you see signs of real effort in finding the right language and visual medium. Do we in critical urbanism take the trouble to do this? We in the humanities and social sciences have never been Latourian in that sense. We've tended not to really work at building coalitions or maintaining coalitions, working on translation and enrolment, worrying about the language of public communication.

COLIN: I guess related to this, we might think of how the urban agenda has taken shape in the United Nation (UN) system, for instance in the Sustainable Development Goals and in the international development debate. Sue Parnell has written about this, and it was clearly a long struggle over many years involving all kinds of networks and coalitions. Now that the city has arrived, so to speak, in urban development, Sue makes the point there that we – distributed scholars of the city – have an obligation to shape what the "urban" means in those discourses, policies, and practices. But I think you're driving at a more fundamental point here.

ASH: I think there's a kind of crisis of confidence. Partly because of the pushback against expertise that populism has brought and partly because of a reticence to oversimplify what are complex and evolving urban processes and phenomena. Yet, in the knowledge vacuum left, with the corporate world and elites having their own way, you sense that urban planners feel perplexed about what to do and how to intervene in the city. So, I think they require – in the clearest and simplest of language – a handful of takeaways, whether that's about developing a master plan, or thinking about zoning practices, or supporting the economy. And on the latter, for instance, often it's not enough for them to hear about the social or circular economy or about micro-enterprises, while the elephant in the economic room remains the privatisation of property, high finance, corporatism, the violence of neoliberalism left unaddressed by our pitch for the socialised economy.

But, as I already hinted, there has to be an openness on the part of policy makers to left-field thinking on cities. Let me give you one illustration. A while ago, I was asked to give a talk on cities to senior UK civil servants. I precirculated the opening chapter of *Seeing Like a City* (Amin and Thrift, 2017), which is not an easy read, in arguing for a 'science' of citiness based on understanding urban sociotechinical systems and their formative influence. I was given five to seven minutes to map out the main ideas and was convinced that the whole affair was a fiasco. Interestingly, though, there followed a lively discussion, followed by an invitation from one of the attendees to talk further because he recognised that the language of relational emergence and non-human urbanism echoed with his own sense of the city as experimental ground. Such moments of interest from senior decision makers in non-staple categories of urban analysis and in more evanescent forms of evidence are rare and could be grabbed in the spirit of developing an open and collaborative dialogue around complex phenomena.

MICHELE: Bent Flyvbjerg (2011), in *Making Social Science Matter*, put together a range of examples on the city and the way in which social sciences could make a difference in cities differently from natural sciences. It's not only about the big data, but it's also about this building of constituencies that you were referring to and engaging with real-life problems and offering a language to speak to the problems themselves. But then as soon as we enter into that, which is a correlation between neoliberal agencies, *and* the state, *and* the local government, we see more often than not that these local agencies are colluding with those forces. Which raises the question: What is the politics of doing these kinds of interventions using the language of complexity? By politics I am talking about the risk of colluding in the production of a form of urbanism that we don't want to be part of. Sometimes constituting an alliance is not the way to go.

ASH: Yes, but I'm not talking about collusion. I'm thinking of how we can make our kind of thinking and the expertise associated with it, and the insight that flows from it, publicly available. Available in ways that become tangible to interlocutors that we want to side with but also, through care taken to persuade intelligibly, those who are caught between lots of different interests and find themselves arbitrating in the way that many public officials do. To find a way of showing that it might be in their interest to intervene in a particular direction because that makes for the better, the healthier, the more democratic city. The task here is to make available new arguments and evidence to help planners interested in change to act differently.

Think how the New Zealand government is beginning to recognise the legal rights of nature. It's an outcome of decadal pressure from many quarters to reconsider indigeneity, respect Maori ways of thinking, bring in the vernacular into the legal and constitutional life, and, finally, be convinced by the Maori that rivers are living entities and therefore need to be protected. Here is an example of intellectuals and activists working with indigenous peoples and cultures to make public alternative cosmologies, continue to press for them, reveal nature and our debt to it in compelling ways, maintain the traction to the point of shifting the mainstream mindset. Without such work, the legislative changes would have stalled, I think. I am deeply interested in how fragile, discrepant, in-between forms of knowledge and action gain traction through a convincing politics of compulsion and amplification.

MICHELE: There is something very interesting here I think, which relates to the temporality of this kind of engagement. The debates we are witnessing both within academia and outside, when it comes to the Left, are attached to heavily, dense identity politics that make everything either being of one colour or another. It's very difficult to sustain the kind of political argument that you're sustaining now because you can be boxed in from the outset within one camp or another. So, the element of temporality matters here because you are talking not just about speaking to a public but about constituting a public that is not there yet.

ASH: That's exactly right, but there is also a journey on the road to the constitution of a different public. You have to make decisions; you have to choose your friends and your enemies, and then maybe you have to make coalitions as well. And sometimes these coalitions might actually be with those you least expect to ally with because in the making of a public, assumptions of what is 'the given' in a counter-public also have to be dispelled. It is an extremely tricky terrain to navigate, but it still needs to be navigated.

A politics of the emergent

COLIN: Okay, great, but there are some who would say that if you really want to effect change, do you not need to get back to some of the big ideas that animated, for example, the Marxist turn in urban thinking, where there was a clearer sense of what justice or equality demanded? There's the coalition; there's the constituting of a public; there's this focus on process-based urbanism, but how do you deal with questions of justice, equality, and development? Big ideas where some on the Left would feel, actually, here I know the tradition, I have my thresholds and yardsticks through which I can judge progress, whereas the form of urbanism we're talking about here is far too uncertain, far too hazy on where it's trying to get to. How do you respond to that claim?

ASH: I think now, because of the insights we've had from post-structuralism, postcolonialism, feminism, and the ecological movement, we can no longer assume that there is a direct one-on-one relationship between ontology and ideology. Or between ontology and normative outcomes and aspirations, with an assumed linearity of political travel. It may be wiser to ask ourselves: How can certain historic values of critical thinking or of the Left – the commitment to fairness, equality, and justice – be facilitated by the processes, disjunctures, and opportunities of the urban ontological?

It may transpire that a politics of justice is well served through careful and strategic interventions in the techno-sphere of the city or through minor or even major adjustments in the allocation of property rights. It may turn out, given the contingencies that arise in most complex systems, that a particular institutional or legal fix is in practice no fix, requiring close attention to emergent developments or to the combinatorial associations that produce ill effects, for example, the quality of air differentially affecting the life chances of a particular social group. These are the sorts of insights that come out of assemblage thinking, which brings different actants-in-relation into political view. Easterling's (2016) work on 'extra-statecraft' working through the hidden machinery of infrastructures is revealing.

COLIN: We might think too here of the work of Casper Bruun Jensen or Penny Harvey on how infrastructure can produce ontological politics because they are pulled into all kinds of relations, sometimes presentable and sometimes surprising, which threw up new political questions (e.g., Jensen and Morita, 2017; Harvey, 2012). Politics unfolds, things change, the debate shifts, new publics spring up, and so on.

ASH: And how do you, then, stay ahead of the game? How do you place cultures of anticipation at the centre? Foresight is really quite central for a new urban politics because what we're saying is that an awful lot in the city is simultaneously repetitive and emergent. This requires forms of intelligence that get close to this ground. Where are the appropriate research centres, scenario-building capabilities and capacities, laboratories of experimentation? These are the questions that a political language of global urbanism might pose.

COLIN: One of the things you have raised about smart technologies and how they monitor, as you have put it, the myriad transactions of phenomena of the city in real time is how they then come to "see" the emergent world. At the same time, the smart city has been the subject of all sorts of critique around what it doesn't see and the forms of knowledge that it might marginalise. And that includes global urbanism in that certain kinds of cities will show up perhaps a little bit less often, or more partially, or are off the map in all sorts of ways. Is there a risk that a politics of the emergent of anticipation of different kinds, via the digital turn, becomes dominated by particular cities and forms of urban imagination?

ASH: Absolutely! But don't blame digital awareness for that, but how it is numbingly used, or even manipulated, for corporate ends. Digitality is immensely powerful in its own right. Think how far and fast participatory mapping from the ground could go by making available real-time data that could be processed for participatory, pro-justice, urban ends. To say that ethnographies are strong enough weapons against digital information is a bit of a delusion. Critical urbanism could expand the repertoire of ground knowing, through cadastral maps made by people themselves, to authenticate or challenge central depositories of smart data. There are so many ways of getting closer and closer to the city and its hidden processes in real time, and there are so many new ways in which sensorial and virtual images of the processes can be now produced. In the multiplication of a whole series of vernacular laboratories – digital, archival, ethnographic, filmic, and more – the available digital data that is locked and manipulated in private or governmental hands, almost always beyond reach, can be placed under critical scrutiny, perhaps also made to work more democratically.

MICHELE: To me, this is really the starting point of the whole of this discussion. The material condition of the digital data today is not accessible. Creating a sense, as social scientists, of the urgency of making this data available matters.

ASH: Indeed, and more. Work is needed to produce maps and monitor the urban assemblage so that the sites of opportunity and the sites of resistance become clearer. The city is opaque and needs to be made visible first. Secondly, work is needed to break into the corporatist machine and its archive, to have a very clear sense of the power dynamics that turn the city into a playground of centralising returns. Third, the bringing together of diverse forms of urban intelligence new coalitions and alliances, horizontally and ground-up, to be made to imagine the other city. In these ways, a knowledge politics of pushback and the opening of new ground is put into place, slowly forcing settled power to shift, give way.

Styles of urban thinking

COLIN: You've spoken and written before about a "style" of urban thought, one of conceptual approximations, of provisional understandings that can be refined, rejected, consolidated, and so on. I wonder whether you'd say urban theory and urban studies are anywhere near this? What do you think needs to change in urban studies and urban theory in order to create that kind of approximating form of urbanism where we can get things wrong and shift things around, and so on because I don't think that is encouraged a great deal; there is a strong element of career stage about it.

ASH: It's a really tough big question. Categorical and canonical thinking is compelling and has the weight of tradition behind it, and it's very difficult to overturn that. But when categorical thinking stops delivering, and those categories don't help us anymore, then there is a little opening for emergent thinking. When ontology itself demands a rethinking of core categories, it may not be a bad thing to have a proliferation of concepts. Critical urbanism has seen a proliferation of concepts in recent years so as to get close to the manifold and always changing city. I don't think that's a problem. The problem has to do with questions of rigour, clarity, testability. The obligation to test the rigour and applicability of new concepts and methods sometimes has been sacrificed in the rush to develop new languages of the urban. If experimental thinking becomes loose thinking, then I think it is right that we are criticised.

COLIN: There is also something about different kinds of rigour here. When you talk about conceptual approximation, for example, I think partly what you're opening up there is a different kind of rigour. One in which, for example, you might go into a field situation with a certain degree of uncertainty and hesitancy over the concepts. This is different from a tighter sense of the concept that has a more rigid internal consistency, with a deeper commitment to that concept. If you set out stating that the concept you are using is deliberately open and reviseable – and certainly the way in which we're thinking about global urbanism is relevant here – then you could well be deemed unrigorous from the start. That's a battle in and of itself.

ASH: Sure. But theory, to me, is simply a testing ground. Theory is not a map of the world, and the moment in which you make theory into the substantiable aim, it can become dogma. Theory is an intellectual opening that requires verification.

References

Amin, A. and Howell, P. (2016) *Releasing the Commons: Rethinking the Futures of the Commons*. London: Routledge

Amin, A. and Thrift, N. (2017) *Seeing Like a City*. Cambridge: Polity Press.

Easterling, K. (2016) *Extrastatecraft: The Power of Infrastructure Space*. London: Verso.

Flyvbjerg, B. (2011) *Making Social Science Matter: Why Social Inquiry Fails and How it Can Succeed Again*. Cambridge: Cambridge University Press.

Harvey, P. (2012) 'The topological quality of infrastructural relation: An ethnographic approach'. *Theory, Culture, and Society* 29, no. 4–5: 76–92.

Jensen, C.B. and Morita, A. (2017) 'Introduction: Infrastructures as ontological experiments, *Ethnos*, 82, no. 4: 615–626

Spivak, G.S. (2005) *Death of a Discipline*. New York: Columbia University Press.

3

DECENTERING GLOBAL URBANISM

An interview with Ananya Roy

In this interview, Ananya Roy offers a series of powerful provocations of how 'global urbanism' might be pursued as a political and theoretical project. She offers a compelling case for how critical urban work might contribute to local activist struggles both by 'doing the homework' that those struggles need and learning from struggles elsewhere. Her arguments depart from a commitment to the feminist, black, and postcolonial traditions, and she raises vital questions around the extent to which urbanists have brought difference to their work in meaningful ways, including in relation to – and perhaps particularly in relation to – global urbanism.

Ananya is professor of urban planning, social welfare, and geography at the University of California, Los Angeles, and director of the Institute on Inequality and Democracy. Her work has been enormously influential in urban research, from questions of informality, poverty, and development to racial banishment and the nature and status of urban theory. Her recent books include *Encountering Poverty: Thinking and Acting in an Unequal World* (2016, edited with Genevieve Negrón-Gonzales, Kweku Opoku-Agyemang, and Clare Talwalker) and *Territories of Poverty: Rethinking North and South* (2015, edited with Emma Shaw Crane). We conducted this interview in early February 2019, not long after Ananya had run a conference on housing justice at the Institute on Inequality and Democracy at the University of California, Los Angeles (UCLA).

MICHELE LANCIONE: If we throw at you this couplet – "global urbanism" – what do you think of it? What are for you the epistemological challenges of thinking urbanism globally?

ANANYA ROY: This is a wonderful opportunity to think about and reflect on some of these issues with the two of you. I don't get to do that very often nowadays. The first piece of this is that I do not want to take lightly the tremendous work that has gone into building global urbanism, as what I would say is a loosely structured field of enquiry. This has been the work done by scholars in different genres, perhaps from different positionalities, with different epistemological commitments and methodologies. But there is also a shared space of debate and knowledge production. One piece that is central to this field of enquiry is a commitment to decentring the production of geographical knowledge and urban studies knowledge. But as we all know, decentring is not as simple as shifting knowledge production to what we might call the Global South. It is instead to be acutely aware of the relationalities

DOI: 10.4324/9780429259593-5

of knowledge production, and to think very much about the relationship between knowledge and place-based power. And we can return to some of that later in our conversation.

So we have to ask: What is that decentring? I also think that this has been a way of recentring, or attempting to centre certain epistemologies and methodologies that are not necessarily at the very heart of canonical urban studies knowledge, be it postcolonial theory, be it feminist thought, be it queer geographies. Or the ways in which some of us care to journey with radical social movements. And what sort of knowledge is then produced, subaltern knowledge if we are to call it that, and here we may want to think with the framework of subaltern cosmopolitanisms. This is for me the second piece of what I would see as the project of global urbanism. And then the third piece of it, is a somewhat tense relationship between global urbanism as a potential field of enquiry, as a loosely structured field of enquiry, and what for a long time has been the study of globalisation. I don't think we have fully had that conversation, but the study of globalisation dominated the social sciences for a substantial period of time, and I feel that global urbanism didn't just take on the global as a label but really recrafted the global in relation to, say, world systems of power, imperialism, and North-South relationalities, and so forth. In urban studies, the global has of course served as a proxy for the functions of capitalism, as in the global cities' schema, but in what we are calling global urbanism, the global also serves as a site of critical enquiry.

MICHELE: If I can just follow up on this? Again, it's a very broad question – but what do you think is specifically *urban* about this?

ANANYA: That's a terrific question, and it lies at the crux of the planetary urbanisation debates that continue to unfold. As you both know, I wrote an essay called "What Is Urban about Critical Urban Theory?" precisely because this question was on my mind. I think of the urban not as the city, and not necessarily as a bounded space, but as a particular set of transformations. So I'm with Brenner and Schmid (2015) on this, that we need to think about urbanisation as a process and as a process that is transforming a whole set of spaces that don't look like a city. I think the early work of David Harvey and Manuel Castells is incredibly useful in thinking about the urban in these terms. But where I strongly disagree with planetary urbanisation is the argument that urbanisation as a process does not have an outside. As many of the postcolonial critiques, including mine, have articulated, what is crucial is the outside. Various strains of feminist, poststructuralist, postcolonial thought insist on thinking about the outside, whether we see it as the periphery, the rural, the agrarian question, the hinterland, the colony. Wherever and whatever that is, that constitutive outside of urbanisation matters to me as much as what *we* constitute as an object called the urban.

MICHELE: Absolutely. So it wouldn't be unfair to say that in order to understand this global urbanism question, we need to get rid of the old categories defining the urban, and we need to look between the cracks in a sense?

ANANYA: My interest has been repeatedly in the ways in which new geographies of the outside are being produced and how they constantly slip out of view. And this is both a set of spaces and a set of subjectivations and subjectifications. I think the danger with global urbanism is that the moment we establish the field of enquiry and its set of conceptual frameworks, we risk that it becomes a very ossified canon. I should clarify that I don't have a problem with producing Theory with a capital T. I disagree here with Jenny Robinson, who has called for theorising rather than theory. I think we should unapologetically produce Theory with a capital T. But the question is what kinds of theory and produced from what sorts of positionalities? And can that theory be attentive to what feminist philosophers have called the standpoint of absence (see Roy, 2016)?

COLIN MCFARLANE: Can the kind of thinking that different theoretical approaches being used at the moment, such as planetary urbanisation or variants of global urbanism, accommodate the question of the outside as you are phrasing it (e.g. Sheppard et al., 2013; Brenner, 2018; Oswin, 2018; Pratt, 2018; Sheppard et al., 2013)? Can there be a coming together of these different approaches?

ANANYA: I am a little tired of rehearsing what feels like old debates. But it's worth stating again. I am committed to studying the political economy of global urbanism. The key question for me that follows is this: What is the place of historical difference in the making of urban landscapes and in the unfolding of urban transformations? I have insisted and want to insist again that historical difference is not about variegated capitalism. It is not about socio-spatial differentiation. It is about understanding the processes of expropriation and extraction through which space has been produced and is lived. I feel that in much of the planetary urbanisation work these processes are elided. They are not in view. And they are not in view at precisely the historical moment when we desperately need to keep these things in view. I thus think that there is a disconnect between the ways in which we talk about space and power in planetary urbanisation or even global urbanism and the very exciting scholarship happening in black geographies. I think that urban studies has a long way to go before we do a better job of studying historical difference. And in explaining the relationship between historical difference and the making and unmaking of space.

MICHELE: Some of the things that you've just said are not too dissimilar to interventions that were produced twenty years ago by the likes of Doreen Massey (1994) and Katz (1994). I wouldn't say that they are forgotten because I don't think they are forgotten, but they are definitely marginalised, assimilated. Why can't we move on with a more progressive sense of theory in urban studies?

ANANYA: Because, to quote Kate Derickson (2017), we belong to a discipline, let's just call it urban studies, that is "unbearably white". It is also an unbearably masculinist and unbearably Eurocentric discipline. Colin introduced me to Aamir Mufti's work on theory cultures, and Aamir is now my colleague here at UCLA. The theory culture of urban studies remains precisely what it was a while ago, when Doreen and Cindi and others were advancing their critiques. So, what I'm really tired of, and the reason I'm not going to Association of the American Geographers' conference (AAG) this year, is that I feel that at every one of these conferences, some of us have to repeat, rehearse, and perform the same postcolonial, feminist, poststructuralist critiques. And as you noted, it gets absorbed and assimilated. So, the piece that I really want to write is on citationary alibis. I'm tired of my work having been turned into an alibi.

MICHELE: Tell us more about that.

ANANYA: What I mean by citationary alibi is that citing me does not make an argument attentive to postcolonial methodologies. Citing me does not make that work attentive to the feminist practice of critique, in the same way that my citing Edward Said or Spivak means very little if I do not know how to practise postcolonial theory and its commitments. I think all our fields, including global urbanism, very easily absorb and depoliticise critique. One of the reasons why I am now immersed in learning from black geographies and in doing scholarship on racial banishment is because it's an auto-critique. I think that Los Angeles forces me, requires me in fact, to challenge the work I've done, to do something else, and to be a student of material that I had not taken up in serious fashion before. I'm very interested, then, in how, instead of building canonical power, we are committed to the terrain of critique, and how, therefore, in advancing theory, we think about postcolonial studies, feminist thought, black geographies not as the margins of geography but as its very centre. I will say that over all of the time that I have been involved in urban studies, those traditions of thought have remained the margins. That marginalisation continues, and those alibis, which are citationary structures, are deepening such marginalisation.

COLIN: You said you're not as inspired by your own previous work. What was it that you would say that you think now you weren't doing? Was it that you weren't doing enough of a certain kind of reading, engaging certain traditions on the margins that you've been bringing into the centres of your work, or was it something more fundamental about how you were thinking about your work as an urban researcher, about your place in relation to your work? What was it that had to change, do you think, for you? And just to follow on from that, is it your argument that scholars in urban studies need to transform the way in which they are theorising, reading, thinking, acting?

ANANYA: Those are excellent questions, and of course as you noted, they're linked. For me personally, precisely at the moment when our ideas become popular, when they become a new common sense, we are required to launch a critique of our own work. This is the case for urban informality. I am proud of that work, but I would hate to see urban informality codified as *the* way of thinking about the law and space and, in particular, about the relationship between state and subjects. So, part of it is simply the ongoing need to pull apart our own work, and that's part of the intergenerational production of knowledge. I'm not interested, for example, in having my students replicate my work and my ideas. I am interested in how I can support them to do radically different kinds of work. With global urbanism, I felt that I was settling into a very comfortable discussion of thinking from the South whether it be worlding cities, whether it be building on the work of other urban studies scholars doing similar work, and what I was doing less of was rethinking the West. I don't think that I did an adequate job of what "rethinking the West" might mean. To me that meant, particularly with the move to Los Angeles, and with the work of the Institute on Inequality and Democracy at UCLA, which I direct, a new rooting in understandings of racial capitalism.

What I am most interested in at the moment is holding in simultaneous view postcolonial feminism, which remains a key inspiration for me, my foundations, and the Black Radical Tradition, and black geographies more broadly. If I were to do a one-liner on the work of the Institute on Inequality and Democracy, I would say that we take a postcolonial approach to global Los Angeles. And I don't use those words lightly – I'm very invested in articulating the theoretical agenda and research agenda around what that might mean. What it means to seriously study settler colonialism in the United States in relation to urban change. I also want to understand the ways in which this is a global project and how from this place of understanding Los Angeles as a postcolony, I might return to what has been called the Global South.

This is not as straightforward as conventional understandings of thinking from the South, but I think that this continues the work of decentring, and of rethinking, the West. This also means that I would like to see many more urban studies scholars think about the places from which they're producing knowledge. I continue to be very troubled by what I see as disembodied knowledge production. All knowledge *is* in fact embodied and embedded, and I'm a little tired of these conceits of theory production that still speak from nowhere and therefore from everywhere. This goes back to the discussion we've had about historical difference. And this is not about reflexivity. This is about taking seriously embodied difference and the sheer epistemic and material violences through which embodiment is produced and which, therefore, cannot be separated out from the knowledge we produce.

COLIN: A lot of the debate around global urbanism has, in terms of the methodologies, been around comparative urbanism, which itself has become almost commonsensical in some areas of urban studies – and established as a way of thinking and talking about global urbanism. Can you say something about our broader question of how one "does" global urbanism from your standpoint based on some of the experiences you've had since you moved to LA because what you're articulating here is very different.

ANANYA: I take my cue from what I would broadly call critical race studies and critical ethnic studies. I think that there is still a vast gap, or disconnect, between urban studies and global race studies and critical ethnic studies. Pulido (2006) was just here for our housing justice conference, and her work in particular points out the need for a critical race studies approach in urban studies. For me, what is at stake is a reformulation of some of the key concepts in urban studies. For example, I've been very interested in rethinking dispossession, not just as moving beyond the usual frameworks of accumulation by dispossession but studying dispossession in a historical context here in the US such that this allows a new sort of *global* conversation around dispossession. So on the one hand, we want to be acutely aware of US imperialism, but on the other hand, if we are able to see the US as one of many settler-colonial societies, then I think we can push back against US exceptionalism, but we can also push back against disembodied, dis-embedded theory production. And think about the very specific forms of dispossession in the US that have always been connected to forms of dispossession elsewhere in the world. At our housing justice conference, Erin McElroy emphasised that we need to think not about universalisms but about connections (see, for example, McElroy and Werth, 2019). And this has really stayed with me.

Many of the social movements we seek to learn from and journey with at the Institute on Inequality and Democracy have already built these connections. These are political connections, and they also happen at the conceptual scale. They happen in how we study financialisation; they happen in terms of how we study the law; they happen in terms of how we study dispossession. I'm very interested in how those concepts – could be the urban – allow us to think not so much comparatively but about the connections. Establishing those connections is of course key, and for me, so many of those connections are established in the two theoretical frameworks that I talked about. For me, this is what postcolonial theory does. It is also what the Black Radical Tradition does. It's why Cedric Robinson's *Black Marxism* and Du Bois's *Black Reconstruction* are foundational texts, precisely because they give us a world-systems understanding of racial capitalism. This is the work we tried to do at our 2017 Race and Capitalism conference where we thought about racial capitalism as transnational histories and global territories. And that, to me, is different from thinking about neoliberalism as a global project. It is also the difference between universalising and generalising. I'm very interested in generalisability, but generalisability is not the same thing as, say, a universal theory of capital, or even a universal history of racial capitalism.

MICHELE: I think both Colin and I particularly agree on this last point. Some of our writing has been precisely about this universalising, generalising. I think a lot of what you said, Ananya, goes precisely into the next question, which is about how you relate global urbanism to ideas of the urban political, and you have been already saying a lot about this. One particular thing that really is a challenge for all of us is trying to experiment with these traditions – the feminist tradition, critical race studies tradition, and grassroots anarchist interventions on the political – in order to *intervene.* Is this how you go about studying the urban political?

ANANYA: The Housing Justice Conference and Network is a good example of this because these are some of the issues we're grappling with. And these are issues I've grappled with since the inception of the Institute three years ago. My terrain is that of scholarship. I'm interested in the relationship between university-based scholarship and movement-based scholarship in relation to the urban political. That is a very specific entry point. The question then becomes what is the scholarship we produce of the urban political and why might that matter? Two things. The first is the question of justice. This goes to why we decided to focus on housing justice in the Network and Conference rather than housing crisis or the housing question. I think there is important work to be done, scholarly work to be done, that learns from, critically reflects upon, convenes, curates, and theorises the proliferation of movements and mobilisations that make up our present historical moment. I feel that university centres and university-based scholars have some resources to be able to do that, but they've got to think about their accountability to movements and communities, and they've got to think about how fraught that shared terrain of scholarship can be. Overall, I think that there is much more work to be done on thinking about the key imaginations and practices of the urban political, struggles for freedom, struggles for liberation, struggles for justice today.

But the other part of this is that there is also work to be done in thinking about the key tropes through which we understand the urban political. Those key tropes and histories return us to that Euro-American centre of urban studies, of what is considered to be urban revolution, of what are the barricades, of who are seen to be the protagonists of struggle. I think that all three of us have tried hard in our work to move away from those histories and tropes. Surely, the barricades of Paris cannot be the only point of reference. History did not start with Occupy Wall Street. Such work has new urgency at this moment. Take for example the ways in which Black Lives Matter has re-shaped understandings of race and violence in the United States but without imposing a uniquely American conceptualisation of race. How do we think about anti-blackness in a global way? How do we think about the incredible mobilisations that connect the favelas of Rio to the inner city of Los Angeles? At the Institute and in the housing justice network, we thus rely on scholars who give us hemispheric understandings of anti-blackness, as well as of freedom.

COLIN: Can you give some examples of work you've been doing in LA and how those conversations with activists in LA around housing justice have been informed by work that's going on in places like Rio? Is there a way in which that kind of global conversation around anti-blackness is formed into the politics you're pursuing with the nexus of the university and these movements?

ANANYA: I've noted before that rather unapologetically, I am an urban theorist. I say this because the movements that the Institute on Inequality and Democracy journeys with and learns from give us homework. And the homework they often give us is to do the history and theory. They don't see my useful role as setting up barricades and marching in the streets. It's one thing to be there in solidarity, but it's another thing to be clear about one's contribution. And this has been a big debate with my graduate students. We just finished something called Activist Graduate School. And the focus of this class was housing justice activism. I had severe disagreements with my graduate students who are appropriately critical and hesitant of academic power in the support of housing justice activism. But I am not hesitant to use and wield academic power. Because I have come to that power via postcolonial feminism. I am, in Spivak's words, "outside in the teaching machine."

I'll give you a couple of examples. I have been doing new work on what I've been calling racial banishment. New geographies of segregation in the United States. Part of this is the criminalisation of poverty, but the key part of this is the public means of expulsion and banishment. And I've been trying to trace the long histories of this banishment but also the new geographies. In that work it became clear to me that many of the movements that I've been studying and working with, like the Chicago Anti-Eviction Campaign or the LA Community Action Network here in Los Angeles, were also using the term of banishment. They were saying that the familiar concept of gentrification, or even of displacement, were insufficient to capture what was going on. And so, with me, as with other academic scholars, they have said, "Give us a robust theory of banishment, trace the history of banishment, show us the mechanisms, do the research".

We need to recognise what that sort of homework means, and we need to recognise that these concepts are not simply produced by us as academic scholars but rather produced in relationship to such struggle. Again, the work that the each of you has done speaks directly to this.

Another example comes from the LA Tenants Union. We are in the midst of quite fierce housing justice struggles here in California, and tenant unions are a manifestation of such energy. The tenant as the protagonist of struggle. The LA Tenants Union has a very interesting conceptual framework of who the tenant is. The tenant is not the renter. For them, the tenant is someone who does not control her or his housing, which I interpret to mean someone who does not own the means of production of housing. This means that tenancy includes the houseless. So, it is very much about rethinking landlordism, rethinking the proprietary prerogatives of property, rethinking who is within this class that we might call tenants. That, to me, is the starting point. It is homework. It gives me a point of departure to return to questions of land and property and tenancy, as well as to the financialisation of housing. And to do so from a vantage point that the LA Tenants Union has laid out – which is their vantage point, but which, for me, is a starting point for history and theory.

MICHELE: In a lot of what you are saying, Ananya, the epistemological questions about, you know, what kind of tradition we come from when we build knowledge are central. But your questions are also methodological. I mean, you are saying, "I'm looking at *this* problem because I see it as a concern coming directly from the tenants, and in that concern, I see my starting point." Methodologically, I find this very, very interesting when we need to think about global urbanism, isn't it?

ANANYA: This is about the practice of postcolonial feminism. This is about the practice of the Black Radical Tradition. We know how difficult that practice is. It's incomplete. Every day, I fail at it. This methodological practice comes from these theoretical commitments, which are ultimately, for me, ethical commitments. But the other methodology that has been on my mind for a while now is Said's notion of contrapuntal thinking. And this goes back to issues of connections and universalisms. When we study the financialisation of housing, or we study the proliferation of housing justice movements, whatever that is, we can't be interested in the universalisms. I'm interested in the connections, but I'm also interested in how Said thinks about contrapuntal method as something attentive to discrepancy. That we are not looking to tell one story. I don't want to tell one unified, universal story about shack dwellers and tenants and mortgage-affected people. Nor do I want to say that they're also different from one another, and they are all place-specific and no generalisations are possible. So, I've been returning to the contrapuntal method to think about how we hold difference in view and what we might learn to pay attention to discrepancy rather than commonality.

COLIN: There's a line in Caldeira's (2017) piece on "peripheral urbanisation" where she talks not about the "juxtaposition of singularities". There's something in that line in thinking as a frame for connecting these discrepancies and similarities across space.

ANANYA: I think the issue is whether it's *similarities* or something else. What is it that we are looking at? What's the translation we strive to achieve, and what are our methodologies of translation? As I've just noted, I am much more interested in discrepancy, and its relationship to historical difference, than in juxtaposition.

MICHELE: The book we're trying to do here is a bit crazy. We have 54 authors from all over the world, and without putting any pressure on you, Ananya – and maybe already thinking about our readership and the authors that are going to contribute to this book – if you want to give us a sense, from your own standpoint, from your own feminist tradition, and also critical race studies tradition, and from your own place, LA: *What is the future of global urbanism?*

ANANYA: In my earlier work, for example the *Poverty Capital* book, I was very influenced by postcolonial feminists like Chandra Mohanty, who have insisted on feminist methodologies for studying the global nature of capitalism. That for me is the imperative of global urbanism. But I am now very interested in what the project of global urbanism might be in what I have been calling the age of Trumpism, as in my essay on sanctuary and abolition in *Society and Space* (Roy, 2019). If this is a historical conjuncture of resurgent right-wing nationalism and organised white supremacy, what then is the urban political? What is the global as a critical site of enquiry? What is the ethic of postcolonial critique?

And for me, there is an ethical commitment to building power, which includes a set of questions. about the role of scholarship in the age of Trumpism. This is about front lines of knowledge and how we shield ourselves from political reality and epistemic violence. But I think that global urbanism, as that loosely structured field of enquiry, also has tremendous potential to address this historical moment in a way that some of the other fields of enquiry in urban studies may not be able to do. And I would love to think about a theoretical research agenda that does that. In my own work, after Trump got elected, and when sanctuary cities become sort of a key site of struggle, I decided to become a student – of border studies, migration studies, and that in turn led me across the Atlantic to the debates in Europe about hospitality, cosmopolitanism, refuge. The global became the threshold of empire, a phrase I use in the *Society and Space* essay.

I hope this is work we can take up in global urbanism. And I think that if we approach it from a global urbanism lens, that work is slightly different from what it looks like in border studies or what it looks like in critical refugee studies. And part of this returns us to the question of the urban political – not simply the city as refuge and sanctuary but how we think about sovereignty, power, belonging, and rightful presence.

References

Brenner, N. (2018) 'Debating planetary urbanization: For an engaged pluralism'. *Environment and Planning D: Society and Space* 026377581875751. doi:10.1177/0263775818757510.

Brenner, N. and Schmid, C. (2015) 'Towards a new epistemology of the urban?' *CITY*, 19(2–3), 151–182.

Caldeira, T. P. (2017) 'Peripheral urbanization: Autoconstruction, transversal logics, and politics in cities of the global south'. *Environment and Planning D: Society and Space*, 35(1), 3–20.

Derickson, K. (2017) 'Urban geography II: Urban geography in the age of Ferguson'. *Progress in Human Geography*, 41(2), 230–244.

Katz, C. (1994) 'Playing the field: Questions of fieldwork in geography'. *The Professional Geographer*, 46(1), 67–72.

Massey, D. (1994) *A Global Sense of Place*. University of Minnesota Press, Minneapolis.

McElroy, E. and Werth, A. (2019) 'Deracinated dispossessions: On the foreclosures of "gentrification" in Oakland, CA'. *Antipode anti*. 12528. doi:10.1111/anti.12528.

Oswin, N. (2018) 'Planetary urbanization: A view from outside'. *Environment and Planning D: Society and Space*, 36(3), 540–546.

Pratt, G. (2018) 'One hand clapping: Notes towards a methodology for debating planetary urbanization'. *Environment and Planning D: Society and Space*, 36(3), 563–569.

Pulido, L. (2006) *Black, Brown, Yellow and Left: Radical Activism in Los Angeles*. University of California Press, Los Angeles.

Roy, A. (2016) 'What is urban about critical urban theory?' *Urban Geography*, 37, 810–823.

Roy, A. (2019) 'The city in the age of Trumpism: From sanctuary to abolition'. *Environment and Planning D: Society and Space*, 37(5), 761–778.

Sheppard, E., Leitner, H., and Maringanti, A. (2013) 'Provinicalizing global urbanism: A mainfesto'. *Urban Geography*, 34(2015), 893–900.

4

HINTERLANDS OF THE CAPITALOCENE

Neil Brenner and Nikos Katsikis

Contemporary conceptualizations of the urban condition emphasize the specific geographies, dynamics and consequences of city-building processes, whether in specific regional contexts or as a national, continental or global aggregation. Such conceptualizations have been pervasive in recent debates on the global urban age and global urbanism, which treat "the city" as the key arena for the major political-economic and ecological transformations of our time. Major disagreements abound regarding how best to understand cities, their spatial patterns, their developmental dynamics, their crisis tendencies and their future potentialities. Nonetheless, from the resurgent universalism of contemporary agglomeration economics and the new "science of cities" to the more contextually embedded explorations of postcolonial, southern and assemblage-theoretical approaches, there is an underlying consensus regarding the basic parameters of the urban question. This consensus can be concisely captured in the following formulas:

urban = city

urbanization = city growth

Today, this methodologically cityist framing of the urban question appears nearly all-pervasive, even among critically reflexive accounts that reject inherited technoscientific, Eurocentric and economistic models, that insist on the situatedness, provisionality and ideological overdetermination of urban epistemologies and that emphasize the uneven, crisis-riven character of urban development. The urban is conceived, at core, with reference to the conditions, morphologies and transformations of the city, a specific type of human settlement or techno-infrastructure that is contrasted to other, putatively non-urban (suburban, rural or wilderness) spaces.[1]

In our recent work, in collaboration with Christian Schmid and other scholars of planetary urbanization, we have questioned this inherited epistemology of the urban.[2] We do not deny the fundamental connection between urbanization and the power of agglomeration (and thus with spaces that have been labeled as "cities"). We argue, however, that the urbanization/city nexus is not logically self-evident. Unlike an analytical statement whose subject and predicate mutually contain one another (for example, "a triangle is a three-sided figure"), the assumption that "urbanization entails city growth" requires careful theoretical clarification and comparative historical-geographical investigation: it is not a logical truism but a synthetic proposition whose content must be systematically explicated. From this perspective, the connection between

DOI: 10.4324/9780429259593-6

urbanization and city-building is not a definitionally pregiven *premise* for urban research but one of its core *problematiques*. Rather than presupposing the outcome (the growth of "the city") of the process under investigation (urbanization) as a logical requirement or definitional necessity, this approach to urban studies asks: how, why and with what consequences do determinate patterns and pathways of city-building flow from historically and geographically specific forms of urbanization? This reframing of the urban question requires not only the interrogation of its historically specific manifestations across spatiotemporal contexts, but a reflexively multiscalar analytic framework that embeds city-building processes within the broader political-economic and environmental geographies that at once support and result from them.

While the notion of the city is likely to remain useful and even necessary as a generic shorthand reference to zones of spatial concentration, densification and intensification (and their juridical expressions in municipal governmental apparatuses), its deployment as a social science concept requires considerable caution, precision, specificity and reflexivity. Too often, folk concepts of the city have been naturalized in the social sciences, in part due to their pervasive deployment by state institutions, corporate alliances and growth coalitions to promote projects of territorial regulation, biopolitical management and economic development. This widespread conflation of categories of practice with categories of analysis has undermined the capacity of urban scholars to interrogate the specific processes and (geo)political strategies through which apparently "city-like" spaces are produced and transformed across time and space. The concept of *concentrated urbanization* breaks from the long-entrenched, *a priori* assumption that urbanization processes result in formally identical, city-like spatial units, directing attention instead to the patterned diversity of such processes, their contextually specific causes, their variegated spatial expressions and their wide-ranging consequences.

This epistemological reorientation is closely connected to a second core research agenda among scholars of planetary urbanization—namely, to consider processes of city-building as being intermeshed with massive transformations of sociospatial and ecological relations beyond the city, across diverse territories and landscapes. While such "hinterland" spaces may not assume city-like morphologies or demographics, their sociospatial relations, land-use arrangements, infrastructural configurations and biogeophysical dynamics are fundamentally shaped and reshaped through their evolving connections to city-building processes, networks and systems. From this point of view, the preconditions and consequences of city-building stretch far beyond the city's immediate environs, at once in political-economic, infrastructural and metabolic terms. The term "hinterland" is used here to demarcate the variegated non-city spaces that are thereby swept into the maelstrom of urbanization, whether as supply zones, logistical corridors, impact zones, sacrifice zones or otherwise. Such spaces include human settlements (cities, towns, villages), land-use configurations (industrial, agrarian, extractive) and nonhuman ecologies (terrestrial, oceanic, subterranean, atmospheric). We refer to explorations of such spaces, and their role in urbanization processes, as engagements with "the hinterland question."

The concept of *extended urbanization* refers to the co-evolution of hinterland zones with metropolitan centers and, more generally, in relation to ongoing processes of concentrated urbanization. Within this expanded conceptualization of the urban question, the non-city preconditions and consequences of city-building are essential and intrinsic, not secondary or extrinsic, dimensions of the urbanization process. Here urbanization is conceived as a dialectic of implosion-explosion: it encompasses not only the moment of spatial concentration (city-building), but equally, the making and remaking of extra-city geographies and ecologies in direct relation to the latter.[3] Thus conceived, city/hinterland dynamics are not a merely contingent articulation among distinct, externally related sites or spaces but represent mutually co-constitutive, dialectically interlinked moments of the urbanization process.[4]

Against the background of these epistemological realignments and conceptual proposals, this chapter revisits the *problematique* of the hinterland question, especially since the accelerated industrialization of capital in the mid-nineteenth century. Across the urban social sciences and design disciplines, the hinterland question is today largely considered secondary or irrelevant to the study of urbanization. The city, its dense socioeconomic networks and its powerful agglomeration economies occupy center stage. We consider this position untenable: from our perspective, the dialectic of city/hinterland relations lies at the heart of the urban *problematique*. And yet, these relations are today undergoing mutations that necessitate not only a repositioning of the hinterland question into the core of urban research and practice but also its radical reconceptualization as such.

Cities without hinterlands?

Prior to the 1970s, the field of urban studies devoted extensive attention to the role of non-city landscapes in the urbanization process. From Johann Heinrich von Thünen's early nineteenth-century model of the relationship between an isolated city and land-use differentiation in its agrarian hinterland, through the early twentieth-century writings of Patrick Geddes, Lewis Mumford and Benton MacKaye on ecological regionalism, up through post-Second World War explorations of central place hierarchies and polarized regional development, city/hinterland relations were widely regarded as constitutive dimensions of the urban *problematique*.[5]

However, during the last half-century, the hinterland has largely disappeared from urban theoretical discourse or has been relegated to mere background status. Under conditions of accelerated geoeconomic integration; splintering national economies; the rollout of neoliberal austerity programs; cascading social, financial and ecological crises; and proliferating local growth initiatives, cities are increasingly viewed as self-propelled economic engines. Within this post-1980s approach to the urban question, the major emphasis is on the internal preconditions, dynamics and consequences of agglomeration. Urbanization is understood as city growth *tout court*—in effect, as *city*ization—rather than as a process that is actively supported by non-city spaces.[6]

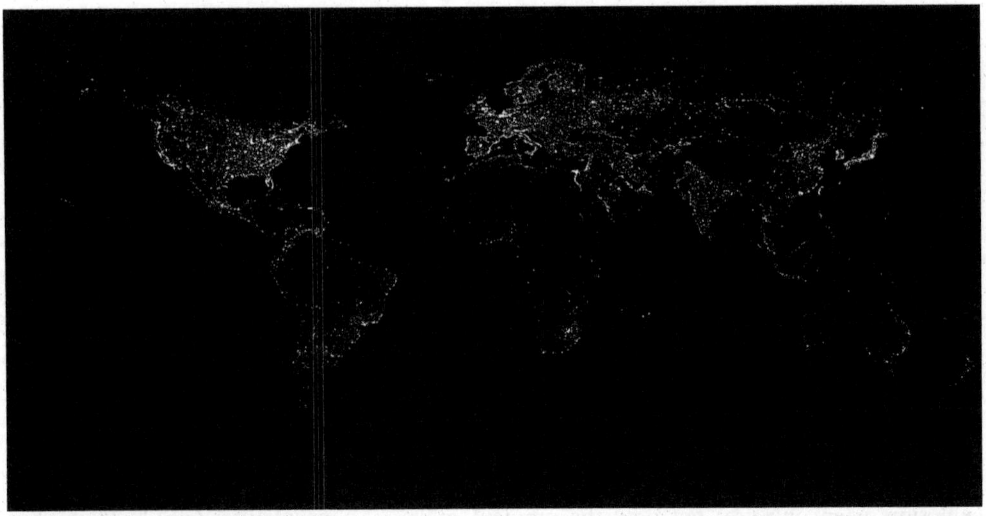

FIGURE 4.1 Nighttime lights of the world. Few images have had a greater impact on contemporary metanarratives of global urbanization than the "nighttime lights of the world" series, initially synthesized during the 1990s in the National Geophysical Data Center (NGDC) in Boulder, Colorado, and subsequently improved through NASA's remote sensing networks (Source: Visualization by the authors based on the publicly available dataset: VIIRS DNB Nighttime Lights Composites, NOAA National Center for Environmental Information (NCEI))

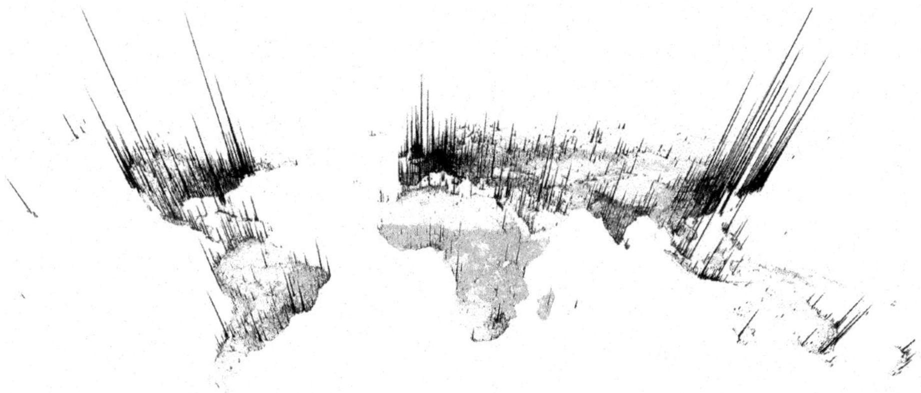

FIGURE 4.2 Spiky world: geographical distribution of global GDP in a three-dimensional perspective. Based on a disaggregation of national GDP data, this visualization represents cities and metropolitan regions as the "spiky" concentration points for major economic activities, reflecting an approach popularized by Richard Florida (2005). Hinterlands—the world's non-city spaces—are correspondingly represented as empty, barren and, by implication, economically marginal (Source: Visualisation by the authors based on the original idea of the spiky world by Richard Florida ("The world is spiky," *The Atlantic Monthly*, October 2005, 48–51). This image is derived from the publicly available dataset: UNEP – United Nations Environment Program, 2012. Gross Domestic Product 2010)

The empty, desolate and isolated condition to which the planet's hinterlands are thereby consigned is starkly illustrated in the iconic image of the world's nighttime lights, in which brightness is treated as a proxy for cityness. This excision of the hinterland's role in urbanization is even more starkly spatialized in the influential concept of the "spiky world," developed by urbanist Richard Florida. Here cities are viewed as the nodal concentration points of global gross domestic product. In both visualizations, non-city spaces appear as barren, depopulated, shapeless voids.

While the roots of this conceptualization predate the 1970s, it was consolidated into a broadly shared episteme of urban studies following the erosion of Fordist-Keynesian, national-developmentalist capitalism. Debates on industrial clusters in the 1980s, global cities in the 1990s, postcolonial cities in the 2000s and more recent assertions of a majority-urban world or "urban age" represent but variations on an underlying vision of cities without hinterlands.

Counterpoint: metabolic urbanization

The major contemporary counterpoints to this hegemonic, city-centric approach to urban studies are associated with various streams of urban ecological thought. Despite their otherwise divergent agendas, these dissident approaches conceive urbanization as a sociometabolic process. From this point of view, cities are supported by diverse metabolic inputs (labor, materials, fuel, water and food), the vast majority of which are produced within non-city zones and, eventually, absorbed back into them as metabolic by-products (waste, pollution, carbon emissions). Such approaches articulate a multiscalar understanding of urbanization that encompasses not only cities and metropolitan regions but also extended landscapes of primary commodity production, logistics and waste management. Metabolic approaches to urbanization thus seek to connect the dynamics of agglomeration to a panoply of non-city geographies—for instance, of land enclosure, population displacement, deforestation, industrial agriculture, extraction, energetics, logistics, waste processing and ecological load displacement.[7]

The contemporary vibrancy of metabolic approaches to urbanization underscores the continued centrality of hinterland questions to early twenty-first century urban studies. These research traditions have contributed fundamental insights that unsettle the myopic narrowing of urban investigations to cities and intercity relations while illuminating the myriad sociomaterial processes through which city development is supported by, and actively coevolves with, non-city spaces. Thus understood, cities are *not* self-propelled. The urban process is materialized within city spaces while invariably exceeding them. City and non-city landscapes are thus dialectically *coproduced* under modern capitalism. The urban *problematique* can only be deciphered adequately through an approach that dialectically connects them at once in social, political, material, infrastructural and ecological terms.

The hinterland enigma

Despite its role in offering powerful scholarly counterpoints to the ideology of the self-propelled city, the bulk of contemporary urban ecological scholarship has confronted the hinterland question only indirectly. While studies of urban metabolism have exhaustively quantified the material and energetic flows that mediate city/hinterland relations, they have tended to bypass the question of how non-city spaces are reconfigured through these mediations. Consequently, the hinterland itself has remained something of a "black box": metabolic flows move in and out, but what actually happens "inside" the box and how the latter has itself evolved are not interrogated.

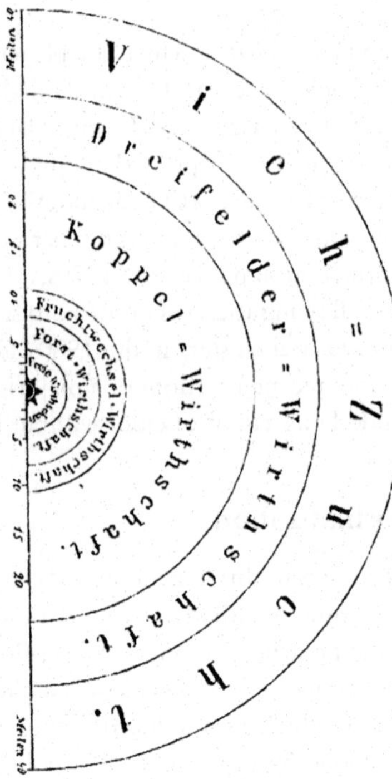

FIGURE 4.3 The hinterland of the "isolated state": Von Thünen's visualization (1826). Johann Heinrich von Thünen's model of city/hinterland relations under mercantile capitalism shaped many subsequent generations of scholarship in urban economic geography. However, except in a few limit-cases of continued, dense metabolic interchange between settlements and their immediately contiguous supply zones, its basic assumptions have been superseded through the forward-motion of capitalist industrialization (Source: Johann Heinrich von Thünen, *Der isolierte Staat in Beziehung auf Landwirtschaft und Nationalökonomie*. Hamburg: Friedrich Perthes, 1826)

The hinterland's internal political-economic operations, land-use matrices, property relations, spatiotemporal dynamics and socioecological crisis tendencies thus remain enigmatic.

Many contemporary urban researchers appear to presuppose a conception of the hinterland that is derived from the mercantile period of capitalist development in which von Thünen constructed his famous account of the "isolated state" (1826).[8] Here the hinterland is territorially contiguous and directly linked to the city, which in turn serves as its market outlet and its manufacturing center. Although commodity production is generalized, there is no structural impulsion to enhance labor productivity or to maximize crop yields. In this model, the non-city zone is, by definition, non-industrial; land-use sorting occurs due to differential transport costs.

The point here is not to assert that contemporary urbanists self-consciously embrace von Thünen's conception of a contiguous, non-industrial hinterland but to suggest that some version of this nineteenth-century model continues to shape our collective imagination of non-city landscapes, which are thereby reduced to an amorphous "ghost acreage" of "emptied spaces, homogeneous blanks yet to be inscribed by human history."[9] As a result, scholars have only rarely sought to decipher the specific patterns and pathways through which hinterlands have been enclosed, colonized and creatively destroyed since the 1850s, even though such transformations have been as far-reaching as those that are commonly ascribed to the crisis-riven remaking of cities' own built environments during the same period. Investigating such mutations will require new conceptualizations of city/hinterland matrices in relation to historical, contemporary and emergent geographies and ecologies of planetary urbanization.[10]

Hinterlands of the Capitalocene

How, then, to decipher the role of hinterlands in supporting and buffering the metabolic dynamics, rifts and crisis tendencies of urbanization under capitalism? This challenge is, on the one hand, a conceptual one insofar as it requires us to rethink the very nature of hinterlands in the age of capital, or "Capitalocene."[11] The hinterland concept is not only a technical term of art within economic geography but also has been thoroughly intermeshed with racialized historical geographies of colonization, land grabbing, enclosure, territorial dispossession and institutionalized violence against indigenous people, enslaved Africans and other subordinated, displaced and "surplus" populations.[12] The exploration of such issues will require rigorously historical analysis since the social, infrastructural, institutional and environmental geographies of "hinterlandized" regions have been dramatically transformed during the process of capitalist industrial development, in close connection to state spatial strategies to shape and reshape patterns of spatial inequality within and beyond their territories. Indeed, much like the term "city," the inherited notion of the hinterland implies a static, context-transcendent universalism that belies the extraordinary variegation and dynamism of the spaces it purports to demarcate while occluding the relations of imperial power, racial domination and unequal ecological exchange through which they are constituted. As such, even within the definitionally specific parameters demarcated earlier, the hinterland concept should be used only with utmost reflexivity and precision, and as no more than a "first cut" towards a broader inquiry into the *problematique* of extended urbanization.

Investigating the hinterlands of the Capitalocene will also require a systematic reengagement with classic questions about cities and empire. Much of the inherited scholarship on such issues emphasizes the role of cities as "spearheads" for colonial territorial conquest, plunder of resources and ideological domination.[13] While our approach affirms the contributions of this classical line of analysis, it also directs attention to the intrinsic connection between accelerated city-building processes under capitalism and what we might think of as the "hinterlandization" of the world—that is, the large-scale enclosure, colonization and operationalization of non-city zones within and beyond capital Euro-American heartlands into subordinate zones of extraction through which successive historical "cheap natures" are appropriated.[14] From this expanded point of view, urbanization involves not only the role of city-building in grounding imperial projects of territorial expansion but also variegated processes of hinterlandization through which non-city zones are operationalized, both economically and ecologically, to support agglomeration-centric processes of geographical transfer of value in both metropole and colony alike.[15] In this sense, the project of superseding narrowly Eurocentric formations of the urban question must not only explore the decidedly *non*-derivative theoretical significance of urbanisms beyond the West but the *coloniality of urbanization* itself—that is, the multiscalar, intercontinental force fields of extraction, surplus value transfer and thermodynamic imperialism that underpin, animate and result from city-building processes under modern capitalism.[16]

Confronting the hinterland enigma is, finally, a challenge that will require critical appropriations of newly available sources of geospatial data, which may offer a powerful basis for investigating the historical and contemporary rearticulation of land uses, built and unbuilt environments and political ecologies around the world.[17] It is not sufficient simply to posit that such non-city "outsides" are constitutively important for city-building processes or to focus on measuring the role of such spaces as "taps" and "sinks" for the metabolic dynamics of capitalist urbanization. While this vast planetary hinterland covers nearly 70% of the earth's terrestrial surface and is densely layered with productive, extractive, circulatory and informational infrastructure, it has remained an obscure background to the study of contemporary urbanization. It is precisely in this sense that the "black box" of the hinterland must be opened and systematically rearticulated to the central agendas of urban studies. What is required is a framework that can connect historically and geographically specific forms of city and non-city space (and associated relations of domination) as coproduced, coevolving moments within the combined, uneven, variegated and crisis-riven world ecologies of capitalist urbanization.

FIGURE 4.4 Agglomerations and the "used area" of the planet in the early 21st century. This map juxtaposes a demarcation of the world's metropolitan agglomerations (red zones) onto a rendering of the total "used area" of the entire planet (black and grayscale). Agglomeration zones constitute only a miniscule percentage of the planet's operationalized landscapes, which are mostly devoted to primary commodity production (agricultural cultivation, grazing, forestry), resource extraction, logistics and waste disposal (Source: Visualisation by the authors based on the following sources and publicly available datasets: European Commission Joint Research Center, 2016, Global Human Settlement Layer; K.-H. Erb, V. Gaube, F. Krausmann, C. Plutzar, A. Bondeau and H. Haberl, "A comprehensive global 5 min resolution land-use dataset for the year 2000 consistent with national census data," *Journal of Land Use Science* 2, 3 (2007): 191–224; Vector Map Level 0 (VMap0) dataset released by the National Imagery and Mapping Agency (NIMA), 1997)

The development of such a framework requires systematic elaboration elsewhere. Here it must suffice to offer some initial generalizations regarding four key mutations of city/hinterland relations that have been particularly pronounced during the last half-century. These relatively abstract propositions are not intended to foreclose more contextually embedded, historically specific lines of inquiry, but to stimulate further reflection, investigation and debate regarding the restlessly churning dynamics of planetary urbanization. We aim not to subsume all aspects of social life in non-city zones under the latter rubric, but rather to mobilize the methodological tactics proposed here as a basis for investigating historically and geographically specific processes of hinterlandization and their far-reaching implications for the political-economic and environmental geographies of capital. For present purposes, we continue to deploy the term "hinterland", even as our analysis demarcates various ways in which its analytic usefulness has been ever more narrowly circumscribed through the vicissitudes of capitalist uneven spatial development.

Enclosure, distanciation and infrastructuralization

First, primary commodity production has been globalized and specialized, causing local, contiguous hinterlands to be enmeshed within export-oriented transnational production networks. Contiguous hinterlands remain important, but are no longer the norm, either in the older industrialized world or in most southern megacities. This implosion-explosion of hinterland zones has been animated by capital's drive to increase labor productivity, to extend interspatial connectivity and to appropriate ecological surpluses, all of which entail the construction of large-scale infrastructural configurations for extraction, production and circulation, as well as the concomitant expropriation of peasant populations from their historic means of subsistence. While such strategies may temporarily boost profits, they also increase the organic composition of capital, as living labor is replaced by machinery, equipment and infrastructure. This leads to the precipitous decline of the non-city workforce ("depeasantization"), the dispossession of formerly agrarian laborers from nonmarket means of social reproduction and their mass expulsion into the squatter settlements, or "slums," of expanding megacities where they forge informal economies of survival as a proletarianized, migratory reserve army of "surplus labor."[18] This demographic "emptying of the

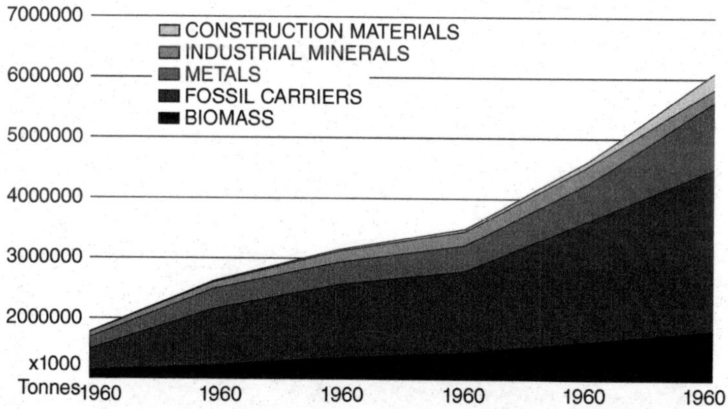

FIGURE 4.5 Growth in global trade of basic materials, 1960–2010. Over the last decade, the global trade in primary commodities—such as agricultural and forestry products (biomass), fossil fuels, industrial minerals, metals and construction materials—has increased more than threefold. This reflects the increasing globalization of hinterland economies (Source: Visualisation by the authors based on data from F. Krausmann, S. Gingrich, N. Eisenmenger, K.-H Erb, H. Haberl and M. Fischer-Kowalski, "Growth in global materials use, GDP and population during the 20th century," *Ecological Economics* 68, 10 (2009): 2696–2705)

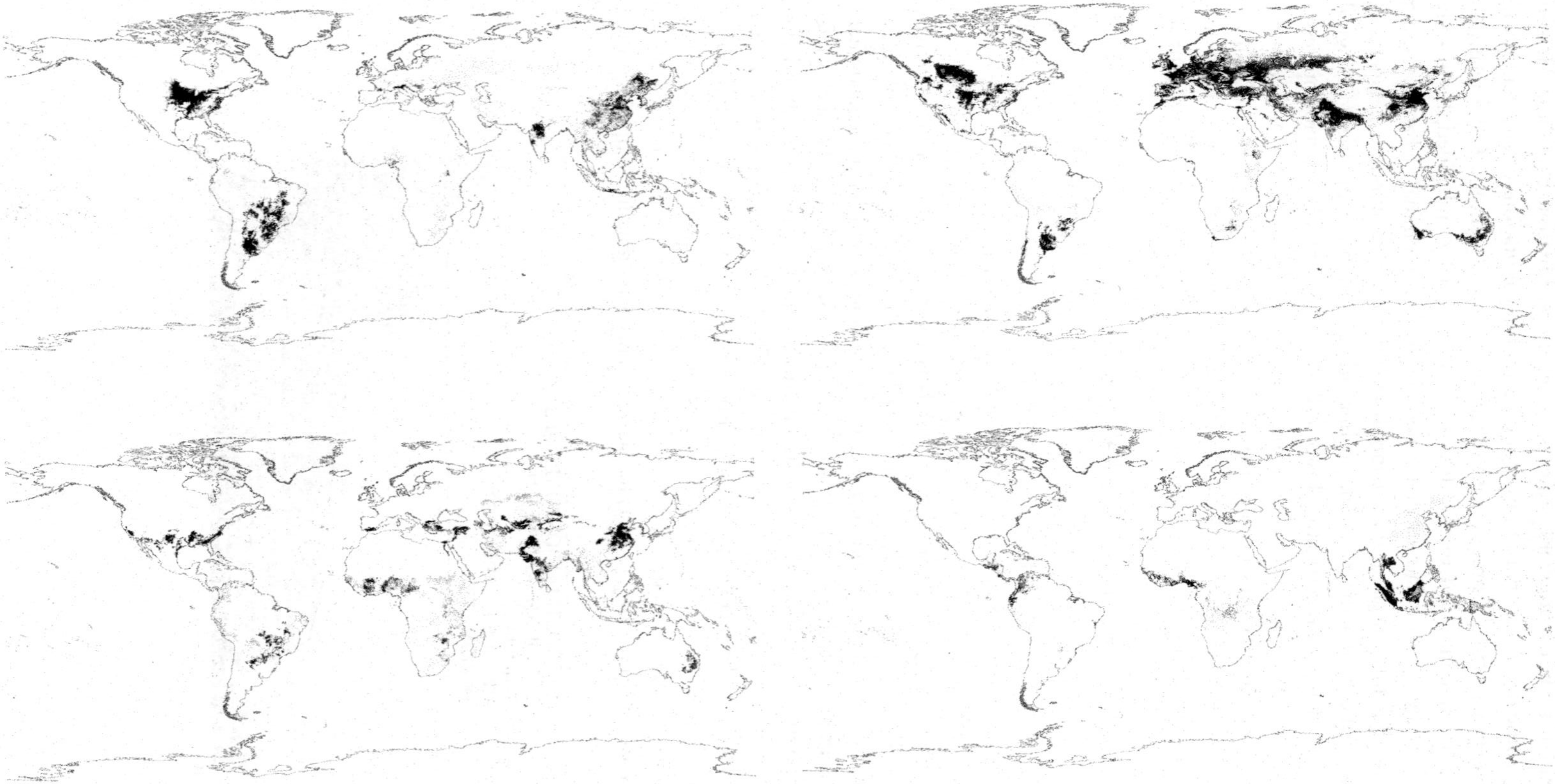

FIGURE 4.6 Worlds of specialized agricultural production, 2000. This map series depicts the geographical distribution of production sites for four of the most globally traded agricultural commodities as of 2000. The grayscale gradient on the maps corresponds to production areas for soybeans (top left), wheat (top right), cotton (bottom left) and palm oil (bottom right) (Source: Visualisation by the authors based on data presented in C. Monfreda, N. Ramankutty and J. Foley, "Farming the planet 2: Geographic distribution of crop areas, yields, physiological types, and net primary production in the year 2000," *Global Biogeochemical Cycles*, 22, 1 (2009): GB1022)

world's countryside" has been inextricably linked to strategies to hollow out, enclose and operationalize the spaces of erstwhile rural life—manifested, for example, in the differential integration of smallholders into cash-crop-based commercial networks; the progressive concentration of agrarian property regimes; the resurgence of plantation agriculture; the large-scale conversion of arable land into enclosed livestock pasture; the establishment of robotized, monofunctional and export-oriented agribusiness landscapes; and massive ecological devastation, as significant territories are transformed into "sacrifice zones" for capital's spatially externalized impacts.[19]

Hinterlands of hinterlands

Second, as they are embedded within global supply chains, inherited hinterlands lose their directional articulation to specific zones of consumption, metropolitan or otherwise. The linear directionality of von Thünen's classic model—in which each hinterland has "its" city and each city "its" hinterland—is thus no longer a reliable guide. The point is not simply that contemporary cities' hinterlands are more distanciated than previously, but that their operational logics, infrastructural configuration, metabolic relays and developmental dynamics have been qualitatively transformed. On the one hand, most of the world's most productive, specialized and export-oriented hinterlands now circulate their outputs to a multitude of metropolitan agglomerations and across the global metropolitan network as a whole. Just as importantly, many zones of primary commodity production are now most directly articulated not to major cities and metropolitan regions but to other operational landscapes of extraction, cultivation, processing and distribution, which are in turn embedded and intermeshed within an intercontinental logistics space. This situation is exemplified in the monocrop soybean landscapes of Amazonia, whose outputs are mostly exported as cattle feed to Chinese livestock hinterlands; in the export of phosphate fertilizer from Central Florida to Brazilian agro-industrial hinterlands; or in the use of hydroelectric dams to power the extractive hinterlands of northern Chile. In this sense, the hinterlands of the Capitalocene are now more tightly articulated to transnational production matrices than to localized relays of consumption.[20]

From formal to real subsumption

Third, most forms of primary commodity production have remained heavily contingent upon the extra-human geographies of the earth system (for instance, soil and weather conditions, water availability or resource deposits) that can only be modified through significant industrial investment (for instance, in fertilizer, greenhouses, irrigation systems, biotechnology and other sociotechnical fixes). Historically, therefore, the industrial operationalization of hinterland spaces has occurred through strategies to establish new resource frontiers and, as the latter are exhausted, through compensatory efforts to intensify techno-extractive logics. In both moments of this process, new industrial infrastructures are established and intensively operationalized before being superseded as successive regimes of accumulation are exhausted. Many contemporary hinterlands, therefore, are no longer zones of mere "formal subsumption" in which inherited socioecological resource matrices and regimes of cheap nature are directly appropriated as commodities for external market exchange. Insofar as the geographies and ecologies of non-city zones are themselves systematically re-engineered to maximize surplus value extraction and accelerate capital accumulation, a "real subsumption" of hinterland spaces appears to be under way: nature is itself increasingly capitalized in a profit-driven terraforming process.[21] In this manner, many erstwhile hinterlands, or parts thereof, are transformed into configurations of large-scale territorial-ecological machinery: capital-intensive, more-than-human infrastructural assemblages oriented towards capital accumulation within a planet-encompassing profit matrix.

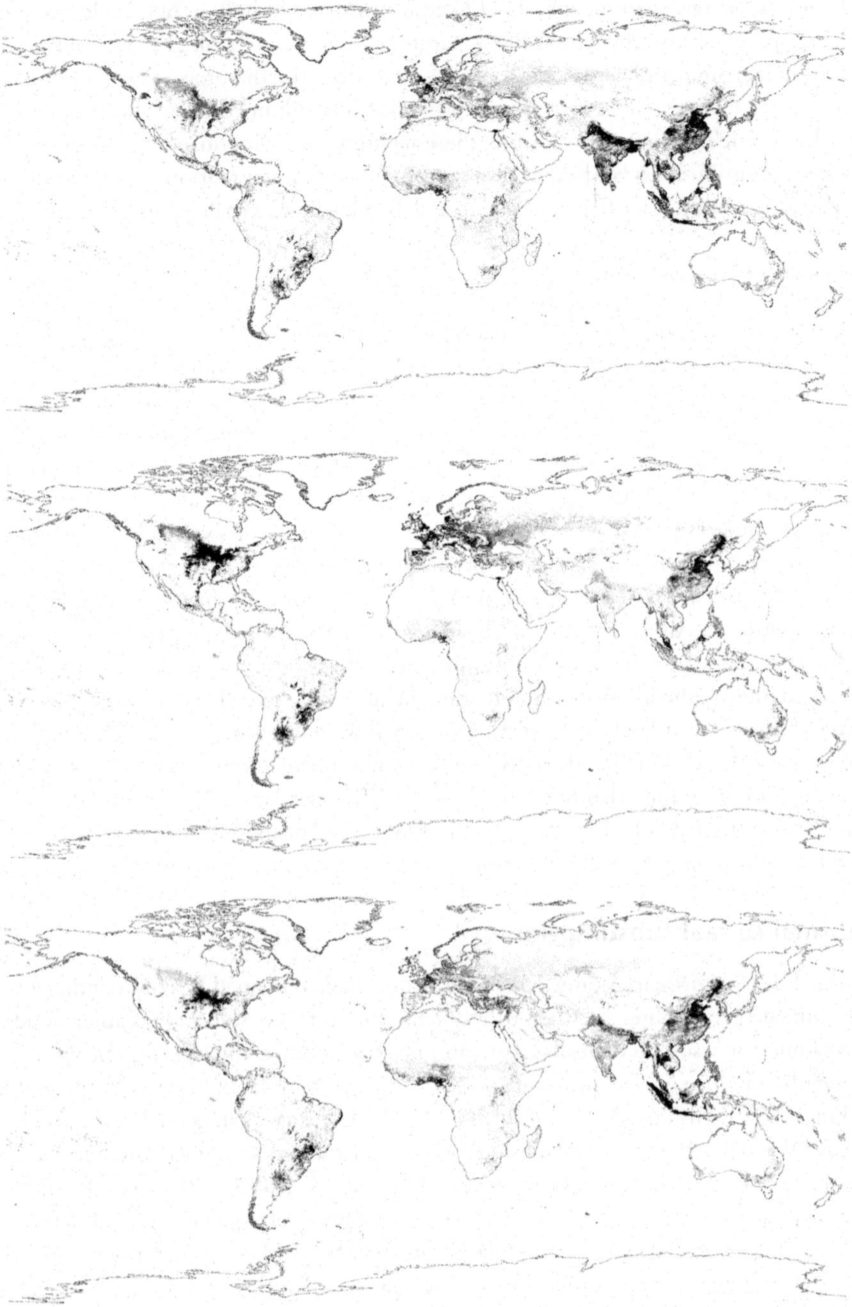

FIGURE 4.7 Hinterlands of hinterlands, 2000. This map series depicts the geographical distribution of cropland areas (arranged, respectively, from top to bottom) dedicated to food, feed, or non-food uses (such as energy and industrial inputs) The bottom two maps can be conceived as representations of hinterlands of hinterlands, insofar as they depict the supply of specific industrial inputs to other hinterlands (for example, cattle feed to livestock production zones, or biofuel to the energy sector) (Source: Visualisation by the authors based on the publicly available datasets: E. Cassidy, P. West, J. Gerber and J. Foley, "Redefining agricultural yields: from tonnes to people nourished per hectare," *Environmental Research Letters* 8, 3 (2013): 034015)

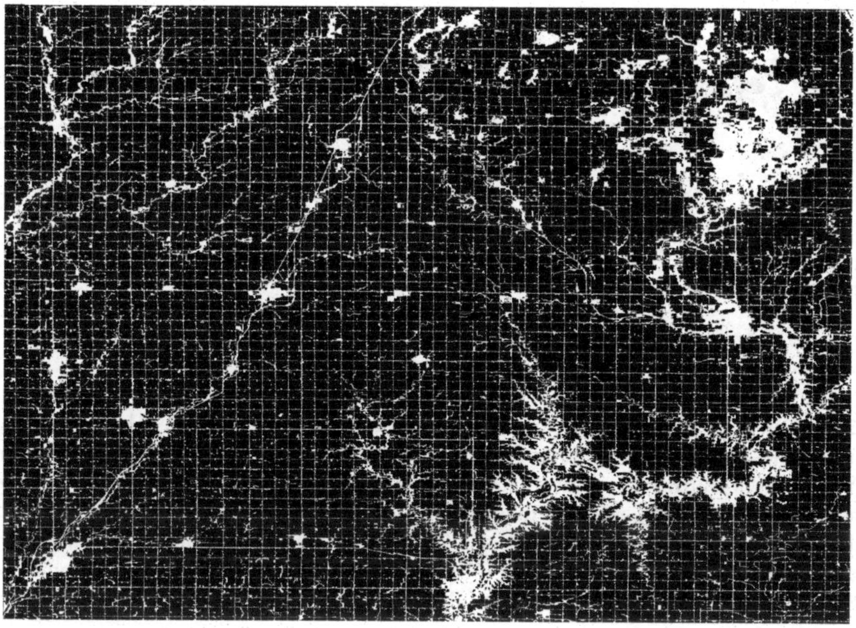

FIGURE 4.8 Mechanized, monoculture landscapes of corn and soybean production in the US Midwest, 2018. Capital-intensive, highly industrialized and densely equipped landscapes of cash-crop monocultures dominate the US Cornbelt, where more than 80% of all land is dedicated to the cultivation of corn and soybeans. The zone of corn and soybean production (depicted in black) is configured among one-mile tiles within a Jeffersonian grid pattern. This permits the maximally efficient operation of agro-industrial machinery. Beneath this terrestrial surface is an extensive subterranean drainage system that supports soil tilling (Source: Visualisation by the authors based on the following publicly available dataset: USDA National Agricultural Statistics Service Cropland Data Layer (2018), published crop-specific data layer, available at https://nassgeodata.gmu.edu/CropScape/ (accessed 06/05/2019))

Metabolic rifts and cycles of creative destruction

The proliferation of specialized, globally interdependent zones of primary commodity production reveals not only the ways in which inherited human and nonhuman landscapes have been commodified, infrastructuralized, capitalized and terraformed, but also the progressive depletion of their capacity to contribute ecological surpluses to sustain the accumulation process as they are subjected to overcultivation, overgrazing, deforestation and generalized toxic contamination.[22] The proliferation of such metabolic rifts further accelerates capital's drive to redesign, terraform and mechanize hinterland geographies, at once through the substitution of manufactured, biotechnological inputs into production and through the construction of colossal techno-infrastructural configurations.[23] The hinterlands of the Capitalocene are, therefore, chronically unstable. As the capitalization of nature proceeds apace and ecological surpluses are exhausted, the resultant metabolic rifts severely destabilize prevalent regimes of accumulation, leaving vast tracts of "dead land, dead water" in which not only economic activity, but life itself, are largely expunged.[24] Consequently, established hinterland infrastructural configurations and political ecologies are rendered obsolete, even though their sociotechnical capacities and resource stocks may have been only partially amortized. This leads to intense struggles over the choreography, form, socioecological impacts and future pathways of territorial transformation across the global metropolitan network and throughout the interconnected landscapes of extraction, production and circulation that underpin transnational production

FIGURE 4.9 Intensity of synthetic fertilizer application (nitrogen) over the global croplands, 2000. This map depicts annual levels and locations of nitrogen fertilizer use through a black-dotted gradient pattern. Since 1950, fertilizer use has increased ninefold, while total cropland area has expanded by less than 30% (Source: Visualisation by the authors based on the following publicly available datasets: N. Ramankutty, A. T. Evan, C. Monfreda and J.A. Foley, *Global Agricultural Lands: Croplands, 2000* (Palisades, NY: SEDAC, 2010); P. Potter, N. Ramankutty, E. M. Bennett and S.D. Donner, Global Fertilizer and Manure, Version 1: Nitrogen Fertilizer Application (Palisades, NY: SEDAC, 2012))

systems. For this reason, any contemporary approach to the hinterland question must consider the systemic political-economic *and* political-ecological vulnerabilities of those non-city spaces that have been forged to support the globalizing, profit-maximizing dynamics of supply-chain capitalism and its coordinating centers in the global metropolitan network.

The hinterland question, reframed

Under contemporary conditions, there is no singular hinterland of "the" city. Instead, non-city productive landscapes have become more specialized, infrastructurally dense and industrially intensive, and they are intermeshed with one another through extended material, operational and informational linkages, as well as through their continuous, but largely indirect, exchanges with (strategic nodes within) the global metropolitan network. However, the operational landscapes of planetary urbanization are hardly a stable foundation for territorial development, social reproduction or political-ecological security. Indeed, even as they support (unevenly distributed bursts of) enhanced industrial productivity and the (crisis-riven) accelerated, long-distance circulation of commodities, the hinterlands of the Capitalocene expose local territories and communities to increasing turbulence, risk and precarity while systematically degrading the ecological preconditions of both human and nonhuman life on a planetary scale.

How, and by whom, has this planetary fabric of urbanization been forged? What are its social, political, institutional, regulatory and ecological preconditions? What are its major contradictions, crisis tendencies and vulnerabilities? Can the massive sociotechnical capacities it has unleashed somehow be harnessed to support more just, democratic, nonviolent, culturally vibrant and ecologically sane forms of collective existence? Are there alternative forms of urbanization, planetary or otherwise, and can their sociometabolic dynamics be reflexively designed, negotiated and institutionalized through political agency? Or will the violent, exclusionary and relentlessly profit-driven illogics of planetary urbanization continue to degrade, erode and destroy the web of life? These are among the most urgent dimensions of the hinterland question in the Capitalocene. They are essential for any critical approach to contemporary urbanization.

Notes

1 Neil Brenner and Christian Schmid, "The 'urban age' in question," *International Journal of Urban and Regional Research* 38.3 (2013): 731–755; Hillary Angelo and David Wachsmuth, "Urbanizing urban political ecology: A critique of methodological cityism," *International Journal of Urban and Regional Research*, 39.1 (2015): 16–27.

2 See, for example, Neil Brenner and Christian Schmid, "Towards a new epistemology of the urban?" *CITY*, 19.2–3 (2015): 151–182; Neil Brenner, "Urban Revolution?," in *Critique of Urbanization: Selected Essays* (Basel: Bauwelt Fundamente/Birkhäuser Verlag, 2016), 192–211; and Nikos Katsikis, "The 'other' horizontal metropolis: landscapes of urban interdependence," in Paola Vigano ed., *The Horizontal Metropolis Between Urbanism and Urbanization* (Berlin: Springer, 2018), 23–46.

3 Henri Lefebvre, *The Urban Revolution*, trans. Robert Bononno (Minneapolis: University of Minnesota Press, 2003 [1970]); Neil Brenner ed., *Implosions/Explosions: Towards a Study of Planetary Urbanization* (Berlin: Jovis, 2013); Neil Brenner, *New Urban Spaces: Urban Theory and the Scale Question* (New York: Oxford University Press, 2019).

4 Martín Arboleda, "Spaces of extraction, metropolitan explosions: planetary urbanization and the commodity boom in Latin America," *International Journal of Urban and Regional Research*, 40.1 (2016): 360–378; Martín Arboleda, "Financialization, totality and planetary urbanization in the Chilean Andes," *Geoforum* 67 (2015): 4–13.

5 Nikos Katsikis, *From Hinterland to Hinterglobe: Urbanization as Geographical Organization*. Doctor of Design (DDes) thesis, Graduate School of Design (GSD), Harvard University (Cambridge, MA: 2016).

6 Edward W. Soja, *Postmetropolis* (Oxford: Blackwell, 2000).

7 The most significant contemporary streams include, among others, historical investigations of city/hinterland relations, such as William Cronon's study of Chicago and the US Midwest in *Nature's Metropolis* (New York: Norton, 1991) or Gray Brechin's investigation of urbanizing California in *Imperial San Francisco* (Berkeley and Los Angeles, University of California Press, 1999); approaches to materials flow analysis by Marina Fischer-Kowalski, Helmut Haberl and their colleagues in the Institute of Social Ecology at Klagenfurt University (https://www.aau.at/en/social-ecology/); the investigation of "teleconnections" through which land use transformations in cities impact land-use change elsewhere developed by Karen Seto and her colleagues at Yale University (https://urban.yale.edu/); and the analysis of urban ecological footprints developed by William Rees, Mathis Wackernagel and their collaborators (https://www.footprintnetwork.org/our-work/cities/). For overviews and detailed citations to these literatures, see Katsikis, *From Hinterland to Hinterglobe.*

8 Johann Heinrich von Thünen, *Von Thünen's Isolated State*, trans. Carla M. Wartenberg, ed. Peter Hall (Oxford: Pergamon Press, 1966 [1826]).

9 Gavin Bridge, "Resource triumphalism: postindustrial narratives of primary commodity production," *Environment and Planning A*, 33 (2001): 2154.

10 Brenner and Schmid, "Towards a New Epistemology of the Urban?"; Katsikis, *From Hinterland to Hinterglobe*; and Álvaro Sevilla-Buitrago, "*Urbs in Rure*: Historical Enclosure and the Extended Urbanization of the Countryside," in Brenner, *Implosions/Explosions*, 236–259. These texts explicate in more detail the specific conceptualization of "planetary" urbanization we are presupposing here.

11 On the "Capitalocene," see Jason W. Moore ed., *Anthropocene or Capitalocene?* (Oakland: PM Press, 2018).

12 The link between hinterlandization—the subordination of non-city space into a production and supply zone for city-based economic operations—and the violent seizure and enclosure of land is evident, albeit not centrally theorized, in Cronon's classic account of Chicago and the "Great West." On the genealogy of the hinterland concept within the study of ports, trade routes and "remote" zones by economic geographers, see Eugene Van Cleef, "Hinterland and Umland," *Geographical Review*, 31.2 (1941): 308–311.

13 On cities as "spearheads" for empire, see Anthony D. King's classic account in *Urbanism, Colonialism and the World-Economy* (London: Routledge, 1990). In King's work, this conception is closely connected to a world-systems conception of spatial polarization under capitalism in which the world economy is coherently divided among core, semi-peripheral and peripheral zones, with each defined as a more-or-less bounded territorial whole. Cities form the mediating links between such zones, the nodes of connectivity through which world-systemic relations of domination are articulated.

14 The concept of extraction is used here in a broad sense, not simply with reference to mining but also as a more general characterization of capital's relentlessly expansionary transformation/appropriation of social and ecological relations to fuel its accumulation process. See Sandro Mezzadra and Brett Neilson, "On the multiple frontiers of extraction: Excavating contemporary capitalism," *Cultural Studies*, 31.2–3 (2017): 185–204. On the appropriation and capitalization of cheap natures during successive

world-ecological revolutions under capitalism, see Jason Moore's seminal account in *Capitalism and the Web of Life: Ecology and the Accumulation of Capital* (New York: Verso, 2015).

15 This formulation provides a basis on which to revisit some of the classic Marxian formulations on the city and the countryside, from Marx and Engels' *German Ideology* (https://www.marxists.org/archive/marx/works/1845/german-ideology/) and Antonio Gramsci's account of the "Southern Question" to Raymond Williams' *The Country and the City* (London: Oxford University Press, 1973), which explicitly connect intra-national city/countryside divides to broader geographies of worldwide imperialism. For further elaborations, see Timothy Brennan, "On the image of the country and the city," *Antipode*, 49, 1 (2015): 34–51; Michael Ekers, Gillian Hart, Stefan Kipfer and Alex Loftus eds, *Gramsci: Space, Nature, Politics* (Malden, MA: John Wiley & Sons, 2013); and Kanishka Goonewardena, "The country and the city in the urban revolution," in Brenner, *Implosions/Explosions*, 218–231. This approach also productively destabilizes traditional, methodologically territorialist models of the capitalist world system as involving a neat division among core, semi-peripheral and peripheral zones. For a critique of the latter in the context of post-1980s patterns of geoeconomic restructuring state rescaling and urban redevelopment, see Neil Brenner, "Beyond state-centrism: Space, territoriality and geographical scale in globalization studies," *Theory and Society* 28.1 (1999): 39–78.

16 Many intellectual resources for such an inquiry can be gleaned from established literatures—for instance, on race, territory and empire; on agrarian environments, deruralization and depeasantization; on the neoliberalization of nature; and on the geographies and spatial politics of extractivism, forestry, fishing and energy. However, their appropriation for the lines of inquiry proposed here is constrained by the continued intellectual parcelization of research among distinct, city-centered or rural- or nature-centered fields of scholarship that are only rarely connected, either in theoretical-epistemological or historical-empirical terms. For a critical reflection on this problem with reference to important links between accounts of extended urbanization and Marxian agrarian studies, see Swarnabh Ghosh and Ayan Meer, "Extended urbanization and the agrarian question: Convergences, divergences and openings," Urban Theory Lab, Harvard GSD, December 2019. For exceptions to this generalization with reference to the study of urbanization and the contemporary global land rush, see Annelies Zoomers, Femke van Noorloos, Kei Otsuki, Griet Steel and Guus van Westen, "The rush for land in an urbanizing world: From land grabbing toward developing safe, resilient and sustainable cities and landscapes," *World Development* 92 (2017): 242–252; and Joshua Leon, "The role of cities in land grabs," *Third World Quarterly*, 36.2 (2015): 257–273. For a pioneering urbanization-theoretical approach to wood and forestry, see Daniel Ibañez, Jane Hutton and Kiel Moe, *Wood Urbanism: From the Molecular to the Territorial* (Barcelona: Actar, 2020).

17 Luke Bergmann and Mollie Holmberg, "Land in motion" *Annals of the American Association of Geographers* 106.4 (2016): 932–956; and Neil Brenner and Nikos Katsikis, *Is the World Urban? Towards a Critique of Geospatial Ideology* (Barcelona: Actar, forthcoming 2020).

18 On the link between depeasantization and slum formation in southern megacities, see Farshad Araghi, "Global depeasantization, 1945–1990," *Sociological Quarterly*, 36.2 (1995): 337–68, and Farshad Araghi, "The great global enclosure of our times: Peasants and the agrarian question at the end of the twentieth century," in Frederick Buttel, Fred Magdoff and John Bellamy Foster eds, *Hungry for Profit: The Agri-Business Threat to Farmers* (New York: Monthly Review Press, 2000), 145–160. See also Mike Davis, *Planet of Slums* (London: Verso, 2005).

19 On the "emptying" of the global countryside and the various pathways of peasant displacement and dispossession, see Araghi, "The great global enclosure," 145, *passim*.

20 Bergmann and Holmberg, "Land in motion."

21 On the application of the Marxian distinction between formal and real subsumption to biogeophysical processes, see William Boyd, W. Scott Prudham, and Rachel Shurman, "Industrial dynamics and the problem of nature," *Society and Natural Resources*, 14.7 (2001): 555–570. On the capitalization of nature, see Moore, *Capitalism in the Web of Life*. For a reflection on terraforming in the Anthropocene, via a discussion of Kim Stanley Robinson's *Mars Trilogy*, see McKenzie Wark, "Molecular red: Theory for the Anthropocene," *e-flux*, 63, March (2015).

22 Moore, *Capitalism in the Web of Life*. See also Farshad Araghi's account of "ecological enclosures" in "Accumulation by displacement: Global enclosures, food crisis and the ecological contradictions of capitalism," *REVIEW*, XXXII.1 (2009): 113–146.

23 David Goodman, Bernardo Sorj and John Wilkinson, *From Farming to Biotechnology: A Theory of Agro-Industrial Development* (New York: Blackwell, 1987).

24 See Saskia Sassen, *Expulsions: Brutality and Complexity in the Global Economy* (Cambridge, MA: Harvard University Press, 2015).

5

MAKING SPACE FOR QUEER DESIRE IN GLOBAL URBANISM

Gavin Brown and Dhiren Borisa

This chapter began as a dialogue between two, very differently positioned, geographers interested in sexual lives in contemporary cities. That dialogue initially revolved around a comparison of the experiences of queer Dalit men in Indian cities that are the focus of Dhiren's work, alongside the European and North American cities studied by Gavin. Out of that initial dialogue, an argument began to distil, which we present here. Our argument is in three parts: first, that the types of urban gay spaces which proliferated in major Australasian, European, and North American cities from the 1970s onwards have become globalised and replicated in cities around the world. These are sites of aspiration and desire for many different groups – both the desire to be part of an imagined global 'gay community' and lifestyle and an aspiration for a form of 'progressive' liberal politics with which it is often (but not exclusively) associated. Our second point is that it is not possible to construct Western-style 'gaybourhoods', pride events, film festivals, or activist circuits in many cities. Equally, even in those cities where they do (precariously) exist, not everyone who desires to be part of them can access those spaces. This does not mean that they do not aspire to be part of global gay modernity, nor that they do not construct (often more ephemeral) spaces imbued with this aspiration elsewhere in the city. The second part of our chapter explores some of the spaces that are created in more peripheral locations in cities around the world. Finally, we recognise that (globally) these opportunities are primarily restricted to those people who already have some familiarity with the vocabulary of identity categories that have circulated from the North Atlantic world over the last century or so. These more globalised lesbian, gay, bisexual, transgender (LGBT) spaces exist in tension with other sexualised subjectivities that do not fit the 'global gay' model. In the final section of the chapter, we step outside the 'global gay imaginary' to take a more pluriversal approach to sexual and gendered difference and to consider some of the spaces that sustain a more diverse range of 'queer' lives in cities around the world.

The big city has been central to the mythologies that sexual and gender minorities create for themselves for more than a century (Abraham, 2009). The rapid expansion of cities in Europe in the nineteenth century created space for sexual minorities to find each other and the anonymity to do so in relative safety (Hubbard, 2012). Writing in a North American context, Weston (1995) associated the proliferation of lesbian and gay communities in the 1970s, with a 'great gay migration' to the coastal metropolises on the continent. Metropolitan urban space was central to how many (North American) lesbians and gay men understood their sexualities and the possibilities for a better life.

DOI: 10.4324/9780429259593-7

Within urban studies, despite the role cities play in terms of sexual possibilities, sexuality has remained an under-researched theme (Hubbard, 2012). Within the geographies of sexualities literature, for too long, the same roster of cities – Amsterdam, London, New York, San Francisco, and Sydney – kept recurring. Slowly, over the last 15 years, the field has expanded to consider the co-constitution of sexualities and urban space in a wider set of cities. Beyond the cities that initially dominated research on sexualities and urban space, Vickers (2010) argues there is a need to understand the ways in which sexual (and, particularly, queer) possibilities are imagined in relation to cities. Sexual tolerance has gone global as a badge of progressive politics, and many world cities have celebrated themselves as inclusive, tolerant spaces. Consequently, the global gay subject is also a classed and racialised subject (Oswin, 2015). Cities also continue to be sites of community building and the contestation of sexual norms. This is as true in postcolonial countries like India (Shah, 2015) as it is in Western Europe and North America. Alongside elite celebrations of tolerance, urban world making is enacted from below through working-class re-imaginings of queer (im)possibilities (Gopinath 2005; Manalansan, 2015).

So much of the debate about planetary urbanisation (Brenner and Schmid, 2015) focuses on the dynamics of contemporary capitalism in driving increasing urbanisation. As Oswin (2018) and others (Hubbard, 2012) highlight, too frequently, debates in urban studies focus singularly on the role that technical composition of capitalism and class dynamics play in shaping cities and urbanisation processes without paying attention to the ways in which sexuality, sexual norms, and desires also shape cities and urban life. This critique notwithstanding, there is a sizeable body of work (D'Emilio, 1983; Duggan, 2002) focused on the experience of the Global North/Minority World, which has considered how changes in the division of labour have created both opportunities for gay people and changes in the dominant expressions of sexual politics.

In contrast to political economy approaches that rely on a strong critique of consumer capitalism or neoliberalism, Nash and Gorman-Murray (2014, 2017) have conceptualised how different generations of queer people, from specific class fractions, come together (with other people and other things) to form sexual assemblages that produce differentiated urban spaces and cultures. We believe there is a need to examine how sexuality (and sexual desire) contributes to shaping cities and urban space around the world. This includes paying attention to how heterosexuality and normative family arrangements shape cities (Oswin, 2019). However, our focus here is on the potential of a 'queerer' global urbanism which attends to the diversity of non-normative expressions of sexuality and gender in shaping cities around the world. It is relatively easy to examine expressions of 'global gay' lifestyles across diverse cities (and this is important), but here we attend to the ways in which other ways of understanding 'queer' desires shape cities and urban life. We feel that this approach, which pays attention to how specific expressions of sexuality and sexual lives arise out of more complex urban assemblages, has the potential to offer more nuanced understandings of the geography of sexualities within cities globally (and in the spirit of provincializing the experiences of northern cities). Methodologically, we see an importance in examining dominant expressions of urban sexual politics and space, as well as those desires which exist more precariously beyond and alongside these normativities.

In the last two decades, an increasing number of city governments around the world have incorporated a celebration of their LGBT populations, spaces, and events in their place-marketing strategies. As sexual politics appears to be becoming increasingly polarised globally, liberal attitudes towards sexual difference (what Puar (2007) referred to as a biopolitical regime of 'homonationalism') have increasingly become entangled with geopolitical struggles and the defence of particular forms of globalised (neo)liberal connections and politics. Even so, as cities like Singapore demonstrate, the liberal model of toleration is not the only one in operation

(Oswin, 2019). It is increasingly important to look beyond the Minority World to examine how particular cities function within this globalised gay imaginary. For example, Yang (2019) has argued that Bangkok increasingly plays a pivotal role, regionally, in the articulation of specifically East Asian expressions of homonormativity.

Looking beyond the global urban gay

Much of the extant geographical work on queer urban lives has focused on charting those city centre neighbourhoods (primarily in the Global North) where gay bars, clubs, and other services have clustered and concentrations of queer residents have sometimes grown up around them (Ghaziani, 2014). As Podmore (2006) highlighted, too many of these studies privileged the spatial expressions of (white) gay male culture and obscured the lives of others. There has been some growth in studies of lesbian and gay life in the suburbs of cities in the Global North (Tongson, 2011; Bain et al., 2018), but these are still studies of relatively privileged and homogenous social groups. So we are interested in thinking, globally, about what happens if we shift our focus from city centre locations to examine the social and spatial peripheries of cities and how queer life is lived within them.

The infrastructure of the city divides and fragments urban life (McFarlane, 2018). This highlights the ways in which inequalities within LGBT populations intersect with other structures of oppression, as well as histories of colonialism and war (Benedicto, 2014; Merabet, 2014). Tucker (2009) has demonstrated how gay life in Cape Town remains fractured by the legacies of spatial segregation and unequal infrastructural provision from the apartheid era. Borisa (2018) has mapped the ways in which caste, class, and religion fracture the experiences of queer men in Delhi; their sexual lives are as spatially segregated as the city they inhabit. Economic resources, access to affordable transport, and the temporal rhythms of working life all have material impacts on who can access 'global gay spaces'. Marginalisation is material, but it is also symbolic and can be internalised. If society repeatedly tells you that you do not belong in the city centre (where globalised gay spaces exist), it is harder to imagine oneself in those spaces and harder to embody the attitudes and aesthetics needed to fit into them.

Two things follow from this. First, it is a mistake to imagine that gay and queer place making does not occur in the peripheries and the margins of the city. Second, to be marginalised and peripheral within the city does not mean that one does not aspire to have access to more globalised gay spaces. We find it telling that in cities like Kinshasa (Hendriks, 2016, 2019) and small towns in South Africa (Reid 2013), gayness is ambivalently articulated through claims to be 'modern'. Social stigma is deflected and negotiated through the performance of demonstrable cultural capital in relation to global styles and fashions. These practices are not necessarily performed in 'classy' globalised gay venues but in quotidian hair salons (Reid, 2013), drag balls (Reid, 2013; Benedicto, 2014), popular dance halls (Hendriks, 2019), and promenading along the seafront (Merabet, 2014). Paying attention to gay lives in (and as) urban peripheries also requires attention to the parks, cemeteries, and public toilets where public (homo)sex occurs. In his requiem for the porn cinemas of Times Square, which were erased by gentrification, Delaney (1999) celebrated such public sex venues as spaces of cross-class and inter-racial contact and encounter.

Much has been written, recently, about the impact of digital and locative technologies, such as hook-up apps, on gay life (Mowlabocus, 2010). These apps form a new technological infrastructure facilitating the socio-sexual interactions of queer people in cities. Given that the functionality of many of these apps includes a relational locative function, they also reshape how people understand the urban landscape (Miles, 2017). They can reveal new forms of propinquity (but that, in turn, can draw painful attention to the ways in which the urban landscape is divided by other infrastructures that produce and reinforce social distance). As Borisa (2018) discovered, in Delhi, lower-caste men often adopt upper-caste names or misrepresent where they are located within the city in order to appear more desirable. By attempting to negotiate caste-based prejudice, as well as access to more globalised gay spaces, such acts reproduce the ways in which the intersections of caste, class, and desirability are mapped onto the physical infrastructure of the city.

Urban queer life is an act of refashioning common-sense truths (about queer people *and* about the city). It is both a city-making and a world-making project where, through their practices and interactions with the city, queer people challenge the dominant ideologies that normalise and discipline them. In an intersectional sense, this is never just about sexuality and gender but about those multiple social locations that queer folk embody and through which they survive their desires (khanna, 2013, 2016). It is the plethora of social locations that make people (un)desirable. Against the coherent mapping of the global gay city which urban theory keeps reproducing there is an inherent messiness to the lives of queer subjects through which they map these cities. These are the cities and urban spaces where queer desires emerge for those trapped within the logics of uninhabitability (Simone, 2019) and barely surviving the city or their own bodies and desires.

Getting pluriversal

In the final section of this chapter, we want to decentre 'global gay' urban spaces to think in more decolonial and pluriversal ways about sexuality and gender (Bakshi et al., 2016; de la Cadena and Blaser, 2018). We draw attention to those ways of doing and understanding sexuality and gender which do not fit within the epistemologies and ontologies of Eurocentric models of LGBT subjectivities. We want to sketch just a few ways in which a variety of more localised or indigenous subjectivities, which modern, globalised LGBT identities "cannot capture" (Di Pietro, 2016: 678), produce different kinds of spaces, and articulate different spatial imaginations, within cities around the world. Frequently, such identities are feared and denigrated by those with access to the "whitened networks of [global] lesbian and gay middle class aspirations" (Di Pietro, 2016: 678) precisely because they are too black/brown/indigenous, too associated with working-class femininities, and too heavily associated with the past.

In India, we might think about the spaces created by *kothi* and *hijras*. *Kothi* is the preferred self-identification of many (working-class) men who have a feminine sense of self and practice 'passive' sexual roles (Boyce, 2007). *Hijras* are gender-transgressing bodies (often considered a 'third sex') who have a history that is framed as predating the medico-juridical recognition of homosexuality and is often rooted in a variety of religious mythologies from across the country (Reddy, 2007). Many *hijras* live in communal houses, organised around a guru (or mother figure) who decides which members should engage in religious ritual or sex work, controls the collective income, and guarantees their well-being. While global HIV prevention funding mechanisms have brought many of these diverse gender expressions together under the rubric of trans*, they are often only treated as local variants of universal identities without interrogating them contextually and in relation to their caste histories (Dutta and Roy, 2014). However, the emergence of an upper-caste Hindu *hijra* woman as a *guru* of a religious *Kinnar Akhara* at the Kumbh (a major Hindu religious gathering) demonstrates how the indigeneity of the *hijra* tradition is being co-opted by reactionary Hindutva nationalism in contemporary India.

Over recent decades, there has been a resurgence in the adoption of the term *takatapui* by many Maori. The term can be literally translated as an 'intimate companion of the same sex'.

By adopting this term, queer Maori resist assimilation into Pakeha identity categories, challenging both the racism and coloniality within mainstream LGBT community contexts and heteronormative assumptions amongst other Maori. To name oneself *takatapui* centres one's obligations and "belonging in the whanau, the iwi and in the ancestral and future generations" (Hutchings and Aspin, 2007: ii). It is an articulation of belonging that decentres the LGBT community and its spaces without entirely rejecting them. It imagines belonging and one's place in the world through a more expansive set of relations across time and space.

To think comparatively, the identification as *takatapui* by some 'queer' Maori draws attention to the multiple idioms through which sexuality and sexual practices are understood in various parts of the world. In the context of South Asia, Katyal (2016) frames this through a logic of 'doubleness'. While homosexuality becomes one idiom through which sexuality is practiced (such as the loosely coined literal translation of the word for mobilisation purposes – *samlaingik* in Hindi), there are many others shaped through the language of habits (*aadat, lat, baazi*), play (*baazi, shauq, masti*), friendships (*dost, saheli*), or of love (*ishq, pyaar*). These idioms allow a simultaneity in understanding sexuality both within the Westernised subjectivities that many aspire to embody and allow for more creative and contextual responses to the limitations on these aspirations produced by the material contexts in which people live. Thus in a context where gendered norms are heavily scrutinised in public, men and boys in India can create space for themselves through the idiom of *masti* (fun), which blurs the boundaries between the homosocial, homoerotic, and homosexual (Boyce and khanna, 2011). In these ways, the urban infrastructure can be manipulated to sustain liveable lives that do not rely on Westernised forms of public visibility (Browne et al., 2019).

There is a danger of presenting these localised identities and practices as distinct and separate from the globalised gay aesthetics that circulate widely through globalised media. They are more hybrid. For example, Gupta (2005) has described how, in the context of a park in Mumbai filled with *kothis* cruising, the intersection of his relatively privileged class position and sexuality gets labelled as '*Englishpur ki Kothi*'. In cities across the world, it is possible to witness processes by which sexual minorities recognise and know spaces that allow them to weave queer desires into the material fabric of the city. These framings can be identitarian, but they are not necessarily so.

Concluding implications

As pictures of 'global gays' parading at pride events circulate so frequently around the globe, there is a danger of overlooking the myriad other queer sub-cultures that coexist in and actively produce urban space in registers that are easily fetishised as an exotic anomaly. In exploring a range of identities and ways of understanding sexuality, which have their origins in the Majority World, our intention is not to essentialise them in any way but to recognise how they might help open up an appreciation of the rich multiplicity of ways in which sexuality shapes urban space and urban life around the world. Our purpose in this chapter has been twofold: first, to reiterate the importance of attending to the ways in which sexuality and desire shape cities, large and small, across the globe, and, second, to argue that there is an urgent need to theorise urban sexualities using epistemologies and ontologies that originate in the Majority World.

Critical scholars of urban life and sexualities need to look beyond the spaces of 'global gay' life to negotiate more decolonial and pluriversal approaches to understanding the spaces that queer desire make in urban spaces around the globe. In doing so, we recommend that they attend to the importance of tracing sexual lives and the spaces they create and adapt. We also recommend taking local and indigenous idioms relating to intimacy and sex seriously (and recognising that they are often more hybrid, contextual, and translocal than appreciated). A 'queerer' global urbanism pays attention to the spaces that are created by these other ways of understanding sexuality, and the ephemeral traces left by them, within, against, and beyond 'global gay space'. It pays attention to how these idioms also produce the city.

References

Abraham, J. (2009) *Metropolitan Lovers: The Homosexuality of Cities*. Minneapolis: University of Minnesota Press.

Bain, A.L., Podmore, J.A. and Rosenberg, R. (2018) '"Straightening" space and time? Peripheral moral panics in print media representations of Canadian LGBTQ2S suburbanites, 1985–2005'. *Social & Cultural Geography*, 21 (6), 1–23.

Bakshi, S., Jivraj, S. and Posocco, S. (eds.) (2016) *Decolonizing Sexualities: Transnational Perspectives, Critical Interventions*. Oxford: Counterpress.

Benedicto, B. (2014) *Under Bright Lights: Gay Manila and the Global Scene*. Minneapolis: University of Minnesota Press.

Borisa, D. (2018) *Imagined Spaces of Freedom: Negotiating Queer Cartographies of Desires in Delhi*, Unpublished PhD thesis, India: Jawaharlal Nehru University.

Boyce, P. (2007), '"Conceiving kothis": Men who have sex with men in India and the cultural subject of HIV prevention'. *Medical Anthropology*, 26 (2): 175–203.

Boyce, P. and khanna, a. (2011) 'Rights and representations: Querying the male-to-male sexual subject in India'. *Culture, Health & Sexuality*, 13 (1): 89–100.

Brenner, N. and Schmid, C. (2015) 'Towards a new epistemology of the urban?' *CITY*, 19 (2–3): 151–182.

Browne, K., Banerjea, N., McGlynn, N., Bakshi, L., Beethi, S., & Biswas, R. (2019) '"The limits of legislative change: Moving beyond inclusion/exclusion to create" a life worth living'. *Environment and Planning C: Politics and Space*. doi: 10.1177/2399654419845910.

D'Emilio, J. (1983) *Sexual Politics, Sexual Communities: The Making of a Homosexual Minority in the United States, 1940–1970*. Chicago: University of Chicago Press.

de la Cadena, M., & Blaser, M. (eds.) (2018) *A World of Many Worlds*. Durham, NC: Duke University Press.

Delaney, SR. (1999) *Times Square Red, Times Square Blue*. New York: New York University Press.

Di Pietro, PJJ. (2016) 'Decolonizing *travesti* space in Buenos Aires: Race, sexuality, and sideways relationality'. *Gender, Place and Culture*, 23(5): 677–693.

Duggan, L. (2002) 'The new homonormativity: The sexual politics of neoliberalism', in R Castronovo and DD Nelson (eds.), *Materializing Democracy: Toward a Revitalized Cultural Politics*. Durham, NC: Duke University Press. pp. 175–194.

Dutta, A., and R. Roy (2014) 'Decolonizing transgender in India: Some reflections.' *TSQ: Transgender Studies Quarterly*, 1(3): 320–337.

Ghaziani, A. (2014) *There Goes the Gayborhood*. Princeton, NJ: Princeton University Press.

Gopinath, G. (2005) *Impossible Desires: Queer Diasporas and South Asian Public Cultures*. Durham, NC: Duke University Press.

Gupta, A. (2005) 'Englishpur ki Kothi: Class dynamics in the queer movement in India', in Narain, A and Bhan, G. (eds.), *Because I Have a Voice: Queer Politics in India*. New Delhi, Yoda Press, pp. 123–142.

Hendriks, T. (2016) 'SIM cards of desire: Sexual versatility and the male homoerotic economy in urban Congo'. *American Ethnologist*, 43(2): 230–242.

Hendriks, T. (2019) 'Queer(ing) popular culture: Homo-erotic provocations from Kinshasa'. *Journal of African Cultural Studies*, 31(1): 71–88.

Hubbard, P. (2012) *Cities and Sexualities*. London: Routledge.

Hutchings, J. and Aspin, C. (eds.) (2007) *Sexuality and the Stories of Indigenous People*. Wellington, Aotearoa New Zealand: HUIA.

Katyal, A. (2016) *The Doubleness of Sexuality: Idioms of Same-Sex Desire in Modern India*. New Delhi: New Text.

khanna, a. (2013) 'Three hundred and seventy seven ways of being–sexualness of the citizen in India'. *Journal of Historical Sociology*, 26(1), 120–142.

khanna, a. (2016) *Sexualness.* New Delhi: New Text.

Manalansan IV, M. F. (2015) 'Queer worldings: The messy art of being global in Manila and New York'. *Antipode*, 47(3): 566–579.

McFarlane, C. (2018) 'Fragment urbanism: Politics at the margins of the city'. *Environment and Planning D: Society and Space*, 36(6), 1007–1025.

Merabet, S. (2014) *Queer Beirut.* Austin, TX: University of Texas Press.

Miles, S. (2017) 'Sex in the digital city: Location-based dating apps and queer urban life'. *Gender, Place & Culture*, 24(11): 1595–1610.

Mowlabocus, S. (2010) *Gaydar Culture: Gay Men, Technology and Embodiment in the Digital Age.* Farnham: Ashgate.

Nash, C. J. and Gorman-Murray, A. (2014) 'LGBT neighbourhoods and "new mobilities": Towards understanding transformations in sexual and gendered urban landscapes'. *International Journal of Urban and Regional Research*, 38(3): 756–772.

Nash, C. J. and Gorman-Murray, A. (2017) 'Sexualities, subjectivities and urban spaces: A case for assemblage thinking'. *Gender, Place & Culture*, 24(11): 1521–1529.

Oswin, N. (2015) 'World, city, queer'. *Antipode*, 47(3): 557–565.

Oswin, N. (2018) 'Planetary urbanization: A view from outside'. *Environment and Planning D: Society and Space*, 36(3): 540–546.

Oswin, N. (2019) *Global City Futures: Desire and Development in Singapore.* Athens, GA: University of Georgia Press.

Podmore, J.A. (2006) 'Gone "underground"? Lesbian visibility and the consolidation of queer space in Montréal'. *Social & Cultural Geography*, 7(4): 595–625.

Puar, J.K. (2007) *Terrorist Assemblages: Homonationalism in Queer Times.* Durham, NC: Duke University Press.

Reddy, G. (2007) *With Respect to Sex: Negotiating Hijra Identity in South India.* Chicago: University of Chicago Press.

Reid, G. (2013) *How to Be a Real Gay. Gay Identities in Small-Town South Africa.* Scottsville, South Africa: University of KwaZulu-Natal Press.

Shah, S.P. (2015) 'Queering critiques of neoliberalism in India: Urbanism and inequality in the era of transnational "LGBTQ" rights'. *Antipode*, 47(3): 635–651.

Simone, A. (2019) *Improvised Lives: Rhythms of Ein an Urban South.* Cambridge: Polity Press.

Tongson, K. (2011) *Relocations: Queer Suburban imaginaries.* New York: New York University Press.

Tucker, A. (2009) *Queer Visibilities: Space, Identity and Interaction in Cape Town.* Oxford: Wiley-Blackwell.

Vickers, E. (2010) 'Queer sex in the metropolis? Place, subjectivity and the Second World War'. *Feminist Review*, 96 (1): 58–73.

Weston, K. (1995). Get thee to a big city: Sexual imaginary and the great gay migration. *GLQ: A Journal of Lesbian and Gay Studies*, 2(3), 253–277.

Yang, Y.-H. (2019) *Travelling for Becoming a 'Normal' Gay Man: Liminality, Sexual Fields, and Gay Taiwanesemen's Sex Tourism to Bangkok.* Unpublished PhD thesis. Durham: University of Durham.

6

SEEING LIKE AN ITALIAN CITY

Questioning global urbanism from an "in-between space" in Turin

Francesca Governa

Introduction

The title of this chapter is borrowed from Amin and Thrift's latest book (2016). Nevertheless, I add the location of Turin to suggest that seeing global urbanism from Italian cities may help to reveal the empirical heterogeneities of relationships between the urban and the global and increase the spaces from which to question urban theories. This is not a new idea. Indeed, in 2011, Roy and Ong published a book titled *Worlding Cities: Asian Experiments and the Art of Being Global.* The aim of the book written by Roy and Ong is to critically describe the "worlding practices" of Asian cities and thereby define a heterodox research project with the goals of disrupting the "normal" maps of global urbanism and showing multiple attitudes towards being global in urban Asia. According to McCann, Roy and Ward (2013), worlding means to challenge the limits of urban theories in order to acknowledge the role of southern cities within global networks of flows and relations; to define a different stance towards the Global South, overcoming the predominance of transnational elites; and to illustrate the ways in which cities are each distinctively global in professional and scientific discourses and practices.

Is it possible to adapt this conceptual device to Italian cities and specifically to Turin? Does Turin experiment with its own specific "art of being global" (Roy and Ong, 2011), and how might this art be used to conceive a more open global urbanism? In order to answer or at least start indicating these questions, I take the apparently "mundane" meaning of worlding – "a plethora of situated experiments that reinvent what urban norms can count as 'global'" (Ong, 2011: 2) – and I refer to a long-standing study developed in Barriera di Milano (Turin), a so-called urban margin located in the north of the city (Governa, 2016; Bonini Baraldi, Governa and Salone, 2019). The fieldwork in Barriera di Milano has been conducted in different periods since 2015 and combines various methods according to a conscious eclecticism (Law, 2004) aimed at avoiding any mechanical application of qualitative methods.

This chapter is a preliminary exploration – from the standpoint of an Italian city – on two points. The first point is to question the consistency of the "areal identification" of urban/city models – that is, as a European city, a Mediterranean city, an Italian city and so on. The second is the need to transcend the dichotomy between Global North/Global South and to understand the relationships between the urban and the global, shifting contemporary geographies of

DOI: 10.4324/9780429259593-8

knowledge and theory production (Sheppard, Leitner and Maringanti, 2013). By doing this, I aim to tackle the slippery notion of "global urbanism" from within: by questioning its makings across geographies, with particular reference to modes of worlding that exceed Anglophone debates while remaining anchored in a (southern) Western experience.

The chapter is organized into three parts. Following the introduction, it briefly reviews global urban research as seeing from Southern European cities. It subsequently discusses the possibility of reconsidering "worlding practices", starting from a so-called urban margin in Turin, an Italian "ordinary city" (Robinson 2006). Finally, it reassesses some research paths that may help us to pursue a "different" urban view – neither global nor local, general nor specific – in order to apply Nancy's (2000) notion of "being singularly plural and plurally singular" to global urbanism.

From the South (of Europe)

There is a rich amount of investigations into the "exceptionality" or, at least, the "specificity" of Southern Europe and its cities, pertaining both to their history and traditions (in which Southern European cities are inserted into the so-called Mediterranean city model, which plays a foundational role as an archetype of urban civilization) and to the impacts of the urban crisis affecting Western countries in the last decade (Leontidou 1990, 1996; Hadjimichalis, 2011; Seixas and Albet, 2016). However, I would like to introduce a deviation from these interpretations, not to deny their value but to insert Italian cities into the global urbanism debate and to overcome the constraints imposed by national or at least areal canons in defying what a city is and does. In which sense, areal canons identify a tradition (or even an identity)? Which is the role of this kind of tradition and identity if not simultaneously to reassure our own certainty and limit encounter, contacts, conflicts and messy relations?

Inserted in the Global North/Global South divide, (most) Italian cities can be considered an "in-between space" (Leontidou, 1996), just like other Southern European and Mediterranean cities. Such a space evades conventional and more or less mainstream theoretical frameworks and conceptual dichotomies: it is probably more the space of "and" than that of "or"; it is a city and a non-city, global and local, rich and poor and so on. These traits do not only pertain to somewhat defined Italian cities but to the "XXI century metropolis" as a whole (Roy, 2009). Indeed, is it enough to be located in Italy? Is there a "model" of an Italian city and, if so, what is it? The Italian Renaissance city, like Urbino or Pienza? The "one company town", like Turin during the Fordist period, considered by Bagnasco in the 80 more similar to industrial cities of the United Kingdom than to "traditional" Italian cities? The "cosmopolitan city", like Palermo, or the "global city", as Milan is now perceived and described?

As postcolonial and feminist critiques have taught us, knowledge production is necessarily situated and demands openness to plural and even contradictory positions (McKittrick, 2006; Roy, 2015; Robinson, 2016). However, the production of urban theory is firmly located in the experiences of North America and Western Europe, with concepts developed in Euro-American urban studies pretending to be applied ipso facto to southern urbanism (Schindler, 2017).

The critiques made by a number of scholars on "the imagined geographies of new subjects of theorization" (Robinson, 2016: 188; see also Mbembe and Nuttall, 2004; Roy and Ong, 2011; Parnell and Oldfield, 2014) highlight the opaqueness of opposition between the Global North and the Global South. In particular, considering the parts of the Western world as if they are all the same is a simplification that must be scrutinized by the same advocates of a critical urban theory. "The West" is heterogeneous and fragmented: there are margins, inclusionary and exclusionary dynamics, centres, as well as a "monist core", constituted by Anglophone, male, white scholars located in just a few Euro-American universities (Leitner & Sheppard, 2016).

Seeing from an "in-between space" – that is to say, simultaneously from the centre (the European Union [EU]) and from the margin (Italy) – moulds this first attempt towards the heterodox and exploratory field of enquiry of global urbanism, seeking to avoid the risk of replicating "a dull and regressive regionalism" (Barnes, 2019: 173).

Worlding Turin?

Turin is not a "global city". Some years ago, in particular around the triumphalist period of the 2006 Winter Olympic Games, the strange idea to regard Turin as a "minor global city" or, as Vanolo (2015) suggests, the hyper-optimistic dream of a "wannabe global city", emerged. Today, any claim that the city participates in a somehow "minor" global championship is over. Since the Olympics, Turin has become smaller (seeing a 2.3% reduction in its population between 2010 and 2016), older (with an ageing index, i.e. the proportion of older to younger people, of 207.7 in 2016, compared to 169.8 in other regions of North-West Italy) and poorer (marked by an unemployment rate of 11.4%, the highest of all metropolitan cities in Northern Italy, as well as an increase in the number of poor people by 46% between 2011 and 2013; Regione Piemonte – CEP, 2018). While Turin is experiencing a severe urban crisis, additionally related to the austerity policies imposed by the EU Stability and Growth Pact and some endogenous weaknesses due to the disappearance (or at least aphonia) of political, cultural and economic elites (Bagnasco, Berta and Pichierri, 2020), some inhabitants and some places are particularly affected by that. More or less extreme forms of marginality unfolding in the everyday life characterize in particular some old working-class neigbourhoods as Barriera di Milano. In the urban historiography of Turin, Barriera di Milano plays a paradigmatic role. Certainly, in just a few decades, it has gone from being a key place in constituting Turin as a city of the working class and socialism (Spriano, 1972) to that of a restless, involuntary laboratory of multiculturalism brought about by globalization.

Starting from the late nineteenth century, Barriera di Milano was the emblem of Turin's industrialization, working-class culture and later resistance to the fascist dictatorship. It was the place where the first Italian trade unions were born, where the Communist Party laid its roots and where destruction and workers' revolts during World War Two all defined a common past, which was maintained until the unrest and contradictions associated with the death of Fordism in the late 1970s. The stigmas of deindustrialization at the turn of the 1990s saw high levels of unemployment, alongside the precarious and at times difficult cohabitation of "local" and "new" residents (coming from all over the world), considerable overpopulation due to the high density of degraded, low-cost housing and poor-quality public services. Not only in official documents but also among the local media and the inhabitants, Barriera di Milano is viewed as the symbol of deprived urban spaces in Turin: it is a poor, problematic, tough and neglected neighbourhood, a place of theft and drug dealing, an unsafe place, a slum.

Today in Barriera di Milano, we can certainly no longer talk of the uniformity of class, nor the cultural uniformity of the social groups that live there and help weave the mix of relations that continuously (re)define daily life. Barriera di Milano is the most avant-garde district in Turin: in the past few years, it has enjoyed a boom in cultural and artistic activities (Bonini Baraldi, Governa and Salone, 2019). At the same time, Barriera di Milano is the most multi-ethnic area of the city. Indeed, in 2011, the concentration of foreign nationals here was around twice the city average. In 2013, the district was home to the largest number in absolute terms of the foreign population compared to other districts and had the highest percentage of foreign nationals within the total resident population (Città di Torino, 2013). As occurred with the migrants from Apulia, Sicily and Calabria in the 1950s and 1960s, today Romanians, Moroccans, Chinese and Senegalese nationals populate the area and, here even more so than in the rest of the city, have proved able to develop their own productive and commercial activities. Indeed, the arrival of foreign immigrants has led to a progressive transformation (or rather complexification) of local trade. Of course, the progressive spread of foreign shops instigates conflicts with the "local" population in a difficult path of meetings and clashes, which becomes especially complex when the presence of small foreign businesses is healthy and competitive in the face of the crisis of traditional commercial and trade activities. The rise and spread of new ethnic activities (bars, restaurants, bakeries, markets) linked to Turin's new urban populations highlight some economic specializations that can foster urban life. As outlined by Lagendijk et al. (2011), ethnic entrepreneurship, especially in the form of retail, plays a key role in strengthening ethnically diverse places from both economic and sociocultural points of view. In Barriera di Milano, old and new social structures live side by side, as do (residual) industrial, commercial and cultural functions. The foreign presence in the daily economy and cultural institutions are made and remade by global relations, both physically (the migrant population arriving in the neighbourhood from different parts of the world) and immaterially (the flow of migrants' remittances to their countries of origin). At the same time, when we enter museums, art galleries, arts and craft workshops and so on, we penetrate different places that ignore the local (neighbourhood) level. What happens inside – not only artistic performances but also consumption and leisure practices – has little or nothing to do with what happens outside. It is more akin to other experiences of the reuse of old industrial spaces for cultural purposes around Europe: it institutes networks and relations at the global level (for example, "artistic residences" host artists from all over the world, and the exchanges of artistic works or experiences create global networks of cooperation) to a greater extent than with the place that physically winds its way around it, as well as to the daily practices that take place within it.

Despite these changes, Barriera di Milano is increasingly depicted in an opaque manner compared to the complexity of what is currently redefining socio-spatial relationships: it is a place far from the centre, whether from a physical point of view (which it certainly is not!) or from social, economic and cultural perspectives; it is a 'problematic' district, where decadence, segregation and social exclusion thrive; it is a non-place, with no quality or history, where the centre keeps what it needs but cannot (or will not) host. Such a viewpoint not only continues to deny the cultural and economic complexity of contemporary Barriera and its multiplicity of practices that constantly redefine the relationships between Turin and the global but reproduces a "pathological" attitude towards urban margins, asserting a fixed, certain and reassuring spatial order that legitimizes urban regeneration strategies as good for everyone (Uitermark, 2014).

Worlding within an ordinary city?

While inevitably simplified, the features of today's Barriera di Milano show how many differences and flows at both local and global levels cross this place (and Turin and all other cities), revealing social stories, economic events and physical transformations that cannot be reduced to outcomes of simple cause-effect mechanisms. Global urbanism is a fuzzy expression that gathers different meaning and sensibilities, a heterodox yet intertwined and exploratory field of enquiry to think with the connections between places challenging questions about the specificity or limited scope of some concepts and about the extent to which it is productive to think with ideas across many different experiences, acknowledging the locatedness of all theoretical endeavours. Inserted in this framework, what can we learn from Barriera di Milano? The first thing is the impossibility to strictly cut the difference between Barriera and the global. Barriera is "global" because the life and the relations that unfold in it daily are not made of locally embedded traditions but of translocal relations that mix and overwhelm in this place. The second thing is that the global is materialized in places and in urban practices unfolding in daily life: global is not only abstract, not only fast, not only economics and finance, not only 'central business district'. As Bunnell and Maringanti (2010: 418) maintain, "ostensibly 'global' research is paradoxically parochial in failing to extend beyond world-city financial-enclave spaces".

"Worlding practices" in Barriera di Milano are both "from below" experiences (Simone, 2001) involving seemingly irrelevant and ordinary practices that are often neglected both by research and urban policies, as well as "institutional" experiences that are led by cultural institutions and involve Barriera di Milano in the network of so-called creative cities. Together, these practices constantly redefine urban-global connections through everyday activities. They constitute a widespread tissue reassembling the complexity of socio-spatial relations, as opposed to being a pure mimesis of contemporary mainstream narratives of what "global" is and could be. The multiplicity of spaces, times and webs of relations that overlap and intersect challenge the very notion that generalizations can be made about what (global) cities are, allowing us to trace the entanglements between objects and bodies, discourses and power, in their contextual deployment in space and time. By drawing attention to largely unseen relations and practices, migration flows in Italian cities could be seen not only as a problem but also as economic opportunities; by highlighting things and processes normally inserted in strict boxes (i.e. the "creative city" model), we can search for the (global) value of cultural practices not only as a deterritorialized recipe.

The "in-between space", both on the macro-regional (Italy) and the urban scale (Barriera di Milano as urban margin), challenges predefined dichotomies based on the geopolitical Global North/Global South and centre/periphery, potentially enabling us to broaden urban theorizing (Lancione and McFarlane, 2016). Indeed, the "in-between space" opens up the possibility to reconsider deprived urban spaces in Southern European cities not (only) as "pockets of poverty", "excluded places" or "spaces of danger and violence" that are normally explained as a result of increasing social inequality, primarily related to global and local economic restructuring processes and defective welfare policies and to move away from the conventionally dualistic framing of urban space. The latter excludes what is out of place (i.e. what is not where "it should be"), hides the antagonist and conflictual dimension and reduces conflict to an archaic anomaly involving only minorities that are particularly needy or that have pathologies that require treatment (i.e. through the shortcut of urban regeneration). Seeing from within an urban margin in an ordinary city, global urbanism is more the urbanism of the middle, of the trivial and the everyday where existence is essentially co-existent, even conflictual. Seeing from the "in-between space", global urbanism is made by various experiences and traditions interweaving different geographies of the making and remaking of geographical ideas and frameworks, overlapping interactions and physical, cultural, epistemological and mental borders.

References

Amin, A. and Thrift, N. (2016) *Seeing Like a City*. Polity Press: Cambridge.

Bagnasco, A. (1984) *Torino: Un Profilo Sociologico*. Einaudi: Torino.

Bagnasco A., Berta, G. and Pichierri, A. (2020) *Chi ha fermato Torino? Una metafora per l'Italia*. Einaudi: Torino.

Barnes, T. (2019) 'The importance of 'being various': A commentary on 'Moving beyond Anglo-American economic geography'. *International Journal of Urban Sciences*, 23 (2), pp. 170–176.

Bonini Baraldi, S., Governa, F. and Salone, C. (2019) '"They tried to make me go to rehab. I said, no, no, no". Representations of 'deprived' urban spaces and urban regeneration in Turin, Italy'. *Urban Research & Practice*, online first (doi:10.1080/17535069.2019.1611911).

Bunnel, T. and Maringanti, A. (2010) 'Practising urban and regional research beyond metrocentricity'. *International Journal of Urban and Regional Research*, 34 (2), pp. 415–420.

Cittá, di Torino (2013) *La Popolazione Straniera nel 2013*, Torino.

Governa, F. (2016) 'Ordinary spaces in ordinary cities: Exploring urban margins in Torino and Marseille'. *Mediterranée Revue géographique des pays Méditerranéens/Journal of Mediterranean Geography*, 127, pp. 109–119.

Hadjimichalis, C. (2011) 'Uneven geographical development and socio-spatial justice and solidarity: European regions after the 2009 financial crisis'. *European Urban and Regional Studies*, 18 (3), pp. 254–274.

Lagendijk, A., Pijpers, R., Ent, G., Hendrikx R., von Lanen, B. and Maussart, L. (2011) 'Multiple worlds in a single street'. *Space and Polity*, 15 (2), pp. 163–181.

Lancione, M. and McFarlane, C. (2016) 'Life at the urban margins: Sanitation infra-making and the potential of experimental comparison'. *Environment and Planning A*, 48.12, pp. 2402–2421.

Law, J. (2004) *After Method: Mess in Social Science Research*. Routledge: London and New York.

Leitner, H. and Sheppard, E. (2016) 'Provincializing critical urban theory: Extending the ecosystem of possibilities'. *International Journal of Urban and Regional Research*, 40 (1), pp. 228–235.

Leontidou, L. (1990) *The Mediterranean City in Transition*. Cambridge University Press: Cambridge.

Leontidou, L. (1996) 'Alternatives to modernism in (Southern) urban theory: Exploring in-between spaces'. *International Journal of Urban and Regional Research*, 20 (1), pp. 178–195.

Leontidou, L. (2014) 'The crisis and its discourses: Quasi-Orientalist attacks on Mediterranean urban spontaneity, informality and joie de vivre', *City: Analysis of Urban Trends, Culture, Theory, Policy, Action*, 18 (4–5), pp. 551–562.

McCann, E., Ward, K., and Roy, A. (2013) 'Assembling/worlding cities'. *Urban Geography*, 34 (5), pp. 581–589.

Massey, D. (2005) *For Space*. SAGE: London.

Mbembe, A. and Nuttall, S. (2004) 'Writing the world from an African metropolis'. *Public Culture*, 16 (3), pp. 347–372.

McKittrick, K. (2006) *Demonic Grounds: Black Women and the Cartographies of Struggle*. University of Minnesota Press: Minneapolis.

Nancy J.-L. (2000) *Being Singular Plural*. Stanford University Press: Stanford.

Ong, A. (2011) 'Introduction: Worlding cities, or the art of being global', in Roy, A. and Ong, A. (eds.) *Worlding Cities. Asian Experiments and the Art of Being Global*. Wiley-Blackwell: Oxford, pp. 1–26.

Parnell, S. and Oldfield, S. (2014) *Handbook for Cities of the Global South*. Routledge: London.

Regione Piemonte – CEP (Conferenza Episcopale Piemontese) (2018) *Chi Offre e Chi Crea Lavoro in Piemonte*. Indagine sulla condizione lavorativa dei giovani piemontesi, Torino.

Robinson, J. (2006) *Ordinary Cities: Between Modernity and Development*. Routledge: London.

Robinson, J. (2016) 'Comparative urbanism: New geographies and cultures of theorizing the urban'. *International Journal of Urban and Regional Research*, 40 (1), pp. 187–199.

Roy, A. (2009) 'The 21st century metropolis: New geographies of theory'. *Regional Studies*, 43 (6), pp. 819–830.

Roy, A. (2015) 'Who's afraid of postcolonial theory?' *International Journal of Urban and Regional Research*, 40 (1), pp. 200–209.

Roy, A. and Ong, A. (2011) *Worlding Cities. Asian Experiments and the Art of Being Global*. Wiley-Blackwell: Chichester.

Schindler, S. (2017) 'Towards a paradigm of Southern urbanism'. *City: Analysis of Urban Trends, Culture, Theory, Policy, Action*, 21 (1), pp. 47–64.

Seixas, J. and Albet, A. (2016) *Urban Governance in Southern Europe*. Routledge: London and New York.

Sheppard, E., Leitner, H. and Maringanti, A. (2013) 'Provincializing global urbanism: A manifesto'. *Urban Geography*, 34 (7), pp. 893–900.

Simone, A. (2001) 'On the worlding of African cities'. *African Studies Review*, 44, pp. 15–41.

Spriano, P. (1972) *Storia di Torino operaia e socialista*. Einaudi: Torino.

Uitermark, J. (2014) 'Integration and control: The governing of urban marginality in Western Europe'. *International Journal of Urban and Regional Research*, 38 (4), pp. 1418–1436.

Vanolo, A. (2015) 'The image of the creative city, eight years later: Turin, urban branding and the economic crisis taboo'. *Cities*, 46, pp. 1–7.

7

THEORISING FROM WHERE? REFLECTIONS ON DE-CENTRING GLOBAL (SOUTHERN) URBANISM

Hyun Bang Shin

One of the major contributions of the postcolonial critique of urbanism is the call for de-centring knowledge production, especially the emphasis on the need for 'theorising' from the Global South so that urban processes embedded therein are not rendered simply as variants of the Global North. What can we learn from this endeavour to re-assert the importance of southern cities, and are there any limitations in our existing practices? Does the focus on a select number of southern cities result in replicating the extant shortfalls of urban studies from the Global North? Are post-colonial and neo-Marxian critiques of urbanism mutually exclusive? In this short chapter, while advocating the current efforts amongst critical scholars to give more weight to the studies of the Global South, I reflect on these questions, introducing some of the latest contributions from scholars working in and on what I refer to as the Global East (Shin et al., 2016) as a deliberate attempt to interrupt the North-South binaries.

Theorising from the global south?

De-centring the production of knowledge on global urbanism involves moving away from over-reliance on Western cities (e.g., London, Chicago and Vancouver) as sources of theoretical inspiration and treating all cities across the globe as 'ordinary cities' placed on the level-playing field (Robinson, 2006; Roy, 2009). Analytical generalisation is pursued by scholars working on the Global South to challenge the existing wisdom on the urban way of life and what urbanisation means to different places and people.

While acknowledging the contributions made by the emphasis on southern cities as new sites of theorising, I posit two shortfalls that can be observed in existing practices. First, a persistent problem is the absence of certain global regions. Critical scholars have been lamenting this absence, voicing their dissatisfaction with the way some regions are under-represented. For example, in a recent international conference on 'the frontiers of the urban' held in November 2019 at the University College London, Oren Yiftachel called for efforts to 'theoriSE', stressing the need of accumulating more work on learning from what he defines as 'global Southeast', which covers the Middle East and Turkey. In another conference on urban geographies of post-communist states held in Kyiv in 2017, multiple sessions were organised to probe 'theorising

DOI: 10.4324/9780429259593-9

cities from the global East'. Here contributors lamented the absence of the former 'second world' whose post-communist transitional urban experiences were missing in the vocabularies of urban scholarship. As the session organiser later reiterated,

> [t]he demise of the Second World's political project – communism – wiped the East off the global map, any distinctiveness of more than 70 years of communist rule erased. The East is too rich to be a proper part of the South, but too poor to be a part of the North. It is too powerful to be periphery, but too weak to be the centre.
>
> (Müller, 2017: 2–3)

Global East is also a term I have elsewhere introduced (see Shin et al., 2016) to use East and South-East Asia as a disruptive means to "problematise the existing common practices of grouping all regions other than Western European and North American ones into the Global South" (ibid.: 456), practices that often render East and South-East Asia invisible. All these emergent efforts add weight to the ongoing project of decolonising urban theories and de-centring knowledge production but at the same time signal the lack of presence of certain regions from the global urban scholarship. There is still more work to be done to re-insert cities outside the West (or North Atlantic) in the global epistemological map of urban theorising.

Second, a trap researchers may often find themselves in is that researchers, willingly or unwittingly, portray a picture of the world that is divided into two binaries: Global North and Global South, which are increasingly limited categories that constrain analytical enquiries. For students of East Asian (and to some extent, South-East Asian, if Singapore is considered) urban studies, there is a recurring – often imposed – dilemma as to where their region fits. As a scholar researching Asian cities, I get to hear occasional murmurs from students who are questioned by their professors about their attempt to use South Korea as a case study to discuss development in the Global South. Similarly, Hae and Song (2019: 10) also bemoan "the invisibility of East Asian societies within the dominant geographical nomenclatures of Global North and Global South". Such invisibility may perhaps stem from the understanding that East Asia's trajectory of urbanisation does not fit into the characteristics of the Global South. Indeed, a study proposing 'a paradigm of Southern urbanism' summarises three distinctive characteristics of cities in the Global South but considers them to be different from cities in 'North Atlantic and Northeast Asia' (Schindler, 2017: 60). In a similar vein, other scholars may hesitate to use Singapore, Taiwan and South Korea as case studies in their research on the Global North: after all, despite their status as high-income countries according to the World Bank, they are 'postcolonial' and non-Western, having experienced urban development trajectories dissimilar to those of Western Europe and North America. In other words, they are inadvertently in interstitial spaces, neither Global North nor Global South.

The rise of East and South-East Asia, to some extent, provides an opportunity to disrupt the North-South binaries, which do not capture the more complex web of spatial connectedness. If the export of urban development experiences occurs from, say, South Korea to Ethiopia, is this South-to-South mobility? For some who do not regard South Korea as part of the Global South, such a policy transfer gets lost in the usual categorisation of Global North and South, but it may present an interesting moment of disruption to the conventional study of global urbanism and expose the increasingly limited utility of the North-South divide.

De-centring and re-centring urban studies?

Ananya Roy (2009: 820) notes succinctly that

> [t]he concern is the limited sites at which theoretical production is currently theorized and with the failure of imagination and epistemology that is thus engendered. It is time to blast open theoretical geographies, to produce a new set of concepts, in the crucible of a new repertoire of cities.

Here the emergence of East and South-East Asian cities has created a niche in the global urban studies, bringing scholarly attention to this part of Asia. The rise of East and South-East Asian cities as a reference point for urban development in other cities of the Global South has resulted in an increasing volume of studies on these cities in recent years (for example, Doucette and Park, 2019; Park et al., 2021). The contribution of studies on Chinese urbanism has been enormous in particular, not only in terms of the scholarly opportunities that were created as China began to tread the path of economic reform and rise to become one of the biggest economies in the world but also in terms of the expanding higher education market. Indeed, as Wu (2015) has noted, "Chinese cities provide a chance to expand the geography of theories".

Re-centring urban studies by focusing on cities outside the North Atlantic is a challenging endeavour in itself. One particular problem perhaps is the hesitation amongst funders and publishers to be open to proposals that nominate a non-Western single site (e.g., a city or a country) from which theorising can occur. Journal editors may also favour a special issue that includes multi-site comparison instead of a collection on a single site. The experience of Jesook Song and Laam Hae, who have recently edited an insightful volume titled *On the Margins of Urban South Korea: Core Location as Method and Praxis*, is quite telling. In their afterword, itself a rare contribution that provides a detailed genealogy of the evolution of their thoughts from their seeding to fruition, they spell out how their previous efforts to pursue a special issue, which was to treat 'urban South Korea' as a single case, were to no avail, an experience that replicates my own from some years ago. One of the reasons journal editors rejected their special issue proposal seems to be that "research on a particular country (i.e., South Korea) was incapable of engaging in transnational analysis" (Song and Hae, 2019: 189). They found it challenging to ascertain that there is "the significance of knowledge grounded in a particular location, especially one in the non-West, because of its supposed limitations in appealing to conception of universality and certain understandings of transnationality" (ibid.). It is interesting to note how this hurdle is less of the case with China studies, as can be illustrated by a number of journal special issues that have focused on China (for example, He et al., 2017; Logan, 2018; Wu and Zhang, 2020), the reasons of which may include the surge of interests in China studies as China has risen to become a powerful economy and how much the academic world, including the publishing industry, has worked hard to capture its market and readership.

Even if it is possible to use a single site from which theorising can occur, there is a risk of giving prominence to a small number of prime cities that already enjoy political, economic and cultural privilege in their host countries. In this way, some places are over-represented as sources of inspiration. For instance, a select number of prime cities situated in India (noteworthy are Bangalore, Delhi and Mumbai), in Africa (notably Johannesburg, Cape Town and, to a lesser extent, Lagos) and in mainland China (so-called first-tier cities of Beijing, Shanghai, Shenzhen and Guangzhou) have risen to become key sites of academic enquiries. The rise of these economies has also helped elevate the positions of these cities, as academic markets expand rapidly in line with their trade volumes in the world economy. The same can be said for South Korea: there is a predominant focus on its capital, Seoul. The focus on a select number of prime cities means that our understandings of urban processes in their countries are largely skewed towards the few cities that frequently feature in research outputs and public debates. In other words, our discussions on urbanism in the Global South and East are largely based on the optics of major prime cities of the region, and this is indeed a mirror image of the global urban studies that have been largely based on a select number of global cities.[1]

Another inevitable shortcoming of such preponderance of studies on prime cities is that smaller and more regional cities are largely hidden from scrutiny, thus remaining in the epistemological blind spot despite calls for more inclusion of cities dropped 'off the map' (Robinson, 2002). Some of the megacities may also remain in the blind spot despite their sizeable population and economic influence in their national territories due to their loose global connections and therefore remaining as 'black hole' cities (Short, 2004). Cities in the blind spot may also result from inadequate research infrastructure and the institutional constraints they face, preventing lone researchers from entering the under-studied sites, and anyone who has succeeded is not able to form a critical mass of scholars that would make the sites more visible in the academic world. Whatever the reasons are, places in the blind spot remain under-researched, which in turn further contributes to their drop from the global epistemological map of what constitutes studies of global urbanism.

There are, of course, benefits of studying such prime cities in a rapidly urbanising world. For one thing, it provides a window of learning opportunity, as the urbanism rooted in such cities may generate a further impetus of changes in the rest of the country, acting as a reference that gets emulated by those whose development lags behind. For instance, the vertical accumulation in South Korea that saw the domination of high-rise condominiums[2] stemmed from Seoul's urban development experiences (Shin, 2011, 2021; Park and Jang, 2019; Sonn and Shin, 2020). China's globalising cities in the eastern provinces, such as Shanghai, Nanjing, Guangzhou and Shenzhen, got 'rich' first in the midst of China's economic reform from the 1990s, thus acting as a source of aspiration for other inland cities to learn from (for example, Chen and Zhou, 2009; O'Donnell et al., 2017). Such understandings promise the unveiling of the nature of Korean or Chinese developmental urbanism as ideology (Shin and Zhao, 2018; Park and Jang, 2019; Shin 2019), which affects cross-regional (and, more recently, transnational) mobility of corresponding urban policies and urbanism (McFarlane, 2011). However, the shortfall is the absence of studies on interstitial spaces and smaller/regional cities; this limitation prevents us from adequately capturing the urban experiences of the majority populations in host countries. To this extent, studies on the Global South/East urbanism may retain a mirror image of the hegemonic global urbanism that has heavily relied upon global cities at the apex of urban hierarchies that are prone to repeating the epistemological chauvinism that Vainer (2014: 53) warned against.[3]

Where the neo-Marxian political-economic perspective meets the postcolonial approach

A simplistic response to address the shortfall as noted in the previous section – that is, the select few cities outside the North Atlantic as the main sites of shaping our understanding of urbanism – may entail undertaking more studies on smaller towns and cities. In this regard, the ongoing studies by critical scholars are welcoming (see Bell and Jayne, 2009). But a more profound means, I would argue, is to situate our enquiries in the web of historic conjunctures and incorporate relational perspectives on spatial connectedness in the narratives of development that cities of the global South/East have treaded. We need to practice "historical geographies of such imaginations and practices of development" (Roy, 2016: 318) in such a way that would allow us to "to move beyond the methodologies of geographic inversion and corrective inclusion" (Roy, 2018: 45). Such an exercise requires us to start from developing greater awareness that global southern cities may be connected and are capable of producing genuine materialisation of such connections "through various combinations of shared colonial histories, development strategies, trade circuits, regional integration, common challenges, investment flows, and geopolitical articulation" (Simone, 2010: 14).

While such perspectives have been central to postcolonial critique of urban studies, this is also not new to the geopolitical economic perspectives advocated by critical geographers working on and in the Global East. From the late 1990s in particular, with the establishment of the 'alternative geographies' group that saw the inaugural conference in Daegu 1999, critical scholars working on urban and regional development in the Global East have been wrestling with the questions of the state, capital and class relations, often from a neo-Marxian perspective that inherited progressive social and labour movement traditions in the region. More recently, noteworthy development includes efforts to de-centre the production of knowledge by incorporating local histories and multi-scalar perspectives to counter the hegemony of Western narratives of capitalism and globalisation. This is particularly pronounced in the critical scholarship that strives to overcome the 'territorial trap' (see Agnew, 1994; Brenner, 1999) and the methodological nationalism, which gives supremacy to national actors, embedded in the dominant narratives of the 'developmental state' (see Glassman, 1999; Hwang, 2016). For example, situating the South Korean economy in the Cold War geography, Glassman and Choi (2014) attribute the success of South Korean industrialisation, and that of the Korean developmental state and conglomerates, in part to the country's incorporation into the US-led, military-industry complex, thus overcoming the limitations of 'national–territorial' and 'state-centric' perspectives. The collection of essays that fed into a recent edited volume entitled *Developmental Cities? Interrogating Urban Developmentalism in East Asia* (Doucette and Park, 2019) also questions the pervasive 'methodological nationalism' and 'methodological statism' in narrating the history of urbanisation in the Global East, calling for more attention to how multi-scalar social forces collide in a relational way to produce the particularist geographies of capital and social relations that cannot be simply confined to a national territory.

The aforementioned neo-Marxian perspective of the geopolitical economy suggests that neo-Marxian and postcolonial perspectives are not mutually exclusive and can be complementary (see also Brenner, 2017). The geopolitical economic perspective also aims to disrupt the conventional narratives of globalisation that considers the developmental experience of the Global South as a deviation of modernity and capitalism rooted in the West, but it does this by retaining its focus on political economy and resistance politics. This is prominent in the work of the aforementioned Song and Hae (2019), who have re-oriented their study to frame their discussions "in transnational and relational terms" while highlighting their "interest in frameworks that challenge Euro-American epistemologies and anglophone hegemony in the field of knowledge production" (2019: 189). Referring to the work of Buckley and Hanieh (2014), Hae (2019: 84) further argues that the project undertaken as part of their edited collection is to "identify old and new forms of domination and subordination that are also connected to the broader global capitalist system, and point to the cracks, ruptures, and contradictions of systems that may open up political spaces for on-the-ground dissident politics". These perspectives are, in fact, speaking directly to what Roy (2016: 207) has reiterated: "It [postcolonial theory] allows me to undertake a political economy attentive to historical difference as a fundamental and constitutive force in the making of global urbanization. I rely on postcolonial theory to think relationally about cities".

Coda: The politics of knowledge production and praxis

I have argued in this chapter that de-centring knowledge production has the potential to run into the risk of re-centring it through the optics of prime cities, which are themselves exercising privileges in their respective countries and regions, thus exhibiting 'epistemological chauvinism'. In other words, studies on global southern urbanism are prone to reproducing another power dynamic within their own geography and susceptible to becoming mere manifestations of "geographic inversion and corrective inclusion" that Roy (2018: 45) has cautioned against.

Critical enquiries into global urbanism require recognition of multi-scalar processes at work and how these processes are situated within historical conjunctures and spatial connectedness, attending to "the global historical processes through which the two are inevitably entangled" (Roy, 2018: 45). Not only is an urbanising space the site of a palimpsest of historically accumulated processes that are both local and trans-local, but it is also the site of competing socio-spatial relations. As Massey (1993: 145) noted in her discussion of the locality,

> Interdependence [of all places] and uniqueness [of individual places] can be understood as two sides of the same coin, in which two fundamental geographical concepts – uneven development and the identity of place – can be held in tension with each other and can each contribute to the explanation of the other.

Contesting global urbanism requires us to regard our own locality – for me, Asia and, more specifically, East and South-East Asia as Global East but more narrowly cities in which I have thus far carried out empirical research – as a starting point of critical enquiry and an ending point of intervention as praxis and practicing such epistemologies without 'essentialising or appropriating' the geographical region under scrutiny (Moosavi, 2020: 19). This entails the retention of a historical and relational perspective, which involves analysing urban processes under observation historically and relationally, exploring particularities without losing the grip of universality.

Furthermore, urban scholarship needs to be more mindful of the politics of knowledge production. Radical scholarship may also be liable to some limitations, such as the exclusion of the voices from outside the Global North,[4] despite its attention to the project of decolonising knowledge production and demolishing the existing hierarchy that builds on social injustice. It would be vital to pursue intellectual decolonisation projects that are inclusive of radical scholars located outside the Global North and to ensure that, as Moosavi (2020) emphatically implores, such projects retain 'epistemological vigilance' while distancing from 'nativist' or 'tokenistic' decolonisation. It is also imperative to support the network of (particularly early career) scholars from the Global South/East whose intellectual endeavours face multi-layered constraints due to censorship, gendered hierarchy, seniority and racial and ethnic inequality.

Here I am particularly inspired by the way scholars on the margin (in the sense that they felt marginal in their pursuit of publication projects as they confront difficulties in securing a contract for a special issue on a regional city/country) were able to challenge the postcolonial scholarship by making use of South Korea as the 'core location', a concept that Song and Hae (2019) have adopted from Baik (2013) in their collective project. These core locations are sites of contestations, the examinations of which would allow observers to reveal the conflicting relations and processes that give rise to such contestations.[5] I would like to see that every student of critical urban studies develops their own core locations that help them practice analytical generalisation *and* pursue activism as praxis. It is valuable to be reminded of late Anne Haila's legacy: "Singapore is the protagonist of my story. It is a case and a comparison, but it also…enables me to generalise. My generalisations concern land, property and land rent" (Haila, 2016: xxiii). Here the act of theorising from outside the North Atlantic or the West is not to involve a mere academic exercise of testing the applicability of concepts and ideas generated in the West.

There is much to learn from the radical scholarship from the Global East and elsewhere that remains in the blind spot. As Hae and Song (2019: 11) argue, "A universal common ground of resistant struggles against unjust capitalist exploitation, dispossession, and expropriation that have erupted across different locales in the world still needs to be identified, explained, and highlighted", an endeavour that has not been attended to sufficiently by postcolonial urban studies. However, fundamentally, praxis rests upon reflection on rich narratives and particularities each locality presents, learning from those whose lives suffer from injustice and inequity. This is where the duty of care and post-research ethics kick in for critical scholars. The study of ordinary cities or Global South/East urbanism is ultimately to give voice to the people in each scholar's core locations. For this, every scholar is to return to their site in the way the late Anne Haila cared for Singapore (or what Singapore did to her). After all, Karl Marx was calling for theories to change the world and the revolutionary thoughts were to be localised: "The philosophers have only interpreted the world, in various ways; the point is to change it".[6]

Acknowledgements

The writing of this chapter has been partly supported by the National Research Foundation of Korea Grant funded by the Korean Government (NRF-2017S1A3A2066514) and the Tackling the UK's International Challenges 2018 grant (reference: IC3\100155) provided by the British Academy. I also thank Laam Hae, Sin Yee Koh, Jesook Song, Charles Stafford and Yimin Zhao for their encouragement and comments on the earlier drafts. I am, however, responsible for any errors in this chapter.

Notes

1 I am, too, guilty of this, having used Beijing, Guangzhou and Seoul as the main sites to theorise from.
2 According to the results of 2018 Census in South Korea, more than three-fifths (61.4 per cent) of all dwellings in the country were apartments (Statistics Korea, 2019).
3 In discussing the decolonisation of urban knowledge, Carlos Vainer (2014: 53) warns against the rise of "epistemological nationalism or chauvinism", that is, "the replacement of a Eurocentric, monotopic epistemology by another one – a global southern one – also mono-topic in nature, though centred instead in Latin America or elsewhere in the periphery". A similar logic can be extended to the way in which a select number of global southern cities possibly produce their own version of urbanism that gives no regard to smaller and peripheral cities.
4 For instance, a recent publication from the *Antipode* journal, *Keywords in Radical Geography: Antipode at 50*, included a wide-ranging group of critical scholars to discuss 50 keywords. While the efforts and interventions by the contributors may be informative and insightful, it comes as a surprise that almost all contributors came from academic institutions based in the Global North (with one exception being a Argentina-based member of a collective).

5 Hae and Song (2019: 13) further note that these core locations act as "nodal points of 'multiple historical/geographical determinations, connections, and articulations' (Hart 2006, 984)" while distancing from privileging "the local scale and difference".
6 Thesis XI in the *Theses on Feuerbach* by Karl Marx. URL: https://www.marxists.org/archive/marx/works/1845/theses/theses.pdf (last accessed 31 May 2020).

References

Agnew, J. (1994) 'The territorial trap: The geographical assumptions of international relations theory'. *Review of International Political Economy*, 1(1):53–80.

Baik, Y.-S. (2013) *Haeksimhyeonjang-eseo Dong-asiareul Dasi Muda: Gongsaengsahoereul Wihan Silcheon-gwaje [Rethinking East Asia in Core Locations: A Task for a Co-Habitation]*. Paju: Changbi

Bell, D. and Jayne, M. (2009) 'Small cities? Towards a research agenda'. *International Journal of Urban and Regional Research*, 33(3): 683–699.

Brenner, N. (2017) 'Debating planetary urbanization: For an engaged pluralism'. http://urbantheorylab.net/uploads/Engaged-pluralism_Brenner.pdf

Brenner, N. (1999) 'Beyond state-centrism? Space, territoriality, and geographical scale in globalization studies'. *Theory and Society*, 28(1): 39–78.

Buckley, M. and Hanieh, A. (2014) 'Diversification by urbanization: Tracing the property-finance nexus in Dubai and the Gulf'. *International Journal of Urban and Regional Research*, 38(1): 155–175.

Chen, X. and Zhou, Z. (eds.) (2009) *Shanghai Rising: State Power and Local Transformation*. Minneapolis: University of Minnesota Press.

Doucette, J. and Park, B.-G. (2019) (eds.) *Developmental Cities? Interrogating Urban Developmentalism in East Asia*. Leiden, the Netherlands: Brill.

Glassman, J. (1999) 'State power beyond the "territorial trap": The internationalization of the state'. *Political Geography*, 18(6): 669–696.

Glassman, J. and Choi, Y.-J. (2014) 'The Chaebol and the U.S. military–industrial complex: Cold War geopolitical economy and South Korean industrialization. *Environment and Planning A: Economy and Space*, 46(5): 1160–1180.

Hae, L. (2019) Against the construction state: Korean pro-greenbelt activism as method. In: Song, J. and Hae, L. (eds.) *On the Margins of Urban South Korea: Core Location as Method and Praxis*. Toronto: University of Toronto Press, pp. 65–88.

Hae, L and Song, J. (2019) Introduction: Core location, Asia as method, and a relational understanding of places. In: Song, J. and Hae, L. (eds.) *On the Margins of Urban South Korea: Core Location as Method and Praxis*. Toronto: University of Toronto Press, pp. 3–20.

Haila, A. (2016) *Urban Land Rent: Singapore as a Property State*. Chichester: Wiley Blackwell.

He, S., Kong, L. and Lin, G.C.S. (2017) 'Special issue: Interpreting China's new urban spaces: State, market, and society in action'. *Urban Geography*, 38(5).

Hwang, J.-T. (2016) 'Escaping the territorially trapped East Asian developmental state thesis'. *The Professional Geographer*, 68(4): 554–560.

Logan, J. (2018) 'Special issue: People and plans in urbanising China'. *Urban Studies*, 55(7)

Massey, D. (1993) Power-geometry and a progressive sense of place. In Bird, J., Curtis, B., Putnam, T., Robertson, G. and Tickner, I. (eds.) *Mapping the Futures: Local Cultures, Global Change*. London: Routledge, pp. 60–70.

McFarlane, C. (2011) *Learning the City: Knowledge and Translocal Assemblage*. Oxford: Wiley Blackwell.

Moosavi, L. (2020) 'The decolonial bandwagon and the dangers of intellectual decolonisation'. *International Review of Sociology*. doi: 10.1080/03906701.2020.1776919.

Müller, M. (2017) 'In search of the global East: Thinking between North and South'. *Geopolitics*. doi: 10.1080/14650045.2018.1477757.

O'Donnell, M.A., Wong, W.W.Y. and Bach, J.P.G. (eds.) (2017) *Learning from Shenzhen: China's Post-Mao Experiment from Special Zone to Model City*. Chicago: The University of Chicago Press.

Park, B.-G. and Jang, J.-B. (2019) The gangnam-ization of Korean Urban ideology. In: Doucette, J. and Park, B-G. (eds.) *Developmental Cities? Interrogating Urban Developmentalism in East Asia*. Leiden, the Netherlands: Brill, pp. 134–161.

Park, S.H., Shin, H.B. and Kang, H.S. (eds.) (2021) *Exporting Urban Korea? Reconsidering the Korean Urban Development Experience*. Routledge.

Robinson, J. (2006) *Ordinary Cities: Between Modernity and Development.* London: Routledge

Robinson, J. (2002) 'Global and world cities: A view off the map'. *International Journal of Urban and Regional Research*, 26(3): 531–554

Roy, A. (2018) Urban studies and the postcolonial encounter. In: Hall, S. and Burdett, R. (eds.) *The SAGE Handbook of the 21st Century City*. Thousand Oaks, CA: SAGE Publications, pp. 32–46.

Roy, A. (2009) 'The 21st century metropolis: new geographies of theory'. *Regional Studies*, 43(6): 819–830

Roy, A. (2016) 'Who's afraid of postcolonial theory?' *International Journal of Urban and Regional Research*, 40(1), 200–209.

Schindler, S. (2017) 'Towards a paradigm of Southern urbanism'. *City*, 21(1): 47–64

Shin, H.B. (2021) Urban transformation "Korean style": Lessons from property-based urban development. In: Park, S.H., Shin, H.B. and Kang, H.S. (eds.) *Exporting Urban Korea? Reconsidering the Korean Urban Development Experience*. Abingdon, Oxon; New York: Routledge, pp. 58–80.

Shin, H.B. (2011) Vertical accumulation and accelerated urbanism: The East Asian experience. In: Gandy, M. (ed.) *Urban Constellations*. Berlin: Jovis, pp. 48–53.

Shin, H.B. (2019) Asian urbanism. In: Orum, A.M. (ed.) *The Wiley-Blackwell Encyclopedia of Urban and Regional Studies*. Wiley-Blackwell. doi: 10.1002/9781118568446.eurs0010.

Shin, H.B., Lees, L., and López-Morales, E. (2016) 'Introduction: Locating gentrification in the global East'. *Urban Studies*, 53(3): 455–470.

Shin, H.B. and Zhao, Y. (2018) Urbanism as a state project: Lessons from Beijing's Green Belt. In: Jayne, M. (eds.) *Chinese Urbanism: Critical Perspectives*. Abingdon: Routledge, pp. 30–46.

Short, J.R. (2004) *Global Metropolitan Globalizing Cities in a Capitalist World*. London: Routledge.

Simon, A.M. (2010) *City Life from Jakarta to Dakar: Movements at the Crossroads*. New York: Routledge.

Song, J.-S. and Hae, L. (eds.) (2019) *On the Margins of Urban South Korea: Core Location as Method and Praxis*. Toronto: University of Toronto Press.

Sonn, J. and Shin, H.B. (2020) 'Contextualizing accumulation by dispossession: The state and high-rise apartment clusters in Gangnam, Seoul'. *Annals of the American Association of Geographers*, 110(3): 864–881.

Statistics Korea (2019) Press release of 2018 population and housing census: Registered census method results. URL: http://kostat.go.kr/portal/korea/kor_nw/1/1/index.board?bmode=read&aSeq=377115 (Last accessed 30 April 2020).

Vainer, C. (2014) Disseminating "best practice"? The coloniality of urban knowledge and city models. In: Parnll, S. and Oldfield, S. (eds.) *The Routledge Handbook on Cities of the Global South*. New York: Routledge, pp. 48–56.

Wu, F. (2015) Emerging Chinese cities: Implications for global urban studies. *Urban Geography*, 68(2): 338–348.

Wu, F. and Zhang, F. (eds.) (2020) 'Special issue: New directions of urban studies in China'. *Urban Studies*, 57(3).

8

GLOBALIZING POSTSOCIALIST URBANISM

Liviu Chelcea, Slavomíra Ferenčuhová and Gruia Bădescu*

Often obscured in discussions of global urbanism, cities found in (ex)socialist countries have been for many decades relegated to their own idiosyncratic category. Socialist and postsocialist cities have been contentious and unstable research objects for a number of reasons (Wiest, 2012; Grubbauer, 2012; Hirt, 2013; Ferenčuhová, 2020; Hirt et al., 2016; Tuvikene, 2016; Gentile, 2018; Müller, 2019). During the Cold War geopolitical division of the world, the cities of the former Soviet Bloc were the subject of ideological stereotypes, as well as the object of idealistic imaginations on both sides of the Iron Curtain. Next to serious attempts to understand the specificity of existing 'socialist cities' that were sometimes followed by hesitations about the analytical relevance of the concept, socialist cities were often subject to patronizing overtones, as well as caricatural popular culture representations of life in 'totalitarian' societies and planned economies. The analysis of socialist cities also over-emphasized homogeneity, under-estimated internal differentiation and limited their comparability to examining the relevance of theories on cities in capitalist economies and to checking for differences (Hirt et al., 2016).

Since the end of the Cold War, cities of the ex-Soviet Bloc have often been subject to teleologies of postsocialist 'transition to capitalism,' which emphasized the emergence, and indeed the necessity, of free-market urbanism and the value of private property, as well as the retreat of the state as part of the imagined road to a propertied civil society. Moreover, the ever-present reference to the past in the concept of the 'postsocialist city' has locked its meaning and trapped it in historical debates about the difference 'from the West' (Wiest, 2012; Hirt et al., 2016; Gentile, 2018; Müller 2018, 2019). Eventually, explicit calls appeared to drop the concept, as it potentially brought more trouble than analytic worth (e.g., Hirt et al., 2016; Gentile, 2018; and especially Müller, 2019). This turn rejects even the topical attempts to redefine or to 'deterritorialize' the notion to make it useful outside the context of cities geographically located in the former Eastern Bloc (Chari and Verdery, 2009; Tuvikene, 2016).

We suggest, however, that these cities offer multiple vantage points to think globally. We understand the project of *global urbanism* mainly as a shift in perspective and in theorizing about cities,

* Slavomíra Ferenčuhová would like to thank Imre Kertész Kolleg at the FSU Jena for providing pleasant conditions for working on the chapter during her fellowship period and to the Czech Academy of Sciences (Lumina quaeruntur, No LQ300282103).

DOI: 10.4324/9780429259593-10

which was inspired by postcolonial and decolonial theory and by a turn to comparative research. This change in perspective has parallels in other fields – for example, in sociology, where inspiration by postcolonial theory stimulated similar criticism of the current state of the art in the discipline as in urban studies (Burawoy, 2016; Go, 2016; Bhambra, 2016; Connell, 2018). The idea behind these shifts across social sciences is, first, to recognize the value of academic knowledge and theory produced in places beyond the West and Global North for what is understood as 'universal' or generalizable knowledge. Second, it leads to taking into consideration the often-unequal historical relations between places as influencing how knowledge has been produced in the past (Bhambra, 2016). In the case of postsocialist cities, adopting a perspective of *global urbanism* means recognizing as globally relevant the urban experience of postsocialism, as it is locally researched mainly in cities in Central and Eastern Europe and Central Asia, but also the multiplicity of urban experiences in producing knowledge about cities, a point raised by decolonial theory. Additionally, it means understanding that knowledge and theories about socialist and postsocialist cities have been produced in specific historical conditions – that of the bipolar world of the Cold War era and by the post-1989 project of transition to liberal democracy and the capitalist market.

Since the inception of questions about the existence and specificity of a socialist and postsocialist city, the answers have changed significantly: from ideas about socialist cities expressing a new societal order by the mid-twentieth century, to differences between, as well as similarities with, capitalist cities (of the then 'First World') mainly in the 1970s and 1980s, to 'convergence' with cities of Western Europe, the precedence of agency over structure and frictions during the 1990s and 2000s. Eventually, scholars expressed doubt about the validity of ontological assumptions about postsocialism during the 2010s, about the usefulness of this body of research for urban theory production and searched for new conceptual inventions (see e.g., Sjöberg, 2014). We do not want to dwell here on the pitfalls of using the term 'postsocialism,' but rather point to new analytics and metaphors helpful in making sense of differences and erasures in the efforts to capture socialist and postsocialist formations. If understanding postsocialism as a temporal and spatial container (legacy) is, indeed, obsolete, what analytic language can we use to capture socialist and postsocialist formations that persist in different parts of the world? In the following section, we indicate three areas where socialism and postsocialism seem likely to endure. We close with some remarks about analytic engagements with postsocialist cities as a form of comparative global urbanism theorization.

The ongoing aftermath of socialism: (Post)Cold War urbanisms, ghosts, intransigent materialities

(Post)Cold War urbanisms

Early theorizations of postsocialism (Verdery, 1996: 4) indicated that the Cold War division between West and East meant a form of 'knowledge and a cognitive organization of the world.' The emergence, transformations and implosion of socialism also meant the transformation of the West. The argument was that, with the Cold War over, we need to think differently about West and East, North and South (Chari and Verdery, 2009). Under the 'post-Cold War Studies' analytic, Chari and Verdery argued that it is time to liberate the Cold War studies from the ghetto of Soviet area studies and postcolonial thought from the space of Third World and colonial studies. Calls for relational geographies when researching 'Eastern Europe' appeared as early as 2005 (Stenning, 2005).

One way to do that is to trace the constitutive effects of Soviet socialism on urbanization outside the Soviet Bloc. Current historical research on state socialism brings forward numerous examples of contacts and exchanges that occurred transnationally during the Cold War era (Pabiracki and Jersild, 2016, Mark et al., 2019). In architecture, urban planning and city building, exchanges of ideas, work and even material objects (such as prefabricated family houses) existed between socialist states and the rest of the world, including the 'developing countries' or the 'West,' and influenced the cities around the world that now still contain traces of these exchanges (Ward et al., 2013; Ward, 2012; Cook et al., 2014; Stanek, 2015, 2019; Ferenčuhová, 2020). Moreover, a paradigmatic example is the federal support of 1950s suburbanization as civil defense in the United States. In addition to well-known factors, the suburban boom of the 1950s and 1960s relied on the permanent Cold War economy and fear of nuclear bombs, which understood the suburbanization of industry and the urban core as civil defense. One editor of the Bulletin of the Atomic Scientists explained in the 1950s that 'displacement is the only measure which could make an atomic super Pearl Harbor impossible' (Tobin, 2002).

Another illustration is Erin McElroy's use of a postsocialist analytic to understand the post-Cold War transformations of San Francisco and Silicon Valley. Silicon Valley emerged in the 1960s as a 'military/education hybridity designed to engineer the "free world"' (McElroy, 2017: 3), yet as the post-1989 federal defense budget diminished, it quickly veered towards financialization and entrepreneurialism. The dot-com invasion of entrepreneurs, Internet industries and 'digital nomads' radically transformed the class and racial geography of San Francisco and Oakland.[1] For McElroy (2017: 8),

> postsocialist Valley geography therefore has not only sublated local urbanities such as San Francisco and Oakland through new technologies of transit, but also former socialist landscapes, signaling new imperial spatiality endemic to the age of global techno-capitalism. The phenomenon of producing Silicon space beyond Valley confines is thus one of postsocialist spatiality, mobility, and global liquidity. This is not simply because prestidigitation has outgrown Silicon Valley space; postsocialist relations inhere globality.

One may think of many other forms of post-Cold War developments that extend beyond the ex-Soviet countries. Tuvikene (2016), for example, reframed the notion of postsocialism as relevant to all instances where urban ideals of a just access to resources, redistribution and sharing of common goods have been abandoned for a market-lead development. A recent article stresses a different, global *in-betweenness* of these cities that, however, is shared with many cities outside the 'East' (Müller, 2018). Furthermore, Müller (2019) suggests that the concept of the 'Global East' should be introduced as an antidote to the omission of the (former) Second World from current critical debates in urban studies and in geography about global power inequalities, as well as relations between the Global North and Global South. Ex-socialist countries are omitted because they are neither former colonies nor colonizing powers, and they are hardly at the extremes of power/powerlessness, economic or political, that are associated with the two categories (ibid.).

Müller's concept designates both the global in-betweenness and a 'political project' (2018: 16). Müller suggests, convincingly only up to a point, that the Global East is not merely a new label for the geographical area and/or for its historical specificity that would replace the adjective 'postsocialist.' The state of in-betweenness and the tendency to be globally invisible is not reserved for the former Second World (although those places *in between* often happen to be the countries of the former Eastern Bloc). The concept, therefore, possibly applies to a variety of places with different histories. As to the second point, the author calls for thinking 'with the East,' for postulating its existence and reclaiming 'a place *and* a voice for it in the world' (Müller, 2018: 16, original emphasis). This second point echoes similar ideas about the concept of the Global South as it is used in 'the southern urban critique' (Lawhon and Truelove, 2019) – that is, as a 'time-limited concept-metaphor,' which is 'of tactical *analytical* and *political* utility in the project of change [sic] knowledge production' (ibid.: 14).

Müller's project is an example of a bold move towards searching for new concepts of wide relevance while they are embedded in the research of particular places and cities – in this case cities in the Global East/postsocialist countries – yet open to critique and revision. This clearly resonates with the project of a 'more global urban studies' and with the comparative urbanism, but it carries with it two immediate dangers. First, the 'East' reproduces squarely the Cold War divisions, just like the spatial container understanding of 'postsocialism' did, despite Müller's redefinition of the concept that gestures towards a more general (nongeographical) meaning. Second, it may postulate an (expected) empirical specificity of the places labeled as the Global East, a contentious point discussed in urban theory also in relation to the category of the Global South (see Lawhon and Truelove, 2019).

Ghosts and zombies of socialism

If socialism and postsocialism are no longer relevant, why is it that the debates for the 2020 US presidential election have been framed around capitalism versus socialism? Why are right-wing thinkers and politicians from Eastern Europe and the United States still passionately fighting communism today, increasingly outside the ex-socialist countries, such as in the European Parliament? Why is 1965 anti-communist legislation upheld in Indonesia three decades after the end of communism and six decades after the onset of the military dictatorship in that country (Anon., 2017)? Anti-communism has been a palpable ideological and material force, inscribed in urbanism, social policies, landscapes and the bodies of denizens of ex-socialist cities (Light and Young, 2015; Lancione, 2019; Martinez, 2018). The post-1989 fight against state socialism and non-state socialism more generally has been at the forefront of capital accumulation in ex-socialist cities. Socialism is well and alive as a zombie in the realm of post-1989 anti-communism. Anti-communism, often materialized in anti-state and pro-business stances and anti-redistribution of wealth, is a hegemonic formation that does not show any signs of weakness. Memory wars have reframed the opposition of people against repressive techniques of control, political violence and dictatorship to an alleged war against everything public, common, collective action and emancipatory ideals (Bernhard and Kubik, 2016). Socialism is a zombie and ghost kept alive through its "strategic essentialization" by the Koch brothers and their libertarian accomplices in the United States, for instance, and the winners of transition in ex-socialist countries (Chelcea and Druță, 2016).

The 'urbanization of anti-communism' has had palpable effects on urban transformation (Golubchikov, 2016; Deoancă, 2017; Chelcea, 2019; Moll, 2019; Atanasoski and McElroy, 2018; Ghodsee 2011, 2017). It may help residents signal rupture and closure with the socialist past, as Fehérváry's (2013) describes in her study of transformation of socialist modernism through colors that depart from the socialist era concrete-gray overtones of housing estates. Anti-communism may show up in the reclusive gated communities in ex-socialist cities, built as an antidote to the alleged 'collectivism' of socialist-era housing estates. Anti-communism has also reshaped cities through housing and urban land property rights restitution policies (Lancione, 2019; Kusiak, 2019). The ghost of socialism may equally deflect questions about the political present by pointing to the socialist past, as Saygun Gökariksel (2019) identified in the spectacles of 'facing history' through open-air exhibitions with large posters, with secret police officers circulating in Polish cities. The socialist era, on the other hand, may also function as a resource against neoliberal normalization (Channell-Justice, 2019: 50; Kurtović and Sargsyan, 2019; Borelli and Mattioli, 2013: 11).

Intransigent materialities

Even the most critical voices of 'postsocialism' ultimately depict socialist urban materiality as difference. The socialist-era housing estates, the high density of the built environment (Hirt, 2013; Marin and Chelcea, 2018), the Cold War urbanization of Siberia (Gentile, 2018: 7; Gerlach and Kinossian, 2016), but also the technological characteristics of urban infrastructures (Collier, 2011), seem to resist the metanarrative and practices of transition from socialism and survive the theoretical extinction of postsocialism. Gentile (2018: 7), for instance, mentions that the Siberian cities cannot follow a postsocialist fix because the 'costs of resettlement would be prohibitive.' Stephen Collier (2011) has described that the inbuilt features of networked infrastructures for heating in Russian cities precluded the introduction of consumer-based remaking of municipal heating in some industrial towns. The material environment of cities built in the socialist era – housing estates with abundant green spaces – is at times perceived as a limiting obstacle and at other times as an advantage in facing new challenges, such as adapting cities to climate change (Bakay, 2012).

Some socio-technical networks have been reluctant to engage in transition; sometimes they have created friction in the circulation of fast policy and fast capital (Golubchikov et al., 2014; Bouzarovski et al., 2016; Bouzarovski et al., 2018). Foregrounding such intransigent materialities is not to focus on the objects of nostalgia, socialist 'legacies' and socialist hangovers but rather on those assemblages that create friction with regional and global flows. While some urban materialities are much more malleable to such normalizations (Fehérváry, 2013), Murawski (2019) suggests the notion of 'still socialist' objects, such as the Palace of Culture in Warsaw, which resists capitalist and anti-communist normalization through its bulky materiality and its surrounding landscape. In terms of postsocialist reshaping of urban materialities, these cities have echoed an array of global processes, mirroring both Global North and Global South urban outcomes from gated communities to gentrified historic centers, to informal urbanism (Hirt, 2013; Tsenkova, 2017). These transformations have been often transnational and transregional, seeing the emergence of new circulations, flows, but also power relations between local elites and foreign actors in regions as diverse as the Americas, the Middle East and East Asia (Bădescu, 2016).

(Post)socialist city debates and the project of global urbanism

The academic project of postcolonial and more global urban studies has been developing extensively for almost two decades and has included scholars from a variety of disciplines and places. A thorough critical debate about parochialism, local embeddedness, limitations of 'international' and 'general' urban theories and inequalities in the academic knowledge production resulted in searching for new ways of thinking about cities, determining new strategies of empirical research and theory-making and establishing connections in scholarly work and dialogue.

The debate about the postsocialist city and the value of its previous research has been, to an extent, inspired or provoked by the attempts to develop postcolonial urban studies and comparative urbanism, which theorizes from 'elsewhere' (McFarlane, 2010; Jazeel and McFarlane, 2010; Robinson, 2016). The uneasiness about the concept of the postsocialist city and the need to revise the category came from within the field as researchers struggled with the concept's limitations (Ferenčuhová, 2016). Eventually, the strategies adopted in the past decade by researchers studying ex-socialist cities included attempts to deconstruct categories, uncover their embeddedness in geographical and historical contexts and reconstruct an analytic language sensitive to socialist formations and their aftermath. The postsocialist city debate, insular as it may seem, has, therefore, been part of the ongoing *global* search to change the ways of creating knowledge in urban studies.

Postsocialist cities have the potential to contribute to the dialogue on global urbanism in three ways. First, they expose, and perhaps epitomize, the neoliberal transformations of the last decades. Their comprehensive transformation of property relations and the state withdrawal from city-making have accounted for a radical laboratory of neoliberal change, often related to a politics of austerity. Deterritorializing the postsocialist condition implies using the lens of a global postsocialist transformation after 1989 for the analysis of processes occurring elsewhere. Second, these cities are part and parcel of the global urban through the flows, circulations and entanglements they mediate. They are often hubs of transnational migration – from Moscow's hyperdiversity in the shadow of empire, to migrants and refugees in Central and Eastern Europe – and exhibit economic relations and flows stemming from the socialist era and beyond (Rogers, 2010; Bădescu, 2016). The postsocialist experience is thus one that highlights the existence of both multiple and entangled modernities (Therborn, 2003; Wiest, 2012). Third, the ongoing empirical research, briefly presented earlier, brings more broadly to the fore several aspects in which it is still reasonable to 'think with the past': historically formed connections across places, obvious continuities with the past and new uses of historical references and past materialities in the present-day.

From an epistemological standpoint, postsocialist cities offer, in a Gadamerian sense, a horizon of knowledge which showcases the existence of urban situations that are both idiosyncratic and globally entangled. The postsocialist experience underlines that a southern and decolonial theory of global urbanism is not only possible but also necessary. Just as the Global South is an umbrella term about a myriad of urban lifeworlds, postsocialist cities add to this diversity of situations but also contribute to understanding the nuances of general processes: intertwined, internationalized neoliberal regimes, forms of resistance, movements towards commons, linkages, flows and transnational mobilities. The vantage point of former socialist cities adds to the postcolonial call to think of concepts and urban theory from elsewhere. Take for instance the work on idiosyncrasies of *gentrification* in the former socialist world, where property restitution, debates about housing justice and high ownership rates complicate the global understanding of the concept (Bernt, 2016).

The politics of global urbanism can also benefit from the (post)socialist experience. If global urbanism means an inclusion of voices to mirror the multiplicity and entanglement of urban experience worldwide, a real engagement with what has been a rather neglected urban form (Gentile, 2018) needs to occur. Conversely, scholars working on ex-socialist cities can continue to implode the area studies container. If, however, the politics of global urbanism relates to a global attempt to link voices for an emancipatory politics of the city, the laboratory nature of these cities both during socialism and after is an enormous resource. From the emancipatory goals and achievements of socialism to the pitfalls of the party-state, coercive methods and top-down city-making during the same era; from the postsocialist erosion of the public, large-scale privatization and dispossession to the contemporary resistance practices and collective action of people, these cities offer invaluable insights for shaping debates for emancipatory global urbanism. Finally, creating a dialogue between urban struggles and forms of resistance across cities can well be a platform for engaged global urbanism.

Note

1 In her subsequent work on 'digital nomads' in a Central European city, McElroy (2019) documents the displacement induced by such powerful urban actors.

References

Anon. (2017) 'Indonesia's Dangerous 'Anti-Communist' Paranoia Rumors of Communist Party Revival Sparks Riot.' *Human Rights Watch*, accessed 14 January 2021, https://www.hrw.org/news/2017/09/18/indonesias-dangerous-anti-communist-paranoia.

Atanasoski N. and E. McElroy (2018) Postsocialism and the afterlives of revolution: Impossible spaces of dissent. In: Pireddu N. (ed.) *Reframing Critical, Literary, and Cultural Theories*. New York and London: Palgrave Macmillan.

Bădescu, G. (2016) '(Post) colonial encounters in the postsocialist city: Reshaping urban space in Sarajevo.' *Geografiska Annaler B: Human Geography*, 98(4): 321–329.

Bakay, E. (2012) 'The role of housing estates' green surfaces in forming the city climate of Budapest.' *Applied Ecology and Environmental Research*, 10(1): 1–16.

Bernhard, M. and J. Kubik (eds.) (2016) *Twenty Years After Communism: The Politics of Memory and Commemoration*. Oxford: Oxford University Press.

Bernt, M. (2016) 'How post-socialist is gentrification? Observations in East Berlin and Saint Petersburg.' *Eurasian Geography and Economics*, 57(4–5): 565–587.

Bhambra, G. K. (2016) 'Postcolonial reflections on sociology.' *Sociology*, 50(5): 960–966.

Borelli, C. and F. Mattioli (2013) 'The social lives of postsocialism.' *Laboratorium*, 5(1): 4–13.

Bouzarovski, S., J. Frankowski, and S. Tirado Herrero (2018) 'Low-carbon gentrification: When climate change encounters residential displacement.' *International Journal of Urban and Regional Research*, 42(5): 845–863

Bouzarovski, S., L. Sýkora and R. Matoušek (2016) 'Locked-in post-socialism: Rolling path dependencies in Liberec's district heating system.' *Eurasian Geography and Economics*, 57(4–5): 624–642.

Burawoy, M. (2016) 'The promise of sociology: Global challenges for national disciplines.' *Sociology*, 50(5): 949–959.

Channell-Justice, E. (2019) 'Thinking through positionality in post-socialist politics.' *History and Anthropology*, 30(1): 47–66.

Chari, S., and K. Verdery (2009) 'Thinking between the posts: postcolonialism, postsocialism, and ethnography after the Cold War.' *Comparative Studies in Society and History*, 51: 6–34.

Chelcea, L. (2019) '*After Engels: Evictions and the Urbanization of Anti-communism*.' *Dialogues in Human Geography*, 9(2): 205–207.

Chelcea, L. and O. Druță (2016) 'Zombie socialism and the rise of neoliberalism in Central and Eastern Europe.' *Eurasian Geography and Economics*, 57(4–5): 521–544.

Collier, S. (2011) *Post-Soviet Social*. Princeton: Princeton University Press.

Connell, R. (2018) 'Decolonizing sociology.' *Contemporary Sociology*, 47(4): 399–407.

Cook, I.R., S.V. Ward, and K. Ward (2014) 'A springtime journey to the Soviet Union.' *International Journal of Urban and Regional Research*, 38: 805–822.

Deoancă, A. (2017) 'Class politics and Romania's 'white revolution.' *Anthropology News*, 58(4): e394–e398.

Fehérváry, K. (2013) *Politics in Color and Concrete*. Bloomington: Indiana University Press.

Ferenčuhová, S. (2020) 'Thinking relationally about socialist cities: cross–border connections in Czechoslovak post-war urban planning and housing construction.' *Planning Perspectives*, DOI: 10.1080/02665433.2020.1844042

Ferenčuhová, S. (2016) 'Explicit definitions and implicit assumptions about post-socialist cities in academic writings.' *Geography Compass*, 10: 514–524.

Gentile, M. (2018) 'Three metals and the "post-socialist city".' *International Journal of Urban and Regional Research*, 42(6): 1140–1151.

Ghodsee, K. (2011) *Lost in Transition: Ethnographies of Everyday Life after Communism*. Durham: Duke University Press.

Ghodsee, K. (2017) *Red Hangover: Legagies of Twentieth-Century Communism*. Durham: Duke University Press.

Go, J. (2016) 'Globalizing sociology, turning south: Perspectival realism and the southern standpoint.' *Sociologica*, 10(2): 1–43. doi: 10.2383/85279.

Golubchikov, O. (2016) 'The urbanization of transition: Ideology and the urban experience.' *Eurasian Geography and Economics*, 57(4–5): 607–623.

Golubchikov, O., A. Badyina, and A. Makhrova (2014) 'The hybrid spatialities of transition: Capitalism, legacy and uneven urban economic restructuring.' *Urban Studies*, 51: 617–633.

Gökariksel, S. (2019) 'Facing history: Sovereignty and spectacles of justice and violence in Poland's capitalist democracy.' *Comparative Studies in Society and History*, 61(1): 111–144.

Gerlach, J. and N. Kinossian (2016) 'Cultural Landscapes of the Arctic: "Recycling" of soviet imaginary in the Russian settlement of Barentsburg, Svalbard (Norway).' *Polar Geography*, 39(1): 1–19.

Grubbauer, M. (2012) 'Toward a more comprehensive notion of urban change: Linking post-socialist urbanism and urban theory.' In *Chasing Warsaw: Socio-Material Dynamics of Urban Change Since 1990*, edited by M. Grubbauer and J. Kusiak, pp. 35–60. Frankfurt: Campus Verlag.

Hirt, S., S. Ferenčuhová and T. Tuvikene (2016) 'Conceptual forum: The 'post-socialist' city. *Eurasian Geography and Economics*, 57(4–5): 497–520.

Hirt, S. (2013) 'Whatever happened to the (post)socialist city?' *Cities*, 32, S29–S38.

Jazeel, T., and McFarlane, C. (2010) 'The limits of responsibility: A postcolonial politics of academic knowledge production.' *Transactions of the Institute of British Geographers*, 35(1): 109–124.

Kurtović, L. and N. Sargsyan (2019) 'After utopia: Leftist imaginaries and activist politics in the postsocialist world.' *History and Anthropology*, 30(1): 1–19.

Kusiak, J. (2019) 'Rule of law and rules-lawyering.' *International Journal of Urban and Regional Research*, 43(3): 589–596.

Lancione, M. (2019) 'The politics of embodied urban precarity: Roma people and the fight for housing in Bucharest, Romania.' *Geoforum*, 101: 182–191.

Lawhon, M. and Truelove, Y. (2019) 'Disambiguating the southern urban critique: propositions, pathways and possibilities for a more global urban studies.' *Urban Studies.* doi: 10.1177/0042098019829412.

Light, D. and C. Young (2015) 'Toponymy as commodity: Exploring the economic dimensions of urban places.' *International Journal of Urban and Regional Research*, 39(3): 435–450.

Marin, V. and L. Chelcea (2018) 'The many (still) functional housing estates of Bucharest.' In *Housing Estates in Europe: Poverty, Ethnic Segregation and Policy Challenges*, edited by D. Hess, T. Tiimmaru and M. van Ham, pp. 167–190. Cham: Springer Nature.

Mark, J. et al. (2019) *1989. A Global History of Eastern Europe*. Cambridge: Cambridge University Press.

McElroy, E. (2017) 'Postsocialism and the tech boom 2.0: Techno-utopics of racial/spatial dispossesions.' *Social Identities*, 24(2): 206–221.

McElroy, E. (2019) 'Digital nomads in siliconising Cluj: Material and allegorical double dispossession.' *Urban Studies.* doi:10.1177/0042098019847448.

McFarlane, C. (2010) 'The comparative city: Knowledge, learning, urbanism.' *International Journal of Urban and Regional Research*, 34: 725–742.

Moll, Ł. (2019) 'Erasure of the common: From Polish anti-communism to universal anti-capitalism.' *Praktyka Teoretyczna*, 31(1): 118–143.

Müller, M. (2018) 'In search of the global east: Thinking between north and south.' *Geopolitics.* doi: 10.1080/14650045.2018.1477757.

Müller, M. (2019) 'Goodbye, postsocialism!' *Europe-Asia Studies*, 71(4): 533–550.

Murawski, M. (2019) *The Palace Complex*. Bloomington: Indiana University Press.

Pabiracki, P. and A. Jersild (eds.) (2016) *Socialist Internationalism in the Cold War – Exploring the Second World.* Cham: Palgrave Macmillan.

Robinson, J. (2016) 'Comparative urbanism: New geographies and cultures of theorizing the urban.' *International Journal of Urban and Regional Research*, 40(1): 187–199.

Rogers, D. (2010) 'Postsocialisms unbound: Connections, critiques, comparisons.' *Slavic Review*, 69, 1. doi: 10.1017/S0037677900016673.

Sjöberg, Ö. (2014) 'Cases onto themselves? Theory and research on ex-socialist urban environments.' *Geografie*, 119: 299–319.

Stanek, Ł. (2019) *Architecture in Global Socialism: Eastern Europe, West Africa, and the Middle East in the Cold War.* Princeton: Princeton University Press.

Stanek, Ł. (2015) 'Architects from socialist countries in Ghana (1957–67).' *Journal of the Society of Architectural Historians*, 74: 416–442.

Stenning, A. (2005) 'Out there and in here: Studying Eastern Europe in the West.' *Area*, 37(4): 378–383.

Therborn, G. (2003) 'Entangled modernities.' *European Journal of Social Theory*, 6: 293–305.

Tobin, K. (2002) 'The reduction of urban vulnerability: Revisiting 1950s American suburbanization as civil defense.' *Cold War History*, 2(2): 1–32.

Tsenkova, S. (2017) *Housing Change in East and Central Europe: Integration or Fragmentation? Abingdon*: Routledge.

Tuvikene, T. (2016) 'Strategies for comparative urbanism: Post-socialism as a de-territorialized concept.' *International Journal of Urban and Regional Research*, 40: 132–146.

Verdery. K. (1996) *What Was Postsocialism and What Comes Next*. Princeton: Princeton University Press.

Ward, S.V., I. R. Cook, and K. Ward (2013) 'Lesson Learning: From Liverpool to Moscow.' *Town and Country Planning*, 82: 344–347.

Ward, S.V. (2012) 'Soviet communism and the British planning movement: Rational learning or utopian imagining?' *Planning Perspectives*, 27: 499–524.

Wiest, K. (2012) 'Comparative debates in post-socialist urban studies.' *Urban Geography*, 33: 829–849.

9

BEYOND THE NOOSPHERE? NORTHERN ENGLAND'S 'LEFT BEHIND' URBANISM

John Flint and Ryan Powell

Introduction

Despite the diversity of scholarship on 'global urbanism', there are several common elements underpinning most definitions. First, the global trend towards an urbanised population through migration and city growth. Second, the increasing complexity of urban flows and interconnections across the world driven by a particular form of globalised capitalism that is, in turn, underpinned by cultural and economic valorisations and imperatives of cosmopolitan and mobility dispositions (Farrugia, 2019). Urban scholarship has further explored the new configurations of spatial juxtapositions, often framed dialectically between constellations of 'global' and 'left behind' cities.

As Farrugia (2019) argues, these developments are shaped by (and themselves inform) global urban imaginations. The contemporary prominence of cosmopolitan complexity and connection within these imaginaries resonates strongly with Wyly's (2015) work examining Teilhard de Chardini's concept of the noosphere. This depicted a planetary super stage of consciousness in which humanity would move to new degrees of organised complexity and reflective consciousness as emergent urban elites and dynamic nodes of global cities are constituted through new circuits of finance, communication and culture in a reconfigured form of 'cognitive-cultural capitalism' (Moulier-Boutang, 2012; Wyly, 2015). This global urban imaginary remains underpinned by the legitimacy of capitalist competition, while the skills and aptitudes valorised are those of creativity, technological advancement, knowledge and diversity. In parallel with new flows of investment between urban nodes, there is a realignment of social networks powered by corporations such as Apple, Google and Facebook (Wyly, 2015). The winners, both individuals and cities, are conceptualised as 'cosmopolitan patriots' (Appiah, 1997) dissolving traditional boundaries of space and sovereignty. Florida (2003) describes these new orientations of technology, talent and tolerance as generating 'a new industrial revolution inside people's heads' (quoted in Wyly, 2015).

Such an understanding generates three problematics. First, beyond the predominant interest in world cities (London being the United Kingdom's example), as Nayak (2019) asks, what is happening among the populations of English places forged in the original industrial revolution, many of whom are claimed to be peripheral to contemporary planetary urbanism? Second, this is not merely a matter of change 'inside people's heads', a phrase that articulates the dominance given to logics, calculation and knowledge by proponents of the new global order. Rather, it is

DOI: 10.4324/9780429259593-11

about wider notions of embodiment and affect in which time (history) and place (geography) are fundamental. Third, the new global patriots, and the dominant world cities and city-regions they reside in, are perpetually juxtaposed against 'the fortress mentality' of borders, conservative traditions, obsolete skill sets and insular identifications of populations and places 'left behind' (Wyly, 2015). This necessitates an analysis of global urbanism that understands populations and places as being in a dialectic relation with one another, occupying particular spaces in a constellation of interlinked positions. In Wacquant's (2016) understanding, the relegation of particular urban sites and populations is always a collective, power-laden and relational process resulting in distinctive socio-spatial formations. That is, places are not simply 'left behind' in peripheries or 'back waters' outside the super-speed highways of global economics. Rather, they always occupy a space and position relative to other urban configurations – discursively, economically, socially and symbolically. Likewise, the dialectics of identification and disidentification (de Swann, 1995) through which people increasingly see themselves as similar to, or different from, others rest on a strong attachment to place and a *group* orientation that is formed inter-generationally in relation to other people and places (Willis, 2014; Farrugia, 2019). Although we focus on Northern England, we build on broader works, such as Farrugia's (2019) on young people in Australia. While the specificity of place is fundamental to our understanding, there are considerable similarities with other 'peripheral' regions in terms of their socio-spatial positioning and the urban imaginaries underpinning this.

In this chapter, we further three arguments. First, an understanding of particular regions or populations as 'left behind' fails to capture the *historical complexity* of global urbanism's impacts. Second, a *revitalized concept of habitus* provides a dynamic mechanism for better understanding the intergenerational, geo-political impacts of deindustrialisation and the centrality of place and affect within them. Third, the framing of particular regions as 'left behind' is constitutive of an urbanism in these regions that is shaped by a *reconfigured imaginary* of their relative relations to the globalised economy, other cities, regions and the nation. Simultaneously, an understanding of their collective relative disadvantage creates tensions between cooperative or competitive tropes within the regions themselves.

Left behind?

The north of England appears as the exemplar antithesis of the noosphere discourses framing the success stories of contemporary global urbanism. The area situated between Scotland and the Midlands of England has a population of 18 million. It includes the major conurbations of Leeds, Liverpool, Manchester, Newcastle/Gateshead and Sheffield, other cities such as Bradford and Sunderland and numerous smaller towns. The economy of the north would rank as the eighth largest in the European Union (EU; IPPR North, 2016). There are important distinctions in their economic histories (for example, Sheffield's prominence as a 'steel city' or shipbuilding in the north-east) but a shared industrial manufacturing and mineral extraction heritage. Although specific trajectories also vary, the region has been transformed by the social, economic and cultural effects of deindustrialisation since the 1970s. Such seismic shocks continue: the decline in GDP in South Yorkshire between 2007 and 2011 was almost as large as that experienced by Greece (Forth, 2017, cited in McKay, 2019).

The discourse of the north of England as a post-industrial space of obsolescence increasingly adrift from the new drivers of economic prosperity and with "unproductive" populations ill-equipped for globalisation, crystallised in the UK 2016 referendum on EU membership. The Brexit outcome was posited as driven by a social and political alienation entrenched in 'left behind' deindustrialised urban areas as their economy and social fabric declined. Jennings et al. (2016) defined such 'backwaters' as being disconnected from the global economy with their populations more likely to be inward looking, illiberal, negative about the EU and immigration, nostalgic, and more English in their identity. Thus areas with a strong tradition of manufacturing employment and those areas with relatively low pay and high unemployment were more likely to vote to leave (Becker et al., 2017). Variations within the north of England, with cities like Manchester voting strongly to remain while towns like Doncaster and Oldham voted to leave, appeared to reinforce the distinction between a confident cosmopolitan urbanism and post-industrial peripheral insularism.

The 'left behind' narrative about the north of England articulated a common understanding of global competition in which connectivity was the decisive factor. However, the posited solution was the forging of a 'northern powerhouse' (with its deliberative evocation of former industrial might) focused on connectivity *within* the region rather than directly to global circuits of wealth creation. There was also a localised intra-British understanding of periphery in which the north was juxtaposed against London and the south-east of England. All of these elements were captured in former chancellor George Osborne's first articulation of the powerhouse idea:

> The powerhouse of London dominates more and more...[we need] not one city, but a collection of northern cities – sufficiently close to each other that combined they can take on the world.
>
> (Osborne, 2014)

The three policy approaches were developing a counterweight to London, agglomeration achieved through infrastructure investment and devolution of power to city-regions. But even here there was an imaginary dominated by the major cities that failed to understand that much economic capacity and opportunity (for example, in advanced manufacturing) was located in the smaller cities and towns and that there were strong and reciprocal interdependencies and connections between the major cities and their surrounding urban settlements (Pike et al., 2016). If the political economy of 'the left behind' may be contested, we also need a stronger understanding of the mechanisms through which a post-industrial urbanism may generate alleged social and cultural alienation, to which we now turn.

The habitus of industrial ruination

'The industrial order exerted an overwhelming influence on the organization and formation of working class lives and spaces' (Emery, 2018: 4; see also Cowie and Heathcott, 2003). This influence was most pronounced in towns and cities dominated by single industries. Places that were defined as 'coalfields', 'steel towns' or 'shipbuilding communities' not only offered a *relatively* predetermined route to employment (especially for young men) but also social organisation and leisure were often intrinsically bound up with those industries. The cataclysmic shift in social and spatial interdependencies was defined by a decline in relative power and a diminution in the importance of these places to the national and global economy. For example, as coal became less important to economic growth, the collective power of coalfield communities also diminished. This is sharply exemplified by the sudden obsolescence of a highly skilled workforce; the eradication of stable, predetermined routes to employment; loss of status (individually and collectively); institutional abandonment; and the almost instantaneous undermining of the social organisation of communities around the extraction of coal.

The impacts of the collapse of these industries are well documented in terms of worklessness and indicators of relative deprivation (Beatty et al., 2017). The longer-term impact of such shocks upon habitus formation, however, remains relatively under-explored (for exceptions, see Evans et al., 1996; Charlesworth, 2000). Indeed, changes in the psychic structure tied to wider social transformations appear to be a sizeable gap in current understandings of contemporary urbanism. Mah (2012) argues that deindustrialisation represents a form of ruination encompassing the cognitive, cultural and material degradation of geographic areas of industrialism that, according to several studies, generates a sense of alienation, loss, dislocation and mourning (Emery, 2018). We suggest that a revitalized sociological understanding of habitus (Silva, 2016) offers a fruitful frame for articulating and grasping the longer-term and neglected impacts of industrialisation on the social self – a self that is always embedded in wider, dynamic social interdependencies and relations, both enabling and constraining (Burkitt, 2016). The embodied and affective dimensions (Edensor, 2015) of what Emery (2018) terms the 'qualitative traumas' arise from what Elias identified as the lag between changing social conditions and the social habitus to be in tune with these changes (Elias, 2000 [1939]; Connolly, 2019).

Extending the concept of habitus as set out by Elias (2000) and Bourdieu (2000; see Paulle et al., 2012) can help reveal the deep interrelationship between deindustrialisation, intergenerational change and territorial stigmatisation, as well as help explain how 'the emotional pull of class loyalties can entangle subjects in the affinities of the past' (Friedman, 2016), even as they are called upon to 'contemplate new schemes for existing, a different way of being because the old habit world, the ethos…has decayed' (Charlesworth, 2000: 83). A revitalized understanding of habitus that goes beyond common critiques of rigidity or plasticity (which arguably miss the point) and moves beyond a narrow focus on individualized class trajectories offers a useful theoretical lens. Such an understanding calls attention to group relations and place-based identifications beyond the individual in facilitating a more explicit micro inclusion of habitus tied to local affordances. Deploying notions of the fragmented habitus over time, and rooted in space, offers a conceptual understanding of these ruptures and the neglected (and often collective) emotional difficulties that they produce (Friedman, 2016; Silva, 2016; Schilling et al., 2019).

The socio-spatial configurations of northern towns and cities were shaped by their locations and topography – for example, the specific relatively isolated geography of mill towns that constituted spatially bound contexts for intergenerational transmission of orientations and cultures. Identities were shaped by the landscape that arose from the industries that made these places, what Emery (2018) describes as urban collages of factories, plants, dockyards, collieries, warehouses and transport networks. Through deindustrialisation, these sites become the visible manifestation of economic decay, the decline of the lives connected to them (Mah, 2010) and their neglect as a symbol of the wider abandonment of these places and populations by a new global urbanism and its political and governing classes.

But industrial ruins constitute only one element of visible economic trauma within a landscape that is continually remade in the sites of production (call centres and distribution centres) and consumption (high streets). These urban localities are still intrinsically linked to the national and global economy – even if they provide precarious low-paid employment – and are the sites of operation of international companies servicing urban marginality (payday lenders, bargain retail stores, betting shops). Just as the landscape was made when these towns and cities prospered as flourishing nodes of a previous global economy, it's remaking in the post-industrial, socio-spatial decline is rooted in the new relations and positions – always within a space of positions – that many northern English urban areas now occupy.

The Victorian town halls, libraries, swimming baths and parks also remind us that the industries that made and defined these localities were coupled with a strong sense of civic pride in local place, whereby a small industrial town would forge an identity in distinction to, and in fierce competition with, other places. Grand town halls were never the preserve of major cities. Even in smaller settlements, such as English coalfield communities, the Miners' Welfare Hall reinforced local pride and was a site for social organisation, education and leisure. Crucially, the mercantile elites, as much as their workers, were rooted in a shared sense of place in which economic surplus was symbolised and aggrandised in a civic architecture and philanthropy physically manifest in the locales of its production. Here was, in the very bricks and mortar of public buildings, a legitimacy and valorisation of industrial endeavour that may be contrasted with the contemporary geographical dislocation between capital, elites and urban marginality (Shaxson, 2011; McKenzie and Atkinson, 2019).

The north of England and reconfigured imaginaries

The ruptured habitus of northern deindustrialisation is situated in a wider reframing of the imaginary position of the north of England within the space of positions of contemporary global, national and regional urbanism (Haughton et al., 2016). As McKay (2019) argues, discontent has begun to incorporate a spatial element in the political moment post–Brexit, although such processes predate 2016. There are attempts to constitute the north of England as a specific and meaningful entity in terms of its political economy (IPPR North, 2016) and its culture and emotional identity (Morley, 2013) – what Cox (2018) terms 'a Northern Mind'. The 'we-identity' of the north is constituted in competition with the south of England (and more particularly with London, which has contrastingly seen a rejuvenated sense of itself as a world city with its own trajectory and international connectivity that looks beyond as much as within the United Kingdom) and with Scotland, again with a reinvigorated sense of nationhood that seeks to reframe the borders of belonging. An increasing sense of Englishness (as opposed to Britishness) has been identified with Leave voters. Thus there is a paradox whereby the decision to remove the United Kingdom from the integrational dynamics of the EU occurs in parallel with the declining sense of Britain as the primary cohesive economic, cultural and political imaginary entity. McKay (2019) found evidence of a relational 'grievance' effect whereby individuals are more negative about community representation when they believe that their local economy is less successful than the national one: "[The perception] that different parts of the country are on different economic trajectories has tangible effects on the sentiment that one's community is ignored" (2019: 9). Thus the 'left behind' increasingly encompasses political and cultural, as well as economic, disempowerment.

But the sense of the north as a collective facing a common challenge as a periphery region in the new global urban order neglects the extent to which rivalries, forged in the industrial revolution, continue. The pervasive narrative of competition between places within globalisation is manifested in the challenges of both establishing northern city-regions and in fostering collaborations between them.

There emerges a new geographical imaginary emphasising distinctions between cities and towns as the new fault line of urban marginalisation. A study by the Centre for Towns (2017) found that town residents were more likely to perceive themselves to be financially less well off than those living in cities and to believe that their locality is currently less central to British society than in the past, and it will be less central in the future. Thus a strong identification with towns in the industrial north increasingly goes hand in hand with a disidentification from elsewhere, particularly where geography serves as shorthand for class (Skeggs, 2004) in the post-mortem discourse of the Brexit outcome.

The alleged 'dislocation' of post-industrial areas is a somewhat misleading term as populations have not been physically relocated, nor subject to gentrification. Rather, they remain in situ in the same locale, although its landscape and social fabric are transformed. It is the position of this place, relative to others in the new spaces of positions within global urbanism, that has radically altered, with the collective trauma of socio-spatial demotion and denigration in turn affecting habitus formation. The shift in the interdependence between northern industrial urban areas and global urbanism is characterised by a decline in relative power: economic, cultural, political and symbolic, which manifests today in a deeply ingrained territorial stigmatisation from the outside and an unravelling of interdependencies and social organisation intrinsically related to a bygone industry. Yet the response to this complex ensemble might usefully be understood as a 'heroic simplification' (Scott, 1988) in which the new orthodoxy of city-region governance at best pays little attention to, and at worst ignores, the reality of post-industrial towns in the north. City-region governance de-centres power nationally but re-centres power regionally within urban centres. Policy initiatives such as core cities, the urban renaissance and inclusive growth models show a consistency in their neglect of peripheral towns and their local practices, histories and relations.

It is not difficult to understand political alienation when we consider the fortunes of places like Barnsley or Middlesbrough, their relative exclusion from city-led growth and opportunity and the disjuncture with the rhetoric of state simplifications. For example, the governmental claim of almost full employment, measured solely by receipt of unemployment benefits, ignores the reality that residents of these towns are far more likely to be claiming sickness and disability benefits or to have dropped out of the benefits system altogether due to punitive conditionality (Beatty et al., 2017), and those in employment are more likely to be in precarious, low-wage jobs. Likewise, discourses of 'inclusive growth' and the promise of a 'northern powerhouse' by their very logic of agglomeration in, and connectivity between, the major cities demote or ignore towns peripheral to this vision. Beyond discourse, the ravages of austerity are far more visible within the post-industrial towns of the north manifested in the standardisation of consumption spaces, disuse and dilapidation on the once vibrant high street.

De Chardini's noosphere appears to predict a contemporary global urbanism characterised by imaginaries of complexity, connectivity and cosmopolitanism. Our thinking on this global urbanism would be enhanced by a research agenda that devotes the same attention to the 'left behind' places and populations within this urban constellation as it does to the iconic world city exemplifiers of the new order. Such an agenda would also recognise the importance of history and sociological processes and how new geographical imaginaries and reconfigured dispositions of attachment and belonging (i.e. changes in habitus formation) are being shaped in these places.

Acknowledgements

Thanks to Jay Emery, Michele Lancione and Colin McFarlane for helpful comments on earlier drafts of this chapter.

References

Appiah, K.W. (1997) 'Cosmopolitan patriots'. *Critical Inquiry*, 23(3): 617–639.

Beatty, C., Fothergill, S. and Gore, T. (2017) *The Real Level of Unemployment 2017*. Sheffield: Centre for Regional Economic and Social Research.

Becker, S. O., Fetzer, T. and Novy, D. (2017) 'Who voted for Brexit? A comprehensive district-level analysis'. *Economic Analysis*, 32(92): 601–650.

Bourdieu, P. (2000) *Pascalian Meditations*. Cambridge: Polity Press.

Burkitt, I. (2016) 'Relational agency: Relational sociology, agency and interaction'. *European Journal of Social Theory*, 19(3): 322–339.

Centre for Towns. (2017) 'Centre for Towns: Launch briefing'. Available at https://www.centrefortowns.org/reports/launch-briefing/viewdocument/5 (Last accessed 25 July 2019).

Charlesworth, S.J. (2000) *A Phenomenology of Working Class Experience*. Cambridge: Cambridge University Press.

Connolly, J. (2019) 'Generational conflict and the sociology of generations: Mannheim and Elias reconsidered'. *Theory, Culture and Society*. doi: 10.1177/0263276419827085.

Cox, E. (2018) 'What now for the North? Realities, myths and intelligence in the northern powerhouse', 9 May 2018. Available at https://www.ippr.org/blog/what-now-for-the-north-realities-myths-and-collective-intelligence-in-the-northern-powerhouse (Last accessed 25 July 2019).

Cowie, J. and Heathcott, J. (2003) *Beyond the Ruins: The Meanings of Deindustrialization*. Ithica, NY: ILR Press.

De Swann, A. (1995) 'Widening circles of identification: Emotional concerns in sociogenetic perspective'. *Theory Culture and Society*, 12(1): 24–39.

Elias, N. (2000 [1939]) *The Civilizing Process*. London: Blackwell.

Emery, J. (2018) 'Geographies of deindustrialisation and the working class: Industrial ruination, legacies and affect'. *Geography Compass*, 13(2).

Edensor, T. (2015). 'The ghosts of industrial ruins: Ordering and disordering memory in excessive space'. *Environment and Planning D*, 23, 829–849.

Evans, K., Frazer, P. and Taylor, I. (1996) *A Tale of Two Cities: Global Change, Local Feeling and Everyday Life in the North of England*. London: Routledge.

Farrugia, D. (2019) 'Class, place and mobility beyond the global city: Stigmatisation and the cosmopolitanisation of the local'. *Journal of Youth Studies*.

Florida, R. (2003) 'Cities and the creative class'. *City and Community*, 2(1): 3–19.

Forth, T. (2017) 'To bring back trust in politics, Britain needs a local measure of GDP', 30 October 2017. Available at https://www.citymetric.com/business/bring-back-trust-politics-britain-needs-local-measure-gdp-3440 (Last accessed 25 July 2019).

Friedman, S. (2016) 'Habitus clive and the emotional imprint of social mobility'. *The Sociological Review*, 64: 129–147. https://doi.org/10.1111%2F1467-954X.12280

Haughton, G., Deas, I., Hincks, S., and Ward, K. (2016). 'Mythic Manchester: Devo Manc, the northern powerhouse and rebalancing the English economy'. *Cambridge Journal of Regions, Economy and Society*, 9(2): 355–370.

IPPR North (2016) *Taking Back Control in the North: A Council of the North and Other Ideas*. Manchester: IPPR North.

Jennings, W., Stoker, G., and Twyman, J. (2016) 'The dimensions and impact of political discontent in Britain'. *Parliamentary Affairs*, 69(4): 876–900.

Mah, A., (2010) 'Memory, uncertainty and industrial ruination: Walker Riverside, Newcastle upon Tyne'. *International Journal of urban and Regional Research*, 34(2): 398–413.

Mah, A. (2012) *Industrial Ruination, Community and Place: Landscapes and Legacies of Urban Decline*. Toronto: University of Toronto Press.

McKenzie, R. and Atkinson, R. (2019) 'Anchoring capital in place: The grounded impact of international wealth chains on housing markets in London'. *Urban Studies*. https://doi.org/10.1177%2F0042098019839875

McKay, L. (2019) '"Left Behind" people, or places? The role of local economies in perceived community representation'. *Electoral Studies*, 60.

Moulier-Boutang, Y. (2012) *Cognitive Capitalism*. Cambridge: Polity.

Morley, P. (2013) *The North (and Almost Everything in It)*. London: Bloomsbury.

Nayak, A. (2019) 'Re-scripting place: Managing social class stigma in a former steel-making region'. *Antipode*, 51(3): 927–948.

Osborne, G. (2014) 'Chancellor: 'We need a northern powerhouse' speech at the Museum of Science and Industry, Manchester, 23 June 2014. Available at https://www.gov.uk/government/speeches/chancellor-we-need-a-northern-powerhouse (Last accessed July 25 2019).

Paulle, B., van Heerikhuizen, B. and Emirbayer, M. (2012) 'Elias and Bourdieu'. *Journal of Classical Sociology*, 12(1): 69–93.

Pike, A., MacKinnon, D., Coombes, M., Champion, T., Bradley, D., Cumbers, A., Robson, L. and Wymer, C. (2016) *Uneven Growth: Tackling City Decline*. York: Joseph Rowntree Foundation.

Schilling, H., Blokland, T. and Simone, A. (2019) 'Working precarity: Urban youth tactics to make livelihoods in instable conditions in Abidjan, Athens, Berlin and Jakarta'. *The Sociological Review*, 67(6): 1333–1349. https://doi.org/10.1177%2F0038026119858209.

Scott, J.C. (1988) *Seeing Like a State: How Certain Schemes to Improve the Human Condition Have Failed*. New Haven, CT: Yale University Press.

Shaxson, N. (2011) *Treasure Islands: Tax Havens and the Men Who Stole the World*. London: Bodley Head.

Silva, E.B. (2016) 'Unity and fragmentation of the habitus'. *The Sociological Review*, 64: 166–183. https://doi.org/10.1111%2F1467-954X.12346.

Skeggs, B. (2004) *Class, Self, Culture*. London: Routledge.

Wacquant, L. (2016) 'Revisiting territories of relegation: Class, ethnicity and state in the making of advanced marginality'. *Urban Studies*, 53(6): 1–77–1088.

Willis, P. (2014 [1977]) *Learning to Labour*. Farnham: Ashgate.

Wyly, E. (2015) 'Gentrification on the planetary urban frontier: The evolution of Turner's noosphere'. *Urban Studies*, 52(4): 2515–2550.

10

FOOTNOTE URBANISM

The missing East in (not so) global urbanism

Martin Müller

> All cities are global, but some are more global than others.

Can you locate the city ofVilnius on a map? If your hand just twitched towards your smartphone, take comfort in that you are not alone. Most people can't. The capital city of Lithuania, a European Union member state, draws a blank even among most global urbanists. Vilnius wanted to end that dire state of affairs in 2018. It launched a marketing campaign that provided some basic geographical education of its whereabouts to the global public. The tagline read: 'Vilnius – the G-Spot of Europe. Nobody knows where it is, but when you find it – it's amazing.' I will leave to the side a rich feminist reading of the mystification of female sexuality and the presentation of a city as inviting male penetration and will just say this much: If you are still wondering where Vilnius is, just follow the convulsive hand in Figure 10.1 to where it clutches the bedsheet.

FIGURE 10.1 Image campaign of the city ofVilnius in 2018: 'Nobody knows where it is, but when you find it – it's amazing' #vilniusgspot (Source: Go Vilnius, reproduced with kind permission.)

DOI: 10.4324/9780429259593-12

But Vilnius is not just a blank space in the mental maps of tourists; it is just as much one in the mental map of global urbanism. This holds true for the history of global urbanism as we know it, from Peter Hall's (1966) *The World Cities* to the more recent postcolonial critique of urban studies. This critique, emerging in the 2000s, takes issue with a situation where a few large cities in wealthy countries are theoretical beacons, but most others, often outside the West, remained shrouded in darkness. Drawing out cities from the shadows and margins of global urban theory-making and putting them centre stage has been the declared mission of this critique (McFarlane, 2008). Scholars have been vociferous about the need for 'new geographies of theory' (Roy, 2009: 819), about making urban studies 'more cosmopolitan' (Robinson, 2002: 549) and about not elevating a limited number of places – Paris, London, Chicago, Los Angeles – to the status of paradigmatic cities for urban theory.

Many people doing research on cities outside North America and Western Europe feel the postcolonial insistence on a more global vision of urban studies, and crucially of urban theory, as a great liberation. I certainly do. It has breathed life into research on cities that had been treated as exotic case studies, and it has encouraged theory-making from places that had hitherto been confined to the margins of urban theory. As a consequence, urban studies has seen influential theorisations emerge from cities such as Jakarta and Johannesburg, Kolkata and Cape Town.

Yet for Vilnius and its brother and sister cities in the Global Easts – from Belgrade to Kyiv, from Istanbul to Teheran, from Almaty to Taipeh and from Moscow to Seoul – the advent and rapid rise of this critique has not much improved the situation. Quite the opposite. If anything, it has made them recede even more deeply into the dark. Global urbanism, then, is much less global than it purports to be: it comes with a particular gaze that makes some cities – in the Global North and, increasingly, in the Global South – more visible and others, in the Global Easts, less so. Or to paraphrase George Orwell, all cities are global, but some are more global than others.

When I write of the Global Easts, in the plural, I refer to all those places that do not easily fit into the categories of Global North and Global South. In particular, this concerns the Middle East, East Asia and the postsocialist East. Often, these are countries and cities that were outside or at the margins of European colonialism. Either they were empires themselves, but not European (such as the Russian, Ottoman, Persian, Chinese and Japanese empires), or they were colonies of, or occupied by, non-European empires (such as much of Eastern Europe, Central Asia, Korea, Taiwan and parts of the Middle East). I use the term 'East' here, as elsewhere (Müller, 2020), not as a determinate geographical descriptor but as a floating signifier that denotes an epistemological project: that of thinking urban theory with and through the locations outside European empires and their former colonies.

The cities of these Global Easts have become mere footnotes to contemporary global urbanism. I mean this metaphorically, although it is often true in a literal sense: cities of the Easts are frequently relegated to the footnotes of books and articles, reduced to mere afterthoughts for the sake of completeness. The footnote, of course, is not where the exciting action happens. It is for tucking away things you think you should mention but deem not important enough for the running text. English playwright Noël Coward put it quite memorably: 'Having to read a footnote resembles having to go downstairs to answer the door while in the midst of making love' (quoted in Grafton, 1999: 70). The Easts are thus the unwelcome disruption when you are busy with more exciting things in global urbanism. And let's face it: for the most part, global urbanism has not chosen to go downstairs.

The goal of this contribution is to rummage through the footnotes of global urbanism in order to reconstruct the gaze of global urbanism and its blind spots with regard to the Global

Easts. For only if one is aware of that gaze – and its silences – can one direct it elsewhere. The analysis in the rest of this piece focuses on one of those multiple Easts: those 30 countries, and their cities, that emerged from the fall of state socialism between 1989 and 1992 to form the post-socialist East. This East includes cities from Wrocław to Vladivostok, from Sarajevo to Samarkand, from Tirana to Tartu, as well as the largest city in Europe, Moscow.

Rummaging through the footnotes of global urbanism

Literature on global urbanism has one particular place reserved for cities of the postsocialist East: the footnote. This is often true enough in quite a literal sense. When in the 1980s Austro-American urban theorist John Friedmann started to think urbanisation in relation to global economic processes with his world city hypothesis, this marked an important opening up of urban studies to cities around the world. The world cities approach explicitly cast its gaze beyond the narrow confines of large metropolises to a world of interlinked cities. Friedmann, however, framed the cities of the (then still socialist) East as a problem in this approach and confined them to the footnotes:

> A *problem* is posed by cities in the actually existing socialist countries, particularly China and the Soviet Union. The question is an empirical one: how and to what extent are their major cities tied into the international capitalist economy and therefore subject to its influence?
>
> (Friedmann and Wolff, 1982, footnote on page 310, emphasis added)

The only time the East occurs in Friedmann's influential paper on the world city hypothesis is, again, in a footnote to one of his tables (Friedmann, 1986: 72). This table, entitled 'The World City Hierarchy,' lists world cities in core and semi-peripheral countries. Its footnote reads, 'Also *eliminated* from consideration were all centrally planned economies which are integrated into the Soviet block' (emphasis added). No mention thus of Moscow or Warsaw and their control functions in the world economy – a *problem eliminated* by definitional fiat in one fell swoop.

In the early 2000s, a critique of the specific geography of the notion of 'world cities' emerged – a geography that favours rich and powerful cities over others, less rich and less powerful, notably those in the Global South (Robinson, 2006). The clarion call was not to privilege certain cities on the basis of their economic power but to widen the empirical embrace of urban studies to work towards creating a more global discipline. This opening would have come at the right time for cities of the postsocialist East. Socialism had dissolved a decade earlier, and cities from Wrocław to Vladivostok were integrating, and being integrated with force, into the global (capitalist) economy and flows of investment, capital, migration and expertise (Pojani and Stead, 2019). More and more local researchers started to publish in English, and Western researchers had much easier access to Eastern cities as field sites than during socialist times. 'The situation appears more propitious than ever, then,' Jennifer Robinson (2006: 93) remarked, 'for an integration of urban studies across long-standing divisions of scholarship, especially between Western and other cities, including "Third World" and former socialist cities.' In theory.

In practice, however, cities of the postsocialist East continued to be consigned to the footnotes. Let us look at how this is the case in three of the most formative interventions for global urbanism: Robinson's (2002) 'Global and World Cities: A View from Off the Map,' Roy's (2009) 'The 21st Century Metropolis: New Geographies of Theory' and Sheppard, Leitner and Maringanti's (2013) 'Provincializing Global Urbanism: A Manifesto.'

Jennifer Robinson's paper is an early critique of the division between urban theory, associated with cities of the West, and development studies, associated with cities in the Third World. Postsocialist cities make a tentative appearance in a footnote (!) on the first page: 'This argument [of the division between Western and Third World cities] would seem to hold also for other cities "beyond the West", such as post-Socialist [sic] cities' (Robinson, 2002: 531). This assumed parallelism is not explored further, although postsocialist cities are mentioned several times later in the paper. Whilst Robinson devotes a later paper entirely to the question of cities of the postsocialist East in global urbanism (Robinson, 2016), this early intervention, together with her book *Ordinary Cities* (Robinson, 2006), mentions the postsocialist East for completeness but does not engage with cities there, and their empirical realities, in any depth beyond postulating an analogy.

The other two pieces (Roy, 2009; Sheppard et al., 2013) pass over the postsocialist East in silence. Ananya Roy argues for the need to multiply the locations from which urban theory is produced and highlights the potential of different world regions to that effect. She comes out in favour of enlisting area studies for forging new geographies of urban theory. Yet in her otherwise comprehensive consideration of different areas – Latin America, South Asia, East Asia, the Middle East, Africa – one area is conspicuously absent: the postsocialist East. That is all the more curious because it was precisely the confrontation between the West and the then socialist East that led to the ascendance of area studies during the Cold War (Engerman, 2010; Sidaway, 2013: 986). Yes, Roy (2009: 821) does mention that her 'coverage of world-areas is thus highly selective and strategic rather than comprehensive'. But it is hard to see what other non-northern areas have been left out, apart from Southeast Asia and non-white Oceania.

Finally, the manifesto for provincializing global urbanism (Sheppard et al., 2013) performs an interesting manoeuvre in which it theoretically leaves the door open for cities of the postsocialist East whilst practically closing it. It does so by calling for challenging mainstream urbanism from the South, which, in a footnote (!), they define as 'those, *everywhere*, whose livelihoods have been made precarious by geohistorical processes of colonialism and globalizing capitalism' (p. 898, emphasis added). This could include the postsocialist East, at least where 'globalizing capitalism' is concerned (where the East stands in the forcefield of colonialism is more complicated, see Müller 2020). Yet, when it comes to putting names and locations to this all-encompassing South, one quickly realises that the manifesto is articulated around the conventional notion of the South as 'the post-colonial societies of Latin America, Asia and more recently Africa and non-white Oceania' (p. 894). This is even more curious knowing Eric Sheppard's earlier interest in socialist cities (Sheppard, 2000). Thus, the cities cited in the manifesto are Jakarta, Lagos, Calcutta [sic], São Paulo, Mumbai, Hongkong, Singapore and Shanghai – all with a history of European colonialism. Cities in the Easts, postsocialist or otherwise, are passed over in silence.

The position of the postsocialist East is not much different in the major textbooks of urban geography. Textbooks are central sources of representing urban studies and a world of cities in all its diversity, both to students and instructors. The global urban critique notably pointed out that Third World cities had been relegated to the obligatory last chapter of textbooks, as special or exotic cases (Roy, 2009). In general, English-language urban geography textbooks exhibit a strong bias towards Anglo America, as a recent analysis demonstrated (Lawhon and Roux, 2019). Let us take a closer look at the two most international textbooks according to that analysis: Jonas et al.'s (2015) *Urban Geography: a critical introduction* and Pacione's (2009) *Urban Geography: a global perspective*. What is the position and role of cities in the postsocialist East in them?

Sadly, the tally is quickly made. *Urban Geography: a critical introduction* does not mention the cities of the postsocialist East even once. There is one mention of 'socialist cities', but only to argue that this label is becoming obsolete with the spread of the term 'global South' (p. 75). *Urban*

Geography: a global perspective has a few pages devoted to the urban form and urban planning in the socialist city. But these cities disappear from view with the collapse of socialism. A mere two paragraphs (p. 185) explain how the postsocialist transition process has turned those cities really into their Western counterparts: the free market has replaced central planning, and suburbanisation, private homeownership and gentrification have taken hold. Postsocialist cities undertake large redevelopment of their downtown areas with private capital, 'like many entrepreneurial cities in the US' (p. 185), and there is a trend towards a '"European" urban residential structure characterised by an affluent inner core and a poor periphery' (p. 185). The postsocialist city seems to return, in this account, to Western capitalist normality, with all difference effaced – an account not shared by many of the researchers working on those cities.

In this footnote urbanism, the postsocialist East is therefore either completely absent in attempts to present a more global picture of urban studies or is confined, often quite literally, to the footnotes. These footnotes come in two kinds. The first could be called the 'idem' footnote, conveying a message akin to 'the same goes for the post-socialist East', without presenting much evidence for this thesis. This idem footnote aligns the East either with the West, assuming a neat transition from socialism to Western-style capitalism (as in Pacione, 2009), or with the South, assuming a similar trope of marginalisation (as in Robinson, 2002). The second type of footnote is the elimination footnote. It mentions cities of the postsocialist East for completeness, but then goes on to disregard, or eliminate, them from further consideration (as in Friedmann, 1986 and Friedmann and Wolff, 1982).

From the footnotes to the main act: 'Nobody knows where it is, but when you find – it's amazing.'

Footnotes are the places where we put information that is too important to leave out, but too trivial to discuss in a meaningful way, lest it disrupt the main argument. The footnote is a side show or afterthought tacked on, sometimes awkwardly, to the lead narrative. It is where we find the cities of the postsocialist East in contemporary global urbanism: their existence is sometimes acknowledged (and sometimes not), but its implications are ignored.

This ignorance has problematic consequences. It creates a blind spot in the endeavour of global urbanism to work towards a more global urban studies. What is more, this is a blind spot that global urbanism is currently only dimly aware of, enthralled as it is by theorising from the South and the hemispheric fallacy that North plus South equals the globe. This footnote urbanism puts all those interested in cities of the Easts, whether students, scholars or policymakers, in an awkward position: How shall one articulate one's thoughts if speaking from what is essentially terra incognita in global urbanism? Who shall one speak to and with what concepts? And who will listen?

For global urbanism to become truly global, it must embrace the Easts. Two things follow. First, global urbanism must come to terms with its Eastern blind spot and move towards an active engagement with the wealth of literature that is out there, whether on the postsocialist or on other Easts. The large and growing body of writing, almost all in English, now comprises several edited volumes, some of them dating back to the 2000s when the first interventions of the postcolonial critique of global urbanism appeared (in addition to the early classic of Andrusz et al., 1996 there are Brade and Neugebauer, 2017; Grubbauer and Kusiak, 2012; Kliems and Dmitrieva, 2010; Stanilov, 2007; Stanilov and Sýkora, 2014; Tsenkova and Nedović-Budić, 2006; Tuvikene et al., 2019). Several excellent monographs present a nuanced picture of postsocialist urbanism, avoiding the twin dangers of assuming full assimilation to the West on the one hand

and exotic difference on the other (Czepczyński, 2008; Fehérváry, 2013; Hirt, 2012). There is a biannual conference on the international urban geographies of post-communist states, colloquially known as the CATference, which draws about 200 participants. Perhaps most importantly in this age of social media, there is a lively Facebook group (Cities after Transition, 2020) and Twitter account (@AfterCities).

A second step ensues. It is about listening and entering into a dialogue. Postsocialist urban research has sought to converse with research on cities in other parts of the globe and contribute to global urban theory. Debates have circled around the practice of comparison (Tuvikene, 2016; Tuvikene et al., 2017; Wiest, 2012), the need to produce and 'export' theory (Gentile, 2018b; Ouředníček, 2016; Sjöberg, 2014), the obstacles to engaging with global urban studies (Ferenčuhová, 2016a, 2016b) and the position of the East in global urban studies (Müller and Trubina, 2020b). Authors have shown chutzpah in proposing ways of speaking back to global urban theory through revising existing (Western) concepts such as gentrification, neoliberalism, urban geopolitics, urban informality and improvisation, mobility and shrinking cities (Gentile, 2018a, 2019; Gentile and Sjöberg, 2019; Haase et al., 2016; Kubeš and Kovács, 2020; Kusiak, 2019; Müller and Trubina, 2020a; Tuvikene, 2018; Tuvikene et al., 2019).

There is no excuse for ignoring cities in the Easts: literature is out there; it is plentiful; it is in English; it has something to say to global urban theory. Entering into a dialogue with the East will add some nuance, some shades of grey (Pobłocki, 2013; Yiftachel, 2009), to the sometimes overwrought debates that emerge from a global urbanism divided into hemispheric binaries of North and South.

Noël Coward may have remarked that engaging with a footnote was akin to having to go downstairs to answer the door whilst in the midst of making love. Cities from the East are banging on that door, loud and clear. If global urbanism chose to go downstairs and answer, it might be surprised to discover yet unknown pleasures. *The Global East: nobody knows where it is, but when you find it – it's amazing.*

References

Andrusz, G., Harloe, M. and Szelenyi, I. (eds.) (1996) *Cities After Socialism*. Oxford: Blackwell.

Brade, I. and Neugebauer, C. (2017) *Urban Eurasia: Cities in Transformation*. Berlin: DOM.

Cities after Transition (2020) Cities after Transition (CAT). Available at https://www.facebook.com/groups/274911812535400/.

Czepczyński, M. (2008) *Cultural Landscapes of Post-Socialist Cities: Representation of Powers and Needs*. Aldershot: Ashgate.

Engerman, D.C. (2010) 'Social science in the Cold War.' *Isis*, 101(2): 393–400. doi: 10.1086/653106.

Fehérváry, K. (2013) *Politics in Color and Concrete: Socialist Materialities and the Middle Class in Hungary*. Bloomington: Indiana University Press.

Ferenčuhová, S. (2016a) Accounts from behind the curtain: History and geography in the critical analysis of urban theory.' *International Journal of Urban and Regional Research*, 40(1): 113–131. doi: 10.1111/1468-2427.12332.

Ferenčuhová, S. (2016b) 'Explicit definitions and implicit assumptions about post-socialist cities in academic writings.' *Geography Compass*, 10(12): 514–524. doi: 10.1111/gec3.12282.

Friedmann, J. (1986) 'The world city hypothesis.' *Development and Change*, 17: 69–83.

Friedmann, J. and Wolff, G. (1982) 'World city formation: An agenda for research and action.' *International Journal of Urban and Regional Research*, 6(3): 309–344. doi: 10.1111/j.1468-2427.1982.tb00384.x.

Gentile, M. (2018a) 'Gentrifications in the planetary elsewhere: Tele-urbanization, Schengtrification, colour-splashing, and the mirage of "more-than-adequate" critical theory.' *Urban Geography*, 39(10): 1455–1464. doi: 10.1080/02723638.2018.1500248.

Gentile, M. (2018b) 'Three metals and the "post-socialist city": reclaiming the peripheries of urban knowledge.' *International Journal of Urban and Regional Research*, online. doi: 10.1111/1468-2427.12552.

Gentile, M. (2019) 'Geopolitical fault-line cities in the world of divided cities.' *Political Geography*, 71: 126–138. doi: 10.1016/j.polgeo.2019.03.002.

Gentile, M. and Sjöberg, Ö. (2019) 'Neoliberalism(s) as a guide to post-wall urban change: Explanation out of the blue?' *Tijdschrift Voor Economische en Sociale Geografie*, 0(0). doi: 10.1111/tesg.12374.

Grafton, A. (1999) *The Footnote: A Curious History*. Cambridge, MA: Harvard University Press.

Grubbauer, M. and Kusiak, J. (eds) (2012) *Chasing Warsaw: Socio-Material Dynamics of Urban Change Since 1990*. Frankfurt: Campus.

Haase, A., Rink, D. and Grossmann, K. (2016) 'Shrinking cities in post-socialist Europe: What can we learn from their analysis for theory building today?' *Geografiska Annaler: Series B, Human Geography*, 98(4): 305–319. doi: 10.1111/geob.12106.

Hall, P. (1966) *The World Cities*. London: Weidenfeld and Nicolson.

Hirt, S. (2012) *Iron Curtains: Gates, Suburbs and Privatization of Space in the Post-Socialist City*. Oxford: Wiley.

Jonas, A.E.G., McCann, E. and Thomas, M. (2015) *Urban Geography: A Critical Introduction*. Oxford: Wiley.

Kliems, A. and Dmitrieva, M. (eds) (2010) *The Post Socialist City: Continuity and Change in Urban Space and Imagery*. Berlin: Jovis.

Kubeš, J. and Kovács, Z. (2020) 'The kaleidoscope of gentrification in post-socialist cities.' *Urban Studies*, early online. doi: 10.1177/0042098019889257.

Kusiak, J. (2019) 'Rule of law and rules-lawyering: legal corruption and "reprivatization business" in Warsaw'. *International Journal of Urban and Regional Research*, 43(3): 589–596. doi: 10.1111/1468-2427.12702.

Lawhon, M. and Roux, L. L. (2019) 'Southern urbanism or a world of cities? Modes of enacting more global urban geographical textbooks, teaching and research.' *Urban Geography*, 40(9): 1251–1269. doi: 10.1080/02723638.2019.1575153.

McFarlane, C. (2008) 'Urban shadows: Materiality, the "southern city" and urban theory'. *Geography Compass*, 2(2): 340–358. doi: 10.1111/j.1749-8198.2007.00073.x.

Müller, M. (2020) 'In search of the Global East: Thinking between North and South.' *Geopolitics* 25(3): 734–755. doi: 10.1080/14650045.2018.1477757.

Müller, M. and Trubina, E. (2020a) 'Improvising urban spaces, inhabiting the in-between.' *Environment and Planning D: Society & Space*, 38(4): 664-681.

Müller, M. and Trubina, E. (2020b) 'The Easts in global urbanism: Views from beyond North and South.' *Eurasian Geography and Economics*, 61(6): 627–635.

Ouředníček, M. (2016) The relevance of 'Western' theoretical concepts for investigations of the margins of post-socialist cities: The case of Prague.' *Eurasian Geography and Economics*, 57(4–5): 545–564. doi: 10.1080/15387216.2016.1256786.

Pacione, M. (2009) *Urban Geography: A Global Perspective*. 3rd ed. Abingdon: Routledge.

Pobłocki, K. (2013) 'Neither west nor south: Colour and vernacular cosmopolitanism in urban Poland.' *Przegląd Kulturoznawczy*, 17(3): 205–215. doi: 10.4467/20843860PK.13.020.2072.

Pojani, D. and Stead, D. (2019) 'When West–East planning policy advice fails to gain traction.' *Journal of Environmental Planning and Management*, 62(8). 1402–1419. doi: 10.1080/09640568.2018.1497586.

Robinson, J. (2002) 'Global and world cities: A view from off the map.' *International Journal of Urban and Regional Research*, 26(3): 531–554.

Robinson, J. (2006) *Ordinary Cities: Between Modernity and Development*. London: Routledge.

Robinson, J. (2016) 'Starting from anywhere, making connections: Globalizing urban theory.' *Eurasian Geography and Economics*, 57(4–5): 643–657. doi: 10.1080/15387216.2016.1271617.

Roy, A. (2009) 'The 21st-century metropolis: New geographies of theory.' *Regional Studies*, 43(6): 819–830. doi: 10.1080/00343400701809665.

Sheppard, E. (2000) 'Socialist cities?' *Urban Geography*, 21(8): 758–763. doi: 10.2747/0272-3638.21.8.758.

Sheppard, E., Leitner, H. and Maringanti, A. (2013) 'Provincializing global urbanism: A manifesto. *Urban Geography*, 34(7): 893–900. doi: 10.1080/02723638.2013.807977.

Sidaway, J. D. (2013) 'Geography, globalization, and the problematic of area studies.' *Annals of the Association of American Geographers*, 103(4): 984–1002. doi: 10.1080/00045608.2012.660397.

Sjöberg, Ö. (2014) 'Cases onto themselves? Theory and research on ex-socialist urban environments.' *Geografie*, 119(4): 299–319.

Stanilov, K. (2007) *The Post-Socialist City: Urban Form and Space Transformations in Central and Eastern Europe after Socialism*. Dordrecht: Springer.

Stanilov, K. and Sýkora, L. (eds) (2014) *Confronting Suburbanization: Urban Decentralization in Postsocialist Central and Eastern Europe*. Oxford: Wiley.

Tsenkova, S. and Nedović-Budić, Z. (2006) *The Urban Mosaic of Post-Socialist Europe: Space, Institutions and Policy*. Physica: Heidelberg.

Tuvikene, T. (2016) 'Strategies for comparative urbanism: Post-socialism as a de-territorialized concept.' *International Journal of Urban and Regional Research*, 40(1): 132–146. doi: 10.1111/1468-2427.12333.

Tuvikene, T. (2018) 'Post-socialist (auto)mobilities: Modernity, freedom and citizenship.' *Geography Compass*, 12(3): e12362. doi: 10.1111/gec3.12362.

Tuvikene, T., Neves Alves, S. and Hilbrandt, H. (2017) 'Strategies for relating diverse cities: A multi-sited individualising comparison of informality in Bafatá, Berlin and Tallinn.' *Current Sociology*, 65(2): 276–288. doi: 10.1177/0011392116657298.

Tuvikene, T., Sgibnev, W. and Neugebauer, C. S. (eds) (2019) *Post-Socialist Urban Infrastructures*. London: Routledge.

Wiest, K. (2012) 'Comparative debates in post-socialist urban studies.' *Urban Geography*, 33(6): 829–849.

Yiftachel, O. (2009) 'Theoretical notes on "gray cities": The coming of urban apartheid? *Planning Theory*, 8(1): 88–100. doi: 10.1177/1473095208099300.

11

COMPARATIVE URBANISM AND GLOBAL URBAN STUDIES

Theorising the urban

Jennifer Robinson

For some time now, postcolonial critiques of urban studies have sought to expand its horizons to respond to the global and globalised nature of the urban and decolonise its concepts and practices (Southall 1976; Mitchell, 1987; Hannerz, 1980; Parnell 1997; Robinson 2011; 2006; Parnell, Pieterse and Watson, 2009; Roy 2009; Wu 2016). In the process, many theoretical and methodological challenges have come to the fore. These have inspired sustained debates, which have to some extent productively stretched the vocabularies and practices of urban scholars (for example, Robinson, 2006; Parnell, Pieterse and Watson, 2009; Roy, 2009; Brenner and Schmid, 2015; Scott and Storper, 2016; Buckley and Strauss, 2016; Bhan, 2019). These debates also expose the ongoing limitations on which voices amongst urban scholars across the globe shape the terms of urban theorising (Parnell and Pieterse, 2016; Ferenčuhová, 2016), even as the leading centres of urban research become more globally dispersed (Parnell and Pieterse, 2014; Revi et al., 2015; Marques, 2020). Nonetheless, the issues presented by a more global framing of urban studies are far from settled – and, indeed, are of long-standing concern (Abu-Lughod, 1976). Alongside the ambition to de-/postcolonialise urban studies and frame the problematic of the urban and urbanisation as "global", open to consideration based on the experiences of any urban settlements and the wide diversity of urbanisation processes across the planet, coming into view are a series of conundrums associated with thinking the urban globally. What/where is the urban? What is the relation between individual cases and wider theoretical propositions? How do we build new insights from distinctive urban contexts? What is the status and scope of ideas about the urban? Whose voices count in framing understandings and concepts of the urban? Writers, of course, bring their various political engagements, practices, vocabularies and philosophies; empirical concerns and theoretical preferences; regional and locational embeddedness; and positionalities to addressing these questions (Angelo and Goh, 2020). But the problematic of theorising the urban, globally, also brings forward shared concerns which urbanists seek to throw light on. More than positionality and theoretical perspectives inform these debates – often a deeply shared commitment to engaged urban practices and issues of social justice inspires writers concerned with the urban world. And, of course, the urban world itself presses on any conceptualisations of the urban.

Starting from the practical challenges of building knowledge of the urban within the rubric of a more global urban studies and in response to the unruly and often divergent nature of

DOI: 10.4324/9780429259593-13

contemporary urbanisation, all urbanists confront the challenge of thinking the urban in a "world of cities" (Robinson, 2011) and thus of building concepts of the urban across difference. The search for a new "geography of theory" (Roy, 2009) to support a postcolonial and global urban studies (Robinson 2006) has inspired a number of scholars to revisit the nature and terms of comparative method (Nijman, 2007; McFarlane, 2010; Ward, 2010; Robinson, 2011; Peck, 2015; Hart, 2016; Leitner, Peck and Sheppard, 2019). Comparativism has been explored as a broad theoretical practice of global urbanism associated with ways of coming to an understanding of and engagement with the urban through a "comparative imagination" (McFarlane, 2010), bringing different urban experiences into analytical conversation (Ward, 2010), and "thinking with elsewhere" (Robinson, 2016). Comparativism also draws attention to processes of conceptualisation as individual cases are drawn into reflections across more than one case (Connell, 2007) and the agency and role of the specific researcher in framing comparative experiments is evident. The agentful researcher, her positionality and trajectories (Ruddick et al., 2018), her engaged commitments in specific urban contexts (Smith, 2009), her inspired conceptualisations and expansive horizons (Zerilli, 2009) are key elements in analyses of the urban. In this vein, comparative urbanism emerges as much more than a method. It indexes a thick and dynamic research practice which calls for careful attention to different contexts, to the diverse positionalities and concerns of researchers, and to the dynamism of concept formation. Comparativism indicates a mode of theory building which sees concepts as strongly revisable in the context of bespoke and natural experiments, rigorous evaluation and ongoing contestation of insights. Different spatialities of the urban convene methodological experiments and enable the conceptual innovations which a comparative imagination invites. At its most expansive, this chapter suggests, comparative urbanism adds up to a theory of the urban. That is, in searching for ways in which a comparative imagination – thinking with elsewhere – might be specifically formatted for global urban studies, a distinctive perspective on the urban emerges.

Here it is important to clarify what a "global" urban studies might entail. Whatever the urban might be thought to be, it is *the possibility to define this from any urban context* which articulates the meaning of "global" urban studies (Robinson, 2016). Thus "global urbanism", far from signposting the possibility of a definitive, universal or generalised account of the urban instead indexes the potential for theoretical and pragmatic practices, insights and orientations towards and within the urban brought forward by scholars or practitioners, residents, artists, policymakers and planners – starting anywhere (Robinson and Roy, 2015). Thus diverse ways of thinking, practice and experience produce and perform the urban (Lefebvre, 2003). Global urban studies convenes theoretical practices that are deeply aware that the urban takes place, is produced and lived within a globalised world of diverse urban contexts and in the midst of a multiplicity of processes of urbanisation. "Global" urbanism then suggests an analytical project of coming to understand the urban across its many formations, from a wide diversity of contexts and experiences, resulting in knowledge which is strongly revisable, open, diverse and non-singular. Whatever the urban might reference within global urban studies, it is not singular or universal but differentiated and fractured, even disjunct – as any critical geographical or spatial understanding of "global" indicates.

Indeed, if we follow Doreen Massey's foundational "global sense of place" argument, the multiple trajectories and interconnections which produce any space as a "simultaneity" (Massey, 2005) are never "global" as opposed to "local". Her still productive formulation portrays a generalised condition of globalised and "stretched out" social relations composed through complex spatialities. In his important contribution, John Allen (2016) unfolds spatial relations as not simply material trajectories or flows but draws attention to the intricate spatialities (topologies) of relating, which compose presence, absence and reach in the midst of a globalised social world.

In the tracks of such spatial imaginations, the open possibility of the urban within the framing of global urban studies is emergent from simultaneously globalised and embedded interconnections. These connections need to be appreciated in their full diversity beyond political economy (Buckley and Strauss, 2016). Also, the submerged, "submarine" or occluded trajectories of black histories, articulations of global history which have been largely written out of urban studies, as critical race studies powerfully insists, frame an urgent agenda for urban studies as for other fields of research (Glissant, 1997; McKittrick, 2011; Simone, 2016; Myers, 2020). To follow Massey (2005), then, urban outcomes, or territorialisations of urbanisation processes, are intimately and intricately produced at the same time as (are the same as) the prolific circulations which amount to globalised, extended processes and the operational landscapes of urbanisation (McCann and Ward, 2010; Brenner and Katsikis, 2020). The stretched, out, circulating, globalising, fragmented empirical formations of the urban represent the core challenge for twenty-first century urbanism in many parts of the world (Murray, 2017; Kanai and Schindler, 2019). These are precisely the grounds for the provocations and questions articulated towards understanding "planetary urbanisation" (Brenner and Schmid, 2015). Any theorisation of the urban in the context of efforts to encourage "global urban studies", therefore, needs to be open to engaging insights from any urban context or any of the dispersed, extended, fragmented territories of contemporary urbanisation (Monte-Mòr, 2014; Brenner and Schmid, 2015; Keil, 2017; Murray, 2017; Kanai and Schindler, 2019; Choplin and Hertzog, 2020). As a theoretical practice, then, urban scholars need to be willing to follow the circulations, extended tracks and trajectories of urban experiences and urbanisation processes and to take account of "whatever" urban emerges (Simone, 2011).

The concept "urban" ("global urbanism" as theoretical and engaged practice in relation to any urban whatsoever) can only ever exist as emergent and multiple, in a state of constant, strong revisability. It is through prolific circulating processes and interconnections, for example, that the urban is composed of *differentiated* (repeated) outcomes. "The urban" emerges through topographical traces and topological foldings and their territorialised assemblages (Jacobs, 2006). Thus conceptualisation of the urban entails the movements needed to track material processes of urbanisation which link many different urban contexts (Robinson, 2018). As socio-spatial formations, the urban also emerges as *diverse*, as outcomes of the specific socio-spatial and locational dynamics associated with urban forms (Scott and Storper, 2015; Schmid, 2015), the lively materialities of "city life" (McFarlane, 2011; Lancione and McFarlane, 2016; Amin and Thrift, 2016), the ephemeral emergences of "people as infrastructure" (Simone, 2018) or city formation as "rumour" (de Boeck and Plissart, 2004). There are a diversity of territorial outcomes and urbanisation processes (Schmid et al., 2018) evident in different regions and urban settlements, yielding sometimes shared features (not necessarily connected to one another) which researchers might draw attention to and which might be worth thinking together to enrich and stretch understandings of each (Ward, 2010; Robinson, 2015).

One further spatiality of the urban is crucial to "global" urban studies: the urban is always *distinctive* – we can come to know it only through its specific outcomes – through its singularities and specificities. In the classic Hegelian triad (universal-particular-individual), the "urban" is the "individual" which always interrupts and presents itself as the unruly ground to any effort to conceptualise. Attending to the individual case study, comparative urbanism signals the potential for starting conceptualisation from specificity (Schmid, 2015). The key insight for our purposes here is that whatever the "concrete totality" of the urban might be, in Lefebvre's (2003) terms, it is a rich complexity, fundamentally unknowable and open. This makes any conceptualisation of the urban (or of any urban outcome) profoundly uncertain, empirically determined and historically located in conditions of ongoing emergence – and therefore contestable. This is a point that Brenner and Schmid (2015, p. 178) and their critics (Ruddick et al., 2018) agree on: "The urban society is thus never an achieved condition, but offers an open horizon in relation to which concrete struggles over the urban are waged. It is through such struggles, ultimately, that any viable new urban epistemology will be forged". Lefebvre proposes, "It [the urban] is form itself, as generator of a virtual object, the urban, the encounter and assembly of all objects and subjects, existing or possible, that must be explored" (2003 [1974], p. 122).

Whatever the "urban" might be, however, it comes into being (always also) as a concept through processes of conceptualisation. The material emergences of urban life (all the possibilities of what the urban might become), entwined with the virtual series of all possible interpretations and imaginations, present themselves for engagement, practice, interrogation and confrontation, generating concepts (Deleuze, 1994; Robinson, 2016). Thus materialities as such provide no solace from the open potential that is the (global) urban, although concepts may be imagined as tightly tied to specific contexts or more wide-ranging (contrast Lancione and McFarlane, 2016; Bhan, 2019). Differentiated, diverse and distinctive. Emergent. Always conceptual. These aspects of the urban emerge from methodological reflections on how we might go about building insights on the urban as part of a practice of "global" urban studies. But such reflections also pose themselves as starting points for conceptualisation of the urban: for "comparative urbanism" as both a theorisation of the (global) urban and a theoretical practice.

A reformatted comparative urbanism – as a practice of global urban studies – builds a view of the urban as emergent from prolific circulations, trajectories, socio-material proximities and associational practices: whatever the urban is or becomes in any given context, it can be seen as an outcome of all these possibilities of urban spatialities of differentiation, diversity and distinctiveness. In this sense, in the interests of critiquing, rebuilding and reinventing the conceptual repertoires of the field, comparative urban practices of "global urbanism" both work with and expand understanding of the complex spatialities of the urban (Robinson, 2020).

"Genetic" grounds for building comparisons refer to situations where the spatiality of the genesis or emergence of the urban itself draws the researcher towards reflections across different contexts and cases. For example, comparative practices work with the profoundly interconnected nature of urban processes and outcomes, following the many urban phenomena which are repeated across different contexts (Robinson, 2016; Hart's, 2016) wider work on relational comparativism has been important here). Thus repeated instances emerge from interconnected processes that might give rise to "any urban whatever" (Maliq Simone, 2011; Jacobs, 2006, 2012). Urban forms and processes are distributed promiscuously across perhaps quite different urban contexts, suggesting the need and potential to think, for example, large-scale developments or satellite cities across a multiplicity of different urban contexts (Robinson et al., 2020). This leaves little scope for a priori segmentation of urbanisation processes or the urban (as in a putative "southern" urban theory – see the critical commentaries and reflections of Yiftachel, 2006; Parnell and Oldfield, 2014; Robinson, 2014; Lawhon and Truelove, 2020; Müller, 2020) and rather speaks to the potential for an expansive and promiscuous geography of the urban, striated in complex ways by difference, inequality and exclusion.

"Generative" grounds or reasons for bringing different cases or contexts into comparison lie in the curiosity of the researcher to understand a phenomenon or problem. Researchers might design comparative experiments which could help them to think in a more focussed way about the puzzles that specific contexts or a number of diverse contexts and their shared features or divergences present to analysis (Robinson, 2015). This might entail seeking to understand the variety of forms of a phenomenon by turning to question the limits of concepts at hand, perhaps their inability to stretch to provide understanding across the different cases. For Deleuze (1994), the field of available concepts constitutes a virtual series, the series of all possible ideas, which goes alongside and works with the virtual series of all the material

possibilities of the emergent urban world which could present themselves to us. These pose questions or "problems" pressing at existing understandings. The intersection of these two series is the generative site in which new concepts might be produced and old concepts reach their limits. Genetic processes and generative explorations intertwine in the practice of comparing. This is a challenging, slow and emergent research practice in which the potential for comparison, the comparator, can be thought of as "assembled" across the elements of cases, wider literature, individual researchers, evidence gathered, interlocutors and not least collaborators, residents and practitioners who have their own productive "wild" comparisons to put into the mix (Clarke, 2012; Deville et al., 2016).

In a reformatted urban comparativism, insights, therefore, emerge from the practices of researchers seeking to invent, expand, enrich and, perhaps, reach the end of the utility of a certain concept. Perhaps concepts might finally be exploded to reveal new starting points inspired by the "difference" drawn off as the concept travels across diverse cases. Thus after tracking across different contexts we would be left with many aspects of phenomena which remain unexplained by available concepts (Robinson, 2016). An agentful researcher might compose a comparative practice inspired by the conceptual challenges she or he faces, to expand insight, test concepts to redundancy and engage with ideas emergent from different contexts (De Boeck and Baloji, 2016).

The important question of who this researcher might be presents itself. In addition to questions of positionality and diversity in scholarship (Ruddick et al., 2018), the dominant figure of the northern or diasporic urban researcher embedded in the demands and vocabularies of the northern academy or northern-centric politics also needs to be addressed (Todd, 2016; Lawhon and Truelove, 2020). Practices and conceptualisations framed by "new subjects of urban theory" (Roy, 2011) have introduced dynamic and hugely important new insights and agendas, transforming the landscape of global urban studies (Smith, 2009; Parnell and Pieterse, 2014; Ren, 2015; Revi et al., 2015; Simone and Pieterse, 2017; Bhan, 2019; Marques, 2020). Much more of such transformation in the authoring of global urban studies is called for.

This constitutes a triple challenge whose resolution is not entirely straightforward. Firstly, and problematically, addressing the uneven presence of different urban contexts in theorisation could be achieved largely through northern-centric and diasporic researchers. Secondly, prioritising insights and concerns of writers from a wider range of languages, backgrounds, scholarly and theoretical traditions and geographical contexts (Sidaway et al., 2014; Bhan, 2019) could productively involve an internal critique of Anglo-American hegemony within northern-centric urban studies (e.g. Müller, 2020). Crucially, supporting different, southern-centric solutions will entail confronting the extreme inequalities in institutional and personal resources amongst urban scholars. So, thirdly, transformations in urban studies will necessitate radically contesting the institutional norms of "international" scholarly production which militate against the emergent conceptualisations of many scholars gaining wider purchase (Ferenčuhová, 2016; Parnell and Pieterse, 2016).

Thus "global urban studies" is not only about disjunctive and new arguments inspired by different contexts or specific urban experiences – although clearly this remains an important agenda. More intractable challenges persist and have barely been addressed despite considerable editorial goodwill. So long as what counts as excellent is embedded with the institutional norms and definitions of northern and well-resourced academies; and so long as claims to conceptual innovation are expressed in certain globally resonant authorial voices (confident, dominating, authorizing, unmarked); and so long as these are enabled through the practicalities of unevenly resourced circuits of knowledge and publishing - "global" urban studies can only be a distorted reflection of these institutional injustices and inherited practices.

Of course, some of this is hard to impact, as it reflects the political economy of uneven development, as well as the extensive neoliberalisation of universities and calls for decolonisation (and more) of the global academy. Not happening tomorrow. But the infrastructure of how global urban studies is practiced and how comparative urban imaginations are mobilised is at least to some extent in the gift of scholars across the world. Wherever we can, the structures of intellectual political economy and practice which sustain exclusions, disfigure urban studies and restrict contributions urgently need to be dissipated. The agenda is extensive (see Jazeel, 2019: Chapter 9, for a valuable summary). This might include some of the following. Changing review practices in all urban journals to embed alternative and decentred expectations for assessing excellence and rigour, originality and other criteria which valorise different forms of knowledge. This should include strongly prioritising detailed empirical research and reports on practice from scholars whose embedded knowledge and long engagement with contexts offers rich insights (Parnell and Pieterse, 2016; Pieterse et al., 2018). Too often such deeply original knowledge and committed urban practice is completely hidden in the extensive but unacknowledged support which southern scholars offer to northern scholars who might secure the time and money to undertake research and write up insights. The infrastructure and under-labouring of highly skilled, innovative and significant urbanists in many different contexts need to be acknowledged in a committed and dynamic programme of changing publishing norms. Researchers who collaborate with institutions and scholars in different contexts need to follow accredited partnerships of equivalence and be pressed at each stage (from grant application to conference presentation to publication) to explicitly acknowledge, reflect in authorship and pay for the contributions of scholars they rely on for their insights and knowledge. Intellectual theft is no basis for "global" urban studies. Scrupulous accounting and accountability for collaboratively produced insights should be a requirement for publication.

In addition, northern urban studies needs to commit to the propagation of modes of theorising which are actively open to being revised and modest in their voice. Wider conceptual claims should always be precise about locational coordinates (both physical and social). And careful, accurate and respectful reading of the contributions of others should underpin critical engagements – or how will marginalised voices in the field find the confidence to promote their insights? Certainly, some of us could write much less and dedicate our energies to the infrastructural and institutional transformations and collegial support for scholarship in less well-resourced contexts, which are the necessary and sufficient conditions for "global" urban studies.

For any theorist of global urban studies, "the urban" poses significant challenges and conundrums as an object of study. This inserts productive dissonances into any attempt to singularise the urban: in its rich fullness and complexity, its uncertain spatialities, the urban is always more than any theorisation of it (Lefebvre, 2003). And in a "world of cities", the urban is always also somewhere else than where we are looking for it (Simone, 2011). Thought of methodologically, these are less *problems* for theory than invitations to crafting conceptualisations from anywhere, and from any starting point, within the full richness of urban life. Comparative urbanism as one kind of "global" urban practice expands the potential for urban scholars, new subjects of urban theory (Roy, 2011), to launch contributions from anywhere and to build collective and expansive horizons of conceptualisation (Zerilli, 2009) from which the urban can be reconfigured within an expanded global field and in which new urban futures can be imagined. Any such conceptualisation, this chapter has argued, will be provisional; revisable; starting anywhere; open to elsewhere; inspired from differentiation, diversity and distinctiveness; and always incomplete. And any such wider programme for global urban studies has a great deal of hard work to do to undermine northern dominance of this agenda and to build the solidarities and institutional networks which facilitate both new geographies and new subjects of urban theory.

References

Abu-Lughod, J. (1976) 'The legitimacy of comparisons in comparative urban studies: A theoretical position and application to north African cities'. Walton, J. and Masotti, L. H. (eds.). *The City in Comparative Perspective: Cross-National Research and New Directions in Theory*. Sage: New York, 17–40.

Allen, J. (2016) *Topologies of Power: Beyond Territory and Networks*. London: Routledge.

Amin, A. (2014) 'Lively infrastructure'. *Theory, Culture and Society*, 31 (7/8): 137–161.

Amin, A. and Thrift, N. 2016. *Seeing like a City*. Oxford: Wiley-Blackwell.

Angelo, H. and Goh, K. 2020. 'Out in space: Difference and abstraction in planetary urbanization'. *International Journal of Urban and Regional Research*. doi:10.1111/1468-2427.12911

Bhan, G. (2019) 'Notes on a southern urban practice'. *Environment and Urbanization*, 31 (2): 639–654.

Brenner, N. and Schmid, C. (2015) 'Towards a new epistemology of the urban?' *City*, 19(2–3): 151–182.

Brenner, N. and Katsikis, N. (2020) 'Operational landscapes: Hinterlands of the Capitalocene'. *Architectural Design*, 90: 22–31.

Buckley, M. and Strauss, K. (2016) 'With, against and beyond Lefebvre: Planetary urbanization and epistemic plurality'. *Society and Space*, 34, 4: 617–636.

Choplin, A. and Hertzog, A. (2020) 'The West African corridor from Accra to Lagos: a megacity-region under construction', in Labbé, D. and Sorensen, A. (eds.). *Handbook of Megacities and Megacity-Regions*, London: Edward Elgar, 205–221.

Clarke, N. (2012) 'Urban policy mobility, anti-politics, and histories of the transnational municipal movement'. *Progress in Human Geography*, 31, (1), 25–43.

Connell, R. (2007) *Southern Theory: The global dynamics of knowledge in social science*. Cambridge: Polity Press.

De Boeck, F. and Baloji, S. (2016). Suturing the City; *Living Together in Congo's Urban Worlds*. London: Autograph ABP.

De Boeck, F. and Plissart, M.-F. (2004) *Kinshasa: Tales of the Invisible City*. Ghent: Ludion Press.

Deleuze, G. (1994) *Difference and Repetition*. New York: Columbia University Press.

Deville, J., Guggenheim, M. and Hrdličková, Z. (eds.). (2016) *Practising Comparison: Logics, Relations, Collaborations*. Manchester: Mattering Press.

Ferenčuhová, S. (2016) 'Accounts from behind the curtain: History and geography in the critical analysis of urban theory'. *International Journal of Urban and Regional Research*, 40(1): 113–131.

Glissant, E. 1997. *Poetics of Relation*. Michigan: University of Michigan Press.

Hannerz, U. (1980) *Exploring the City: Inquiries Towards an Urban Anthropology*, New York: Columbia University Press.

Hart, G. (2016) 'Relational comparison revisited: Marxist postcolonial geographies in practice'. *Progress in Human Geography*, 39, 5: 647–657.

Jacobs, J. (2006) A geography of big things, *Cultural Geographies*, 13(1): 1-27.

Jacobs, J. (2012) Commentary: Comparing comparative urbanisms. *Urban Geography*, 33, 6: 904-914.

Jazeel, T. (2019) *Postcolonialism*. London: Routledge.

Kanai, J. M., and Schindler, S. (2019) 'Peri-urban promises of connectivity: Linking project-led polycentrism to the infrastructure scramble'. *Environment and Planning A: Economy and Space*, 51(2), 302–322.

Keil, R. 2017. *Suburban Planet: Making the World Urban from the Outside In*. Cambridge: Polity

Lancione, M. and McFarlane, C. (2016) 'Life at the urban margins: Sanitation infra-making and the potential of experimental comparison'. *Environment and Planning A*, 48 (12): 2402-2421.

Lawhon, M., & Truelove, Y. (2020) 'Disambiguating the southern urban critique: Propositions, pathways and possibilities for a more global urban studies'. *Urban Studies*, 57(1), 3–20.

Lefebvre, H. (2003 [1974]) *The Urban Revolution*. Minneapolis: University of Minnesota Press.

Leitner, H., Peck, J. and Sheppard, E (eds). (2019) *Urban studies Inside/Out: Theory, Practice and Method*. London and New York: Sage.

Marques, E. (ed). (2020) *The Politics and Policies of the Urban in São Paulo*. Oxford: Wiley-Blackwell.

Massey, D. (2005) *For Space*. London: Sage.

McCann, E. and Ward, K. (2010) 'Relationality/territoriality: Toward a conceptualization of cities in the world'. *Geoforum*, 41: 175–184.

McFarlane, C. (2010). 'The Comparative city: Knowledge, learning, urbanism'. *International Journal of Urban and Regional Research*, 34(4): 725–742.

McFarlane, C. (2011) 'On context: assemblage, political economy and structure'. *City*, 15: 375–388.

McKittrick, K. (2011) 'On plantations, prisons, and a black sense of place'. *Journal of Social and Cultural Geography*, 12(8): 947–963.

Mitchell, J. C. (1987) *Cities, Society, and Social Perception: A Central African Perspective*, Oxford: Clarendon Press.

Monte-Mòr, R. L. 2014. What is the Urban in the Contemporary World? In *N. Brenner*, ed. *Implosions / Explosions*. Berlin: Jovis, pp. 260–267.

Müller, M. (2020) 'In Search of the global east: Thinking between north and south'. *Geopolitics*, 25: 3, 734–755.

Murray, M. (2017) *The Urbanism of Exception: The Dynamics of Global City Building in the Twenty-First Century*. Cambridge: Cambridge University Press.

Myers, G. (2020) *Rethinking Urbanism: Lessons from Postcolonialism and the Global South*. Bristol: Policy Press.

Nijman, J. (2007) 'Place-particularity and "deep analogies": A comparative essay on Miami's rise as a world city'. *Urban Geography*, 28(1): 92–107.

Parnell, S. (1997) South African cities: perspectives from the ivory towers of urban studies, *Urban Studies*, 34, 891-906.

Parnell, S. and Oldfield, S. (eds) (2014) *Handbook for Cities of the Global South*. London: Routledge.

Parnell, S. and Pieterse, E. (eds) 2014. *Africa's Urban Revolution*. London and New York: Zed Books.

Parnell, S. and Pieterse, E. (2016) 'Translational global praxis: Rethinking methods and modes of African urban research'. *International Journal of Urban and Regional Research*, 40(1): 236–246.

Parnell, S., Pieterse, E. and Watson, V. (2009) 'Planning for cities in the global South: An African research agenda for sustainable human settlements'. *Progress in Planning*, 72: 233–240.

Peck, J. (2015) 'Cities beyond compare?' *Regional Studies*, 49(1): 183–186.

Pieterse, E., Parnell, S. and Haysom, G. (2018) 'African dreams: Locating urban infrastructure in the 2030 sustainable developmental agenda'. *Area Development and Policy*, 3: 2, 149–169.

Ren, J. (2015) Gentrification in China? in Lees, L., Shin, H. and Lopez-Morales, E. (eds.). *Global Gentrifications: Uneven Development and Displacement*. Bristol: Policy Press, pp. 329–348.

Revi, A., Jana, A., Malladi, T., Anand, G., Anand, S., Bazaz, A. et al. (2015) *Urban India 2015: Evidence*. Bangalore: Indian Institute for Human Settlements.

Robinson, J. (2006) *Ordinary Cities: Between Modernity and Development*. London: Routledge.

Robinson, J. (2011) 'Cities in a world of cities: The comparative gesture'. *International Journal of Urban and Regional Research*, 35: 1–23.

Robinson, J. (2014) New geographies of theorising the urban: Putting comparison to work for global urban studies, in S. Parnell and S. Oldfield (eds.) *Handbook for Cities of the Global South*. London: Routledge, pp. 57–70.

Robinson, J. (2015) Comparative urbanism: New geographies and cultures of theorising the urban, in Robinson, J. and Roy, A. (eds.) *International Journal of Urban and Regional Research Debates and Developments Symposium; Global Urbanisms and the Nature of Urban Theory*. 41, 1: 187–199.

Robinson, J. (2016) 'Thinking cities through elsewhere: Comparative tactics for a more global urban studies'. *Progress in Human Geography*, 40(1): 3–29.

Robinson, J. (2018). "Policy Mobilities as Comparison: Urbanization Processes, Repeated Instances, Topologies." *Revista de Administração Pública*, 52, 2: 221–243.

Robinson, J. (2020) *Comparative Urbanism: Tactics for Global Urban Studies*. Oxford: Wiley-Blackwell.

Robinson, J. and Roy, A. (2015). 'Introduction: Global urbanisms and the nature of urban theory', in Robinson, J. and Roy, A. (eds.) *International Journal of Urban and Regional Research Debates and Developments Symposium*, 41, 1.

Robinson, J., Harrison, P. and Wu, F. (2020) 'Comparing business models of urban development corporations: Shanghai, London and Johannesburg', *Progress in Planning*, https://doi.org/10.1016/j.progress.2020.100513.

Roy, A. (2009). 'The 21st century metropolis: New geographies of theory'. *Regional Studies*, 43, 6: 819–830.

Roy, A. (2011) 'Slumdog cities: Rethinking subaltern urbanism'. *International Journal of Urban and Regional Research*, 35: 223–238.

Ruddick, S., Peake, L., Tanyildiz, G. S., & Patrick, D. (2018) 'Planetary urbanization: An urban theory for our time?' *Environment and Planning D: Society and Space*, 36(3): 387–404.

Schmid, C. (2015) Specificity and urbanization: A theoretical outlook, in Basel, E.S., Diener, R., Herzog, J. et al. (eds.), *The Inevitable Specificity of Cities*. Zurich: LarsMuller Publishers, pp. 287–307.

Schmid, C., Karaman, O., Hanakata, N., Kallenberger, P., Kockelkorn, A., Sawyer, L., Streule, M. and Wong, K. P. (2018) 'Towards new vocabularies of urbanization processes: A comparative approach'. *Urban Studies*, 55(1): 19–52.

Scott, A. and Storper, M. (2015) The nature of cities: the scope and limits of urban theory, *International Journal of Urban and Regional Research*, 39(1): 1–16.

Sidaway, J.D., Woon, C.Y. and Jacobs, J.M. (2014) 'Planetary postcolonialism'. *Singapore Journal of Tropical Geography*, 35: 4–21.

Simone, A. (2011) 'The surfacing of urban life'. *City*, 15(3–4): 355–364.

Simone, A. (2016) It's just the city after all! *International Journal of Urban and Regional Research*, 40(1): 210–218.

Simone, A. (2018) *Improvised Lives: Rhythms of Endurance in an Urban South*. Cambridge: Polity.

Simone, A. and Pieterse, E. (2017) *New Urban Worlds: Inhabiting Dissonant Times*. Cambridge: Polity

Smith, L.T. (2009) *Decolonizing Methodologies: Research and Indigenous Peoples*. London: Zed Books.

Southall, A. (1973) 'Introduction', in A. Southall (ed.) *Urban Anthropology: Cross-Cultural Studies of Urbanization*, New York: Oxford University Press, pp. 3–14.

Todd, Z. (2016) 'An indigenous Feminist's take on the ontological turn: "Ontology" is just another word for colonialism'. *Journal of Historical Sociology*, 29(1): 4–22.

Ward, K. (2010) 'Towards a relational comparative approach to the study of cities'. *Progress in Human Geography*, 34(4): 471–487.

Wu, F. (2016) 'Emerging Chinese cities: Implications for global urban studies'. *The Professional Geographer*, 68: 2, 338–348.

Yiftachel, O. (2006) 'Re-engaging planning theory? Towards "south-eastern" perspectives'. *Planning Theory*, 5: 211–222.

Zerilli, L. (2009) 'Towards a feminist theory of judgement'. *Signs*, 34(2): 295–317.

PART III

Everyday global urbanisms

12

GLOBAL URBANISM INSIDE/OUT

Thinking through Jakarta

Helga Leitner and Eric Sheppard

Introduction

In megacities across the postcolony, towers of glass and steel have come to signify ascendancy to global city status. In Jakarta, the urban majority (Simone, 2014) lives in the shadow of its rapidly burgeoning skyscrapers in largely auto-constructed informal settlements of widely varying housing quality, referred to as kampungs. Informal settlements such as kampungs (containing 70%–80% of Jakarta's residents, Kusno, 2019) support dense social networks, sociability and distinct cultural and commoning practices, providing essential support systems and a safety net for residents' livelihood practices. In this chapter, we suggest that the informal settlements so common in cities of the postcolony are not simply being transformed by neoliberal global urbanism but are places from which it can be challenged.

Proponents of global urbanism see informal settlements as spaces of arrested development: as an obstacle preventing 'modernizing' southern cities from realizing global city aspirations (World Bank, 2009). From the perspective of Jakarta's developers, they are an impediment to converting valuable urban and peri-urban land to its highest (most profitable) and so-called best use. From the perspective of Jakarta's kampung residents, the drivers of neoliberal global urbanism (Sheppard et al., 2013)— global discourses and policies shaping global city aspirations, the imaginaries of profit-seeking developers and finance capital and the demand by emergent middle classes for modern residential, commercial and recreational spaces — are more or less invisible. Yet they are a very real force: kampungs are replaced with high-rise apartment buildings and office towers; walls appear, enclosing and separating kampungs from the formal city; and residents' livelihood practices are transformed and often undermined. However, as we discuss in this chapter, kampung residents and their allies do not necessarily embrace global urbanism but are fighting for a right to the kampung as a viable mode of living in contemporary Jakarta, recreating it in new locations or in some cases in situ, reaffirming the value of spaces of alterity that enable alternatives to global urban imaginaries and practices

In this chapter, we draw on our multi-year and multi-sited research on urban land transformations in the Jakarta metropolitan area to advocate for taking kampung alternatives seriously.[1] We examine, first, how global urbanist discourses dismiss kampung living as not fit for purpose in a global city, legitimizing the eviction and displacement of kampungs and their residents.

DOI: 10.4324/9780429259593-15

Second, we discuss how kampung livelihoods can be seen as distinct and valuable alternatives to global urbanism. Finally, we highlight kampung agency and its potential for preventing the global homogenization of urban life.

Global urbanist discourse: Othering kampungs

Before and since independence, kampungs have been represented and looked down on as undesirable spaces – riven with poverty, deviance, criminality, refuse, disease and disorder. While the term 'kampung' translates roughly from the Malay as 'village'—resonating with rural forms of life—in recent years, elite discourses have increasingly turned to calling Jakarta's kampungs slums (Irawaty, 2018). Originating in English cities, the term 'slum' has long been redeployed across the postcolony to describe, pathologize and denigrate poor neighborhoods as places in need of being cleaned up or cleared out. This representation has received a renewed boost through the United Nations (UN) Cities without Slums program, endorsed nationally and internationally as part of the UN Millennial Development Goals (United Nations Human Settlements Programme, 2003). We concur with Alan Mayne (2017) that the continued deployment of the word 'slum' is problematic; it is high time to retire 'slum' because it serves to reinforce and reinscribe negative stereotypes of peoples and places. Apparently, Mike Davis, the author of *Planet of Slums* (Davis, 2006), has also agreed to retire the term (Mayne, 2017).

The representation of informal settlements as slums, eyesores antithetical to the modern, global city, has justified a variety of external interventions. In Jakarta, and across cities in the developing world, government policies to address so-called pathologies have included improvement programs designed to upgrade material and infrastructure conditions (e.g., the Kampung Improvement Program in Indonesia), the commodification of kampung environments and life—such as promoting kampungs as tourist attractions and, last but not least, forced evictions as part of anti-slum campaigns. DKI Jakarta government public order ordinances, introduced in the late 1980s and reaffirmed in the early 2000s, declared settlement to be illegal in designated locations – e.g., along railroad tracks, within 10 meters of rivers and water bodies, in parks and green spaces and under flyovers and bridges – areas where the poorest of the urban majority have sought shelter. These ordinances formed the basis for designating the kampungs occupying such spaces as illegal, in turn legitimating city-initiated razing of selected kampungs and eviction of their residents (Leitner and Sheppard, 2018). Thousands of residents have been evicted since the early 1990s from such so-called illegal kampungs by the public order police, residents who then could only watch bulldozers destroy their former homes (Lembaga Bantuan Hukum Jakarta, 2017).

Since the end of Suharto's New Order Regime, the other distinct, albeit less visible and often overlooked, vector erasing kampungs has been market-driven negotiated displacements. This has affected, again, thousands of kampung residents from central and peri-urban Jakarta's legal kampungs (Leitner and Sheppard, 2018). Beyond the community-dissolving temptation of life-altering windfall cash (the equivalent in some cases of 40 years of work), negotiated displacements are reinforced by representations even of relatively well-off kampungs as arrested development. This distinction between legal (tolerated) and illegal (erased) kampungs, imposed by the Jakarta government, has had far-reaching consequences for the inhabitants of kampungs: whereas residents of illegal kampungs have been facing forced evictions and fighting to get compensation for being unhoused, residents of legal kampungs with some form of land rights were able to negotiate with developers a price for their land in compensation for their displacement.

Kampung lived realities

Jakarta's kampungs date back to Dutch colonial control over what was then Batavia. The Dutch planned a European city of thoroughfares, canals, public buildings, businesses and residential districts. Within its interstices, rural to urban migrants developed auto-constructed settlements that became known as urban kampungs. This is where many of them are still found in present-day DKI Jakarta, with their residents living a hybrid lifestyle in which cultural practices brought from elsewhere in Indonesia encounter global urbanism. In peri-urban Jakarta, with its proliferating new towns and integrated developments (Herlambang et al., 2019), even as some kampungs are being destroyed, new ones are constantly emerging.

The global urban imaginary is one in which land tenure aligns with Western norms of private ownership (freehold, leasehold or strata title), spaces of work are separated from those of social reproduction through zoning laws, with the various zones of the urban land nexus (Scott and Storper, 2015) connected through dedicated pathways, road and rail infrastructure. Applying this to Jakarta, residents would live in landed houses, high-rises and quasi-gated integrated developments; consumption would gravitate to shopping malls and online outlets; and work would be carried out in offices and factories. Urban morphology would conform to the hegemonic model of elite subdivisions and high-rise living, and urban growth would be guided by comprehensive plans and zoning regulations. Jakarta's urban minority conforms to this imaginary, residing in often gated enclaves and subdivisions and consuming and recreating in the ever-proliferating shopping malls and entertainment complexes.

The realities of kampung life are quite different, however. First, kampungs have an organic morphology reflective of their auto-construction. Kampung housing ranges from the opportunistic use of found materials, graduating to bricks and tiles as residents acquire income, with building codes rarely adhered to. The kampung also houses places of worship – mostly mosques, tiny stores (*warung*), informal businesses and social spaces connected by winding pathways that are challenging for nonresidents to navigate. Electricity is now broadly available, but this is not the case for water, sanitation and solid waste disposal. Housing is low rise but high density (at times exceeding densities in Manhattan), and residents have differentiated land rights. These range from a lack of any land rights, to rights conferred by local district-level (*Kelurahan*) officials, to a variety of formal rights regulated by the National Land Agency (*Badan Pertanahan Nasional*), including use and occupancy rights (*Hak Pakai*), the right to build (*Hak Guna Bagunan*) and freehold (*Hak Milik*). Collective spaces and pathways are lively, dedicated to the pursuit of everyday life; people hang around outside their often-cramped homes, socializing, celebrating weddings, gossiping and learning from one another (Figure 12.1).

Kampungs have been quiet neighborhoods until the relatively recent proliferation of motorbikes, with pedestrians, hand carts and pedicabs (and now motorbikes) threading their way along pathways that generally are too winding and narrow for cars and trucks (which are relegated to the kampung periphery and thoroughfares).

Second, kampung livelihoods are characterized by informality. Indeed, kampungs constitute the geographical base for informality in Jakarta – an extensive informal housing market and informal economy. They provide shelter – mostly auto-constructed – to residents who work in the informal sector to survive in the city, and offer affordable housing for low-wage workers employed often close by in the formal sector (Santoso et al., 2009). Kampungs house small factories that hire local residents and markets serving both the local population and beyond (Santoso and Irawati, 2015). They are also places from which mobile, itinerant street traders spread across

FIGURE 12.1 Kampung street life (Source: Authors.)

the city to sell their wares, with kampung residents providing services as maids, security guards, etc. In these ways, kampungs provide cheap labor, food and services beyond their boundaries to the rest of the city, as well as cheap rental housing and food for formal sector workers (Santoso et al., 2009; Kusno, 2020).

Third, kampung life is about conviviality, community and mutual aid. There is a strong collaborative and collective aspect to urban kampung life: when a family runs into financial trouble – e.g., through the loss of work or unexpected medical bills – it is normal for neighbors to help out. Precarity means that margins are small and disaster only a step away, so the same could happen to anyone. Many kampungs also have distinctive cultural and linguistic identities because chain migration often gathers migrants from a particular locality to this one place in Jakarta. Residents thus share a place-based identity that extends to the reproduction in Jakarta of preexisting social and political structures.

As we have learned from our intensive interviews, kampung residents across gender lines and different ages value many attributes of kampung living, which they see as not only enabling them to live in the city but also vital to their quality of life. Even residents who sell out relocate to another kampung instead of buying into the condominiums to be built on their land. Residents value the sociability, mutual aid and close-knit sense of community of the kampung, even as they aspire to the cleanliness of, and utilities provided in, more formal settlements. This was beautifully expressed by a teenage girl in one of our focus groups: 'My dream is to combine the quality of a rumahan [housing in planned development] with the life-style and neighborliness of the kampung' (Leitner and Sheppard, 2018: 447).

Kampungs also have much to offer the broader city. Kampungs provide places of refuge from the crowded, noisy and congested spaces of the formal city. Further, in an era when fostering urban environmental sustainability and resilience are policy priorities, kampungs and their residents have a much lower carbon footprint than the city built under global urbanist imaginaries. Kampungs' residents use significantly less energy; air conditioning is rare, and car ownership is low.

It does not follow, however, that kampungs are some halcyon space of the good life where everyone gets along. First, kampungs range enormously in quality: First-generation immigrants seeking a foothold in the city with little to no resources find themselves often in desperately poor and unhealthy environments. By contrast, residents who have lived in Jakarta longer, for up to three generations, have auto-constructed pleasant and green spaces characterized by established networks of collaboration, such as savings circles (arisans), where many residents are able to live comfortably. Such inter-kampung inequalities are reinforced by unequal legal status, as discussed earlier. The most precarious kampungs are designated as illegal, subject to eviction because the land they occupy has been deemed nonresidential, whereas better-off kampungs, long tolerated by the municipal state, are designated as legal. Further, residents' practices are shot through with self-interest and power hierarchies (Simone, 2014); strongmen (preman) use threats and violence to rule the roost; patriarchy can be the order of the day; residents seek to take advantage of capitalist market exchange, also monetizing the commons. In short, however much they enable more-than-capitalist livelihoods, kampungs are not simply spaces of commoning.

Agentic spaces of alterity

Kampungs are shot through with individual and collective agency. Abdoumaliq Simone (2010, 2014, 2018) has demonstrated the prevalence of such agency—not just within and across kampungs but also connecting such local economies to informal trading practices of global scope. As our research also has shown, kampung residents continually invest sweat equity and any disposable income they have into improving their dwellings and maintaining kampung spaces and communities, in some cases over several generations. Indeed, kampung residents, particularly women and youth, highly value the sociality of the kampung and the closely knit social networks and support systems, continuing to participate even after being displaced. Former residents regularly returned to the kampung, visiting friends left behind, especially during the first year after displacement. One group of families, now dispersed, continued monthly meetings of their savings group (arisan). Displaced residents' commitment to and positive valuation of kampung living is evident also in their desire to relocate to another kampung; in one (peri-urban) case, they collectively rebuilt their kampung at a new location. Kampung livelihood strategies also quietly encroach (Bayat, 2000) on the rest of the city, disrupting its rhythms, as informal traders set up stands on sidewalks or informal motorbike taxis jam thoroughfares—providing services that the urban minority also takes advantage of.

Kampung residents actively resist and contest any threat of eviction, deploying a variety of anticipatory and reactive strategies, often coevolving with one another. Anticipatory politics seeks to proactively head off evictions. Kampung residents set about documenting their legal status – for example, by claiming that past relations with city officials and services amount to de facto acknowledgment of a kampung's legitimacy. They have engaged in counter-mapping; working with residents of Bukit Duri, the Ciliwung Merdeka nongovernmental organization (NGO) documented their history of settlement, the auto-constructed built environment and how kampung-based economic activities benefit the city more broadly (Padawangi et al., 2016). Under the leadership of Gugun Muhammad, residents of Kampung Tongkol in North Jakarta took collective action to move homes back from the riverbank, clean up the river and green the riverbanks, also experimenting with alternative housing designs, to demonstrate to the city how kampungs can conform with public order restrictions and thereby avoid eviction (see Figure 12.2).

FIGURE 12.2 Kampung Tongkol: Rebuilding homes away from the riverbank (Source: Authors.)

Other strategies are more reactive, with kampung residents challenging eviction notices and resisting the destruction of their homes by the public order police. Such strategies have included legal action, as when residents of Kampung Bukit Duri challenged eviction notices in the courts, using land certificates to bolster their claim of legal occupancy status (Irawaty, 2018). In Kampung Pulo, violence ensued in 2015 when kampung residents attempted to prevent destruction of their homes by the public order police without success. Similarly, residents in Kampung Akuarium on the north shore of Jakarta in 2016 were at first unable to prevent evictions. However, some families returned to reoccupy and reclaim their kampung space and associated livelihoods, building makeshift dwellings on the rubble (Figure 12.3).

The reoccupation of Kampung Akuarium, coinciding with pending gubernatorial elections, enabled anti-eviction and housing justice activists to negotiate and sign a legally binding political contract between representatives of Akuarium and 16 other kampungs and the then-candidate for governor, Anies Baswedan. The contract included certain commitments and promises by Anies, including legalization of land rights and financial support for kampung upgrading, in exchange for their electoral support. After his election as governor, Anies initiated a Community Action Plan program for these kampungs, and consultants were hired to assess the needs and provide plans for kampung rehabilitation.

FIGURE 12.3 Reoccupancy of Kampung Akuarium (Source: Authors.)

Many forms of kampung agency have been supported by NGOs (Winayanti and Lang, 2004) and grassroots organizations, such as the Jakarta Legal Aid Institute, the Urban Poor Consortium and Ciliwung Merdeka, supplemented by a network of architects, artists, scholars, lawyers, students and journalists of diverse backgrounds, mobilizing claims of expertise to enhance the voice of kampung residents and drawing attention to their needs and desires. For example, the design of *rusunawa*, the mid- to high-rise replacement public housing built by the city for evicted kampung residents, offer post-eviction compensation but do not provide room for informal-sector economic activities or spaces of sociability, a hallmark of their previous kampung living. In response, architects, working with kampung residents, developed alternative housing designs for 'vertical' kampungs that would enable residents to pursue such practices (Irawaty, 2014; Padawangi, 2018). With a few minor exceptions, none were taken up either by the local or national government.

Concluding reflections

In this chapter, we argue that kampungs, and other such informal settlements across the urban postcolony, should not be dismissed as failing to adhere to global urbanist norms. Instead, flipping the comparative script, these can be seen as the basis for alternative forms of urban housing and living that hold up a critical lens to those norms and are worth taking seriously in their own right.

Kampungs are distinct urban environments that allow for a mode of urban living that stands in sharp contrast to the values and forms promoted by neoliberal global urbanism. They are distinctive, organically auto-constructed neighborhoods whose low-rise and high-density urban morphology does not conform to mainstream city planning dictums; kampung life is sociable, convivial and based on expectations of mutual aid; kampung politics is redolent of forms of governance that exceed those of the national and municipal state; kampung households have a foothold in both the formal and informal economies. Instead of the spatial separation of production, consumption, politics and social reproduction expected under global urban imaginaries, these coexist within the kampung. Importantly, kampungs should not be idolized: they house competition as well as cooperation, exploitation as well as emancipation, coevolving with the more formal city surrounding them. Yet kampungs are where more-than-capitalist urban lifeworlds are most likely to be found. Beyond this, taking kampungs seriously as an alternative to the global urbanist norm can also challenge northern urban theory (Leitner and Sheppard, 2018).

Jakarta's kampungs are threatened by dissolution: While some kampungs flourish as informal spaces of capitalist opportunity (especially in peri-urban areas), residents in central city locations are rapidly being displaced. They occupy land that developers and the city seek to assemble to realize the rent gap and pursue global city aspirations. Negotiated displacement, driven by the dissolving power of money, is pushing kampung residents further out. Evictions, currently in the name of ecological security (flood mitigation), unsuccessfully seek to push the most marginalized populations into poor quality rusunawas or out of the city altogether. As middle-class families move into planned developments that wall them off from nearby kampungs and thus isolate them from the urban majority, and kampungs are rebranded as slums, it has become increasingly commonplace for elite and middle-class voices to dismiss kampung residents as lazy, backward and out of place in a world-class city.

Yet kampung residents are not passively accepting dissolution; they actively contest it. This includes everyday grassroots contestations: relocating to another kampung instead of buying an apartment in the new high-rise building replacing the informal settlement, sometimes even relocating whole kampung communities and recreating kampung environments at the destination: the quiet encroachment of the ordinary. But it also includes contested politics, long-standing struggles by residents aided by pro-poor NGOs to challenge and offer alternatives to state-initiated evictions. This extends beyond questions of compensation and replacement housing to legal cases, negotiations with the governor and the development of alternative, housing designs. In these ways, residents seek to assert the broader right to inhabit the city on their own terms—to have a say over shaping urban environments that support the modes of urban living they value. Beyond Jakarta, annual urban social forums have been convened in Indonesian cities for the past six years, inspired by the world social forums, under the motto "another city is possible". All these activities seek to advance alternative urban imaginaries to those of global urbanism from above.

Such struggles for alternative urban imaginaries that underwrite citizens' claims for a right to the city on their own terms are happening across the globe. This amounts to a grassroots approach to global urbanism from below, enacting an alternative to the dominant neoliberal global urbanism that is based on the demand that not only the elite but also ordinary citizens should be able to realize their ideals of urban dwelling, habitation and living. It gives voice to all those who inhabit the city and confronts attempts to homogenize cities worldwide in the name of neoliberal norms, lobbying for valuing a diversity of urban lifeworlds. After all, it is precisely this kind of diversity and heterogeneity that long has made cities and urban life attractive, stimulating and replete with new possibilities. This can only be realized, however, through addressing the increasing inequalities across different lifeworlds. Thus global urbanism from below endorses and is committed to making more socially, environmentally and spatially just and sustainable cities.

Note

1 The fieldwork included individual intensive interviews with developers of large-scale development projects, focus groups and interviews with current and former residents in central and peri-urban kampungs and interviews with residents in these development projects, local officials and other participants

in the real estate sector. These were complemented by field observations in kampungs and development projects and document analysis (including newspaper articles, documents on spatial planning, land laws, relevant government policies and developers' prospectuses). This research has been undertaken in collaboration with colleagues from Tarumanagara University in Jakarta, including Jo Santoso, Liong Ju Tjung, Herlambang Suryono, Miya Irawati, Rully Mardona, Wita Simatupang, Melinda Martinus and Wahyu Astuti, as well as graduate students at the University of California, Los Angeles (Emma Colven, Dian Irawaty, Dimitar Anguelov and Sam Nowak).

References

Bayat, A. (2000) 'From "dangerous classes" to "quiet rebels": Politics of the urban subaltern in the global South'. *International Sociology*, 15, 3, 533–557.

Davis, M. (2006) *Planet of Slums*. London: Verso.

Herlambang, S., Leitner, H., Tjung, L., Sheppard, E., and Anguelov, D (2019) 'Jakarta's great land transformation: Hybrid neoliberalization and informality'. *Urban Studies*, 56(4): 627–648.

Irawaty, D. (2014) Kampung Deret di Timur Jakarta. https://mkusumawijaya.wordpress.com/2014/03/12/kampung-deret-di-timur-jakarta/ Accessed December 12, 2019.

Irawaty, D. (2018) *Jakarta's Kampungs: Their History and Contested Future*. Los Angeles: UCLA MA thesis.

Kusno, A. (2019) Middling urbanism and the contradictory space of the kampung in Indonesian capitalism. Available at https://www.platformspace.net/home/middling-urbanism-and-the-contradictory-space-of-the-kampung-in-indonesian-capitalism (accessed November 11).

Kusno, A. (2020) 'Middling urbanism: The megacity and the kampung'. *Urban Geography*, 41(7): 954–970.

Lembaga Bantuan Hukum Jakarta (2017) *Seperti Puing: Laporan Penggusuran Paksa di Wilayah DKI Jakarta tahun* 2016. Jakarta Legal Aid Foundation.

Leitner, H. and Sheppard, E. (2018) 'From Kampungs to condos? Contested accumulations through displacement in Jakarta'. *Environment and Planning A*, 50(2): 437–456.

Mayne, A. (2017) *Slums: The History of a Global Injustice*. London: Reaktion Books.

Padawangi, R. (2018) Excavating the ruins of aspirational urban futures in Bukit Duri, Jakarta. In: Bunnell, T. and Goh, D.P.S. (eds.), *Urban Asias: Essays on Futurity Past and Present*. Berlin: Jovis, pp. 205–218.

Padawangi, R., Turpin, E., Prescott, M., et al. (2016) 'Mapping an alternative community river: The case of the Ciliwung'. *Sustainable Cities and Society*, 20: 147–157.

Santoso, J., Febrina, A. and Ferry-Cuellar, M. (2009) *The Fifth Layer of Jakarta*. Jakarta: Graduate Program of Urban Planning, Tarumanagara University.

Santoso, J. and Irawati, M. (2015) *The future of the traditional market and its importance to develop the global competitiveness of the city based on its local-specific potentials*. Korea: Paper presented at the *Conference of International Forum of Urbanism in Seoul*.

Scott, A. J. and Storper, M. (2015) 'The nature of cities: The scope and limits of urban theory'. *International Journal of Urban and Regional Research*, 39(1): 1–15.

Sheppard, E., Leitner, H. and Maringanti, A. (2013) 'Provincializing global urbanism: A manifesto'. *Urban Geography*, 34(7): 893–900.

Simone, A. (2010) *City Life from Dakar to Jakarta*. London: Routledge.

Simone, A. (2014) *Jakarta: Drawing the City Near*. Minneapolis, MN: University of Minnesota Press.

Simone, A. (2018) *Improvised Lives: Rhythms of Endurance in an Urban South*. Nerw York: John Wiley & Sons.

United Nations Human Settlements Programme (2003) *The Challenge of Slums: Global Report on Human Settlements*. Nairobi: UN Habitat.

Winayanti, L. and Lang, H. C. (2004) 'Provision of urban services in an informal settlement: A case study of Kampung Penas Tanggul, Jakarta'. *Habitat International*, 28(1): 41–65.

World Bank (2009) *World Development Report 2009: Reshaping Economic Geography*. Washington, DC: The World Bank.

13

TIWA'S MORNING

Grace Adeniyi-Ogunyankin and Linda Peake

It had been a long night.[1] Tiwa had only managed to sleep for a few hours. It'd been like this for 5 months and 21 days, the time that had passed since she had taken out her loan. Yes, she was counting. Before this, she'd been sleeping for at least four hours per night, when she only had to contend with external noises – generators, local churches' night vigils and grinding machines. The grinding machines, especially, gave her headaches. But now, she had anxieties in addition to the headaches.

Tiwa tossed and turned, trying not to wake her two daughters who slept on the double bed with her. She glanced at them and smiled when she saw Suliat, her eldest, who was visiting from university – her first year.[2] She remembered jumping, shouting, singing and dancing when the admission list was released. "Mo rire, mo rire, mo rire gba o" [I found favour, I found favour, I received favour] was one of the many songs she sang that day. That day, she caught a glimpse of her future, and it briefly erased the pains of the struggles of the past 11 years. Her husband had kicked her out.[3] He'd also ensured that she no longer had access to her shop. She lost the majority of her customers. She almost lost her children.

She frowned at the memory as she got out of bed, careful to avoid stepping on her son who was snoring loudly on a mat beside the bed. She looked around at their one room. Just one room for all four of them. Yet she took pride in keeping it orderly. She really wanted to live in a flat, not a 'Face me, I face you'.[4] At this rate, she'd probably never live in a flat. She thought about how difficult it was to make money nowadays. She was grateful for the *gbomu le lantern* microfinance loan she took out—but it was now also causing her so much anxiety.[5] She'd been borrowing money from friends and neighbours to cobble together her weekly payments.

And another payment was due tomorrow morning. Instead of spending the day at the shop today, she would go see one of her clients, Mrs. Akinwumi, at Secretariat (the Oyo State government offices) to ask her for a larger deposit; then she'll go to Bisi's hair salon in Dugbe and ask her for the ₦1,000 she lent her a month ago. She would also pay visits to her friends and acquaintances at Adelabu market to borrow money and collect the ₦100, the ₦200 – here and there – that she had lent out.[6]

She went to the drum where they stored the water they had collected yesterday and poured some in a bucket for her bath. She poured a little into a pot to boil so that she could feel a little

DOI: 10.4324/9780429259593-16

warmth while bathing. Afterwards, she dressed and woke up Suliat, Funmilayo and Hakeem. She asked them to clean the room and fetch water before they left. She told them that they could make *pap* for breakfast and that they should go to Iya Tosin's shop and ask her to give them three Indomie on credit for lunch.[7] She told Suliat to oversee her tailor's shop and do some work while Funmi was at school and Hakeem at his apprenticeship (waiting for Suliat to finish university so that it could then be his turn).

Tiwa picked up her bag and quickly checked her appearance in the small rectangular mirror with green plastic framing on the wall. She was wearing her favourite Ankara [waxed cloth] skirt and blouse. She applied Vaseline on her lips and wished she could afford makeup.

She said goodbye to her children and started to walk towards the Challenge bus stop. As usual, she was annoyed by how difficult it was to walk on the rough unpaved road. It was so uneven, containing a series of little hills and valleys. Arriving at the Challenge bus stop, she glanced at the digital billboard that had been placed there a few years ago. She'd noticed that such changes had started to take place in Ibadan since Ajimobi became governor. She wondered why these changes weren't benefitting her.[8]

She paused for a moment as she considered her transportation options. She could take the *keke* [tricycles], taxi, bus or *okada* [motorcycles].[9] Since she had much ground to cover, the *okada* was really her best option, although more expensive. She quickly did some calculations in her head. All she had left on her was ₦680. In the end, she chose to go with the bus. She climbed in and immediately regretted her choice—the bus was virtually empty. The annoying thing about these drivers is that they don't leave unless the vehicle is full. After waiting for about five minutes and with only one additional passenger, she climbed out. The conductor abused her. She abused him back. She had no time to waste. If she didn't have enough money to pay tomorrow, the debt collectors would come to embarrass her.[10] She weaved in and out of the crowd and luckily sighted a *keke* that was going in the direction she wanted. She ran to it and climbed in.

"30 Naira", said the driver.

"*Iro ni*. Na 20 Naira I go pay", she replied. He complained about how business was bad and that she was stopping him from eating by wanting to pay a low fare.

She also complained saying, "As you dey see me, I dey chop one meal since two days ago. Abeg have mercy. It will be well with you".

He reluctantly nodded, affirming his agreement with her price.

Tiwa greeted the rest of the passengers. She adjusted herself a few times before she got comfortable. She soon got lost in the passing scenery, marveling at the contradictory changes in the city. In the past ten years, there had been an increase in the number of financial institutions; there were now shopping complexes and various businesses, including real estate offices—one could rarely rent without an agent anymore.[11] There were fewer street traders; they were increasingly becoming displaced in attempts to make the city cleaner.[12] Ibadan certainly looked richer. And she knew people were doing well. But she also knew that a lot of people like her were barely getting by. Tiwa thought sadly of her friend, Dunni, whom she was sure died from *ironu* [depression] after she could see no way out of her situation.[13] It had all started when Dunni borrowed money to pay exam fees for her children.

She got to Secretariat about 40 minutes after leaving Challenge, only having to change vehicles once and went to her customer's building. She quickly searched for Mrs. Akinwumi's number on her mobile.[14] That was also when she remembered that her battery was very low. They'd not had electricity for two days.[15] She checked her bag to ensure that she had brought her charger with her. Hopefully, she'd be able to charge it in Mrs. Akinwumi's office. She walked down the corridor and knocked on Mrs. Akinwumi's door.

"Come in", Mrs. Akinwumi said. Tiwa turned the knob and pushed the door open. She greeted Mrs. Akinwumi deferentially.

"Good morning, ma", she timidly said. She cleared her throat. Today was not the day to act shy. She had to be bold. She cleared her throat again – and stood up straight. "Please, ma, I hope you don't mind, ma. I'm here to ask you to increase your deposit amount for the clothes you will collect in two weeks' time."

Mrs. Akinwumi seemed to look annoyed by the request. "Ah ah. But the amount I paid is what we agreed on".

Tiwa nodded. "Yes, ma, you're right. But I should have asked for at least fifty per cent of the money, especially since you're ordering many things".

"Ehn, are you a novice? Should you not know by now how much you require?"

Tiwa tried to stand her ground. "Yes, ma. I'm sorry, ma. Normally, I have little more extra money so that the deposits are enough to get by".

"Ehn and na my fault abi?" Mrs. Akinwui queried.

Tiwa hated begging. But she had no choice. "No ma, but I dey beg you. One mother to another. I no dey do this kind thing before. I just really need the money".

Mrs. Akinwumi nodded. "You know government never paid our salary. It's by the grace of God and my husband's business that we're getting by".

As Mrs. Akinwunmi reached for her bag in the seat beside her to bring out her wallet, Tiwa tried hard not to roll her eyes. She knew Mrs. Akinwunmi wasn't suffering. She lived in a GRA![16] Mrs. Akinwumi brought out two ₦1,000 notes and put them on the table. "There take it. Don't pull this nonsense again".

Tiwa was a little incensed that Mrs. Akinwunmi didn't even ask her the amount, though she should have known. She had hoped to at least get ₦4,000. Tiwa took the money from the table, thanked Mrs. Akinwumi and said she would see her in two weeks. Mrs. Akinwumi nodded and motioned her hand to dismiss her. Tiwa walked out trying hard not to slam the door.

Tiwa stormed out of the building and walked back to the entrance of Secretariat.[17] She hailed an *okada*. On her way to Dugbe, she saw Operation Burst workers terrorizing some street traders.[18] She noticed one boy trying to run away with his wheelbarrow of watermelons, but it appeared too cumbersome for a fast getaway – he eventually left it behind and kept running. The *okada* zoomed by the scene, and she hoped the boy made it out – she felt bad about the loss of his produce and wheelbarrow. She alighted in front of Bisi's salon and paid the *okada* man ₦150. She entered the salon to hear a very heated debate about Ajimobi. Some of the women seemed to be arguing that he was trying too hard to compete with Lagos. Another woman said that she was frankly happy that Ibadan was cleaner and safer than before.

Tiwa looked around; she didn't see Bisi. She looked at one of the hairdressers and asked, "Where your oga?"

"Madam no dey for salon today o".

Tiwa was vexed. Bisi had promised that she would meet her at the salon. She called Bisi – but she did not pick up the call. She noticed her phone was about to die. She'd been too flustered to charge it for a few minutes at Secretariat. Tiwa kissed her teeth and let out a huge sigh. She turned around and exited the salon angrily. She'd wasted money to come here in the hopes that she'd be able to get her ₦1,000 back. And she definitely had too much dignity to go back to the salon to charge it. She sighed again as she walked towards the taxis to Adelabu. The whole ride there she was frustrated with her life. Her dream for her children to be well educated and the increasing cost of living in Ibadan had put her in this mess. She kept borrowing money. And some customers would promise to pay and never pay or not pick up what she'd finished sewing because they could not pay the balance. She hated that the microfinance loan interest was so high and the weekly payments so taxing. As she got out of the cab to grab a *keke* to go into Adelabu, she heard someone shouting her name. "Iya Suliat! Na you be dis? My money nko?" Tiwa inwardly groaned. She had been hoping to avoid Iya Kayode until next week.

"Shebi you dey say you go pay me quick, quick", accused Iya Kayode.

"Abeg Iya Kayode. Make you no vex. As you dey see me now now, my body dey inside cloth – and that is all", Tiwa explained.

"Na only you? All of us na hand to mouth o", retorted Iya Kayode.

"Okay, okay", muttered Tiwa. "Make I give you five hundred Naira today and one thousand Naira next week. I go come for your shop this evening. I promise". Iya Kayode looked at Tiwa and nodded her head in agreement.

"Make you no disappoint o".[19]

Tiwa nodded as she rushed off to get a *keke*. She really hoped that her debtors and borrowers would come through. She reached inside her bag to check the time on her phone. But the battery had finally died.

We provide no ending to this tale of a few hours in one day of Tiwa's life, leaving the reader to imagine how this stylising of telling a story speaks to a global urbanism of the working poor that both resonates with multitudinous lives across the 'Global South' and which at the same time is grounded in particular histories and geographies. It is, of course, a feminist perspective on global urbanism, one that illustrates the analytical potential of an orientation on Tiwa's morning, on her words, feelings, movements and actions—of a weaving together of the weft and warp of the emotional and material into the temporalities and spatialities of the every day that she inhabits. To understand how the everyday realm of the working poor has been permeated by global financialisation, we follow Tiwa through the bustling neighbourhood of Challenge, in Ibadan, in her efforts to secure monies to meet the weekly payments on her microfinance loan. We refer to everyday urbanism as the daily experiences and agency of the urban majority—the working poor—through their cultural practices of living and modes of relating to each other in a context in which their lives are increasingly seen as expendable. With the erosion of state provision and accountability, individual entrepreneurship and personal responsibility to secure daily life have become normalised. And as ongoing processes of dispossession and alienation render social reproduction—the context within and through which political subjectivity is configured—increasingly financialised and precarious, the emotional, mental and physical strains on bodies are increasingly brought into focus. Thus while Lefebvre (1947) saw the everyday as a potential site of radical transformative politics, we show how, in the twenty-first century, the dominant effect of anxiety and the preponderance of depression can render this potential less determinative, albeit that in this time period of global urbanism, the lives of the working poor are connected via the everyday struggle for survival made possible through the working of the different kinds of social relations played out through 'people as infrastructure' (Simone, 2004).

In this chapter, we have aimed to illustrate the points of convergence between the overarching reach of capital—both global and local—in Ibadan and its insinuation into Tiwa's tenuous attempts to make a living. Since the 2007/08 global economic crisis, African cities have increasingly been seen as the 'last frontier in the global economy' (Wilson Center, 2013 https://www.wilsoncenter.org/event/africa-the-last-frontier-why-it-matters-the-global-economy). And while it has been projected that African economies will continue to experience economic growth and an increase in the numbers of the middle class as a percentage of the population, concerns about the viability of such Afropolitan futures abound. As such, African politicians and other key stakeholders have focussed on neoliberal urban projects to present African cities as secure, world-class and cosmopolitan to strengthen their viability in the competition for foreign direct investment (Adeniyi-Ogunyankin, 2019). As a result the global enters into Tiwa's way of being in the city in various ways—materially, affectively, emotionally, behaviourally, ideologically—through the neoliberal financial strategies of prioritising the needs of capital over urban residents that result in: increasing sociospatial inequality, including the widespread economic practices of multilateral organisations and financial institutions of providing high-interest microloans (primarily) to women; imported governance mechanisms that support the former; the crisis in social reproduction that these processes engender and their impacts on physical, emotional and mental health; technological and financial developments that have enabled the spread of mobile phones and subsequent changes to modes of communication; and the globalisation of production networks that enable the importation of cheap food. Welcome to Tiwa's morning.

Acknowledgements

This account draws on participant observation and interview-based data from fieldwork conducted between June and August 2018 by Grace Adeniyi-Ogunyankin, Sylvester Nmormah and Kehinde Bello in Challenge, a central neighbourhood in Ibadan, in Oyo State, Nigeria. The data were collected as part of the feminist GenUrb research project (Urbanisation, Gender and the Global South: A Transformative Knowledge Network) funded by SSHRC grant # 895-2016-0007.

Notes

1 Tiwa's loan not only alters her daily activities; it also preoccupies her thoughts day and night. And yet feminist (and other critical) urban scholars have made little inroad into investigating the temporal dynamics of everyday life, whether by studying time itself, attending to the temporal aspects of social relations, or attending to the temporal assumptions underlying analytic frameworks. In relation to the latter, urban studies has classically engaged with chronological and linear notions of time and temporality, with the important exception of Lefebvre's (2004) notion of 'rhythmanalysis'. This addresses bodily and cyclical conceptualisations of temporality in which time is understood as an historically and geographically specific social product that shapes the urban through temporal experiences, relations, regimes, institutions and imaginaries of urban futures. While Tiwa's urban future can be read positively through her 'being and becoming' in the plans she is putting in place for her children to succeed, it also includes being 'stuck in time', waiting indefinitely for a future that might never materialise.

2 The education of children is highly valued, both for itself and as an investment in the future. Although the cost of education is free (up to junior secondary school in Oyo State), education has been financialised: there are extra, and essential, costs incurred of uniforms, books, school lunches, transportation and other ancillary costs. The latter include the virtually involuntary after school lessons, the institutions of which are a prominent part of the landscape in Ibadan. These costs of 'free' education and the lack of faith in the quality of state education leads those who can to put their children in private schools, the caliber of which, and the fees for which, also vary (see This Day Live: https://www.thisdaylive.com/index.php/2019/10/30/changing-the-narrative-of-oyo-state-education-sector/).

3 Displacement in urban theory has largely been restricted to processes of mass displacement on national or neighbourhood scales, such as in cases of forced migration or gentrification. There has been far less focus on the bodily scale, on individual women, often accompanied by children, who consistently face displacement through male violence. Neither are data available on the number of women like Tiwa, displaced from their homes and often, by default, their livelihoods. Yet without consideration of these displaced, the full complexity of displacement and of post-displacement strategies for survival cannot be captured nor accounted for.

4 'Face me, I face you' refers to a common building design of one- to two-bedroom units, the doors of which face each other across narrow corridors. The bathroom and kitchen are usually located at the back of the building and shared by all tenants. The lack of privacy in such cramped conditions can commonly lead to outbreaks of violence and a colloquial naming of such units as 'Slap me, I slap you'.

5 *Gbomu le lantern* – which translates as putting one's breast upon a lit lantern – is an acerbic generic term commonly applied to microfinance loans with high interest rates. The term is metaphorically indicative of the physical and emotional pain that women must undertake to ensure the daily and generational reproduction of family life. Tiwa's loan is a solidarity group loan in which a number of women jointly hold a loan, serving as each other's guarantors. Social pressure is often applied to women to make up the necessary numbers needed for these groups, meaning these loans are both associated with being a source of *itesiwaju* [progress] and a curse, promoting sometimes reluctant and precarious entrepreneurial subjectivities. Such loans, whether from local non-governmental organisations, such as Grooming (https://groomingmfb.com), with an interest rate of 15%, or from international institutions like La Fayette (https://www.oikocredit.coop/what-we-do/partners/partner-detail/29276/la-fayette-microfinance-bank-limited) are endemic in the Global South (Roy, 2010), permeating women's everyday lives.

6 ₦1,000 is equivalent to approximately GBP2.0 or US$2.75 (as of February 2020).

7 *Pap*, maize porridge and Indomie, an Indonesian brand of noodles, are both regularly consumed as part of Tiwa's family diet. Quick and easy to make, the noodles stand in for more nutritious meals that Tiwa does not have the time to cook when her loan takes her in search of money. Buying food on credit, one meal at a time, allows her to know that at least on this day, she and children will eat. Without this informal source of credit, Tiwa would have to choose between feeding her children and meeting her loan repayments.

8 Ibadan, Oyo State, 80 miles north-east of Lagos, is Nigeria's third-largest city with a population of 3.4 million (United Nations, 2018). An important transit point between the coast and interior and between western African countries, it is growing rapidly (with an urbanisation rate of 3.2%) (United Nations, 2018). Seyi Makinde, the current governor of Oyo State, and Abiola Ajimobi, the governor from 2011 to 2019, are proponents of neoliberal governance and have pursued policies that are reconstituting space in Ibadan by: reducing public ownership through the privatisation of state companies, primarily through public-private partnerships, for investments in luxury shopping centers, hotels and housing estates; expanding private investment through deregulation, with, for example, discounts of up to 90% on land acquisitions by certain businesses; attempts to privatise public services, such as public education; and increasing securitisation (see LP Oxford Business Group: https://oxfordbusinessgroup.com/overview/opening-doors-investment-state-set-serve-new-investors-it-streamlines-procedures-and-improves). These urban renewal policies strategically address the international 'good governance' agenda by focussing on making the city more friendly for investment and promoting private sector–led development (Roelofs, 2019).

9 Although slightly more expensive than the ubiquitous minibuses, *kekes* and *okadas* are both commonly used forms of transportation that allow their users to quickly circumnavigate the traffic-clogged streets of the city.

10 Many microfinance banks in Ibadan rely on the economy of shame (Karim, 2011) to ensure payment. A key disciplinary mechanism of *gbomu le lantern* loans is the public theatre of shaming. It is culturally taboo to be publicly shamed, thus the thought of defaulting is anxiety inducing, as one can become a target for ridicule and lose respect.

11 The financialisation of the home – whether of homeownership or of rental accommodation – is rife in Ibadan (and other Nigerian cities). It is common to have to pay a two-year deposit to rent even a 'face me, I face you' (Dabara, Ojo and Chiwuzie, 2012).

12 Displacement in the name of state 'beautification' schemes speaks to one of the "transformations wrought by neoliberalism – whereby the welfare of disadvantaged populations is recused from the space of the social" (Feldman, Menon and Geisler, 2011: 9). It is one means whereby the urban working poor—Tiwa amongst them—are no longer assumed necessary to the place-making project of the nation state.

13 The global increase in levels of depression is an outcome of the daily struggle for economic survival in systems where social safety nets are disappearing—whether state or family provided. Depression is now the leading cause of disability worldwide (WHO, 2019) and has significant repercussions for women's

ability to continue to socially reproduce the same quality of life, whether through the mundane rhythms of the everyday or through their subversion. Such 'psychic incompleteness' speaks to the need to engage the intimate, the 'psychological dispositions' (Simone and Pieterse, 2017: 156) that permeate everyday life and influence social change in analyses of the urban.

14 It is not only Tiwa's relationship with a place and people but also with material objects within it that secure her livelihood. Despite Tiwa's inability to secure savings and having to live most days hand to mouth, her mobile phone is neither a frippancy nor a luxury good but a necessary tool to secure work and payments.

15 Electricity blackouts are a common aspect of daily life in Challenge. With state withdrawal from basic needs provision, and the privatisation of the electricity sector, there is a lack of political will to increase capacity. Given the booming generator and inverter businesses, it is the urban poor who are mostly affected, speaking to a politics of distribution that exacerbates the difficulties of securing the means to reproduce everyday life (see also Adeniyi-Ogunyankin and Buckley 2017).

16 Government Reservation Areas (GRAs) were key spatial markers of colonial urbanism, a colonial legacy of a dual city still prominent in Ibadan.

17 The macro- and microaggressions of everyday economic violence, each new relation of individual debt, not only increase economic insecurities but also take a toll on mental and emotional health, as evidenced by Tiwa's anxiety, precipitated by the persistent scramble to enable her to make her weekly repayments, alongside her frustration, annoyance and sadness with the indignities and difficulties of everyday life.

18 Operation Burst is a joint security task force (with officers from the Nigerian Army and Navy, the Special Response Squad of the Police and the Nigeria Security and Civil Defence Corps). They are tasked with securing 'peace' and security in the state, especially for the purpose of making investors feel more at ease about doing business in Oyo State, thereby securing the reproduction of capital at the expense and persecution of the urban poor.

19 Tiwa's encounter with Iya Kayode is indicative of the daily circulation of very small amounts of money amongst networks of poor working women, an example of what Simone (2018: 406) refers to as a "collective effort that...sustain[s] the capacity for...ways of life to take place". Such efforts exist alongside the formal microloans, as well as the more regularised but still informal modes of saving or gaining access to credit based on trust, such as *Esusu* and *Ajo*, in which friends, co-workers, associations (e.g., tailors) make financial contributions over a defined time period, with one person drawing the money at each collection (Osabuohien and Ola-David, 2018).

References

Adeniyi-Ogunyankin, G. A. (2019) '"The city of our dreams": Owambe urbanism and low-income women's resistance in Ibadan, Nigeria'. *International Journal of Urban and Regional Research*, 43(3): 1–19.

Adeniyi-Ogunyankin, G. A., and Buckley, M. (2017) Incremental and instant urbanisation: Informal and spectacular urbanisms, in Bain, A. and Peake, L. (eds.), *Urbanisation in a Global Context,* pp. 70–86. Oxford University Press, Toronto.

Dabara, D. I., Ojo, O. and Chiwuzie, A. (2012) 'An examination of the tenancy agreement as a shield in property management in Nigeria'. *International Journal of Business Administration*, 3(4): 54–66.

Feldman, S., Menon, C. C. and Geisler, G. A. (2011) A new politics of containment, in Feldman, S., Menon, C. C. and Geisler, G. A. (eds.), *Accumulating Insecurity: Violence and Dispossession in the Making of Everyday Life*. University of Georgia Press: Athens, pp. 1–23.

Karim, L. (2011) *Microfinance and Its Discontents: Women in Debt in Bangladesh*. University of Minnesota Press: Minneapolis.

Lefebvre, H. (1947) *The Critique of Everyday Life*. Verso: London.

Lefebvre, H. (2004) *Rhythmanalysis: Space, Time, and Everyday Life*. Continuum: London.

Osabuohien, E. and Ola-David, O. (2018) Esusu (Nigeria), in Ledeneva, A. (ed.), *The Global Encyclopaedia of Informality: Understanding Social and Cultural Complexity*, Vol. 2, pp. 56–69. UCL Press: London.

Roelofs, P. (2019) 'Beyond programmatic versus patrimonial politics: Contested conceptions of legitimate distribution in Nigeria'. *Journal of Modern African Studies*, 57(3): 415–436.

Roy, A. (2010) *Poverty Capital: Microfinance and the Making of Development*. Routledge: London.

Simone, A. (2004) 'People as infrastructure: Intersecting fragments in Johannesburg'. *Public Culture*, 16(3): 407–429.

Simone, A. (2018) The majority-world and the politics of everyday living in Southeast Asia, in Hall, S. and Burdett, R. (eds.), *The Sage Handbook of the 21st Century City*, pp. 399–415. Sage: London.

Simone, A. and Pieterse, E. (2017) *New Urban Worlds: Inhabiting Dissonant Times*. Polity Press: Cambridge.

United Nations, Department of Economic and Social Affairs, Population Division (2019) *World Urbanisation Prospects: The 2018 Revision (ST/ESA/SER.A/420)*. United Nations: New York.

United Nations, Department of Economic and Social Affairs, Population Division (2018) *The World's Cities in 2018—Data Booklet (ST/ESA/SER.A/417)*. United Nations: New York.

World Health Organisation (WHO) (2019) Fact sheet: Depression. Accessed November 20 https://www.who.int/news-room/fact-sheets/detail/depression.

14

"OUT THERE, OVER THE HILLS, ON THE OTHER SIDE OF THE TRACKS"

A horizon of the global urban

AbdouMaliq Simone

Too late for modernity

In this brief chapter, I want to consider exposure to an "out there" as a key aspect of the global urban. Not as a constitutive outside nor space beyond actual or potential colonization or settlement. It is neither something beyond history nor human apprehension. Rather, the "out there" I have in mind is more a matter of how attention is paid in the present, and to what. It is an orientation to a beyond very much in the midst of things or, conversely, a background that provides neither context nor content. It is an "out there" that both proceeds and precedes modernity's dependence on anti-blackness and thus is the destination of abolition—that unanticipated opening in the backwoods, the underground railroad that speeds off the tracks, that something which appears as an aperture out of nowhere, as everyone makes a run for it.

In work that I have done in the past decade on notions of "extended urbanization", much attention has been paid to how urban residents pay attention to the shifting terrain on which they repose their everyday lives. In the expansive transitions from urban cores to urban peripheries, to the valorization of itineraries over that of emplacement, what is it that a person pays attention to in a world where so many things can be attended to; how does one create a space of operation when life can theoretically be moved and oriented in so many different directions. Some theorists have talked about the horizon of the *blur* as the axiomatics of capital continue to churn an expansiveness of spaces across different modalities, as well as the wounding, puncturing and fracturing of space at the same time (Moten, 2017; Saldanha, 2019). Deleuze and Guattari (1987) have long pointed to the constant working out, circumvention and disattention to the tensions and implications generated by capital on the earth and its constantly contingent natures.

On the one hand, the prolific and homogenizing formatting of residency that characterizes much of contemporary urban development institutes a form of retention whereby residents, both near and those who are far and share a similar mode of habitation, come to see that they are actually in the "same place". This approximates what Stiegler (1998) had in mind when he talks about mnemotechnical systems and the synchronization of individual minds. But on the other hand, residents bring with them the residues of all the multifarious, messy situations that they have left behind. The newly formed residency then acts in a way to work out or at least elide the conundrums entailed by these residues now settled in close proximity to each other.

DOI: 10.4324/9780429259593-17

But in the scores of interviews I have conducted with residents of vast apartment blocks, cheaply constructed pavilions, self-constructed makeshift settlements, dormitories and hostels across expanding peripheries in Southeast Asia, they profess to know that the real action is always elsewhere, away from them. They are impelled toward the action but also deterred from fully reaching it. What they pay attention to and are inclined to do may traverse along well-worn grooves and sediments, but these are usually sustained only through unexpected veering, as well as inexplicable starts and stops, as each avoidance or deterrence of threat or discomfort exposes the person to new uncertainties. So instead, they commonly invoke the something "out there", some horizon or background that provides tacit organization for their veering, haphazard and, sometimes, systematic circulations across wider expanses of urban space in search of opportunity. Rather than having specific destinations in mind, or any kind of incremental trajectory of continuous attainment or improvement, the objective is to find a way to keep going, to make accessible what for now is simply a diffuse sense of an "out there."

Such sensibilities would seem to diverge from notions of urban modernity with its promises of eventual development for all. For urban modernities concerned the ways in which space and time were organized in rational ways that reflect linear progressions, differentiated functions, ordered relations of cause and effect and the configuration of circulatory systems that placed landscapes, resources and productivity in relation to each other. Still, cities were haunted by something "out there", beyond the capacity to control, beyond the ordering of sovereignties that attempted to identify that outside. Beyond the incursions of potential enemies, disease vectors or wayward natural occurrences, this "out there" pointed to the unanticipated implications of modernity's own accomplishments. The capacity of the urban to interrelate different ways of life and materialities generated unanticipated excesses and trajectories that sometimes escaped the ability of governments to fold in disorder as an instigation to continuous progress (Koselleck, 2018).

The black outdoors

The elaboration of black life perhaps signals the ambiguities of this "out there" as both an aspect and divergence from modernity, simultaneously "outdoors" from its convictions but also, as a locus of exit, a part of its apparatus at the same time (Wynter, 2003, Moten, 2018). In her work on the plots of plantation life as an integral aspect of an economy of sustenance based on attunements and reciprocities with the slave quarter's physical surrounds, Wynter demarcates an interstitial position through which slaves could demonstrate their humanity differently. McKittrick (2013) picks this up and extends this practice to the urban centers where tending to the very physicality of bodies and the grit and messiness of streets trafficked by these bodies generated an economy of sustenance as blacks were largely kept out of "official" urban life.

The constitution and appropriation of blackness have been at the heart of the development of modernity in its continued extensions of imperialism and coloniality (Hanchard, 1999). The capacity of the human to operate according to the maximization of its position required a notion of free will, of the ability to act freely among otherwise constraining interdependencies (Hartman, 2007). This freedom necessitated relegating certain bodies to the status of property capable of circulating only through the transactional circuits of economic exchange and valuation (Wynter, 2003; Bogues, 2012). Yet as the work of Michele Wright (2015) points out, the inauguration of blackness ushered in a time that did not only progress in terms of the linear, capitalist temporalities it facilitated.

Blackness remains in a process of both constant repetition but also reinvention, sometimes prophetic, sometimes that of quantum entanglements. As such, blackness might posit modes of human generation not dependent upon the primacy of the time of the Middle Passage. As Moten points out, it does not consent to being a single being (2018). It is something *disobedient*. It is a time that correlates past, present and future, needs to be recuperated outside the terms of judicial resolutions (Thomas, 2016) and directed toward the repair of a humanness fundamentally broken. It is broken in its very self-reflection, as living out the idea of itself as sufficiently detached from the world to be its orchestrator (Mbembe, 2017).

Hortense Spillers (2003) has talked about the complicated strategic choices facing black people in the Americas in terms of ensuring their endurance. Blacks could insist upon their humanity in contexts in which this humanity was structurally foreclosed, where the insistence would be construed as evidence of disobedience or the very absence of humanity but which, nevertheless, in this assertion of will despite the odds and consequences could be construed by a black self, with no official recognition, as evidence of being human.

In contrast, indifference to the value of a self-formed human subject could be manifested in the capacity of the black body to extend itself into the very surrounds, terrain and materiality of its limited world of operations. Here an extraordinary attunement to the operations of the earth and its varying atmospheres and ways of being signaled a detachment from the need to be human. Here the processes of social reproduction were experienced in concert with the rhythms of other forms of liveliness (King, 2017). What can such extensionality as a black practice indicate to us today in a world of extended urbanization? How can disobedience and extending into the world be simultaneously coupled as a critical urban practice in light of the possible end of urbanity itself?

As urbanization becomes more extensive and extended, it would also seem to be moving in the direction of an "out there", taking on the risk of what Lauren Berlant (2016) calls constant interruptions. From the elongated circuits of interminable migration, to the diffuse and nearly all-encompassing insecurities that make up the raison d'etre of gated communities, to the vague aspirations of those willing to jettison relatively secure positions in the urban core to inchoate futures of urban hinterlands, to the febrile imaginations of those living in carceral conditions, often without discernible walls, the "out there" is not only an object of longing or dread but also an increasingly active participant in the here and now.

Exposure to time

And what about the "now"? This presence of the "out there" raises the question, "what time is it?" How do we think about this time if "out there" is another one? Well, the answer is that increasingly, we don't know, as time is urbanized itself—rendered more complex, numerous and all over the place, despite all the efforts to settle this question, what time is it, once and for all. In the prolongation of settler mentalities that not only attempts to affix an "out there" to a hierarchy of value and use but also colonize the very possibilities of speech beyond soliloquy, the plantation has returned full force, folding in heterogeneous logics of land use to corporate monocropping and the circulation of goods and knowledge to logistical systems (Li, 2018). At the same time, where the plantation may have once "set the clock", established a form of homogenous time through which human efficacy was to be measured and to which everyday rhythms of what bodies could do with each other were regulated, it is not clear that such temporal discipline works anymore, if ever.

If on the one hand the simultaneous erasure of time's heterogeneity and its particularization through intensified global spatial division appears to be a key characteristic of contemporary urbanization, the extension of urbanization also underlines the simultaneous existence of many temporalities. Here different ways and rhythms of doing things; transacting, buying and selling; making and distributing; and deliberating and deciding are both *exposed* and *extended* to each other. Rather than a multiplicity of times being subsumed into a standardized version, across much of the South, discrepant temporalities tend to coexist, even if the terms of coexistence tend to disallow the capacity of any one of them to posit its own trajectories of implication (Povinelli, 2016).

Yet the time of bazaar, the festival, the factory, the neighborhood, the coordinates of modernity, the time of extended family and kinship relations, the time of religious devotion, the time of diurnal and nocturnal markets, the time of administrative bureaucracies all coexist. But they do so almost as "out of body" experiences – i.e., not in the form through which they had been known. They coexist not as individual tracts, not as the rhythms of autonomous worlds but as pressure points that avail any operation to the exigencies and operational practices of the other. Each becomes its own "out there."

Take for example the bazaar, that multifaceted commercial system that integrated individual merchants and trades; that provided credit to those unable to access formal banking mechanisms; that mobilized political sentiment; that charted out specific geographies of articulation; that shifted resources across various kinds of social, geographic and religious ties; that established the price of things based on considerations that far exceeded those of supply and demand; and that shaped the structure and settlement of built environments continues to thrive. In a world of logistics, just-in-time production and constantly recalibrated commodity chains, the bazaar now thrives not on the basis of what it was and the implications of its own logics and operations but because those very logics and operations offer a resource to so-called modern economies and to advanced logistical operations when those economies and operations run into difficulty when they confront choke points or blockage (Nielson et al., 2018).

There is just enough of a solid, definitive boundary to enable the ongoing recognition of the coherence of a specific time but a boundary that leaks or that can be "reversed engineered" into incorporating or adhering to discrepant times when need be. Here extensive urbanization is not the unfolding of a single temporal format, not the imposition of a standardized time—although there may indeed be elements of such a process—but rather the extending of diverse temporalities to each other. Here is an urbanization of temporality itself, of a process of switching back and forth, changing temporal gears, accelerating and slowing down as a means of diversifying the rhythms of enactment so as to complexify the sensory field of urban life, to engender a wider range of implications and behavioral possibilities and, as such, modalities of valuation. In the seemingly most marginal worlds, such as the Chad Basin, transactions may simultaneously entail the barter of animals, the free-for-all distribution of common goods, insurance swaps, satellite imaging, intuitive maps of ancient trade routes, raiding, derivative trading on handheld computers, bulk deliveries of rare earth minerals, fake property titles and leveraged trading of oil platforms and mines (Brachet and Steele, 2015). At least at the temporal level, the out there is in here, and vice versa.

Itineraries of exposure

Key to this issue is the notion of exposure. Not only the exposure of bodies to a growing number of vulnerabilities, not only exposure in terms of the value of whatever assets an individual might have in their pockets or portfolios and not only exposure in terms of the capacity of individual thoughts and feelings to be read, marketized and used to develop more proficient control systems *but also exposure to an "out there"*, something beyond the familiar trappings of the human, beyond the trap of being human, beyond the entrapment of populations having been convinced that becoming properly human is something worth pursuing.

Given these different connotations of exposure, that the outside to be exposed to may not be emancipation or availing of rights to all. Rather, the "out there" may simply be something within the most banal of landscapes, within miles of faceless apartment blocks, within long stretches of mass-produced storage facilities, cheap hotels, strip malls and faded industrial zones. That "out there" might be rather the possibility of an "anything whatsoever", a mode of visibility where it is difficult to tell for sure what something is or what something contains or embodies.

That "out there" may best be seen as a compression of multiplicities, a capacity for anyone to read something about themselves into a particular situation. Rather than having a specific destination in mind or an idea of fulfillment or accomplishment, many urban residents, particularly those dwelling at the peripheries, would rather opt for a kind of maximum exposure, not to the truth of a particular situation—even as obsessions with religiosity are growing—but to an absence of clarity embodied in a background of prolific details not yet, or if ever, organized into a coherent narrative of development, of what it means to be human.

Here exposure is to the possibilities of auto-construction—but not auto-construction of a home, a community or a way of life but rather more so to the viability of making up itineraries as one goes along (Bhan, 2019). The details of daily itineraries are viewed less as indicative of specifically defined futures that would suggest clear courses of action based on what people knew from their prior residential situations, but entities still in motion, yet to be "settled" within any framework. Instead of provoking stasis and an inability to take decisive courses of action, individual enactment in an environment of detached details is seen as a function of paying attention to a larger picture, itself something vague in people's minds but something "out there" which is determinant, vital.

I am particularly interested in this shifting sensibility about and relation to an "out there" because my long-time preoccupation has been exploring the processes through which the so-called majority world—that oscillating amalgam of working classes, the poor, the lumpenproletariat, the vaguely middle class—attempted to construct viable urban lives within conditions that were at once rigidly formatted and open-ended. In such processes of what is popularly known as auto-construction, the volatility of a larger city, characterized by sometimes violent political contestations, the absence or constant revisions of regulatory frameworks or the unpredictability of essential provisioning, was compensated for through a time of increments. Here a steady beat of small changes, of hundreds of initiatives responding to each other, of call and response produced confidence in what was to come.

What was important was a sense of forward momentum, that the efforts were worth something, not necessarily in monetary terms but in the ongoing capacity to build relationships with the "surrounds"—with other inhabitants, human and nonhuman—that afforded a range of different possibilities for action. When conditions don't seem to be taking you anywhere, where you constantly battle to keep your head above water and where most of the efforts you make, both individually and as a part of a larger collective, at best only manage to repair breakdowns of all kinds, then trying to make a particular way of life that endures doesn't really matter. It doesn't really matter no matter how much it is familiar or embodies cherished memories or attainments. This, then, is the irony of making auto-constructed districts: they became indifferent to their own accomplishments, and they sometimes so eagerly jumped for opportunities to acquire new assets, new property, new lifestyles, if the price was right.

What was important, though, was the capacity to keep going. What was productive about many instances of self-constructed urbanization was the way in which the things that were built could be translated into each other in many different ways. Housing, work, sociability, caretaking, service provisioning and livelihood were all connected to each other. A house wasn't simply a house; a place of work wasn't simply a workplace; a religious institution wasn't simply a church, temple or mosque; a neighbor wasn't simply a neighbor; water, sanitation and power weren't simply material resources coming out of the blue. They were woven through each other, continuously combined in different ways.

In all of these efforts, there was often no clear destination in mind. Whatever occurred never exceeded all of the rudimentary, unsubsumable fragments of effort, inclination and passion. Of course, such districts were subject to seemingly debilitating racial anxieties, as race was a critical barometer in measuring what kinds of opportunities and resources were to be available for whom and in signaling what kinds of bodies were both exempt and ineligible for specific adherences to codes of conduct that emphasized respectability, discipline, doing the right thing. Even as auto-constructed districts curated atmospheres of freewheeling improvisation and everyday vernaculars replete with ribald intensity, propriety through property, the property of owning oneself and caring for its transactional value was at the heart of much of postcolonial governmentality.

As some districts transitioned into those of a middle class, affective intensity was to be traded for a more disciplined pursuit of well-mannered accumulation, playing by the rules, postponing immediate gratification for long-term visions of individual success. It was important to demonstrate increasing distance from communities of the poor, as well as one's eligibility to be noticed and taken seriously by an urban elite. In this way, households were managed as if they were really in "the middle of things", that through hard work, as opposed to inherited privilege, demeaning physical labor or trickery, an emerging middle class would constitute the necessary moral underpinning of urban growth cemented in the acquisition of property and propriety. Here propriety was attainable through considering one's lifetime as a property to be cultivated, a value to be maximized (Blomley, 2016).

Whether marked by a middle-class narrative of self-possession, of accumulation through doing the right thing or the reworked solidarities of a working class barely working, each appears increasingly tempered by an insouciant orientation to outcome, an indifference that has come to characterize the decisions of many former residents of auto-constructed districts as they spread across the vast peripheries and hinterlands of urban regions. Even though failure has a debilitating cost, and households may rightly worry about wrong moves, a sense of provisionality seems to burst through the most persistent anxieties. They are taking their chances in settings where there is often nothing clear to latch on to, where the chief concern seems to be finding a place to "park"—one's belongings or aged parents. The concern is not so much to "settle" as it is to take care of the essential pragmatics in an affordable way so that one can tend to the real business of circulating across the region, betting on being at the right place at the right time.

Inhabiting a place becomes less important than unfolding multiple possible configurations of concretizing shelter and income streams, fulfilling household and extended family obligations and ways of imagining the future. It entails assessments about where and how opportunities are moving across the urban region, under what conditions and what then are various routes available to access them. It is a strategy based on mobility, of generating spaces of inhabitation through movement—a practice that has long been the purview of the poor.

But these actions also reiterate in a more general way the extent to which demarcations and apprehensions of space are a result of intersecting itineraries that enfold specific sites as ever-changing passageways of arrival and departure, articulated to various elsewheres in ways that are always being changed by the itineraries themselves. How itineraries with their concomitant sensibilities, backgrounds, services and goods converge and part; how they work themselves around each other with different proportions of contact; and how they sense the implications of each other's presence or absence all imbue a space with a certain solidity that enables specific rhythms of repetition and adjustment that enable the space to assume a particular character.

Again, this is a matter of exposure, of itineraries exposed to each as people circulate through and around each other. Whereas apparatuses of urban policing seem to have foregone attempting to keep bodies in place, despite the persistence of segregation and increasingly privatized space, circulation is increasingly subject to tracking, controlling and targeting. Despite this capacity to target bodies in movement, the exposures of itineraries to each other—and the experiences and knowledge of those who deploy such itineraries—is perhaps more likely among "people on the move" than it is by developing more stable places for people to work and reside. All crossings do indeed take place in a place, yet as those in movement cross, bringing their histories, aspirations and capacities with them, exposing them to both familiar and unfamiliar situations, the "out there" is brought into the ambit of each person's daily operations. In the midst then of multiple "out theres" different for every person, the actions of those crossing are not undetected but perhaps imbued with enough ambiguity to enable continuous passage onto somewhere else, some unexpected terrain or disposition.

References

Berlant, L. (2016) 'The commons: Infrastructures for troubling times'. *Environment and Planning D: Society and Space*, 34(3): 393–419.

Bhan, G. (2019) 'Notes on a southern urban practice'. *Environment and Urbanization*, 31(2): 639–654.

Blomley, N. (2016) 'The territory of property'. *Progress in Human Geography*, 40(5): 593–609.

Bogues, A (2012) 'And what about the human? Freedom, human emancipation, and the radical imagination'. *Boundary, 2* 39(3): 29–46.

Brachet, J. and Steele, J. (2015) 'Fleeting glory in a wasteland: Wealth, politics, and autonomy in Northern Chad'. *Comparative Studies in Society and History*, 57(3): 723–752.

Deleuze, G. and Guattari, F. (1987) *A Thousand Plateaus* (2nd ed.). Minneapolis, MN: University of Minnesota Press.

Hanchard, M. (1999) 'Afro-modernity: temporality, politics, and the African diaspora.' *Public Culture* 11 (1): 245–268.

Hartman, S. (2007) *Lose Your Mother: A Journey Along the Atlantic Slave Route.* New York: Farrar, Straus, and Giroux.

King, T.L. (2017) 'Humans involved: Lurking in the lines of posthumanist flight'. *Critical Ethnic Studies Journal*, 3.1: 162–185.

Koselleck, R. (2018) *Sediments of Time: On Possible Histories.* Stanford: Stanford: University Press.

Li T.M. (2018) 'After the land grab: Infrastructural violence and the "mafia system" in Indonesia's oil palm plantation zones'. *Geoforum*, 96: 328–337.

Mbembe, A. (2017) *A Critique of Black Reason.* Durham, NC; London: Duke University Press.

McKittrick, K. (2013) 'Plantation futures'. *Small Axe*, 17(3): 1–15.

Moten, F. (2017) *Black and Blur.* Durham, N.C., London: Duke University Press.

Moten, F. (2018) *Consent Not to Be a Single Being.* Durham, NC; London: Duke University Press.

Neilson, B., Rossiter, N. and Samaddar, R. (Eds.) (2018) *Logistical Asia: The Labour of Making a World Region.* Singapore: Palgrave.

Povinelli, E. (2016) *Geontology: A Requiem to Late Liberalism.* Durham NC; London: Duke University Press.

Saldanha, A. (2019) 'A date with destiny: Racial capitalism and the beginnings of the Anthropocene'. *Environment and Planning D: Society and Space.* https://doi.org/10.1177/0263775819871964.

Spillers, H. (2003) *Black, White, and in Color: Essays on American Literature and Culture.* Chicago: University of Chicago Press.

Stiegler, B. (1998) *Technics and Time 1: The Fault of Epimetheus.* Stanford, CA: Stanford University Press.

Thomas, D. 2016) 'Time and the otherwise: Plantations, garrisons and being human in the Caribbean'. *Anthropological Theory*, 16(2–3): 177–200.

Wright, M. (2015) *Physics of Blackness: Beyond the Middle Passage Epistemology.* Minneapolis; London: University of Minnesota Press.

Wynter, S. (2003) 'Unsettling the coloniality of being/power/truth/freedom: Towards the human, after man, its overrepresentation—An argument'. *The New Centennial Review*, 3(3): 257–337.

15

CONSTRUCTING THE SOUTH-EAST ASIAN ASCENT

Global vertical urbanisms of brick and sand

William Jamieson, Katherine Brickell, Nithya Natarajan and Laurie Parsons

Introduction

Across South-East Asia, the explosion of construction is driving record demand for building materials to support the literal ascent of its cities. Bringing together our research on brick and sand, in Phnom Penh and then Singapore, this chapter looks up from their reclaimed land and skyscrapers to question the vertical urbanisms of global city-building. While cities in the region 'are increasingly imagined as global frontiers of urban studies in the twenty-first century' (Bunnell, 2017: 9), the chapter grounds and embodies such scholarly proclamations through the less visible lives of those bound in their making and 'call[s] into question the universals of global urbanism' (Sheppard et al., 2013: 1954) through the fraught production of its material form.

Namely, our research explores the lives of debt-bonded labourers in Cambodia, responding to the extraordinary demands for bricks and those of villagers in coastal areas of the country who have experienced forced eviction, the sand beneath their feet transported to Singapore to enable land reclamation for the high-rise developments the city-state has become so synonymous with. Global cities, as told through brick and sand, are assemblages of international financial and material circuits that mark global urbanism from 'above' and 'below'. The individual examples of Phnom Penh and Singapore show their profitable role in examining these dynamics. Yet taken together, they also enable analysis of 'inter-referenced urbanism' (Shatkin, 2011: 79). Singapore has been described as the 'benchmark to which many Southeast Asia cities aspire' (Moser, 2012: 182), and this is an aspiration which is being pursued by Phnom Penh through the brick-making industry supplying the material components of its vertical drive for global city status (Brickell et al., 2018). Yet Singapore's development and land reclamation are contingent on sand dredged from Cambodian coasts to enable their own feverish expansion. Building materials and the stories behind them, we argue, offer a neglected vantage point from which to study the constructive, but also destructive, human impacts of global city-building.

Brick

Cambodia is in the midst of a construction boom. The building of office blocks, factories, condominiums, housing estates, hotels and shopping malls is pushing its capital city upwards. But this

DOI: 10.4324/9780429259593-18

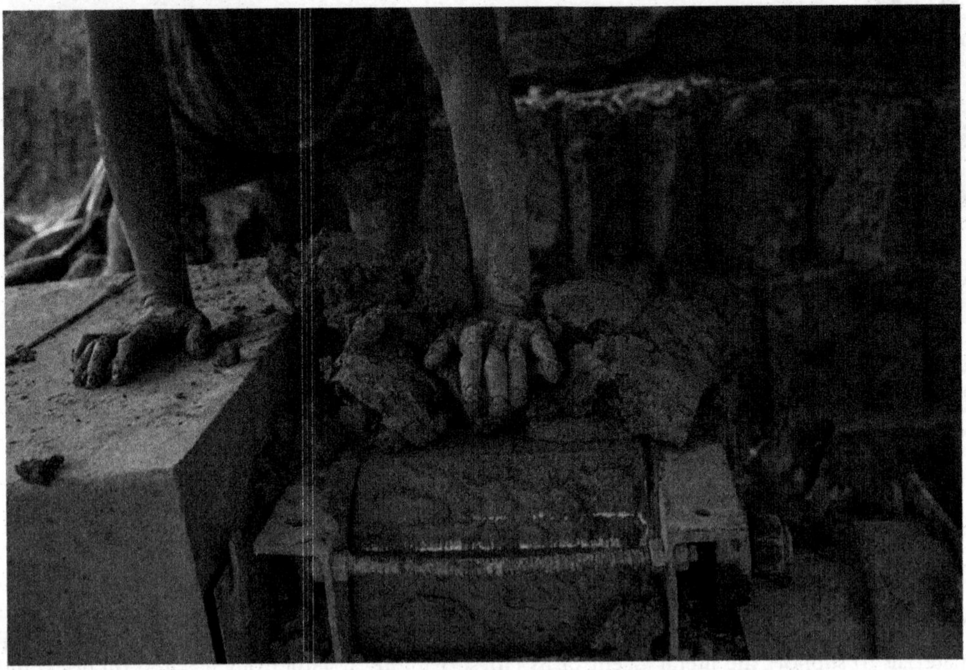

FIGURE 15.1 Boran feeds clay into a manual brick-moulding machine, Cambodia (photograph by Thomas Cristofoletti, copyright Royal Holloway, University of London)

vertical drive into the skies, and the country's status as one of Asia's fastest-growing economies, hides a darker side to Phnom Penh's ascent. Building projects demand bricks in large quantities, and there is a profitable domestic brick production industry supplying them. A national census has established that there are 10,217 people, of whom there 3,937 currently aged under 18, resident in Cambodia's brick kilns (Parsons et al., 2020). This industry relies upon a multigenerational workforce of adults, like Boran (Figure 15.1), and children trapped in debt bondage – one of the most prevalent forms of modern slavery in the world (ONHCR, 2016). The exploitation is a violation of national laws[1] and international human rights treaties, which Cambodia is a signatory of (LICADHO, 2016).[2] Blood bricks raise the question, who is the city built for? And whose lives are being sacrificed in the long shadows of Phnom Penh's peaks and penthouses?

Described as 'lurching upwards' (Nam, 2017), Phnom Penh has taken a rapid vertical turn over the past two decades. Yet as recently as the late 1970s, the city was emptied of its population by the Khmer Rouge regime (1975–1979). Following the genocide's official end, Cambodia endured over a decade of further political instability, during which time Phnom Penh was re-populated. A paucity of state policy and planning in this period meant that the rules of resettlement were forged organically (Fauveaud, 2014). In 1989, the Cambodian government enacted land titling rules, marking the beginnings of a broader structural transformation in Cambodia towards a market economy (Hughes, 2003). The introduction of titling led to rising land prices, speculation and a flurry of foreign investment in land and buildings over the next two decades as the state underwent a process of economic liberalisation (Springer, 2010). This was first ushered in by the United Nations (UN) under the 'United Nations Transitional Authority in Cambodia' (UNTAC) and then taken forward by what is now the ruling Cambodian People's Party (CPP) under Prime Minister

Hun Sen. The drive for foreign investment in construction began in the UNTAC period in 1991 when thousands of UN and non-governmental organisation (NGO) staff from around the world arrived in the capital city looking for high-end residences and leisure spaces.

Today, the construction boom continues unabated, buoyed in particular by an expanding demand for luxury gated complexes for Cambodian and foreign elite and low-cost housing for thousands of migrant workers in the garment and services sectors (Paling, 2012). Sustained increases in foreign capital investment, combined with a proliferating domestic capitalist elite, are forging new projects across Phnom Penh (Fauveaud, 2014). From 2006 to 2016, the construction sector grew by an annual average of 17%, and it represented the largest contributor to growth in 2015 (Durdyev et al., 2016). Indications suggest that the rate of growth in the sector may be slowing down. Yet in 2017, Cambodia saw a 36% increase in imports of building materials and a 27% increase in approved construction projects (ADB, 2018).

While modern slavery has been highly publicised in the building of mega-sporting event structures across low- and middle-income countries, from Qatar to Brazil, in Europe, the construction sector ranks second after the sex industry in terms of the prevalence of labour exploitation (Crates, 2014). On these counts, the construction industry has been termed 'bipolar' for the stark contrast between the 'inspirational buildings' that are constitutive of it and the 'systemic exploitation of millions of migrants' who are sacrificed in their construction (CIOB, 2015). The use of blood bricks reveals how labour exploitation is a much wider problem embedded in the operations of the construction industry before even reaching the construction site.

FIGURE 15.2 The Peak, Phnom Penh (photo by K. Brickell, January 2019)

Tracked from brick kilns to construction sites and, in some cases, from construction sites back to brick kilns where labour practices were verified, we identified the use of blood bricks in the foundations of eight developments in Phnom Penh for which we examined the marketing information and investment profiles. This included a mix of condominiums, shops, restaurants, hotels and office blocks. In doing so, the research highlighted that bricks moulded by debt-bonded labourers were embedded in some of the most prestigious construction projects across the city. This includes The Peak (Figure 15.2), marketed as 'an oasis of luxury in the clouds',[3] which will embolden its new residents to 'embrace spectacular'. The Peak is joined by The Penthouse Residence, with a '360 degree panoramic view of the city skyline';[4] Highland Condo, with 'an open sky bar that allows you to do stargazing at night!';[5] and One Park, 'a New World within the City'.[6] The Elysee (Figure 15.3), with its replica Arc de Triomphe, is pitched meanwhile as 'Little Paris'.[7] Phnom Penh's 'Diamond Island' (Koh Pich), made of reclaimed land, is home to The Elysee, along with Phnom Penh's most expensive and ostentatious building projects. Several architecturally evoke the mimicry noted in 'trans-urban imaginaries' (Bedi, 2016) across world cities. They also emulate, as Roy and Ong (2011: 16) qualify, more than 'concrete, glass, and steel; urban modelling also involves the mimicry of the neoliberal packaging of international glamour, talent, and entrepreneurialism that promises to animate a moribund metropolis'.

The eight developments were mostly domestically funded, but evidence of foreign investment from the United Kingdom, United States, Norway and Singapore was identified in one case. This included a well-known pension investment company, government wealth fund and a host of global investment management corporations. That the Singaporean company was a property development company reflects the growing significance of South-East Asian urbanism in terms of the circulation of capital and expertise (Shin, 2019). Those involved speak to the 'financial-urban conjuncture'

FIGURE 15.3 The Elysee development, Phnom Penh (photograph by Thomas Cristofoletti, copyright Royal Holloway, University of London)

(Sheppard et al., 2015: 1957) that is so often discussed in writing on global urbanism; the intensity of speculative real estate development in Phnom Penh finds roots both internationally and domestically, distorting the demand that drives the precarious production of construction materials and contouring the architectural forms of these developments towards imagined buyers seeking sound investment opportunities. While it is important to pay close attention to the financial ties complicit in these developments, the basis of their built forms reveals their complicity with inhuman labour practices at the point of production. Blood bricks offer a neglected focal point from which to understand people's lives behind the domestic and global growth story of Cambodia's capital city. After all, the price of all this altitude is paid by those at the bottom. Market forces operating in the absence of enforced legal frameworks have seen the most vulnerable slip into inescapable debt, creating a pool of resources that have been greedily subsumed into the national project of construction. The very earth that thrusts Cambodia's growth materially skywards, now baked into brick form, is that which once belonged to the disenfranchised and indebted.

Sand

The humble brick, in constructing the South-East Asian ascent, is accompanied by the grain of sand. A key ingredient of concrete and reclaimed land, the Asia-Pacific region's appetite for construction aggregated grew fivefold over the past two decades (Schandl et al., 2016). This intense and accelerating consumption has been dogged by reports of corruption, illegality and dispossession from India's sand mafias, to small-scale sand mining in Myanmar and illegal cross-border sand trade between Vietnam and Cambodia. In South-East Asia, sand has acquired a covertly geopolitical profile, largely owing to the demand of one single city-state: Singapore. Between 2006 and 2016, Singapore was the largest importer of sand in the world by five times, quite an achievement for a country of only 724 km^2 (Data.gov.sg, 2019). However, 143 km^2 of its total area has been reclaimed over six decades, and it has been estimated that 80% of that land has been built with imported sand (Topalovic et al., 2013). To give some idea of the quantities of sand involved, it has been estimated that around 37.5 million cubic meters of sand are needed per kilometre of reclaimed land (Milton, 2010). Using this as a rough metric, this would mean about 4 billion cubic meters of foreign sand has become Singapore's territory.

Initially constructed for housing and infrastructure in the '60s and '70s, the geography of Singapore's prosthetic territory has diversified considerably since then, with globally recognisable tourist attractions like the Gardens by the Bay and the Marina Bay Sands located on reclaimed land. These sites, and that of the greater Marina Bay, the product of another immense land reclamation project forming the core of Singapore's Global City brand, all lie on sand extracted from rivers, coasts and seabeds in Malaysia and Indonesia. Further from the centre, land built from sand still underpins the infrastructural and logistical core of Singapore's economy, from the petrochemical hub of Jurong Island, to Changi Airport, Port Pasir Panjang and the future Tuas 'megaport'. There are ongoing reclamations until 2030 and plans to reclaim beyond that. But stable sources of sand have become harder and harder to come by.

Malaysia banned sand export to Singapore in 1997, eventually causing the price of sand to temporarily spike by 5,000% in 2003, the same year Indonesia banned sea sand export to Singapore, the primary infill material of Singapore's reclamation projects. Both the Malaysian and Indonesian supply chains were plagued by irregularities, forged documents, misdirected shipments, bribed officials and regulations waived. Indonesian islands were dredged from the surface of the sea, allegedly, including the partial dissolution of Pulau Nipah, an Indonesian island that formed a maritime boundary with Singapore, which prompted a total ban on sand exports in 2007. That same year, Singapore, for the first time, imported sand from Cambodia: a consignment of 127,000 tonnes – two or three bulk carriers worth – then over 3.5 million tonnes in 2008 and 13 million tonnes in 2009. Between 2007 and 2017, Singapore imported 80 million tonnes of sand from Cambodia, entirely from the remote coastal province of Koh Kong.

From the outset, the dredging operation mirrored the steep levels of elite capture that characterise other extractive industries in Cambodia, with all of the sand-dredging concessions controlled by two CPP senators. Environmental Impact Assessments (EIAs) were scant, and when they materialised, they seemed to have been copied and pasted from other EIAs; with no baselines taken, the true impact of the dredging is difficult to ascertain. Along rivers and coasts, sand and silt form the margin between land and water, as well as mediate the velocity of the water. This emergent and complex role means that the sudden extraction of hundreds of tonnes of sand from a river system will have an effect both upstream and downstream; dredging can increase the velocity of a river so drastically that it precipitates erosion so sudden that houses by rivers in Cambodia and Vietnam have tumbled into the river. The silt that gets stirred up affects the nutritional content of the water, so ecosystems become uninhabitable, which not only affects biodiversity but also any communities that depend on fishing. The village of Koh Sralao saw almost 20% of its households leave by mid-2010 because of the decline in fish and crab stocks, as well as a subsequent rise in debt within the community (Marschke, 2012).

According to a 2010 report by investigative NGO Global Witness, the Singapore Embassy in Phnom Penh was involved in issuing the dredging licenses, though the Singaporean government denied this. The sand mining would continue for another six years, unabated despite the multiple bans on varieties of sand and on locations of concession sites in 2009, 2015 and 2016. The Singaporean government always maintained their high standards for their licensed suppliers and its international reputation as a regional leader in sustainability and transparency, as well as the necessity of compliance with regulations and laws for suppliers in source countries, but the steep asymmetry between the technocratic city-state of Singapore and the kleptocratic Kingdom of Cambodia ensured that the sand kept flowing. However, in 2017, after a sustained campaign by activist group Mother Nature, Singapore decided to stop importing Cambodian sand, and Cambodia banned all kinds of sand export permanently. The permanent ban coincided with the revelation that, while Singapore had declared a total import of 73 million tonnes of sand from Cambodia between 2007 and 2015, Cambodia had only declared total exports of 2.7 million tonnes. This shocking disparity, as well as the long-term, viral activist campaign that explored the possibility of suing the Singaporean government over its illegitimate consumption of Cambodian sand, managed to apply enough pressure and stir up enough coverage for the sand to stop. However, this has only prompted Singapore to swap suppliers, with Malaysia now selling its sand to Singapore, as well as Myanmar and the Philippines opening their frontiers to Singapore. Its exceptional demand for sand has anticipated the emerging sand crises, as multiple reports point towards the unsustainable extraction of construction materials, with South-East Asia leading the way. Sand can help us understand how these ubiquitous construction materials and practices, such as concrete and land reclamation, collide with local socio-economic and ecological particularities and reveal the tensions inherent to global urbanism's unsustainable (re)production. As the demand for sand rises, the sand trade has become increasingly transnational, yet sand has remained a precariously constructed resource. While its cheap and plentiful flow is critical for these construction practices to become feasible, only particular kinds of sand are suitable for construction.

Focusing on the flows of sand that facilitate the construction practices of global urbanism reveals the metabolic rift between ubiquitous aspects of its built form, such as concrete, and the ecologies and geomorphologies which that same flow forms part of. These rifts are patched up by precarious supply chains, and the tension between cheap price and widespread consumption was resolved through the opening of new frontiers and the burgeoning black-market trade in sand. Tracing the consumption of sand brings new questions of political ecology and economy to global urbanism: each cubic foot of concrete or reclaimed land is forensically interconnected with its extractive frontier, its murky supply chain potentially disrupting coastal or fluvial metabolisms and the ecologies and livelihoods that depend on them. Understanding the uneven 'choreograph[ies] of sand in place-making across different but interconnected temporalities' (Kothari and Arnall, 2019: 8) forces us to question global urbanism and its construction practices and pay attention to how its extractive frontiers are formed and shifted. As urbanisation continues to intensify in South-East Asia, construction markets and fluvial ecologies become linked in dysrhythmic synchronicity, requiring political and economic intervention to ensure that the sand keeps flowing. Looking more closely at the extractive practices that underlie the built form of global urbanism shifts the focus from 'methodological cityism' to the geographically fragmented processes of urbanisation through which cities become materially and economically coherent (Angelo and Wachsmuth, 2015)

Through sand, the planetary totalities of global urbanism become porous: the contagious surfaces of the global city crumble to reveal the processes of enclosure and dispossession they are built on, as well as the murky supply chains which stitch them together.

Conclusion

While Singapore and Phnom Penh are not natural comparisons, as one is a city-state and the other is a capital city of diverging political and developmental trajectory, both of their practices of city-making point towards the friction that lies beneath the surfaces of global urbanism. Cambodian bricks and sand, whether consumed in Phnom Penh or Singapore, transect the power geometry of global urbanism with the socio-ecological gristle of their material production, grounding its spectacular financialised verticality in the unsustainable extraction of labour

FIGURE 15.4 The proposed Riviera development, Phnom Penh (photo by K. Brickell, 2012)

and resources. More curiously, Singapore's conspicuous architectural brand of global urbanism has become a talisman for growth, as evidenced by the Riviera planned development in Phnom Penh, a budget facsimile of the Marina Bay Sands with its triptych of towers connected by a rooftop garden (Figure 15.4).

In this chapter, we have explored brick and sand as material embodiments of the global urban age, a 'refashioned globalism' (Smith, 2002: 430) told through international processes of production and extraordinary urban growth in South-East Asia. Its cities are assemblages of international financial and material circuits that are produced from 'above' and 'below'. Debt-bonded brick labourers and villagers evicted from coastal areas of Cambodia are bound in this 'below', a corporal and earthy hallowing out of human life and land which problematises both Phnom Penh and Singapore as gleaming beacons of 'development' and progress.

Notes

1 Debt-bonded labour is illegal under Cambodia's Law on the Suppression of Human Trafficking and Sexual Exploitation (2008) Articles 10 and 12. Child labour is also illegal in Cambodian Labour Law (1997), Article 177.

2 Article 3 of the Palermo Protocol (2000) sets out the internationally accepted definition of trafficking – namely, 'the recruitment, transportation, transfer, harbouring or receipt of persons, by means of the threat or use of force or other forms of coercion, of abduction, of fraud, of deception, of the abuse of power or of a position of vulnerability or of the giving or receiving of payments or benefits to achieve the consent of a person having control over another person, for the purpose of exploitation' (United Nations, 2000). Debt bondage is illegal under the Protocol, even if the victim initially consented but was later trapped by fraudulent debt or other coercive means.

3 Oxley-Worldbridge (2018) *'The Peak'*. [Online] Available from https://www.oxleyworldbridge.com.kh/the-peak (Accessed 20 August 2018).

4 Realestate.com (2018) *'The Penthouse Residence'*. [Online] Available from https://www.realestate.com.kh/the-penthouse-residence/25812 (Accessed 20 August 2018).

5 Highland Cambodia (2018) *'About Highland'*. [Online] Available from http://highlandcambodia.com/index.php?op=about-index&id=3 (Accessed 20 August 2018).

6 Editorial (2017) *'One Park: A New World within the City'*. [Online] Available from https://www.realestate.com.kh/news/one-park-a-new-world-within-the-city/ (Accessed 08 August 2018).

7 Morton, E. 'Little Paris comes to Koh Pich'. *The Phnom Penh Post,* 13 January 2014. [Online] Available from https://m.phnompenhpost.com/business/little-paris-comes-koh-pich (Accessed 20 August 2018).

References

ADB (2018) *Asian Development Outlook 2018: How Technology Affects Jobs*. Manila: Asian Development Bank. Available from: https://www.adb.org/sites/default/files/publication/411666/ado2018.pdf

Angelo, H. and Wachsmuth, D. (2015) 'Urbanizing urban political ecology: A critique of methodological cityism'. *International Journal of Urban and Regional Research*, 39(1): 16–27.

Bedi, T. (2016) 'Mimicry, friction and trans-urban imaginaries: Mumbai taxis/Singapore-style'. *Environment and Planning A*, 48(6): 1012–1029.

Brickell, K., Parsons, L., Natarajan, N. and Chann, S. (2018) *Blood bricks: Untold stories of modern slavery and climate change from Cambodia*. Royal Holloway University of London. Available via: https://www.projectbloodbricks.org/publications

Bunnell, T. (2017) 'Introduction: Futurity and urban Asias'. In: Bunnell, T. and Goh, D.P.S. (eds.) *Urban Asias: Essays on Futurity Past and Present*. Jovis: Berlin 9-20.

CIOB (2015) *Modern Slavery: The Dark Side of Construction*. Bracknell: The Chartered Institute of Building.

Crates, E. (2014) *Construction and the Modern Slavery Act: Tackling Exploitation in the UK*. Bracknell: The Chartered Institute of Building.

Data.gov.sg. (2019). 'Total land area of Singapore'. Available from https://data.gov.sg/dataset/total-land-area-of-singapore.

Durdyev, S., Omarov, M. and Ismail, S. (2016) *SWOT Analysis of the Cambodian Construction Industry within the ASEAN Economic Community*. Seville: Proceedings of the 28th International Business Information Management Association Conference.

Fauveaud, G. (2014) 'Mutations of real estate actors' strategies and modes of capital appropriation in contemporary Phnom Penh'. *Urban Studies*, 51(16): 3479–3494.

Global Witness (2010) *Shifting Sand: How Singapore's Demand for Cambodian Sand Threatens Ecosystems and Undermines Good Governance*. London: Global Witness. Available from https://www.globalwitness.org/en/archive/shifting-sand-how-singapores-demand-cambodian-sand-threatens-ecosystems-and-undermines-good/.

Hughes, C. (2003) *The Political Economy of Cambodia's Transition 1991–2001*. London: Routledge.

Kothari, U. and Arnall, A. (2019) '*Shifting sands: The rhythms and temporalities of island sandscapes*'. *Geoforum*. Online before print.

LICADHO (2016) *Built on Slavery: Debt Bondage and Child Labour in Cambodia's Brick Factories*. Phnom Penh: LICADHO.

Marschke, M. (2012) *Life, Fish and Mangroves: Resource Governance in Coastal Cambodia*. Ottawa: University of Ottawa Press.

Milton, C. (2010) 'The sand smugglers'. *Foreign Policy* (blog). Available from https://foreignpolicy.com/2010/08/04/the-sand-smugglers/

Moser, S. (2012) Globalisation and the construction of identity in two new South-East Asian capitals: Putrajaya and Dompak. In Chen, X., and Kanna, A. (eds.). *Rethinking Global Urbanism: Comparative Insights from Secondary Cities*. London: Routledge, 169–189.

Nam, S. (2017) 'Phnom Penh's vertical turn'. *City*, 21(5): 622–631.

ONHCR (2016) Debt bondage remains the most prevalent form of forced labour worldwide–new UN report. Available from http://www.ohchr.org/EN/NewsEvents/Pages/DisplayNews.aspx?NewsID=20504&LangID=E.

Paling, W. (2012). 'Planning a future for Phnom Penh: Mega projects, aid dependence and disjointed governance'. *Urban Studies*, 49(13): 2889–2912.

Parsons, L., Chann, S. and Ly Vouch, L. (2020) *A Census of the Cambodian Brick Industry: Population, Geography, Practice*. Phnom Penh: Solidarity Centre.

Roy, A. and Ong, A. (2011) Worlding cities, or the art of being global. In Roy, A., and Ong, A. (eds.), *Worlding Cities: Asian Experiments and the Art of Being Global*. Wiley Blackwell: Chichester, 1–26.

Schandl, H., Fischer-Kowalski, M., West, J., Giljum, S., Dittrich, M., Eisenmenger, N. and Geschke, A. (2016) *Global material flows and resource productivity: Assessment report for the UNEP International Resource Panel*. United Nations Environment Programme: Nairobi, Kenya.

Shatkin, G. (2011) Planning privatopolis: Representation and contestation in the development of urban integrated mega-projects. In Roy, A. and Ong, A. (eds.), *Worlding Cities: Asian Experiments and the Art of Being Global*. Wiley Blackwell: Chichester 77–97.

Sheppard, E., Gidwani, V., Goldman, M., Leitner, H., Roy, A. and Maringanti, A. (2015) 'Introduction: Urban revolutions in the age of global urbanism'. *Urban Studies*, 52(11): 1947–1961.

Sheppard, E., Leitner, H. and Maringanti, A. (2013) 'Provincializing global urbanism: A manifesto'. *Urban Geography*, 34(7): 893–900.

Shin, H. B. (2019) Asian urbanism. In Orum, A. M. (ed.), *The Wiley-Blackwell Encyclopedia of Urban and Regional Studies*. Oxford: Wiley.

Smith, N. (2002) 'New globalism, new urbanism: Gentrification as global urban strategy'. *Antipode*, 34(3): 427–450.

Springer, S. (2010) *Cambodia's Neoliberal Order: Violence, Authoritarianism, and the Contestation of Public Space*. Abingdon: Routledge.

Topalovic, M., Knüsel, M., Jaggi, M. and Krautzig, S. (2013) *Architecture of territory – hinterland: Singapore, Johor, Riau*. ETH Zürich DArch.

16

NAIROBI CITY, STREETS AND STORIES

Young lives stay in place while going global through digital stages

Tatiana Thieme

What makes a city, a street, a life worth telling about? What localised struggle or daily feats become a matter of shared concern or come to matter to a wider (let's say global) audience? The role of digital life over the last decade, since the start of my ongoing ethnographic engagement in Nairobi, has increasingly extended the temporality and spatiality of urban ethnographic research. One no longer really 'leaves' the field. Having interlocutors who can reach out and be reached via WhatsApp or Facebook adds a particular form of punctuated access to selected (and often highly curated) windows into the lives of our interlocutors and friends from the field. With the added ethical and personal responsibility that this entails in terms of mutual boundary setting and maintenance of ties, this extension of fieldwork through 'globalised' digital platforms shapes our views of our key interlocutors' progression and plans in their lifeworlds, beyond our *time* 'in the field'.

Ten years ago, towards the end of my PhD fieldwork, I collaborated with a local crew of young filmmakers to shoot a short low-budget documentary film we called *Story Yetu* [Our Story]. On the fifth day of production, I met three young boys, barely 18 years old at the time. They said they were 'artists' and spent much of their day in a small container that had been repurposed as a makeshift recording studio in one of the most under-resourced corners of one of the largest and oldest informal settlements in Nairobi: Mathare. In Nairobi, over 60% of residents live in low-income neighbourhoods, where 80% of residents are under the age of 35, a majority of whom are formally unemployed but economically active. Most low-income youth in Nairobi refer to this activity as 'hustling', spending their days *making* work on the edges of the urban economy (Thieme, 2018). Alongside the *hustlers*, some youth bear witness to the Nairobi youth hustle, using creative repertoires to tell stories about the everyday 'hustle' and 'ghetto life'. The three youth I met during the filming of *Story Yetu* were the latter: witness bearers, storytellers. And their stories, over time, became their hustle.

Urban ethnography in the twenty-first century poses a host of epistemic, ontological and methodological conundrums. This chapter interrogates some of these conundrums by focussing on a group of youth with whom I collaborated during the production and recording of that short documentary film and accompanying hip-hop soundtrack in 2010. In a mode of experimental ethnographic reflection, I write about these three youth from a peculiar positionality. Invested in longitudinal ethnographic research,[1] I have continuously returned to my field site over the past

DOI: 10.4324/9780429259593-19

decade and kept in close contact with my key interlocutors. Several have become dear friends and homestay hosts when I spend time in Nairobi. This chapter builds on my ethnographic insights but focusses not on my key interlocutors whose lives I have come to know well. Instead, I depict in semi-stylised form the trajectory of three young people whose lives I have, since those days of music co-production, mostly followed from afar through the prism of social media. This chapter reflects on the particular spatialities of fieldwork and forms of 'telling young lives' (Jeffrey and Dyson, 2008) that might emerge when one's window into these lives is digitally mediated. As ethnographers returning to a field site over time, what can we learn and say about young lives whose intimate vacillations we cannot claim to know but whose carefully curated digital profiles we can consider a form of meaningful auto-portraiture against the backdrop of neighbourhoods and other youth geographies we have come to know something about?

This chapter focusses on these three storytellers from Mathare, whose portrayals of their lifeworlds through social media canvases have increasingly used digital modes of self-fashioning and storytelling to link local and previously unseen ordinary happenings to wider expressions of youth urban culture and belonging. In conversation with this volume's collective reflection on the merits and reach of the conceptual frame, 'global urbanism,' the chapter explores the interplay between the conceptual, imagined and spatial reference points of what might be regarded as 'global' in relation to the performative registers and resonances of digitally mediated hyper-local youth stories aiming to reach multiple precarious elsewheres (Weiss, 2009; Ntarangwi, 2009; Kidula, 2012; Rollefson, 2017). Given that the 'global' is a highly relational and situated concept, this begs the question, 'global from whose point of view?' As I reflect on my years in (and out of) Nairobi, thinking and writing about young lives that are simultaneously cut off from the mainstream city while shaping its peripheral and popular spaces, 'global urbanism' offers an empirical and methodological opening that sets a particular ethnographic positioning in conversation with moments and frames that might be imagined as 'global' to some and 'local' to others. But it is the entanglement of scales – the place-based stories of shared struggle and global imaginaries of shared solidarity – that both give sense to and trouble the notion of 'global urbanism'.

Growing up (digitally) global

There have been long-standing debates about what constitutes *culture* and how to go about interpreting and writing about cultural forms (Geertz, 1977; Clifford and Marcus, 2010). In recent decades, postcolonial scholarship has greatly nuanced the understanding of cultures, examining the hybridities and interconnections of cultural formations marked by layered histories of colonialism, anti-colonial struggles and postcolonial diasporic encounters (Bhabha, 2004). In geographical studies focussed on youth cultures, Massey writes about the 'spatial openness' of urban youth culture, positing that all youth cultures are 'hybrid cultures' (Massey, 1998: 122–123). Youth popular culture in African cities provides a rich repertoire of 'intercultural and transnational formation' that shapes particular forms of 'black vernacular culture' (Gilroy, 1993: ix), reflecting an 'inescapable hybridity and intermixture of ideas' (ibid: xi).

In Nairobi, youth popular culture is expressed through their everyday urban vernacular of Kenyan Swahili known as *Sheng* (Githiora, 2018) and through the Kenyan hip-hop genre of *Genge* that emerged in the late 1990s, mirroring some of the US African American hip-hop styles and Jamaican ragga to form what would come to be known as *muziki wa vijana* (music of the youth) (Ntarangwi, 2009). Both the creative performative practices of street linguistics through the constant wordplay of *Sheng* and *Genge's* 'youth music' have formed a vibrant urban youth culture that relays the everyday shared struggles of youth born and raised in Nairobi's largely marginalised neighbourhoods, facing a constellation of paradoxical economic, environmental and social realities: on the one hand, young lives are cut off from formal institutional support (education, health care and employment opportunities), but at the same time, they are the sources of investment and attention by the myriad improvement schemes purporting to tackle the 'youth crisis' (Thieme, 2015).

Over the past two decades, the concomitant rise of hybrid development schemes targeting 'youth entrepreneurs' and accelerating access to and use of Information Communication Technology (ICT) services have meant that the 'slum' neighbourhoods of Nairobi have become simultaneously under-resourced in terms of basic infrastructure and services but also the grounds for various social innovation experiments, visits from non-governmental organisation donors running youth programmes and, of course, researchers. In a city of protracted uncertainty and adversity (Cooper and Pratten, 2015) and 'hustle economies' (Thieme, 2018), the art of the spoken word and the skill of telling a good story has become an important part of the cultural repertoire of urban youth who may never leave 'their ghetto' but project imaginaries of shared cosmopolitanism with 'global' youth (Weiss, 2009). It is against the backdrop of inner-city under-investment (Parnell and Pieterse, 2014), structural injustice and youth poverty that creative genres of storytelling across urbanscapes have not only emerged from the streets but also spoken to other youth across multiple elsewheres (Massey, 1998; Katz, 2004).

American hip-hop, UK grime, East African *Genge* and European hip-hop have all become manifestations of hyper-localised sounds rooted in the street stories of youth marginalised by formal institutions, spreading their rhymes through diverse scales – performing for each other on the street corner, posting videos and perhaps reaching a wide ('global') audience in some cases (Rollefson, 2017). Writing about inner-city black youth in East London, White (2020: 40) writes, 'For black youth in particular, the contemporary music scene is a site of emancipatory disruption where it is possible to take on a new identity as an artist, a performer, or an entrepreneur'. Similarly, Fredericks argues that hip-hop in Senegal has become a way for youth to assert their citizenship and agency by engaging with representative politics through subversive musical soundscapes, contesting authoritarian elders and rule (Fredericks, 2014). In Nairobi, *Genge* hip-hop tracks are rehearsed on the neighbourhood streets with references to local tragi-comedic happenings and then played on the informal minibuses known as *matatus*. The street banter becomes material for this creative youth practice, and the streetscapes of the corrugated makeshift dwellings and loitering corners serve as the backdrop of DIY music videos featuring what the artists refer to as 'ghetto life'. These prosaic scenes of everyday life turned 'ghetto' inside out, exposing the 'hood as dwelling place' that is at different times either carceral, over-policed or ignored (Richardson and Skott-Myhre, 2012) but also a place of humanitarian spectacle and intervention. Thus for Mathare youth versed in hustling their way through these different forms of ghetto life, this dwelling place also becomes a stage to tell 'their' story. Inflected with stylised autobiographical content meant to reflect shared struggles and solidarities, the tracks are rooted in place but meant to travel, even if youth themselves cannot (Kidula, 2012).

In the absence of mainstream basic services and formal employment opportunities, Mathare youth have had to carve out creative routes to access or distribute both basic services (like electricity and water) and access 'nice to haves' (from Timberland shoes to a smartphone) that equipped local struggles with global consumer cosmopolitanisms (Ntarangwi, 2009; Weiss, 2009). Along Juja Road framing one of the sides of Mathare Valley, the façades of street-side shops are fashioned with a combination of painted logos of international brands, such as Coca-Cola, and stylised portraits of popular culture icons. *Matatus* driving by feature graffiti puns reflecting popular cultural references of the day. In 2016, the year that the American film *Straight Outta Compton* came out 30 years after the famous N.W.A. album's release, I spotted a *matatu* leaving Mathare branded with the sign 'Straight Outta Dumpsite'.

When I first started doing fieldwork in Nairobi in 2005, many young people had access to a 'brick' phone, contributing bits of pay-as-you-go phone credit to a shared device, and some even had a Yahoo or Hotmail email address they would check every few weeks at the local Internet café. When I later returned in 2009 to conduct ethnographic fieldwork, the young people who would become my key interlocutors all had a mobile phone of their own, and some had a Facebook account. Throughout the last decade, as I've returned almost yearly to reconnect, my key interlocutors have all gotten smartphones and become super connected to digital platforms. The spaces in popular neighbourhoods where youth were seen as 'idle' a decade ago if they were on the jobless corner are now 'connected' even during their states of waiting and *kuzurura* (hanging about, loitering). Hunched over their devices, Nairobi youth in these neighbourhoods may not have access to a safe and clean toilet, but they have access to mobile banking, WhatsApp and the labyrinth of the Internet. During my last trip, a young man named Felix was explaining that even if you can't finish high school because you don't have school fees, at least now 'we have Google'.

Many youth living in Mathare Valley are rooted in place, unable (or unwilling) to imagine living and working elsewhere. But some project themselves and their neighbourhood onto an imaginary global stage. Their *global* comes in the form of digital platforms that have increasingly become part of their everyday grammar and embodied experience since the first decade of this century as the digital revolution in East Africa rapidly penetrated even the most under-resourced communities. Smartphones are now in youth's pockets and in the palms of their hands, social media profiles are carefully curated and animated through daily selfies and 'ghetto news' items to remind their 'followers' from near and far that they have local stories to tell to (awaiting) global audiences.

While development practitioners speak of material poverty and the lack of educational facilities in these neighbourhoods, young people who have grown up with smartphones have shaped a kind of leapfrog urbanism using global brand names as a synonym for various forms of connection, education and information that they can access from the street stoop. Just as they say, 'Now with Google I can teach myself about anything,' in the 'Education' section of their Facebook profiles, they humorously and proudly state, 'Ghetto University'. Together, the 'Google' global and the 'ghetto' local are enmeshed to produce a kind of street-based knowledge that is embedded in local social realities on the one hand and inflected with global echoes on the other (Weiss, 2009). Google has become a kind of emic term for the endless online labyrinth of information that transcends class and economic status as the cost of smartphones and data bundles reaches the most under-resourced communities, and this allows youth from these neighbourhoods not only a window with a view of the rest of the world but also, crucially, access to a stage on which they too can perform and curate their cosmopolitan selves. In a lifeworld where a large number of teenagers do not finish secondary school due to a complex number of extenuating circumstances, and where the prospects of waged employment are scarce, the younger youth of Mathare have differentiated themselves from their older 'brothers', the youth I first met over a decade ago. With a smartphone, some smarts and some gall, some of the younger generation of youth are trying to uberise their way out of joblessness while retaining their street credibility and 'ghetto' roots.

Nairobeez: From neighbourhood street rhymes to city street tours

Lazima hustle,
Tumia muscle,
Kutafuta riziki

(you need to hustle
use muscle
to make a living)

Mashanti, 'Everyday' 2010

Consider the particular subjectivities and performative reach of a particular kind of the now widely familiar and globalised phenomenon of digital life – the 'selfie'. For Mathare youth, the selfie can serve as an activist mode of auto-portraiture and storytelling. It displays a counter-representation of the 'Kenyan slums' that have for decades been subject to Afro-pessimistic portrayals of uneven development, informality, unregulated housing markets and municipal under-investment. This section explores the temporal and spatial registers of the stylised digital auto-portraiture of three young friends, Cheddaz, Kissmart and Donga, whom I met in 2010 during the filming of a documentary called *Story Yetu*, in collaboration with Ghetto Films Trust, when we recorded a track titled 'everyday'. These three youth in their late teens were from one of the poorest sub-neighbourhoods of Mathare, known as Kosovo, infamous for its illegal brewing activities along the river.

When I met Cheddaz, Kissmart and Donga, they were spending their days around the makeshift recording studio in their neighbourhood where they worked on their rap rhymes and performed for each other and anyone who cared to listen. Their lyrics and beats were raw and real, but this local street poetry coming from the marginalised voices of youth was at the time largely unheard beyond the place youth call their *baze*. In Nairobi's popular neighbourhoods, the *baze* is the *Sheng* expression connoting a physical space often situated in a semi-public non-descript part of a neighbourhood where childhood friends gather. It becomes a meeting point, but as a *Sheng* expression, it also carries plural connotations of place-making, belonging and the making of claims. The *baze* can turn into kind of site-specific performative space to assert one's turf, as well as confirm one's presence in the *mtaa*. As key interlocutors have explained, it is a place to be found and to be seen, where the local news of the day is told and shared. It is a space for passing the time and marking one's connection to everyday street culture while evoking the palimpsests of youth imaginaries.

Over the last decade, these three friends built their music, visibility and a local storytelling business thanks to the rapid expansion of ICT across the city, including Mathare. Though formally unemployed, these youth built their own hustle portfolio: They became a three-member boy band known amongst their social media circle as *Nairobeez*, whose access to smartphones and affordable data bundles enabled them to regularly upload video selfies onto Facebook and later Instagram. They performed stories from the streets of Mathare that did not seek to sanitise the everyday realities but rather point out the beauty at the corrugated interstices, and they also regularly posted footage from the recording experiments to share the 'behind the scenes' creative process – exposing the creative backstories of their 'ghetto greenroom'. But this public performance still had limited reach or direct income opportunities, so in order to subsidise their music enterprise, they eventually connected with a wider base of potential customers through the digital platform of Airbnb. With the support of a social entrepreneur passionate about youth and popular culture, they co-produced a narrative that would appeal to Nairobi visitors interested in *seeing* the city beyond its familiar tourist landmarks. The three boys from Kosovo connected their local hustle with the digital economy, marketing themselves as three former street kids with in-depth local knowledge of the city centre and its difficult-to-reach peripheral spaces. So when *avant-garde* tourists on a budget and with a taste for grit log on to the Nairobi Airbnb page, they will come across the local "Nai Nami – Our Streets. Our Stories" private walking tours, where tickets can be booked for £29 per person. This clever form of hustling found a bridge between local lived realities and a global audience keen to take part in *off-piste* city tours tailored to visitors keen to seeing the city differently, through the eyes of youth with street-oriented expertise and experience.

In Mathare, youth hustle to 'get by' and get things done (Thieme, 2018), continuously combining a kind of hunt for ways to stay both economically active and credible on the streets. This involves a particular performance where 'making a living' might appear to contribute to a common cause or be read as an expression of community activism. While many youth living in Mathare have over the past three decades engaged in the homegrown residential garbage collection economy as their stepping-stone into working life, a growing number of younger youth who have grown up digitally connected have used *Genge* hip-hop to be both protagonists and activists of 'ghetto life,' exposing the shared struggles and solidarities of everyday street life in Mathare in solidarity with other global 'ghettos' (Rollefson, 2017).

Consider the tag line 'our street, our stories,' particularly the possessive pronoun. It serves as an appropriation, a kind of 'taking back' of the peripheral urbanisation (Caldeira, 2017) that has persisted since the colonial period. Under colonial rule, Nairobi could not have been *their* city. In the early postcolonial era, their neighbourhood was one of the 'slums' in need of upgrading, clean up, policing, and it was under certain affirmative lights regarded as a 'self-help' city (Hake, 1977). To call it 'ours' was an important declaration of belonging – 'we belong' to these streets (in the Central Business District (CBD), the main city), and 'these streets and stories' belong to us. In other words, they placed themselves at the centre of the story, the story of Nairobi as it is today, as it is being frequented today by fellow age-mates and other visitors wanting to learn about the *real* Nairobi, the *real* economy (MacGaffey, 1991), but also rendered highly curated and stylised in its performance.

They perform ordinary everyday life as a kind of magic realism: rendering the ordinary a cosmopolitan ghetto style, and conversely, what might have seemed unattainable becomes banal. Here there is a connection between the modes of cultural production that took place in the United Kingdom with grime music (Hancox, 2019) and that of *Genge* hip-hop in Nairobi (Kidula, 2012; Ntarangwi, 2009). These genres of hip-hop emerged in concert with the advent of digital platforms and social media. Young aspiring artists did not get 'signed' by producers but rather hacked their way onto the virtual stage through free-to-upload platforms, such as YouTube. They didn't need to sanitise their tracks; they could 'spit rhymes' that retained their subversive edge, and they didn't speak about the ascent to fame and success but instead provided a real-time, graphic novel–style musical expression of 'road life' (Bakkali, 2019), retaining the raw and grit of everyday violence on the street and the gall of youth who have nothing to lose and a story to tell.

Nairobeez used their *mtaa* as their backdrop, stage and turf from which they produced video selfies. To an extent, they 'bluffed' (Newell, 2012) about exploits, but never about riches. They did not paint a picture of hard-core struggles against historical dispossession and segregation the way other hip-hop artists of the Black Atlantic have (Gilroy, 1993; Rollefson, 2017). Nairobeez did something different: they did not beautify the ghetto; they rendered the ghetto beautiful, just as it was. Its aesthetics were boasted as 'ghetto style', the creative restyling of *deep Sheng* generated new ways of constantly re-describing and re-affirming their attachment to place, particularly the importance of *this* place. It was fashioned as home, and the *baze* became a hyper-localised expression of global youth culture in its hybrid, spatially open, unfinished forms (Biehl and Locke, 2017; Gilroy, 1993; Massey, 1998).

> The worlds on edge, and the open-endedness of people's becoming, is the very stuff of art.
> (Biehl and Locke, 2017, p. ix)

On one occasion, Donga posted a selfie with his mother, who some twenty years ago would have been a teen mother like so many young women in Mathare (Thieme, 2016). Donga's mother looked tired and didn't smile for the selfie with her son, but she did stare back at the phone camera, acknowledging the moment. She was between morning chores, and the corrugated metal shack they live in is featured in the back, with a line of laundry hanging, baying across the frame. This photo, posted on Instagram and Facebook, could be read as a curation of 'this is what it is'. It was not like Tracey Emin's *The Bed* because nothing about that 'ghetto' scene was meant to appear like a mess or unkept. It exposed on the one hand the mundane hardships of life in a neighbourhood where even the prosaic cup of *chai* requires hours of labour and material cost, and where youth are increasingly targets of police brutality and extra-judicial violence (van Stapele, 2016) and systemic under-employment. Thus these millennial self-proclaimed 'ghetto boyz', who have grown up in Nairobi's popular neighbourhoods with a smartphone in their pockets, know how to reach a world beyond their streets while appearing ever-more tied to, and proud of, their *baze* from where they project shared struggles and solidarities with fights against (not only coronavirus), especially, the 'hunger virus', and since June 2020, localising their allegiance to the global movement against police brutality against young black bodies by posting selfies with makeshift masks that read, 'Kenyan Black Lives Matter'.

Note

1 I have worked and conducted research in Kenya since 2005, with a concentrated period of ethnographic research between 2009 and 2010 and subsequent trips in 2011, 2012, 2016, 2017 and 2019.

References

Bakkali, Y. (2019) 'Dying to live: Youth violence and the munpain'. *The Sociological Review*. https://doi.org/10.1177/0038026119842012

Bhabha, H. K. (2004) *The Location of Culture* (2nd ed.). Routledge: London, New York.

Biehl, J. and Locke, P. (2017) *Unfinished: The Anthropology of Becoming*. Duke University Press Books: Durham.

Caldeira, T. P. (2017) 'Peripheral urbanization: Autoconstruction, transversal logics, and politics in cities of the global south'. *Environ Plan D*, 35, 3–20. https://doi.org/10.1177/0263775816658479

Clifford, J. and Marcus, G. E. (Eds.) (2010) *Writing Culture: The Poetics and Politics of Ethnography* (2nd ed., 25th anniversary ed.). University of California Press: Berkeley Los Angeles London.

Cooper, E. and Pratten, D. (Eds.) (2015) *Ethnographies of Uncertainty in Africa, Anthropology, Change, and Development*. Palgrave Macmillan: United Kingdom. https://doi.org/10.1057/9781137350831

Fredericks, R. (2014) '"The old man is dead": Hip hop and the arts of citizenship of Senegalese youth'. *Antipode*, 46, 130–148. https://doi.org/10.1111/anti.12036

Geertz, C. (1977) *The Interpretation of Cultures* (new edition). Basic Books: New York.

Gilroy, P. (1993) *The Black Atlantic – Modernity and Double Consciousness* (1st ed.). Verso Books: London.

Githiora, C. (2018) *Sheng: Rise of a Kenyan Swahili Vernacular*. James Currey: Woodbridge, Suffolk.

Hake, A. (1977) *African Metropolis: Nairobi's Self-Help City* (1st ed.). Sussex University Press: London.

Hancox, D. (2019) *Inner City Pressure: The Story of Grime*. William Collins: London.

Jeffrey, C. and Dyson, J. (2008) *Telling Young Lives: Portraits of Global Youth*. Temple University Press: Philadelphia, PA.

Katz, C. (2004) *Growing Up Global* (1st ed.). University of Minnesota Press: Minneapolis.

Kidula, J. N. (2012) The local and global Kenyan rap and hip hop culture, in *Hip Hop Africa: New African Music in a Globalizing World*. Ed. Eric Charry. Indiana University Press: Indiana, pp. 171–186.

MacGaffey, J. (1991) *The Real Economy of Zaire: The Contribution of Smuggling and Other Unofficial Activities to National Wealth*. Pennsylvania: University of Pennsylvania Press.

Massey, D. (1998) Spatial Construction of Youth Cultures, in Skelton, T., Valentine, G. (Eds.), *Cool Places: Geographies of Youth Cultures*. Routledge: London, pp. 121–129.

Newell, S. (2012) *The Modernity Bluff: Crime, Consumption, and Citizenship in Côte d'Ivoire*. University of Chicago Press: Chicago.

Ntarangwi, M. (2009) *East African Hip Hop: Youth Culture and Globalization* (1st ed.). University of Illinois Press: Urbana.

Parnell, S. and Pieterse, E. (2014) *Africa's Urban Revolution* (1st ed.). Zed Books Ltd: London.

Richardson, C. and Skott-Myhre, H. A. (Eds.) (2012) *Habitus of the Hood*. Intellect Books: Bristol.

Rollefson, G. (2017) *Flip the Script*. University of Chicago Press: Chicago.

van Stapele, N. (2016) '"We are not Kenyans": Extra-judicial killings, manhood and citizenship in Mathare, a Nairobi ghetto'. *Conflict, Security & Development*, 16, 301–325. https://doi.org/10.1080/14678802.2016.1200313.

Thieme, T. (2016) 'The ghetto will always be my living room' in Hustling and Belonging in the Nairobi Slums, in *Rethinking Life at the Margins: The Assemblage of Contexts, Subjects, and Politics*. Chapter 7. https://doi.org/10.4324/9781315606118-16.

Thieme, T.A., (2015) 'Turning hustlers into entrepreneurs, and social needs into market demands: Corporate–community encounters in Nairobi, Kenya'. *Geoforum*, 59, 228–239. https://doi.org/10.1016/j.geoforum.2014.11.010

Thieme, T. A. (2018) 'The hustle economy: Informality, uncertainty and the geographies of getting by'. *Progress in Human Geography*, 42, 529–548. https://doi.org/10.1177/0309132517690039.

Weiss, B. (2009) *Street Dreams and Hip Hop Barbershops: Global Fantasy in Urban Tanzania*. Indiana University Press: Bloomington.

White, J. (2020) *Terraformed: Young Black Lives in the Inner City*. Repeater Books: London.

17

RETHINKING GLOBAL URBANISM FROM A 'FRIPE' MARKETPLACE IN TUNIS

Katharina Grüneisl

Introduction

The entrance to the *Bab El-Falla* market in the historic inner-city of Tunis is congested with handcarts selling oranges, with elderly vendors offering fresh herbs on cardboard boxes and shoppers trying to push into the main market street with heavy use of elbows. Further up the *Bab El-Falla* street, butcher shops and fruit and vegetable stalls give way to shops and stalls selling second-hand women's underwear, sports equipment, shoes and bags, or what is more generally referred to as '*fripe*' in Tunisia. Trade in imported second-hand garments and other used objects dominates the *Bab El-Falla* market today and more than 200 active trading spaces[1] make it one of the capital's most prominent fripe shopping destinations. Customers from across Tunis come to the *Bab El-Falla* market, especially on Tuesday and Saturday mornings when traders disclose new fripe merchandise.

This moment of disclosure is referred to as bale opening, or '*halen al-bala*' in Tunisian Arabic. Bales are the tarpaulin bags in which different types of imported fripe materials are packaged after having been graded in sorting factories. The bales arriving in the *Bab El-Falla* marketplace are labelled, indicating product type and a value category, but the exact content of the non-transparent tarpaulin packages remains unknown to both traders and customers. This uncertainty, combined with the promise of finding the best pieces when given the chance to inspect the bale first, provides the basic rationale for participating in the bale opening. As the fripe has developed into a popular consumer choice across different age groups and social classes in Tunisia, attending bale opening events in the capital city's ubiquitous fripe marketplaces has become an integral element of many people's weekly routines.

This chapter takes the urban situation of the bale opening as a starting point for exploring how collective encounters with diverse second-hand objects mediate articulations of 'global forms' in the city. Discarded in Europe and North America as 'excess matter' (Gidwani and Reddy, 2011: 1649), many imported fripe materials lack the stable descriptors of value and meaning that habitually guarantee the seamless transfer of standardised commodities from one context to another. Therefore, diverse processes of valuation have to precede the exchange of fripe objects in the Tunis marketplace, and the ritualised performance of the bale opening, which confronts traders and consumers with continuously changing flows of transnationally mobile

DOI: 10.4324/9780429259593-20

objects, constitutes one such moment of valuation. In the unfolding negotiations, fripe objects are incorporated, adapted or reinvented as 'contingent global forms', with the outcome always dependent on the distinct socio-material constellations of the bale opening. Building on cultural economy approaches that have relativized notions of the commodity as a stable global form (Gregson and Crang, 2015; Miller, 1998), this chapter focusses on the contingent remaking of the fripe in situated processes of cultural production in a Tunis marketplace.

Investigating the contingent emergence of the fripe as a global form in the bale opening demands a rethinking of agency in the making of diverse global urbanisms in contemporary Tunis. Urban marketplaces like *Bab El-Falla*, which are constituted by global flows and simultaneously constitutive to the dense socio-spatial relations that assemble the city, offer a privileged terrain for attending to global urbanisms as necessarily plural and rooted in specificity (Robinson and Roy, 2016). 'Global urbanism' has often been approached from the vantage point of allegedly universal global forms, reaching from urban mega-projects, to urban policies or finance mechanisms that circulate between cities in a quest for world-class city-making (Goldman, 2011; Roy and Ong, 2011; Watson, 2014). In this chapter, it is precisely the uncertainty surrounding the fripe as a circulating global form that becomes the starting point for rethinking where and how global urbanisms emerge and who takes agency in shaping diverse understandings of the global in contemporary Tunis. This opens up two particular lines of thought. First, an empirical focus on the fripe emphasises the role of seemingly banal, material circulations in triggering negotiations over connections and differences in a globalised world, resulting in the contingent production of new global forms in ordinary spaces of the urban economy. It thus draws attention to the role of things in establishing global linkages, here between distant locations of production, discard and sorting in the Global North and South (Brooks, 2015). Second, it sheds light on global urbanism, not as a hegemonic project emanating from world-class city-making, but it approaches the 'global' as an inherently contested category that emerges from situated urban constellations and is thus necessarily plural and always provisional.

In the first section, this chapter analyses the coordinated mechanisms of valuation that requalify imported fripe materials as commodities in the Tunisian context and culminates in the strictly timed rhythms and collective performances of the bale openings in Tunis marketplaces. The second section then homes in on the complex negotiations for which the bale opening sets the stage: processes of incorporation, adaptation and reinvention determine the contingent character of the fripe as a global form and illustrate how urban marketplaces become the sites of encounters with 'elsewhere'. The ensuing cultural translation work opens up new connections but also reasserts difference and thus comes to determine articulations of 'the global' at the urban micro-level.

The bale opening: Mediating encounters with contingent global forms

The collective urban event of the bale opening sets the frame for negotiating the fripe as a contingent global form and thus becomes a crucial site for the making of diverse global urbanisms. From the arrival of the fripe in the Tunis container port, various processes of valuation are designed to return the imported materials to their 'commodity situation' (Appadurai, 1986: 13). The bale opening constitutes the culmination of a sequence of different moments of valuation and marks the collective encounter with globally circulating fripe materials in public urban space. The rituals and performances that frame interactions during the bale opening help constitute a temporary form of the urban public that collectively navigates the "sticky materiality of practical encounters" through which global connections become tangible and intimate (Tsing, 2011: 1).

The fripe materials that arrive in the Tunis port have been "cast out in a particular space-time" (Gidwani and Reddy, 2011: 1636), either as charitable donations or as unsold or damaged stocks of fast-fashion labels in Europe and North America. Unlike other commodities for which value after production is stabilised through precisely defined properties and legally binding product standards (Çalışkan and Callon, 2010: 6), fripe imports lack such stable descriptors in their circulation from one value regime and context of use to another. Container loads of fripe materials, usually comprising a heterogeneous assemblage of clothing donations, unsold stocks and often trash materials, are transported from the container port to one of the 47 sorting factories that operate in Tunisia today.[2] Specialised in the systematic grading and packaging of imported fripe materials, these factories not only sort garments and used objects into distinct value categories for the Tunisian market but also prepare a part of the fripe materials for re-export, both onward to sub-Saharan Africa or back to Europe.[3] In Tunis, most packaged fripe merchandise is stored and redistributed from the wholesale quarter *Zahrouni*, with different price segments and types of fripe merchandise being delivered to the highly specialised and differentiated commercial landscapes that have emerged in the Tunisian capital city in recent years. Especially through a largely unchecked increase in fripe imports during the temporary power vacuum after Tunisia's 2011 revolution, the small North African country has today become one of the largest fripe importers on the African continent,[4] a development that has translated into a rapid proliferation of fripe marketplaces in the capital city.

Each fripe market in the capital city of Tunis follows a distinct, habitually biweekly, rhythm of bale opening. Albeit unwritten, the temporalities of bale openings are strictly adhered to in different marketplaces and determine the rhythms of delivery and payment between sorting factories, wholesalers and retailers in the fripe business. Accordingly, the working routines of market traders are organised around the biweekly turnover of merchandise, and customers time their visits to different markets, and thus their patterns of mobility across the city, according to the bale openings. The rhythm of the bale opening has thus become a structuring element not only for the fripe economy but also for urban life, determining people's consumption routines, strategies of navigating the city and weekly family outings.

The joint rhythms of the bale opening are crucial to generating the socio-material densities required to constitute a public urban event. In the *Bab El-Falla* market, newly delivered fripe bales are disclosed to crowds of customers and curious spectators on Tuesday and Saturday mornings according to a strictly adhered to theatrical performance. Rehearsed interactions between fripe traders, customers and spectators give the bale opening the character of a collective performance that is open to public participation, yet also functions on the basis of intimate webs of relationships, insider codes and favours between traders and loyal clients. Tacit rules govern the co-production of value and meaning around heterogeneous fripe materials during the bale opening, and the type of product disclosed in part determines the unfolding of the event: while some bales contain carefully graded garments of comparable product type and value for which fixed prices can be set, others are filled with heterogeneous objects that remain incomparable and whose value and meaning must be established in drawn-out negotiations.

For bales graded according to garment type and value category in the sorting factories, the bale opening process habitually occurs in the following manner: A trader first displays the closed bale on an empty table positioned in front of a shop space and waits for customers to position themselves around the bale. Once an atmosphere of anticipation is palpable around the stall and a sufficient number of people have gathered, the trader lifts up a knife as a symbolic starting sign and cuts the plastic cord of the bale. While unfolding the tarpaulin, laying bare the folded and pressed clothes, the trader announces a standard price for the garments contained within the bale and begins to pull them apart one by one. None of the customers touch the clothes before the trader, and the process of opening is often intentionally prolonged to enhance theatrical tension.

Only once the trader throws garment items in the middle of the table, can those surrounding the stall pick them up, and rapid hand movements decide who is the first to inspect a clothing item. The opening of garment bales often unfolds in concentrated silence, and communication is limited to routinised gestures and rapid transactions.

In stark contrast, the disclosure of bales with more heterogeneous arrays of garments, or of tarpaulin bags containing an eclectic mix of second-hand objects, results in tedious processes of negotiation and collective valuation. In such instances, the sheer heterogeneity and often-alien character of fripe materials contained within the same bag render uniform price setting impossible. Vague sorting categories for fripe objects, such as 'accessories' or 'kitchenware', often contain a wide array of objects with uncertain origin, identity and use value that can only be requalified through collective processes of guessing, researching and creative reinvention. Around a market stall specialised in fripe toy items in *Bab El-Falla* for instance, the process of bale opening tends to drag on over the entire morning. Many of the toys that are being pulled from the tarpaulin bag carry inscriptions in foreign languages or are completely unknown in the Tunisian context, rendering the collective moment of opening – and the ensuing debate about value and use – a precondition to transaction and exchange. The stall's location right by the market entrance enhances the public character of the event, inviting passers-by to participate and complement the knowledge of loyal customers who bring chairs from a nearby café to take comfortable front-row positions.

Repeated interactions over time engender an intense sociality that turns the bale opening into a popular pastime, described by some female customers in *Bab El-Falla* as "*an addiction*" or "*a weekly sport*".[5] The tacit rules and codes of behaviour that frame interactions during the bale opening set it apart from other urban situations, turning it into "a stage on which people can assume different personas" and where "the theatrical and the carnivalesque" (Gregson and Crewe, 2003: 61) enter everyday practices of consumption. People thus often attend bale openings simply as curious spectators or commentators, without the intention of purchasing. The line between distant observers following the bale opening like a dramatic street performance and participants taking an active role in the collective performance thereby often blurs, guaranteeing the public character of the event and its continuously evolving socio-material constellations.

Incorporation, adaptation and reinvention: Remaking global forms in the marketplace

The public event of the bale opening functions like a stage for the "cultural work of valuation" (Kirsch, 2013: 436) that vendors, shoppers and spectators perform around diverse fripe objects. Its socio-material density is crucial for generating situated encounters that open up new connections between the marketplace and a distant 'elsewhere' but can also cause moments of rupture that articulate difference. Homing in on the negotiations emerging around market stalls trading in heterogeneous fripe objects, such as children's toys, sports equipment and exotic dress, this section disentangles processes of incorporation, adaptation and reinvention that define the fripe as a contingent global form.

Most second-hand objects that are pulled from the tarpaulin bags require a degree of "consumption work" (Hansen, 2000: 257) to be incorporated into new contexts of use in Tunisia. This is so because fripe garments and objects often carry "thick layers of inscription" (Appelgren and Bohlin, 2015: 160), obliging consumers to engage in tedious processes of sorting, testing and evaluation as part of their everyday consumption practices. Material inscriptions like traces of wear and tear or faulty cuts render apparent the complex "economic and social biographies" (Kopytoff, 1986: 67) of different fripe objects and add layers of complexity to performances of valuation during the bale opening. Routinised probing and inspection underline the need for an explicit and embodied engagement with the material qualities of individual fripe objects and link everyday fripe consumption to experience, tacit knowledge and a collective repertoire of skill and practice.

In addition to material traces of previous use or damage, the fripe objects carry multiple layers of immaterial inscriptions, reflecting consumption habits, stylistic preferences and cultural idioms of a distant 'elsewhere'. Collective processes of valuation in the bale opening thus result in "subtle and discriminatory processes of incorporation and rejection" (Miller, 1998: 18) that produce contingencies in the ways in which global fripe circulations enter local contexts of use and consumer cultures. The direct link between accelerating trend and consumption cycles in Europe and North America (Ellen MacArthur Foundation, 2017: 3) and the fripe merchandise arriving in *Bab El-Falla* complicates processes of incorporation. Hamdi, who owns a stall specialising in second-hand kitchen utensils in *Bab El-Falla*, complains that the merchandise he buys from wholesalers is becoming increasingly unpredictable and that he runs ever-greater risks by "buying bales blind" (Brooks, 2012: 231). For example, Hamdi recounts finding himself swamped with drinking bottles and takeaway thermo coffee mugs in different shapes, colours and sizes. Every new delivery from the wholesalers contains various drinking containers, and the bottles and mugs are accumulating on the shelves of his small shop space. As Hamdi explains, most inhabitants of Tunis abstain from drinking tap water and coffee is consumed in small glass or carton cups, so the Western takeaway containers have little use in the local context. In order to incorporate new merchandise into local consumer habits, Hamdi relies on inventive sales strategies – for instance, by joining forces with a nearby sports equipment store that has a younger clientele with an interest in plastic drinking containers.

While some fripe materials can be incorporated into local use contexts through inventive sales strategies, other objects can only be requalified through the collective moment of valuation in the bale opening. Dense interactions between objects, traders, customers and spectators open up possibilities for the adaptation or complete reinvention of fripe objects in the Tunisian context. On one Saturday morning, Khaled, who sells second-hand sports equipment, pulls a dark leather bag for golf clubs from the tarpaulin bag. Perplexed by the strange shape and material of the object, he passes it on for inspection to the dense crowd surrounding the stall. Divergent interpretations are offered for what the object might have originally been, and a young man consults his smartphone to prove the hypothesis that the bag belongs to a set of hunting equipment. An elderly man decides he could use the container to store his fishing equipment, but disagreement arises about its price as the high-quality leather is unsuitable for wet equipment. By creatively adapting a seemingly useless object like the golf bag to the new context of use, the bale opening situation allows objects to "become something else" (Kirsch, 2013: 438) without undergoing any major physical alteration. While in some cases, knowledge of distant places through the experience of "*l-a'isha barra*" (life abroad) confers interpretive authority to bale opening participants, imaginaries for creative adaptation or complete reinvention often trump the value of 'correct information' on original contexts of use.

Incorporation of the fripe into local contexts of use often occurs over time, as it necessitates incremental processes of appropriation of seemingly stable global forms. Innovative combinations of diverging stylistic preferences or adaptations of garments from elsewhere to local norms and practices of dress thereby often produce new and hybrid cultural forms. Second-hand Indian saris, which were first imported as parts of second-hand donations from European and American cities with sizeable Indian diasporas, provide a striking recent example. Over the past three years, fripe traders report, Indian saris rapidly gained popularity in Tunisia, as young women began to wear the saris as part of their wedding outfits, consisting of multiple dresses for different parts of the festivities that usually stretch over several days. As a result, specialised shops for second-hand Indian saris have now proliferated in fripe marketplaces like *Bab El-Falla* and imports of second-hand Asian dress from the United Arab Emirates cater to the rapidly growing demand. Worn in combination with traditional Tunisian dresses, scarves and headwear, the saris have morphed into a new kind of cultural form through their adaptation to religious norms of 'modest dress' and their gradual integration into local consumer cultures. The circulation of second-hand saris is accompanied by the ever-more prominent circulation of imagery from Bollywood movies and Indian series that have gained a large following in Tunisia in recent years. The dress thus functions as a 'global form' – providing desirable linkages to a culture that remains exotic and fascinating to many Tunisians – and simultaneously provides an affordable way of complementing wedding outfits in a way that has become locally acceptable.

Despite the unparalleled creative energy unleashed in many bale opening events, collective valuation also frequently fails and fripe objects consequentially remain 'out of place' or are discarded. Not only physical traces of wear and tear but also immaterial inscriptions or symbolic layers of meaning can prevent fripe objects from entering local registers of value. Hakim, who owns a small shop selling accessories in *Bab El-Falla*, hides away certain fripe objects that he has screened out from the deliveries he receives from wholesalers. As he explains, the objects could "*shock clients*", as he deems them culturally or religiously unacceptable in the Tunisian context. Next to jewellery with satanic symbols, Hakim keeps a large box filled with little owl figurines in the back of his shop. He explains that the owl represents *"'ramz mta'a al-nahs' (a bad omen) in Tunisia"*. Unable to resolve the clash of meanings – here between a symbol of wisdom in Europe and a bad omen in Tunisia – Hakim recounts how innumerable similar moments of failed translation have often forced him to discard part of his merchandise. Rather than establishing new connections between a distant 'elsewhere' and the *Bab El-Falla* fripe market, fripe objects like the owl figurines reinforce feelings of alterity. Failed moments of valuation render apparent the fripe's contingent status as a 'global form' and transform the 'consumption work' performed in the bale opening into a 'consumer politics' that can become the starting point for articulating difference. The refusal to incorporate particular fripe objects thereby often becomes politically significant, triggering debates about identity, social norms or diverging living standards. The quotidian consumer politics of the fripe are thus multi-layered, as they lie embedded in the deep global inequalities structuring the second-hand economy and simultaneously respond to local political concerns and the immediate calculations of the Tunis marketplace. As sites of both connection to and rupture with 'elsewhere', the fripe marketplaces blur clear-cut dichotomies between 'the local' and 'the global', remaking scales and allowing both notions to take on new significance.

The outcomes of the negotiations that unfold during the bale openings are always in part unpredictable and can result in immediate incorporation or rejection, as well as gradual processes of adaptation or reinvention. In the 'friction' (Tsing, 2011: 3) between unstable global forms and situated urban constellations, hybrid cultural forms emerge, altering material practices, habits of dress or stylistic preferences. Over the years, collective processes of cultural production in the bale openings have also heightened the overall acceptance of the fripe as a consumer good. From a poverty handout, the fripe marketplace has today become a popular destination for urban dwellers across all social classes[6] and fripe shopping is considered "*a particular Tunisian skill (...), not something anyone would be ashamed of*".[7] It is this prominence that qualifies the fripe marketplaces as key sites for deliberating the continuous remaking of understandings and experiences of 'the global' in contemporary Tunis.

Conclusion

As recurrent public events taking place all across Tunisia's capital city, the bale openings mediate "messy and surprising encounters across difference" (Tsing, 2011: 3) and thus render tangible and meaningful connections to a distant 'elsewhere' in ordinary consumption spaces of present-day Tunis. The rhythms, socio-material densities and rehearsed performances of the bale opening form the necessary backdrop for complex processes of collective valuation that incorporate, adapt or reinvent the fripe as a contingent global form. The particular moments of friction that emerge in the bale opening posit urban marketplaces as important sites of cultural production that "sublate general forms back into specificity" (Miller, 1998: 185) and thus articulate global forms with situated consumption practices, stylistic preferences and registers of value.

Rethinking global urbanism from the perspective of the bale opening challenges the notion of a singular and hegemonic 'global urbanism'. Backed by international finance, and in the case of Tunis investment from the Arab Gulf states, this hegemonic 'global urbanism' has proliferated as a powerful urban development imperative, as is visible, for example, in the Tunis lakefront developments that have transformed parts of the city since the mid-2000s (Barthel and Signoles, 2006). This chapter has instead foregrounded diverse "genres of global urbanism" (Robinson and Roy, 2016, 183), some of which mirror and some of which contest dominant imaginaries and aspirations associated with 'the global city' in the singular. The Tunis fripe marketplaces, and more specifically the events of the bale opening, allow for a close-up engagement with the highly contingent production of 'global forms' at the urban micro-level. Quotidian processes of cultural production in everyday spaces of the economy thus provide a fruitful starting point for deliberating how 'the global' is articulated with, and thus becomes a constitutive part of, the socio-material relations that constitute the contemporary urban.

Notes

1 This number fluctuates, and four counts of active trading spaces in *Bab El-Falla* in October and November 2018 produced an average of 218 active trading spaces.
2 The regime of Zine El-Abidine Ben Ali in Tunisia accommodated the rapidly expanding global second-hand trade through preferential legislation in 1995, authorising 50 fripe sorting factories to operate in Tunisia.
3 Used garments are often re-exported to sub-Saharan Africa and rag materials to Asia, while industrial cloth and other recycling materials are exported to Europe, alongside high-end vintage fashion.
4 Data from the Observatory for Economic Complexity on used clothes imports in Africa between 2015 and 2017 (https://atlas.media.mit.edu/en/visualize/tree_map/hs92/import/show/all/6309/2015/), accessed July 2, 2019.
5 Conversations during three accompanied bale opening events with different women's groups in *Bab El-Falla* in June and July 2018.
6 According to an unpublished survey of the National Consumption Institute of Tunisia in 2018, 87% of young women and 72% of young men buy regularly or exclusively from the fripe marketplace.
7 Interview with a female shopper in *Bab El-Falla* who attends bale openings once a week, Tunis, May 15, 2018.

References

Appadurai, A. (Ed.) (1986) *The Social Life of Things: Commodities in Cultural Perspective* (11. Print). Cambridge: Cambridge University Press.

Appelgren, S. and A. Bohlin (2015) '"Growing in Motion". *Culture Unbound, Linköping'. University Electronic Press*, 7: 143–168.

Barthel, P.-A. and P. Signoles (2006) *Tunis en projet(s): la fabrique d'une métropole au bord de l'eau*. Rennes: Presses universitaires de Rennes.

Brooks, A. (2015) *Clothing Poverty: The Hidden World of Fast Fashion and Second-Hand Clothes*. London: Zed Books.

Brooks, A. (2012) 'Riches from rags or persistent poverty? The working lives of secondhand clothing vendors in Maputo, Mozambique'. *TEXTILE*, 10(2): 222–237.

Çalışkan, K. and M. Callon (2010) 'Economization, part 2: A research programme for the study of markets'. *Economy and Society*, 39(1): 1–32.

Ellen MacArthur Foundation (2017) *A new textiles economy: Redesigning fashion's future*, 28 November 2017, https://www.ellenmacarthurfoundation.org/publications/a-new-textiles-economy-redesigning-fashions-future.

Gidwani, V. and R. N. Reddy (2011) 'The afterlives of "waste": Notes from India for a minor history of capitalist surplus'. *Antipode*, 43(5): 1625–1658.

Goldman, M. (2011) 'Speculative urbanism and the making of the next world city: Speculative urbanism in Bangalore'. *International Journal of Urban and Regional Research*, 35(3): 555–581.

Gregson, N. and M. Crang (2015) 'From waste to resource: The trade in wastes and global recycling economies'. *Annual Review of Environment and Resources*, 40(1): 151–176.

Gregson, Nicky, and Louise Crewe. 2003. *Second-Hand Cultures*. Materializing Culture. Oxford: Berg.

Hansen, K. T. (2000) 'Other people's clothes? The international second-hand clothing trade and dress practices in Zambia'. *Fashion Theory*, 4(3): 245–274.

Kirsch, S. (2013) 'Cultural geography I: Materialist turns'. *Progress in Human Geography*, 37(3): 433–441.

Kopytoff, I. (1986) The cultural biography of things: Commoditization as process, in *The Social Life of Things: Commodities in Cultural Perspective*, edited by Arjun Appadurai, 64–91. Cambridge: Cambridge University Press.

Miller, D. (1998) Coca-Cola: A black sweet drink from Trinidad, in *Material Cultures: Why Some Things Matter*, edited by D. Miller, 169–187. Chicago: University of Chicago Press.

Robinson, J. and A. Roy (2016) 'Debate on global urbanisms and the nature of urban theory: Debates & developments'. *International Journal of Urban and Regional Research*, 40(1): 181–186.

Roy, A. and A. Ong (Eds.) (2011) *Worlding Cities: Asian Experiments and the Art of Being Global. Studies in Urban and Social Change*. Chichester, West Sussex; Malden, MA: Wiley-Blackwell.

Tsing, A. L. (2011) *Friction: An Ethnography of Global Connection*. Princeton: Princeton University Press.

Watson, V. (2014) 'African urban fantasies: Dreams or nightmares?' *Environment and Urbanization*, 26(1): 215–231.

18

LIMINAL SPACES AND RESISTANCE IN MEXICO CITY

Towards an everyday global urbanism

Alicia Lindón

This text examines the urban life of Mexico City in terms of spatial experiences that occur during the mobility of everyday life (Seamon, 2015) and the singularities acquired by the navigation of popular urban areas (McFarlane and Silver, 2017). This approach aspires to understand the city as fragments of unstable, discontinuous lived spaces in movement. The dense fragments of urban life under analysis may pertain to Mexico City but, despite their singularity, they have links with phenomena found in other cities. In other words, the analytical approach is one of global urbanisms (GU) seen by the subjects themselves, referred to herein as everyday global urbanism (EGU), and it embraces the spatial behaviours of the everyday city, with all their emotions and performativities – in short, the lived city. EGU focusses on microscales, even when small spaces operate in association with other more extensive spaces. These small spaces represent dynamic fragments of the city, but some of these fragments present a degree of stability, as they are thoroughly institutionalised, while others are extremely unstable. This particular perspective delves into a part of the city that is configured by fleeting, interstitial, liminal spaces that are inhabited by subjects in positions of subalternity and form part of the heterogeneity and complexity of the city today.

A focus on EGU in the margins and interstices of urban life is not at odds with "GU" – indeed, on the contrary, the exploration of only partially visible EGU is considered capable of enriching GU. It is worth remembering that urbanism can be defined as both a discipline that studies the city and as a profession that designs, organises, creates and intervenes in the city and its life. So, the global attribute provides, above all, an intensification of spatial connections (both near and far), a spatial-temporal simultaneity (Haesbaert, 2010) and an exacerbation of technology. GU is thus the study of – and intervention in – the city, with an emphasis on the simultaneity, interconnection and technologisation of urban life. These global characteristics do not fit, however, into the lives of a city's inhabitants or into micro-spaces in the same manner: GU is reconfigured in cities' different local and lived contexts, and it therefore, emerges through singular means as multiple EGU.[1] As EGU is derived from the repetition of certain practices (routinisation) in urban space by anonymous city dwellers – i.e., certain forms of social action within the city – it is not so far removed from the classic description of urbanism of the city as lifestyle (although it could be redefined today as "lifestyles", due to the plurality of the social subjects now living in cities).

DOI: 10.4324/9780429259593-21

As EGU directs the analytical focus onto the way people act in space, it seems appropriate to link these behaviours with actors' biographical time and with institutionalised projects (in the widest sense of the term) in which they become involved, as well as with the spatial forms of the city in which their actions unfurl.

The global nature of EGU can act on various levels. On the one hand, everyday spatial behaviours are impregnated with technology and therefore with all its inherent connectivity and simultaneity; on the other hand, the global dimension also embraces the many socio-cultural contexts in which the meaning of everyday behaviours is configured. Subjects experience a range of situations in these various socio-cultural contexts, which are thus interconnected via their biographies. For example, harassment of women on public transport usually gives rise to spatial behaviours that reflect other women's responses in distant places and can be reconstituted close to hand through social networks. Finally, it should be noted that the global condition of everyday urbanism can also be read in terms of the tension between a threatening alterity and a more trustworthy type of otherness, although urbanism usually rejects spatial segregation. The scope of GU undoubtedly extends beyond these limits, but this text confines itself to these particular considerations.

This line of thinking therefore maintains that the study of the city (as undertaken by urbanism) should involve understanding citizens' spatial experiences and that a city's design (similarly undertaken by urbanism) has a direct bearing on those who live in it. At the same time, people's spatial behaviours throw light on the city that urbanism seeks to understand and intervene in. Urbanists tend to assume that their work makes interventions in the city, but by the same token, its inhabitants also make interventions by dwelling in the city. These interventions are not always identified and interpreted, however, especially when they are fleeting and scarcely visible or tangible, and this is particularly true when they are made by subjects in a position of subalternity. This text focusses on them.

More specifically, two types of typical everyday situations on public transport will be considered: firstly, the configuration of liminal male spaces within the city's circulatory space marked out by the underground train system (the "Metro") and, secondly, female spatial experiences on daily journeys on public motor transport. This focus on spatial experiences on public transport reflects the central role of such everyday journeys in Mexico City, despite the spread of long-distance means of work and study and the rise in online shopping. Furthermore, both these windows on city life show that everyday journeys in today's urban world represent more than just routes, origins, destinations and specific technologies of transport and communication that structure the city. These journeys are made up of spatial experiences in which the city's inhabitants come into play. These experiences contribute to a daily making and remaking of the city through practices that serve to appropriate places and inscribe events from subjects' lives onto individual places.

Both these types of everyday situations confront and enrich the more technical and economics-driven viewpoints that are usually used to study contemporary cities: liminal spaces reveal everyday strategies to subvert the uses and functions established for specific areas in the global city. These strategies revolve around the intervention of corporeity and emotions. As Tôrres Gomes have observed, liminal spaces make manifest the various forms adopted by the living city, as opposed to the fixed city of regulations and urban functions (2009:100) established for abstract social stereotypes but not for flesh-and-blood inhabitants. Furthermore, the female spatial experiences on public transport examined herein show that, as a lived space, the spatially extended global city interconnected by systems of transport and communication can also diminish, threaten and restrict some of its inhabitants.

Male liminal spaces

Liminal spaces can be considered micro-territories or islets that become gateways to places different from the setting in which they originated. Sometimes they are fleeting; sometimes they endure; they are partially visible and extremely varied as regards the type of activities that they propitiate (crime, sex, entertainment, etc.). They are differentiated from their setting by the activities conducted within them, which allow participants to configure and appropriate these liminal spaces, making them their own by putting a location to unconventional use. Manuel Delgado (2007) has characterised such liminal or interstitial[2] spaces as fissures in a social tissue in the process of fracturing.[3]

These micro-territories are configured in various places within the city. Here the focus is on circulatory spaces in Mexico City and, more particularly on liminal spaces that have been created within the public Metro system for the sexual practice known as cruising (Turner, 2004) – i.e., looking for and finding anonymous gay sex in public spaces. The Metro has given rise to various liminal spaces of this type, sometimes in specific carriages of trains (particularly the last one, otherwise known as the "happy box"), especially on lines 1 and 2, as well as other spots within the system's infrastructure, such as specific corridors and stations (or parts of stations). These micro-territories are places on a living scale that briefly turn into liminal spaces for cruising. Their liminality integrates both the sense of an islet which subverts the established usage and the dimension of a threshold which, once crossed, brings a subject to a different space. And in terms of time, this liminality also corresponds to small fractions of time that are repeatedly recreated in various locations. In the case of the liminal spaces inside the Metro carriages, their temporality is linked to the time taken by a train to cover the distance between two stations: in other words, cruising follows the rhythm established by the speed of the Metro.

These liminal spaces are configured on two acts: that of the performativity of masculine bodies, through the dramatisation of roles (gestures, glances, movements) and, more immediately, that of the practices enacted therein. These liminal micro-spaces configure an insideness (Relph, 1976: 49–55) that is demarcated in two main ways: performatively and through the handling of the visibility/invisibility enabled by bodies and personal objects.

These liminal spaces seem to challenge the functionality of the transport infrastructure. More specifically, by establishing interstices for public male sexuality, they challenge the functionality of circulatory spaces designed to facilitate daily journeys on a massive scale, largely in conditions of anonymity.

The challenges to circulation represented by these spaces are only partial, however, as they occur within a favourable context of elements typical of the metropolitan lifestyle. These include the indifference shown to other people by regular travellers on public transport (and on the Metro in particular), along with a marked tendency towards self-absorption. Another factor that serves this liminality is the extremely high density of the circulatory spaces, resulting in physical proximity that leads travellers to put a great social distance between themselves and their neighbours. Another point that needs to be examined is the dependence of this liminality on objects present within the circulatory space and their location within that space, as it needs the inhabitants of Mexico City to travel every day with numerous personal objects required for a long day at work or in an educational establishment. So, this practice of carrying objects around can be explained by the length of everyday journeys within an extremely large city, and the inevitable consequence is the use of large backpacks to hold all these objects. The flipside of carrying this baggage is the fact that it offers the opportunity to manipulate objects and create a partial invisibility, behind which liminal micro-spaces of sexuality can be established within the circulatory space.

These liminal micro-spaces are phenomena that elude urban studies (Simone and Pieterse, 2017), although they only challenge the functionality of the circulatory space in an improvised manner, as they are configured intermittently as a result of the partial invisibility afforded by everyday objects and the generalised indifference and self-absorption of passengers on public transport. In short, they seem to be examples of the emerging forms of socio-spatial reproduction in the high-connection global city, albeit with subjects unconnected with their close neighbours but connected with other individuals with similar interests.

In other words, these situations transgress – from positions of subalternity – the prevailing urban dynamics but they do not seem to represent ways of reproducing the city socio-spatially in accordance with an urban life that provides everyday richness,[4] in the Lefebvrian sense; instead, they contribute particularities that transgress notions of functionality in an interconnected city. Nevertheless, the ephemeral, improvisational quality of these liminal spaces does lead to a destabilisation of the received wisdoms of urban logics, such as the purpose of everyday journeys being the arrival at a destination (e.g., a workplace, educational establishment or home). And this disruption of received wisdom gives rise to an alternative meaning of the everyday journey and the places involved, configured by the adventures that they can propitiate and the emotions that they can incorporate.

In this type of situation, everyday spatial behaviours (which give rise to specific EGU through repetition) transgress the functionality of GU, although they are remarkable for doing so with the temporality proper to the GU that is being challenged (ephemeral time, associated with contemporary acceleration). In other forms of social and spatial transgression, it is usual to encounter some of the features of the global viewpoint that is being rejected, as in the case of movements advocating slowness.

In these situations, technology is usually appropriated as a mechanism for visually recording experiences that are enjoyable or challenge the urban order. Meanwhile, social networks propitiate face-to-face encounters on journeys, even on a routine basis. These strategies are repeated on different journeys, with different subjects, and they thus become an EGU rather than being confined to unique experiences.

Another feature of these transgressive spatial behaviours is the creation of behavioural barriers that separate people with similar characteristics – and this is a way of reproducing social distances and forms of segregation between those who partake of this transgression and those who reject it.

This transgressive EGU has many facets – such as the functionality of the city – and it can be seen in young subjects on the lookout for fun and various sexual practices who also need to make daily journeys on public transport, as well as those who find an opportunity to make them.[5] The conditions pertaining to the subjects configure specific biographical situations. One of these conditions – the need/opportunity to move around every day on the transport system – gives rise to the intersection of the biographical circumstances of these young people with the institutional project of the transport system and the spatial forms used to construct this system. It is this intersection that gives rise to spatial behaviour that evolves into an EGU that seeks to be transgressive but has also configured itself within the context of this social group and the aforementioned circumstances.

Female spatial experiences on everyday journeys

In large contemporary cities like Mexico City, long daily journeys on public transport[6] and in private cars eat up a substantial portion of the day. In the case of public transport, the vehicle becomes a complex place in motion that gives rise to various components of global urban life, including asymmetrical power relationships, a range of otherings, a feeling of risk and fragility and an unequal appropriation of the public space. Everyday journeys on public transport are therefore not merely a transferal from one place to another via a particular means of transport since they also provide the context for innumerable confluences of spatial experiences.

Spatial experience is a multidimensional concept, but the following analysis views it from two specific dimensions: spatial orientation and routine practices. The former can be understood as a human being's cognitive capacity, manifested as the ability to recognise an environment and organise movements within it. Spatial orientation, therefore, gives rise to a primordial organisation of living space.

In terms of the female spatial experience of travelling within Mexico City on public motor transport – particularly for those young and middle-aged women who do so repeatedly because of their work or studies (and constitute the majority of female passengers) –, spatial orientation fluctuates between the indoors of the means of transport and the outdoors through which the urban space is crossed. In Mexico City, the indoors of public transport usually presents a high density of occupation, which restricts views of the neighbourhoods and streets on the route outside, or at the very least fragments them into discontinuous images. Sometimes visibility is totally lost for long periods, although it is more common to find an alternation between complete lack of vision and partial views outside.

Apart from the partial invisibility of the outdoors, the high density of travellers indoors on the public transport used by these women sets up undesired but unavoidable proxemics, as well as anonymous and threatening othering. This closeness oversteps the culturally accepted physical distances between strangers, as other bodies stand within women's narrow kinesthetic space. This unwanted invasion of kinesthetic distance is not only rejected and inevitable but also activates a state of spatial alert because it puts the women's very body into play, due to the transgression of social conventions as regards distances between strangers and the possibilities of harassment or pilfering of their belongings.

For these women, undesired contacts and occurrences amount to what Bégout calls "the experience of strangeness" (2005:26–27); this constantly reminds us that everyday life includes the unknown, which is disconcerting and calls certainties into question. Strangeness is thus linked to proximities that threaten the body itself. In this context, the other becomes a danger, and so female spatial experiences on public transport are imbued with hostility.

Being inside a public transport vehicle becomes an experience of private insideness that includes strangeness and unwanted proxemics, as well as an almost obligated state of alert. This experience of insideness is dense (Seamon, 2008), due to the accumulation of the factors involved and the negative evaluation of all of them: the sense of otherness being something dangerous, particularised distrust of close neighbours, sense of harassment, theft of personal belongings, violation of personal space, lack of views outside, etc. One traditional but still current female strategy in the face of these experiences is to board public transport equipped with small sharp objects, such as pins, that can be used as protective shields and as a mechanism for keeping other people at bay when the proxemics move from a latent to a manifest threat. This practice is complemented by the possibility that a woman can make a visual record with her own mobile phone. These female strategies in a hostile space-movement are derived from a fusion of the latest technological devices (such as a smartphone) and traditional resources (such as small pointed objects). Both are used as defensive tools for transiting a space that is hostile to women and is controlled symbolically by some men.

This type of spatial experience of the insideness of public transport is also intermittently connected to the outdoors, which provides women with spatial orientation due to various circumstances, such as a change in the landscape outside occasioned by the journey; the alternation of views and lack of views outside, according to variations in the density of occupation within the vehicle; and because the desired destination is outdoors, and this creates a need for visual contact in order to recognise it, almost always involving the heterogeneity inherent in the city's spatial forms. Spatial orientation is also sought outside because, in a situation of unwanted and dangerous insideness, outdoors represents a desired spatial expansion and a liberation from undesired insideness. This sense of freedom usually manifests itself when it is time to disembark from the vehicle in a micro-experience of liberation, however fleeting and ephemeral.

Within this tension in the orientation between indoors and an outdoors that intermittently appears and disappears, indoors comes out on top because its undesired proxemics and strangeness threaten a woman's very body and instil an immediate sense of distrust. Any attention to outdoors becomes secondary, even though the reason behind the journey, the destination, is outside, as is the fantasy of liberation, and the urban heterogeneity that can offer places of interest is similarly outside the vehicle.

Another dimension of the spatial experience of everyday transportation is the practices involved. It should be remembered in this respect that everyday practices develop through sets of spatially anchored micro-acts ordered in specific sequences that make up each subject's daily spatial-temporal trajectory (Pred, 1981, 1984). Micro-practices represent embodiment within the subject (Buttimer and Seamon, 1980): owing to the repetition that they have undergone, they have become encrusted in the body as capacities to perform certain movements in specific ways. Organised sets of embodied movements not only make it possible to carry out certain practices but these also acquire corporeity by forming part of a process by which bodies communicate something to other people. These embodied, routine micro-practices, made up of body movements and gestures, are associated with practical know-how that has resulted in embodied habits and makes it possible to perform everyday tasks in, for example, the work, domestic and educational fields, as well as in the process of travelling.

The female experiences considered herein involve constricting choreographies in these hostile space-movements. These women board a public transport vehicle, and their routine begins with the manipulation of the largest objects that they are carrying as protective shields between their own bodies and other people. They then look for a place inside for their bodies and subsequently perform *in situ* a series of body movements designed to achieve the most comfortable position possible in an unaccommodating context. Both the search for a niche for their bodies and their movements aimed at physical comfort are restrictive: the strategy involves reducing their bodies due to a lack of physical space and thus reducing the proxemics. The refunctionalising of personal objects as protective shields is an expansion of the boundary of the female body, albeit one devoid of physical sensitivity. Women emphasise the difference between their body practices in this setting and male practices, which involve expansive movements that use the body to occupy the largest amount of physical space possible.

As these routines configure the body into strategic niches inside a vehicle, they also therefore constitute, symbolically, constitute forms of resistance against the hostility of the threatening otherings that occur in everyday spatial mobility within a territorially widespread city.

These female spatial experiences demonstrate a particular way of making and remaking a life on the urban margins (Lancione and McFarlane, 2016) – and not just in terms of territories on the metropolitan periphery or empty areas within the city.

These experiences can occur in this type of territory, but the particular focus here is the reconstruction of life on the margins of urbanity by women who have established themselves in the space of public transport (due to their participation in the labour market, in education, in various commercial practices, in leisure activities, etc.) and have partially unyoked themselves from their domestic spaces, while also being obliged to adopt everyday interstitial practices in order to cohabit that notoriously hostile public space.

This type of urban experience demonstrates that women's widespread immersion in circulatory public space makes them part of GU (due to their considerable everyday mobility and active participation in labour markets, educational activities and, more generally, other smaller-scale fields of everyday life) and allows them to know more territories and inhabit the heterogeneity of the contemporary city. This situation does not, however, exclude a defensive cohabitation with highly traditional, hostile otherings. Ultimately, this presence of women in the public space of today's interconnected city is not always accompanied by an appropriation of that space, as they sometimes inhabit it with a sense of being out of place or of being in an alien space.

This presence could be interpreted as progress for women with respect to their right to the city or their participation in some kind of civic urbanism due to their immersion in circulatory space, but in fact, this is only partially true, as women experience the city as harassment, threat and hostility. This experience leads them to put up frontiers behind which they protect themselves; these can be fragile (a backpack) or immaterial (social distance), or they can be frontiers that reduce the visibility of their bodies (shrinking). The repetition of these behavioural strategies configures them into an EGU. The EGU of these women cannot dispense with either the demarcation of frontiers and distances or the insertion of barriers into the micro-spaces in which they move or the stigmatisation of the other: the EGU established in the circulatory space segregates, differentiates and demarcates, although, paradoxically, the GU from which this EGU emerges would be opposed to segregation and spatial exclusion.

These female spatial behaviours also fall back on technology – a characteristic of GU – to appropriate it as a recording mechanism that provides evidence of the harassment experienced. Similarly, social media are used along the same lines to broadcast the harassment and hostility that these women undergo onboard a vehicle.

These experiences can be understood in the context of the dense intersection occurring at a particular time between the public transport project established in the city (with its required configuration of circulatory spaces), the biographies of the women who live through this type of experience (always set within the need to travel to develop their work-related, educational or miscellaneous life projects) and the simultaneous appropriation of this circulatory space by other social subjects within this micro-space and circulatory time period.

Some final reflections

GU – the study of and intervention in the city, as well as globalised and interconnected urban life – usually constructs urban narratives removed from the living practices of a city's inhabitants (even when these narratives are in alignment with various theories and urban concepts), and it also makes interventions in urban space following criteria divorced from the everyday experiences of its inhabitants. One of the ways of ensuring that GU does not lose touch with the constant hubbub of urban living is by integrating the everyday. In this way, GU becomes less clearly defined as it presents itself in numerous EGU. This is the analytical logic underlying this text in which GU is transfigured into two EGU, but there could have been many more. This theoretical and methodological challenge gives rise to a panorama that is always open because the everyday unfurls in an infinite number of settings, and so there will always be EGU that go unstudied. The transfiguration of GU into various EGU is usually distinguished by the loss of some of GU's lines of force and the emergence of others (which may be directly opposed to the original ones).

This text has thus sought to examine some fleeting, spatially liminal and partially visible EGU because these usually provide a means to confront the reproduction of the city. The examination of a host of small-scale urban situations makes it possible to understand the different ways in which GU is appropriated by a city's inhabitants in their spatial practices. It is through everyday experiences that GU acquires singularity (or specific features) and new meanings in line with subjects' aspirations and limits, as well as with the situated nature of everyday practices.

At times, the various EGU can configure themselves on the basis of a search for adventure, fun or less clearly defined objectives, and at other times on the basis of the rationalities of means and ends. The liminal spaces of male sexuality examined herein integrate the search for adventure and fun, as well as affectivity (albeit ephemeral). In contrast, the female experiences demonstrate an EGU structured in accordance with the logic of means and ends: a long spatial journey is required to arrive at a workplace (or educational establishment).

The repetition of spatial behaviours gives rise to the configuration of various EGU. These are usually of a hybrid nature: some dimensions can be seen in global terms (as in the case of high everyday spatial mobility), while others are combined with traditional behaviours or other types, such as the combination of smartphones and sharp objects. The liminal spaces of male sexuality analysed here incorporate an ephemeral temporality corresponding to both the transport system used and the acceleration typical of global cities, but they also bestow a new meaning on the last carriage in the Metro by pursuing fleeting adventure. The female experiences examined, meanwhile, demonstrate female resistance to a hostile circulatory space in which women use the latest technology to record and bear witness to the hostility present therein; and, at the same time, they also face up to harassment with traditional tools, such as small, sharp objects, in order to keep threatening alterities at a distance. They confront and refuse to accept their situation via female spatial behaviour of a traditional kind, such as shrinking their bodies so as to occupy less space, and, at the same time, they make use of another behaviour of a global kind, such as technological recording and posts on far-reaching social networks. In other words, they configure a hybrid EGU that accepts differences and the insertion of protective barriers while also defending their right to the city and a civic urbanism free of harassment and hostility.

This hybrid nature is also evident in the female experiences studied through the incorporation of major everyday journeys (such as particular forms of the high spatial mobility proper to GU), but the women did not appropriate the places or manage to get to know the circulatory space – i.e., they presented the habitual female lack of knowledge of spaces remote from women's primary space.

Both of the situations on public transport examined herein show that, although Mexico City now appears to be totally interconnected, the physical proximity between its inhabitants does not represent any connection (or if it does, it is usually under a premise of violence). Amin and Graham (1999) have stated that places that are physically close tend to be relationally distant and relatively disconnected. The urban hyperconnectivity of today also involves this type of disconnection. An examination of emblematic spatial experiences in Mexico City, such as those analysed therein, shows that something similar occurs between subjects. One of the challenges facing our dispersed cities is a continued exploration of the urban fragments that some subjects manage to appropriate, thus inhabiting the richness of everyday urban life and configuring EGU that do not segregate and exclude, free from asymmetries that represent forms of domination.

These various EGU – along with many others not discussed here – can possibly be understood in greater depth if, apart from integrating the appropriation of the general prescripts of GU and aspects of particular situations, the biographies of the subjects involved are taken into consideration, as these tend to form the basis of connections with a range of places, social groups, projects and values. The biographies of the city's inhabitants, along with the memories that they carry within them, provide a means of integrating density into the everyday.

The liminal spaces and spaces of resistance covered here have become increasingly relevant to any deciphering of the contemporary city through the prism of a GU of the everyday because they reveal "lines of fire", reappropriations and resignifications of the formal, globalised city that sometimes anticipate unconsolidated forms of urban living or reflect surviving forms from the past or unexpected appropriations.

These unexpected fragments do not always constitute forms of everyday richness but at the very least they challenge global unidimensionality: ultimately, the recognition and analysis of dense, liminal fragments demonstrate the analytical potential of topological perspectives that explore the processes whereby biographical events are inscribed onto places and meanings, onto the values and memories accumulated in everyday spaces. It is no coincidence that topological factors have been so overlooked for so long. This oversight can be explained by the predominance of chorological approaches, which explore the fragmentation (McFarlane, 2018) of the extension, urban continuum and territory of cities rather than the fragments themselves.

These various EGU contribute to the city's continuous self-reproduction along the lines established by GU, while also giving rise to micro-territories within the city in which some subjects challenge the functionality or the prescripts implemented for spatial appropriation, and they thus create innovations. One of the paradoxes of these micro-territories of resistance and innovation is that they tend to be invisible (Nogué, 2006) to the other people who pass through them, because they do not possess the spatial awareness to see and decipher them, due to either indifference or symbolic barriers.

The text above examines only two types of liminal and hybrid everyday situations that occur in Mexico City, although they may also be found in other cities. There is still a need to examine other dense urban fragments, but the analysis herein does enable us to ask ourselves whether it is possible for GU to be singularised into EGU without any segregation or exclusion, any distrust of the other or any symbolic demarcations of lived spaces or insertion of partially visible barriers when cities are made up of a considerable multiplicity of subjects, encounters and spatial behaviours.

Notes

1 GU becomes singular (an intersection between the social and the individual) along lines similar to those found in disasters (an earthquake, a war, an epidemic): the event (or, more exactly, cluster of events) acquires specificities in different places and for different subjects. In other words, the experience of the event is not the same for every subject or in every place. Bearing in mind that GU is not dealing with one unique event that occurs at a given time but with chains of events that are prolonged over time, it is clear that its singular expressions will always be multiple.

2 Strictly speaking, a liminal space is not the same as an interstitial one. The former is situated on the borders of various phenomena, whereas the latter is a hollow or empty zone that can remain undisturbed for varying lengths of time. Both are capable of transmuting, however, as interstices can be found on borders (Águila-Flores, 2014).

3 In this respect, Delgado (2007:121) draws on Deleuze's "general theory of the crack", with its references to sudden losses of balance or fractures.

4 Lefebvre's richness of everyday life is associated with daily cyclical rhythms removed from the temporal linearity of rationality. Accordingly, the integration of playfulness and festivity into the everyday can represent richness because these factors do not operate within the linearity of rationality and technology (Lefebvre, 1981:18–22) and involve temporal rhythms which give rise to discontinuities that can trigger other social forms. Lefebvre observed that play, festivity and traditional rituals lose their transformative and enriching capacity when they are totally separated and differentiated from the everyday since the

possibility of renewing social life is thus taken away. They can only maintain their revolutionary capacity when they form an intrinsic part of the everyday.

5 The difference between a need and an opportunity to undertake these journeys lies in the fact that the former refers to journeys made with the specific aim of arriving at a particular place, albeit with an openness to adventure, while in the case of opportunity, adventure is the sole purpose.

6 The distances travelled usually range between 18 and 25 miles each way, with one or two changes in the means of transport. The average daily travelling time is five hours.

References

Águila-Flores, J-L (2014) *Espacio intersticial: Surgimiento y transformación, Caso: Tonalá, Jalisco en México*. Sevilla: Universidad Internacional de Andalucía.

Amin, A. and S. Graham (1999) Cities of connection and disconnection, in Allen, John, Massey, Doreen and Pryke, Michael (eds.), *Unsettling Cities*. London: Routledge/The Open University, pp. 7–53.

Bachelard, G. (2000) *La poética del espacio,* México: Fondo de Cultura Económica.

Bégout, B. (2005) *La découverte du quotidien*, París: Allia.

Buttimer, A. and D. Seamon (1980) (ed.) *The human experience of space and place,* London: Croom Helm.

Delgado, M. (2007) *Sociedades movedizas, Sociedades movedizas: Pasos hacia una antropología de las calles.* Barcelona: Anagrama

Haesbaert, R. (2010) *Regional-Global: dilemas da região e da regionalização na Geografia contemporânea,* Rio de Janeiro: Editora Bertrand Brasil

Lancione, M. and C. McFarlane (2016) "Life at the urban margins: Sanitation infra-making and the potential of experimental comparison". *Environment and Planning A: Economy and Space*, 48 (12): 2402–2421.

Lefebvre, H. (1981) *Critique de la vie quotidienne.* Vol. III: De la modernité au modernisme (Pour une métaphilosophie du quotidien). París: L'Arche Editeur.

McFarlane, C. and J. Silver (2017) "Navigating the city: Dialectics of everyday urbanism". *Transactions of the Institute for British Geographers*, 42(3): 458–471.

McFarlane, C. (2018) "Fragment urbanism: Politics on the margins of the city". *Environment and Planning D: Society and Space*, 36: 1007–1025.

Nogué, J. (2006) *La construcción social del paisaje*. Madrid: Biblioteca Nueva.

Pred, A. (1981) "Social reproduction and the time-geography of everyday life, *Geografiska Annaler, Serie B". Human Geography,* 63(1): 5–22.

Pred, A. (1984) "Place as historically contingent process: Structuration and the time-geography of becoming places". *Annals of the Association of American Geographers,* 74(2): 279–297.

Relph, E. (1976) *Place and Placelessness*. London: Pion.

Seamon, D. (2008) "Place, placelessness, insideness, and outsideness in John Sayles' sunshine state". *Aether: The Journal of Media Geography*, 3: 1–19.

Seamon, D. (2015) Lifeworld, place, and phenomenology: Holistic and dialectical perspectives, in T. Collins and N. Cronin (Eds.), *Lifeworlds: Space, Place, and Irish Culture*. Ireland: Cork University Press.

Simone, A. and E. Pieterse (2017) *New Urban Worlds: Inhabiting Dissonant Times*. Cambridge: Polity Press.

Tôrres, A. and E. Gomes (2009) "Espacios liminares – el frecuente enfrentamiento entre formas, contenidos y procesos de y en la gestión socioambiental del espacio urbano". *Revista de la Facultad de Trabajo Social UPB*, 25(25): 90–105.

Turner, M. (2004) *Backward Glances: Cruising Queer Streets in London and New York*. London: Reaktion Books.

Wright, J. K. (1947) "*Terrae Incognitae*: The Place of Imagination in Geography". *Annals of the Association of American Geographers,* 37: 1–15.

19

DEATH AND THE CITY

Necrological notes from Kinshasa

Filip De Boeck

Intro: Housing in heaven

During masses and prayer group meetings in the omnipresent Pentecostal-charismatic churches of Kinshasa, the capital of the Democratic Republic of the Congo (DRC) and the ethnographic backdrop of this chapter, the faithful are standardly urged by their preachers to make gifts to God. These acts of gift-giving (with gifts commonly known as *nsinani*, *mabonza* or *dîme*) refer to the Biblical idea of giving a tenth of your possessions up to God. The believers (*bandimi*) 'sow' these gifts or 'seeds' so that God can reward them tenfold. In this way, people 'sow' watches, jewelry, diamonds, money, and even cars to obtain a miracle from God: a job, a marriage, healing, children, prosperity, and, very often, a better house. In the latter case, preachers often tell their followers that

> God uses these gifts to build villas for us in Paradise. Those who do not contribute to God's work, and who refuse to give in church, will find themselves in an unfinished house when they arrive in Heaven! Therefore, I beg you, give freely. To give is to prepare your house in Paradise. To give is like sowing the seed. By giving to your preacher in church you will be able to harvest later. Giving is keeping.
>
> (De Boeck & Plissart, 2004: 202)

What this example in the context of the DRC's capital tells us is that, even though we know that gifts are always embedded in broader structures of obligation, the 'free' and spontaneous character that is supposed to underpin the very notion of the gift itself has come to be redefined as a calculated act in which 'giving' has become 'keeping'. Or more correctly, perhaps, the example shows us how 'gifts' have become hedges, investment positions that hold a future promise (in this case in one's afterlife) of house ownership for today's impoverished home- and houseless citizens. This 'hedging' thus already implies the limitation or qualification of one's housing dream by the living conditions and temporal exceptions that are constantly imposed upon urban residents in this kind of city environment. While the possibility of accessing decent housing emerges as a kind of anticipation inscribed in the present (cf. Nielsen, 2011; de Abreu, 2013), for the majority of Kinois who are permanently barred from accessing the kind of villas that spring up in Kinshasa's new gated

DOI: 10.4324/9780429259593-22

communities and satellite 'cities', the materialization of this housing dream is relegated beyond one's own death, and death itself becomes a condition to access housing and the city itself.

Ever since Jane Jacobs' classic *The Death and Life of Great American Cities* (1961), much has been written about the decline and death of cities, about urbicide and violence against cities, but far less about actual death as a producer of the city or bio- and necropolitics (Mbembe, 2003) as the motor of global contemporary urban transformations that shape up around what Martin Murray has recently termed 'urbanisms of exception' – i.e., a wide array of 'landscapes of accumulation' (Searle, 2016) that form zones of exclusion, allowing wealthy elites, whether linked to political power centers or large corporate entities, to carve out new urban zones in which urban life can unfold in ways that separate and set apart these elites from the vast majority of people who reside in the city and will never be able to access these new urban enclaves (Murray, 2017).

Anthropologist and social theorist Ernest Becker once conceived of capitalism as a contemporary search for the Holy Grail: immortality itself (Becker, 1973). Similarly, Baudrillard (1976) understood capitalism as resting on an obsession with the abolition of death. For him, capitalism tries to abolish death through accumulation, warding off ambivalence (associated with death) through value (associated with life). This, of course, is bound to fail, as African interpretations of capitalist accumulation obtained through the work of zombies put to work in the plantation have intuitively captured (cf. Geschiere, 1997: 156–158). Similarly, if a cultural fear of death is systemic in global capitalism, as Becker and Baudrillard argue, then the neoliberal cityscapes projected onto the African city (in this case Kinshasa) constantly point to the contrary: how global capitalism feeds off, and is constructed through, the very idea and tangible presence of death. The financial technologies underpinning neoliberal global urbanization mechanisms inventively use speculative and fictitious capital as a fetish construct in predatory cycles of accumulation by dispossession, destruction, death and resurrection that, as Harvey has argued so well, are at the heart of global urbanism's financial motor (cf. Harvey, 2013: 27ff).

By looking at how global urbanism is 'provincialized' (Sheppard et al., 2013) in Central African urban settings, this contribution offers some ethnographic examples of the ways in which current urban transformations and their financialization in Kinshasa are to a significant extent mobilized and operationalized through a 'thanatopolitics' (Murray, 2006) – i.e., a particular form of necropolitical governance of life and death that seems to form an inextricable part of global urbanism's nervous condition.

Displaced urbanism, urbanism of displacement

In Kinshasa, exclusive new urbanisms often take the form of a 'modernization as spectacle' (Bloom, Miescher & Manuh, 2014) in which various modes of specular modernity feed a powerful political theatricality and imaginary of the 'neoliberal city' (cf. De Boeck, 2011). This 'neoliberal city' refers to a particular kind of 'modern' urban reality and experience: a city where the policies and development plans are market-driven, where the production of space is based on the logic of increasing profits, with accompanying processes of gentrification and exclusion of the poorer urban masses from the urban core, while the social character and regulating role of the state is downplayed due to an increasing privatization of services (see Bayat, 2012). Often, the newly emerging segregationist infrastructural spaces that are called 'cities' operate as zones of what Easterling calls 'extrastatecraft', a portmanteau term that describes the "often undisclosed activities outside of, in addition to, and sometimes even in partnership with statecraft" (Easterling 2014: 15).

In the DRC, the coalition and collusion between statecraft and extrastatecraft to promote the emergence of the neoliberal city was fully realized during the reign of Joseph Kabila (2002–2018).

During the Kabila years, state propaganda presented the neoliberal city as an ideal model to make Congo 'the mirror of Africa' soon again, and that idea was sold in the form of a 'hypnotizing' spectacle (Figure 19.1).

The government promoted this new urban dream and the forceful promise made by this 'aesthetics of arrival' on the global stage (Kaur and Hansen, 2016) through a sophisticated ocular ground: images and billboards in strategic public spaces throughout Kinshasa, conjuring up particular urban imaginaries and forms of (globally exchangeable) urban infrastructure made visible through the workings of specific forms of speculative capital, such as venture, hedge, and vulture funds, as well as other investment vehicles that constitute the dark money of the global "shadow banking system".

During his campaign for the 2006 presidential elections, president Joseph Kabila skillfully instrumentalized these capital flows generated by the private sector to give form to this global neoliberal urban imaginary in his *5 Chantiers* (Five Public Works) program. The *5 Chantiers*, which were rebranded later on as 'the Revolution of Modernity', presented a recovery plan for Congo, summarizing the former president's pledge to modernize education, health care, road infrastructure, housing and access to electricity throughout the DRC and especially in its urban centers.

The project that illustrated the revolution of modernity most fully was the much-debated *Cité du Fleuve* construction project which, as the developer's website states, 'began as a dream in 2008'. *Cité du Fleuve* is currently being built on land reclaimed from sandbanks and swamps along the Congo River. In its final phase, the *Cité du Fleuve* is projected to be 600 hectares in size and to include over 200 villas and 10,000 luxury apartments, 10,000 offices, a marina, schools, cinemas, restaurants, and conference rooms. It will be connected to the rest of Kinshasa by two bridges and a transit road. Self-sufficient in water and electricity supply, *Cité du Fleuve* aims to offer potential buyers a luxurious lifestyle and secure land titles. According to the developers' website, *Cité du Fleuve* will thus provide a "standard

FIGURE 19.1 Developer's 3D representation of the new 'Hypnose' building, inaugurated in Lubumbashi by Joseph Kabila in June 2018

FIGURE 19.2 Still image from promotional video for La Cité du Fleuve (reprinted with kind permission from Robert Choudury, General Director of La Cité du Fleuve)

of living unparalleled in Kinshasa and will be a model for the rest of Africa"; it will be a haven of "modernity, sobriety, friendliness and the pleasure of living," "a white water lily," "a satellite floating on the riversides of the majestic Congo river" that will "showcase the new era of African economic development" and constitute Kinshasa's major "tourist focal point" (Figure 19.2).

But in spite of its glossy, bright aesthetics, *Cité du Fleuve* is not, by any standard, a real and living city. On the contrary: in its preoccupation with security and cleanliness, it aims to keep the real city at bay. The utopian idea of a tabula rasa, the desire to start from scratch and leave former imperfections behind, has always been at the heart of much modernist urban planning, but the materialization of these modernist dreams, whether in Le Corbusier's Chandigarh or Niemeyer's Brasilia, always involved an endeavor to include different social classes in an attempt to offer an alternative city model that reached beyond the strictures of class and caste (Holston, 1989; Kalia, 1999). Contrary to this original inclusive ideal of the modernist city, satellite cities that are currently being constructed or planned in Kinshasa[1] and elsewhere in Central Africa seem to have given up completely on this ideal of a broader collectivity. Conceived as huge gated communities that are run by means of specific bylaws, the new segregationist cities for the emerging African middle class, the supposed beneficiaries of Africa's new economic growth, profoundly redefine the meaning of the current urbanscape. *Cité du Fleuve*, for example, even though it is most uncertain if it will ever emerge as planned, is already redefining what is Kinshasa's center and edge.[2] Consequently, it reflects many of the ideas behind concepts such as the 'charter city' (Fuller & Romer, 2012) – i.e., a special urban reform zone that allows local governments, in partnership with global investors, to adopt bylaws and new systems of rule. At the same time, *Cité du Fleuve*, although predicated on notions of class rather than race, also replicates the segregationist model of *Ville* and *Cité* that proved so highly effective during the Belgian colonial period. It is clear that the islands will become, at least in the minds of their makers, the new *Ville* while the rest of Kinshasa, with its ten million inhabitants, will be redefined as its periphery.

City as cemetery

The city yet to come always seems to imply the necrology of the existing city. The new city reconceptualizes the city of Kinshasa itself as marginal *terrain vague*, as there where the city no longer is, a peripheral space outside of the new normative urban system or, as Kinois themselves say, a 'cemetery city' (*cité cimetière*), a city of zombies where 'the living and the dead have become the same' (*biso nyonso bato moko*). And in many cases, this has to be taken quite literally: in and around the – often vast – cemeteries that are located within the city of Kinshasa, the living and the dead increasingly live in close proximity and constantly dispute each other's right to occupy urban space. Under the pressure of an unstoppable, rapid urban sprawl, many cemeteries are invaded by urban dwellers looking for space to settle. As a result, it has become very common to see housing structures being built on top of existing graves, thereby further blurring the dividing line between the living and the dead (cf. De Boeck, 2016).

These conceptual moves and forms of what can be described as notional urbicide have turned the city as a whole into an abandoned cemetery and have prepared the urban ground for all kinds of – often violent – erasures and displacements. During the Kabila years, the 'revolution of modernity' program, epitomized by the *Cité du Fleuve*, went hand in hand with a large-scale politics of destruction and erasure of 'illegal' public and private spaces and infrastructures: houses, bars, 'informal' agricultural spaces and fields along boulevards and avenues, sidewalk activities (for example, ambulant restaurants, better known as *malewa*), informal cemeteries, and often also the very bodies of those the government sees as illegal occupants of the city, such as the women who run the sidewalk *malewa*'s, street children and members of *kuluna* street gangs (cf. Geenen, 2009; Intudi and Liwerant, 2017).

The same politics of erasure seems to be followed by Kabila's successor, Felix Tshisekedi, who took power in January 2019 after much-contested elections. In September 2019, under the auspices of the president, the new governor of Kinshasa, Mr. Gentiny Ngobila, launched *Operation Kin Bopeto*,[3] a large five-year-long operation to clean up the city. In the same week that the provincial government announced this 'sanitary operation', the newly appointed federal minister responsible for Urbanization and Habitat, Mr. Pius Muabilu, with a full TV crew in tow, showed up at the doorstep of one of Kinshasa's most enigmatic buildings, a ramshackle 12-floor-high skyscraper to order the closure and demolition of the building and arrest its sole inhabitant, a local 'Facteur Cheval' who auto-constructed his tower in defiance of every existing building and planning regulation.[4]

The tower quickly became the icon of the new governmental urban sanitation policy, as shown in a photograph used in a news report by Congo Live TV.[5] In it, President Tshisekedi is photoshopped alongside the building that Congo's government wants to demolish and 'burn down' to the ground in order to be replaced by a modern high-rise building, the figure of the new southern global urbanism that the DRC's government wants to promote.

Cemetery as city

In all of this, the pivotal point for the production of the 'new urban' seems to be the figure of *death*. As already mentioned, nowhere is this fight between two urban visions and realities – between the image of the global neoliberal city promoted by local elites and global investors and the much pauperized but very real city – more pronounced than in and around Kinshasa's cemeteries. Most of the city's cemeteries date back to the colonial period, and over the past three decades, most have been closed by the city authorities for the simple reason that they were considered too full, or on hygienic grounds, or for reasons of public order. The official abandonment of these burial grounds, however, never stopped the inhabitants of Kinshasa from continuing to bury their dead there. The retreat of the urban authorities also opened up these burial grounds to urban dwellers looking for space or a livelihood. The cemetery itself became a fallow city where void and absence also provide promise and opportunity for the millions of subaltern urbanites who are excluded from the new urban dream as exemplified by *Cité du Fleuve*.

Not only is the urban authorities' withdrawal from urban cemeteries exemplifying the way in which the city as a whole is increasingly abandoned by the ruling elites, thereby turning it into a cemetery for the living,[6] but there also is something deeply ironic about this actual retreat of the state from the existing city and the government's dream to reinvest in the new city because this dream of the new neoliberal urban platform is often financed through special opportunity funds commonly referred to as 'vulture funds'. Debt vulture hedge funds prey on the world's poorest countries, buying up their sovereign debt cheaply on capital markets, where such distressed debt trades significantly below its face value, and then seeking to recover the full amount, often through litigation in courts in countries such as Britain or the United States in an attempt to enforce payment of the full value of the debt.[7] In the past two decades, the DRC has been targeted many times by investment funds making an effort to collect debts in this way.

Part of the start-up capital for the construction of 'new cities' in Kinshasa undoubtedly derives from such debt vulture hedge funds, in which case the new city is literally feeding off the corpse of Kinshasa and its pauperized precariat, the immediate by-product of this capitalist globalization. And in the case of the *Cité du Fleuve*, 'corpse' seems to be much more than a mere metaphor. Mr. Hilary Duckworth is the president of Hawkwood Management, an investment group with a diversified fund with close attention paid to risk (Duckworth in an interview in Hedgeweek, January 19, 2007); Mukwa Fund, which is financed by Hawkwood Fund Management, is the fund responsible for building the *Cité du Fleuve*. Through H. Duckworth's brother, Russell Duckworth, the Hawkwood Fund Management has a strong leg in the British banking world and was known until 2012 for its vulture fund technology (in London and after 2011 in Jersey), but his investment group, operating out of Zambia and primarily investing in agricultural and real estate projects, allegedly also made a significant part of its money from the development of new cemeteries and burial grounds in and around cities such as Lusaka and Harare. Monies like these are also used as start-up capital for the construction of new cities in the DRC.

Necropolis – the new global southern urbanism?

That brings us back to the prayer session with which I opened my text. The preacher's sermon that I recorded in the early 2000s has turned out to be rather prophetic in that, due to the increasing material ruination of the city and the (financial or otherwise) difficult materialization of the new city, the most tangible form of urban renewal and upscaling does indeed seem to take place 'in heaven' or, more precisely, in the new cemeteries for the urban middle and upper middle class that, thanks to recent public-private initiatives, are springing up at the edges of Kinshasa. The most emblematic one in this respect is the burial ground that is officially known as 'Necropolis – Between Earth and Heaven' (ETEC: *Entre Terre et Ciel*), consisting of 17 hectares of land on the eastern outskirts of Kinshasa (Figure 19.3).[8]

Preoccupied with cleanliness, order, and security to the same extent as the *Cité du Fleuve*,[9] the site, apart from being advertised as a tourist attraction (again not unlike the *Cité du Fleuve*) promises to offer an 'urbanized' environment (cf. Figure 19.3), 'a veritable city for the dead' ('*une véritable ville des Morts*'), with a clear and readable ground plan, with streets and avenues that have names such as *Rue des rêves* ('Street of dreams'), with 22 neighborhoods, such as *Pegasus, Origa, Atlas, Naos, Constellation, Mercuria, Artiste, Vega, Columbarium*, that, like gated communities, are each destined for specific and exclusive publics (in the neighborhood *Artiste*, well-known artists and musicians, such as Papa Wemba, are buried, while other neighborhoods are reserved for prominent politicians, businessmen, etc.). All these neighborhoods have luxurious tomb-like villas that, like houses in 'real' cities and unlike many houses in Kinshasa, have addresses and house

FIGURE 19.3 "Necropolis cemetery: Urbanized, flowery, securized" (photo by Sammy Baloji, Kinshasa, May 2015)

FIGURE 19.4 The entrance to the Necropolis cemetery: "Nothing reveals better the soul of a country than its cemeteries" (photo by Sammy Baloji, Kinshasa, May 2015)

FIGURE 19.5 View of the Necropolis cemetery (photo by Sammy Baloji, Kinshasa, May 2015)

numbers. Prices of funerals indicate, however, that *Necropole* and similar cemeteries are not within the financial reach of most Kinois. In 2015, the cost of a burial in Necropolis could reach anywhere between US$1,900 and US$6,400 (Figures 19.4 and 19.5).

Conclusion: The eschatological orientation of southern global urbanism

As the aforementioned illustrates, the neoliberal imagination that underpins new forms of urbanization in the Global South constructs capital accumulation through new landscapes of urban real estate made possible through the instrumentalization of death. On its own, this is perhaps nothing new. As Harvey's theorizations of urbanization explain, the whole history of the very idea of creative destruction is predicated on a cyclical movement between creation and death. But what seems to become more outspoken here is the increased intensity of this 'thanatopolitics' mobilizing the possibility of urban renewal. No longer does this politics even bother to occult or disguise the predatory nature of the construction of vast urban terrains of accumulation by dispossession. The possibility of these terrains to be shaped no longer rests upon the assumption of death and dying as a necessary cyclical precondition for the possibility of urban resurrection and renewal. The very space of death, the cemetery, is no longer only material, infrastructural, and conceptual ground to contain, discard, and exclude a vast majority of 'surplus' urbanites who are condemned to death by the ruling elites controlling the neoliberal politics that give form to the segregationist structures of a city such as Kinshasa. Instead, for the new urban elites as well, the cemetery has become the very model and locale for the realization of the ideal of the new urban norm and form: it is in the new necropolises that spring up alongside the city that a global neoliberal urbanism is realized most fully, even though those who can afford it will only be able to access this new urban ideal after their own deaths. In this sense, for both those with and without a right to the city, the thanatopolitics underpinning the neoliberal imagination of a new southern global urbanism relegates the collective prospect of the city yet to come to an eschatologically oriented theological future: that of the afterlife.

Notes

1 Other new 'cities' under construction or projected in and around Kinshasa include *Cité Joseph Kabila Kin-Oasis*, which aims to integrate 182 apartments and 33 villas in the municipality of Bandalungwa, *Cité du Millennium* (Millennium City) in Mitendi and *Cité de l' Espoir* (City of Hope) in Mikondo. In March 2020, the new Tshisekedi government announced plans for the construction of another new city in the vicinity of Kinshasa, inspired this time by the global southern urbanism of *Diamniadio Lake City*, a futuristic township approved by the Senegalese government.

2 Its construction was considerably slowed down for a while because of cash flow problems after the 2008 and 2013 banking crises. The project was supposed to be completed within eight years, a time span that is likely to be expanded much further into the future. The partially constructed new city already seems to face problems with the stability of the engineered islands on which it is built, and it was completely flooded for the first time in late 2019. Also, in November 2019, the offices of the *Cité du Fleuve* management were sealed by the urban authorities because *Cité du Fleuve* did not succeed in paying US$600,000 of overdue taxes to the city.

3 Bopeto: Lingala word meaning purity, cleanliness, hygiene. On the launching of *Opération Bopeto*, see the following coverage by Radio France Inter at http://www.rfi.fr/afrique/20190925-operations-net-toyage-kinshasa-kin-bopeto (retrieved on October 5, 2019).

4 Elsewhere, I devoted part of a chapter to this remarkable example of a vernacular architecture that exemplifies the local appropriation of the modernist skyscraper model, an architectural form that epitomizes both the promises and aesthetics of an older colonialist modernity and the new urban ideals found in places such as Dubai, Doha, and other hotspots of the new urban Global South (see De Boeck and Baloji, 2016: 216ff). We also authored *The Tower. A Concrete Utopia?*, a 70-minute-long film about this building and its maker (Cf. De Boeck, 2015; see for the trailer https://vimeo.com/258620047).

5 See the Congo Live TV news report of September 29, 2019, on YouTube at https://www.youtube.com/results?search_query=ayii+fatshi.

6 This was again made abundantly clear in the Kinshasa-style coronavirus disease 2019 (COVID-19) lockdown strategy that became effective in the second half of April 2020. The urban authorities decided to quarantine the central business and residential city district of Gombe (also known as *la ville*, the former 'white' heart of the city), leaving the rest of the city to fend for itself.

7 In the same way, vulture funds immediately started to 'feast on' troubled company debt caused by the global COVID-19 crisis (Saigol & Bucak, 2020).

8 Between 2013 and 2015, photographer Sammy Baloji and I did field research at several sites in and around Kinshasa. Baloji's photographs of the *Necropole* cemetery included here were produced within the framework of this collaboration.

9 On its website, ETEC advertises that "you will be assured of a high quality funerary service, with guaranteed security, maintenance and respect for the tombs" (*ETEC vous assure un service funéraire de qualité et vous garantit la sécurité, l'entretien et le respect des lieux de sépultures).*

References

Baudrillard, J. (1976). *L'échange symbolique et la mort.* Paris: Gallimard.

Bayat, A. (2012). 'Politics in the city inside-out'. *City*, 24 (2): 110–128.

Becker, E. (1973) *The Denial of Death.* New York: Simon and Schuster.

Bloom, P. J., S. Miescher and T. Manuh (2014) *Modernization as Spectacle in Africa.* Bloomington: Indiana University Press.

de Abreu, M. J. A. (2013) 'Technological indeterminacy: Medium, threat, temporality'. *Anthropological Theory*, 13(3): 267–284.

De Boeck, F. (2011) 'Inhabiting ocular ground. Kinshasa's future in the light of congo's spectral urban politics. *Cultural Anthropology*, 26(2): 263–286.

De Boeck, F. (2015). A concrete utopia. Notes on a video-installation by Sammy Baloji and Filip De Boeck, in M. J. Holm and M. M. Kallehauge (Eds.), *Africa. Architecture, Culture, Identity.* Copenhagen: Louisiana Museum of Modern Ar, pp. 84–88.

De Boeck, F. (2016). 'Corpus vile. Death and expendable youth in urban Congo, in V. Das and C. Han (Eds.), *Living and Dying in the Contemporary World: A Compendium*. Berkeley: University of California Press, pp. 743–769.

De Boeck, F. and M.-F. Plissart (2004) *Kinshasa: Tales of the invisible city*. Ghent/Tervuren: Ludion/Royal Museum of Central Africa.

De Boeck, F. and S. Baloji (2016). *Suturing the City. Living Together in Congo's Urban Worlds*. London: Autograph ABP.

Easterling, K. (2014). *Extrastatecraft. The Power of Infrastructure Space*. London/New York: Verso.

Fuller, B. and P. Romer (2012). *Success and the city. How charter cities could transform the developing world*. A MacDonald-Laurier Institute Publication. Online publication at: https://www.macdonaldlaurier.ca/files/pdf/How-charter-cities-could-transform-the-developing-world-April-212.pdf. Accessed on August 30, 2018.

Geenen, K. (2009). 'Sleep occupies no space': The use of public space by street gangs in Kinshasa. *Africa*, 79(3): 347–368.

Geschiere, P. (1997). *The Modernity of Witchcraft. Politics and the Occult in Postcolonial Africa*. Charlottesville/London: University Press ofVirginia.

Harvey, D. (2013). *Rebel Cities. From the Right to the City to the Urban Revolution*. London/New York: Verso.

Holston, J. (1989). *The Modernist City. An Anthropological Critique of Brasilia*. Chicago/London: The University of Chicago Press.

Intudi, K. K. and S. Liwerant (Eds.) (2017). *Violence urbaine et reaction policière à Kinshasa (RDCongo): Sens et non-sens*. Louvain-la-Neuve: Academia/L'Harmattan.

Jacobs, J. (1961). *The Death and Life of Great American Cities*. New York: Random House.

Kalia, R. (1999). *Chandigarh: The Making of an Indian City*. London/New York: Routledge.

Kaur, R. and T. B. Hansen (2016). 'Aesthetics of arrival: Spectacle, capital, novelty in post-reform India'. *Identities*, 23(3):, 265–275.

Mbembe, A. (2003). 'Necropolitics'. *Public Culture*, 15(1): 11–40.

Murray, M. J. (2017). *The Urbanism of Exception. The Dynamics of Global City Building in the Twenty-First Century*. Cambridge: Cambridge University Press.

Murray, S. J. (2006). 'Thanatopolitics: On the use of death for mobilizing political life'. *Polygraph*, 18: 191–215.

Nielsen, M. (2011). 'Futures within: Reversible time and house-building in Maputo. *Anthropological Theory*, 11(4): 397–423.

Saigol, L. and S. Bucak (2020). '*Vulture funds prepare to swoop in and feast on troubled company debt*'. *Market Watch*, April 17, 2020 (retrieved on April 18, 2020, from www.marketwatch.com/story/vulture-funds-prepare-to-swoop-in-and-feast-on-troubled-company-debt-2020-04-15).

Searle, L. G. (2016). *Landscapes of Accumulation. Real Estate and the Neoliberal Imagination in Contemporary India*. Chicago: The University of Chicago Press.

Sheppard, E., H. Leitner and A. Maringanti (2013). 'Provincializing global urbanism: A manifesto. *Urban Geography*, 34(7): 893–900.

20

PATHWAYS TOWARD A DIALECTICAL URBANISM

Thinking with the contingencies of crisis, care and capitalism

Suraya Scheba

In a recent article in *Society and Space*, Bhan et al. (2020) focus on the intersection between the coronavirus disease 2019 (COVID-19) pandemic, southern urbanisms and collective life. They argue for

> collective life as an analytic that keeps the focus on the ways in which the urban majority is trying to survive and cope within structures of inequality that now bear both the new imprint of COVID-19 while equally holding the continuities of older forms of distancing and exclusion.
>
> (Bhan et al., 2020)

In this chapter, I am indebted to the important disruptive efforts of these and allied urban scholars who have worked to contest a northern-centric vision of urbanisation and decentre geographies of knowledge production (Sheppard et al., 2013: 894, 897). An important element of this work is the concern to trace everyday life and associated practices through adopting an ethnographic attentiveness (Bhan et al., 2020, Myers, 2011; Pieterse, 2011, Caldeira, 2014; Graham and McFarlane, 2014, etc.). More recent advances have moved toward a concern to trace the logics, processes and entanglements that shape urban peripheries, in a comparative sense, arguing for "a bold characterization of modes of the production of space" in the cities of the Global South (Caldeira, 2017: 3).

I locate this piece in support of these efforts toward comparative and cumulative sense-making, as it is easy to see how it would advance urban scholarship theoretically, methodologically and in terms of policy impact (Pieterse, 2011; Caldeira, 2017). However, I argue for an explicit centring of relational tracings of the production of urban space, as these transverse multi-sited and systemic relations (Cirolia and Scheba, 2019). More specifically, what I have in mind is a dialectical approach to thinking the sites and interconnections of urban spatial relations, inspired by both Gillian Hart (2018) and Bertell Ollman's (2003) careful exposition of Marxian dialectical thought. To quote Ollman (2003), on dialectics as a philosophy of internal relations,

DOI: 10.4324/9780429259593-23

> Dialectics restructures our thinking about reality, by replacing the common-sense notion of "thing"…with notions of "process", which contains its history and possible futures, and "relation", which contains as part of what it is its parts with other relations.
>
> (Ollman, 2003: 11)

My argument then is in support of a dialectical orientation to apprehending the urban and grappling with future possibilities that thinks 'sites' and 'systems' as emergent and interconnected. This is an approach that explicitly moves beyond binary debates over 'situated' or 'global' urbanism and toward a *'dialectical urbanism'* that brings an attentiveness to the contradictions and inherent fragility of city-making as emerging through multiple intersecting processes (Hart, 2018). Hence, to borrow from Bhan et al. (2020), this is an analytic that is concerned with tracing the intricate and intimate processes of collective life-making as unfolding in a co-constituted relation to the contingent mechanics of racialised capitalism.

In advancing this dialectical analytic, I take further inspiration from feminist scholars who challenge the notion of the 'economic' as a separate domain of production that exists outside of spaces of social reproduction or what could be termed 'collective life'. Instead, housework, child-rearing, schooling, acts of care and building homes and neighbourhoods are understood as connected to and essential to the functioning of capitalism (Fraser, 2014: 61). These sites are understood as mutually transformative. This body of scholarship is equally important for challenging notions of capitalism as a totalizing logic that moves across space and time in an immutable form. Instead, systemic and structural analyses are undertaken but through tracing intimate networks of human and non-human relations to grasp the unstable, contingent networks of capitalism that surround us. Central to this approach is a concern to explore the intimacy and interconnectedness of practices of life-making and production that cut across multiple social and spatial arenas; and to surface the labor of constructing apparent structural coherence (Bear et al., 2015b).

In what follows, I will present two short narratives on modes of the production of space in Cape Town intensified by COVID-19, to illustrate the contours of my argument.

Infrastructures of inequality in the Cape Town water crisis

In work carried out with Dr Nate Millington on the 2015–2018 Cape Town water crisis, we caution against the uncritical celebration of Cape Town's water crisis management approach (Millington & Scheba, 2021). In following this argument, it is firstly important to recognise that crisis is a general condition of South African water governance, resulting in sustained inequity. However, the threat of 'Day Zero' was recognised and named a crisis when water access to the already included was threatened. At the same time, the daily struggle experienced by residents of informal settlements, backyard dwellings and low-income communities, was neither acknowledged nor resolved. Instead, we argue that it was intensified as a consequence of the dominant crisis-management response. This embodied struggle was captured in an ethnographic study carried out in Beacon Valley, Cape Town, by a student on this project in 2019. The study reveals how living with a water management device results in residents constantly weighing up various activities contributing to a physical and mental load. This takes the form of calculating usage, forgoing a bath or a cup of tea and engaging in time- and labour-intensive practices, including the use of rudimentary infrastructures for greywater collection and reuse. Furthermore, in households with free-flow connections, water debt is a constant source of anxiety, with the fear of eviction for those living in state housing (Byrnes, 2019).

Importantly, this account cannot be disconnected from the crisis-management interventions promoted by the municipality, including the remaking of Free Basic Water (FBW) and a tariff restructuring. These resulted in a shift away from the universal toward the targeted provisioning of FBW only to those registered as indigent, the associated conditional installation of water management devices and tariff increases beginning at the lowest tariff step that effectively increased the price of water for low-income consumers to make up for reduced consumption levels amongst high-volume users. In the context of a drought crisis, these interventions were designed to avoid a financial crisis for the municipality, ensuring that the city's model of managing water remained financially solvent. These measures, influenced by the concern for municipal financial resilience, had a direct bearing on household indebtedness or disconnection for the urban poor, experienced in embodied ways as a mental and physical load, as well as calculations and negotiations over basic conditions of life.

Hence the crisis management response at the municipal level and the intensified everyday struggle of residents to access water are mutually constitutive and need to be understood through this relational tracing. With the onset of the COVID-19 pandemic, it was no longer possible to deny or normalise the more sustained 'Day Zero' of the marginalised. As lockdown drew closer, the Department of Human Settlements, Water and Sanitation introduced water trucks and tanks across the country. This rapid response is significant, considering that "for 25 years, the state has failed to provide adequate water to millions of people living in shack settlements. Suddenly, in a matter of days, taps have been fixed or installed and water tanks provided" (Newframe Editorial Team, 2020a). However, these deliveries have not resulted in universal access and continue to be faced with challenges, including tender-related corruption and claimed budgetary constraints. At the same time, the water management practices intensified by the drought have not shifted and continue to impose indebtedness and disconnection on low-income households across the city.

Policing order: Land occupation and eviction

The tension in the water sector between municipal commitments to realise equitable service provision and simultaneous drives for fiscal austerity are evidenced again in the struggle to access land and housing. In this case, many remain excluded despite several state housing policies concerned with realising access. In the Western Cape province, with Cape Town as the largest city, 600,000 people are on the housing waiting list, with an average waiting period of 15–17 years (Mafolo, 2020). The consequence is that many urban dwellers access land through other forms, with land occupation as a central practice. However, land occupation is largely criminalised, resulting in punitive and, at times, violent state responses, justified through a framing of land occupiers as disorderly 'queue jumpers' who are unwilling to wait for the state subsidised housing that may eventually materialise (Oldfield & Greyling, 2015; Levenson, 2018).

The predominant response to land occupations has come in the form of state-sanctioned evictions emerging as a normalised practice. In the context of COVID-19, the City of Cape Town (although not alone in this practice) has continued to demolish homes constructed on occupied land, despite a national moratorium on evictions. This has resulted in violent clashes between city law enforcement, including the Anti Land Invasion Unit (ALIU), and land occupiers and protesters (Mafolo, 2020, Farr & Green, 2020). The practice has been criticised, but the moment that resulted in widespread condemnation was the shockingly violent eviction, in July 2020, of Bulelani Qolani by four members of the ALIU. He was removed naked from his shack in Khayelitsha, a peripheral neighbourhood in the city. William Shoki (2020), writing for *Africa Is a Country*, brilliantly explores the connections between this apparently singular act of violence and a systemic order that involves a continuous labour of 'normalisation'. It is worth quoting at length:

> What distinguished this moment of evictions from all the rest that South Africans are used to is that Qholani was naked – and what are usually unnoticed acts of ordinary cruelty became a recorded episode of spectacular dehumanization.... In South Africa, we have historically never been able to think of land as anything but a commodity, a means for facilitating the exploitation of labor (by accommodating and reproducing that labor), or something to be exploited itself (aided by that exploited labor). Any "unproductive" use of land—used as simply a site for a home, or as a site where one could independently produce their own subsistence –had to come to an end.... The apartheid regime thrived on this kind of mystification, of obscuring unequal power relations in the guise of an instrumental rationality which makes us concerned with the proper processes of things and not what ends they are serving. But as we now understand, this isn't exceptional to apartheid, it is how capitalism functions. And contemporary South African capitalism learned from its predecessors an effective regime of exclusion, a means of determining which lives matter and which ones don't.

The power of this reflection is to locate an apparent act of individualised violence within a larger rationality, historically located and defined by practices of exclusion, where life is devalued relative to what is deemed the 'productive' use of land. The struggle to access land and housing in post-apartheid South Africa continues to reflect the ways in which race, class and space intersect. It is also a struggle that intersects with both state and private-sector interests. The point is supported by the response of high ranking municipal politicians to the imminent intensification of austerity and anticipated impact on housing delivery as a result of the 2020/2021 adjustment budget. The adjustment has halved the Human Settlements budget in the city, impacting on the capacity to respond to the housing crisis. This is accompanied by an increase in land occupations in the last few months due to the loss of household income during lockdown (Mafolo, 2020). The response of mayoral committee member for safety and security J. P. Smith to this budget reduction was to enquire about the availability of additional funding to assist with removing structures during land occupations. Adding that "it's essential that additional private security costs are covered" (Mafolo, 2020). From this it can be seen how the prioritised considerations revolve around a technical rationality focused on the proper processes to follow to protect private property through an increase in funding for demolition and securitisation. Also evidenced is the the labor entailed in normalising dominant structural relations through the centering of technical (budgetary) considerations. This is despite a clear incapacity on the part of the state to deliver housing at scale and an inability of residents to afford housing due to rising unemployment and economic distress, further intensified by covid-19.

Multi-sited relations of indebtedness

The scale of this distress, is captured in a study conducted in July 2020, almost five months after South Africa's lockdown began. It paints a devastating picture of the estimated effects of lockdown and the pandemic on employment and welfare outcomes in the country.[1] The findings indicate an 18% decline in employment and confirms that pre-existing inequalities have grown. Furthermore, the pandemic is undeniably gendered, with women having borne the brunt of job and income losses and largely assuming the increased burden of care work of children and the elderly (Webster, 2020). It is also clear that hunger is widespread, again impacting women more, as they sacrifice their own bodies to feed their children first.

Despite this grim outlook, there is an increased push for austerity at the national level, notably in the emergency budget developed to give effect to the R500 billion stimulus package announced by the president on 21 April 2020. However, instead of offering a lifeline, analysts have argued that the budget has translated into a brutal austerity response that fails to take forward the rescue package. In presenting an analysis of the budget, the Institute for Economic Justice argues,

> By embarking on this brutal programme of cutting back on resources available for delivery of basic rights to the population, the government is in violation of its constitutional obligations to progressively realise the socio-economic rights, which are outlined in the Constitution.... There is a strong view therefore in civil society that the constitutionality of this Budget needs to be tested in court.
>
> (Coleman, 2020)

It is anticipated that the emergency budget will adversely impact basic infrastructure and service provision. This will undoubtedly have devastating effects at the household level. Furthermore, South Africa's turn to the International Monetary Fund to finance part of the expenditure and shortfalls arising from the economic crisis has prompted concerns that

> the Treasury and hawks in the economic establishment are trying to use the crisis to justify the imposition of austerity...what Naomi Klein calls the Shock Doctrine – which is to use the crisis as a cover to impose right-wing doctrinaire measures on the public under the cover of claiming to represent a solution to the crisis.
>
> (Coleman, 2020)

In concluding, I want to return to the initial concern with 'collective life' as an analytic. Bhan (2019) has recently drawn on the figure of 'repair' to capture the set of practices of collective life emerging in response to conditions of sustained crisis in southern urbanism, further intensified by the pandemic. Significant in this account is the suggestion that repair is an act of immediacy, a necessary response that does not apply to a specific actor and assumes a cyclical temporality. This concept is intriguing for what it potentially gestures to both as a response to an immediate state of breakdown and as an aspiration, suggesting future possibilities (Jackson, 2015). However, as I have attempted to show here, repair practices exist in relation to and often serve to mitigate the effects of structures of inequality. Furthermore, repair as an assemblage of practices is not limited to the spaces of collective life. Instead, it is also the work of state and guided by differing notions of what requires fixing. This second notion of repair frequently assumes the form of a 'techno-managerial fix' to crisis. I will discuss each of these points in turn below before concluding on the value of a dialectical orientation to thinking the relationship between crisis and repair in urban scholarship, and advancing a third notion of repair, as 'radical repair'.

Beginning with practices of repair in collective life-making, Rebecca Solnit (2020) argues, "I sometimes think that capitalism is a catastrophe constantly being mitigated and cleaned up by mutual aid and kinship networks". In the South African case, I have argued that the pandemic response has drawn tighter connections between the worlds of finance, state austerity measures, infrastructural (dis)investment and the intimate ways in which these arrangements assume embodied effects in the spaces of everyday life. In this case, the work of repair has assumed an increased urgency as a practice of mitigation in relation to this structural violence, where households are also expropriated and indebted spaces. Furthermore, as an 'assemblage of practices', repair as care work is often offered as under-paid or unpaid labour in everyday life and is frequently gendered, with women bearing the increased burden of sustaining life (Fraser, 2014; Webster, 2020). Here repair is vital as a survival and coping mechanism but equally can serve regressive ends contributing to forestalling "the need for deep structural changes in the foundations of contemporary life" (Millington, 2019). Instead, it can contribute to the ongoing maintenance of wealth inequality, displacing essential care work into the under-paid spaces of social reproduction.

Secondly, repair practices are not limited to the spaces of collective life. Instead, divergent assemblages of repair are also mobilised by states and private actors as solutions to crisis. The accounts I have provided reveal the ways in which these practices can be given effect through an instrumental rationality and draw connections between finance and indebtedness that move all the way through, connecting global finance, national and local states and households (García-Lamarca & Kaika, 2016; Montgomerie and Tepe-Belfrage, 2017). These techno-managerial fixes as apparent acts of repair are, ironically, acts of violence in that they intensify inequality and abandonment, thereby necessitating the immediate labour of repair at the scale of collective life – emerging to fill the cracks in state neglect.

Hence as we enter an uncertain future, defined by austerity and intensified exclusion, my argument is for the necessity of rethinking our analytic to apprehend both collective life and the multiple 'elsewheres' that land in place. I contend that a dialectical orientation holds analytical possibility by offering pathways into viewing the fragility of structuring relations despite apparent coherence. This effort is essential in a world defined by multiple crises, as it supports an analysis of the constitutive relations of crisis, and thereby opens up prospects for more radical notions and practices of repair. With the latter assuming a more expansive form, moving beyond mitigation, whilst potentially seeded therein, and explicitly rejecting solutions that hold in place the existing order of things. A 'dialectical urbanism' offers an analytical dynamism to thinking both the horror and the hope that resides in the present and opens up possibilities for a more radical politics of repair that can sever its relationship to sustaining structural violence, moving toward more radical structural reconfigurations.

Note

1 The findings were released as a set of research papers about data collected during the National Income Dynamics Study – Coronavirus Rapid Mobile Survey (Spaull et.al, 2020)

References

Bear, L., Ho, K., Tsing, A. and Yanagisako, S. (2015a) 'Generating capitalism'. *Theorizing the Contemporary, Fieldsights*, March 30. https://culanth.org/fieldsights/series/generating-capitalism.

Bear, L., Ho, K., Tsing, A. and Yanagisako, S. (2015b) 'Gens: A feminist manifesto for the study of capitalism'. *Theorizing the Contemporary, Fieldsights*, March 30. https://culanth.org/fieldsights/gens-a-feminist-manifesto-for-the-study-of-capitalism.

Byrnes, K. (2019) Water access practices as a barometer for contested citizenship in Beacon Valley. Honours Thesis, University of Cape Town.

Bhan, G. (2019) 'Notes on a southern urban practice'. *Environment and Urbanization*, 31(2): 639–654.

Bhan, G., Caldeira, T., Gillespie, K., Simone, A. (2020) 'The pandemic, southern urbanisms and collective life'. *Society and Space*. Available at https://www.societyandspace.org/articles/the-pandemic-southern-urbanisms-and-collective-life.

Caldeira, T. (2014) Peripheral urbanisation. London (Paper presented at the LSE Cities Public Lectures, 23 October 2014).

Caldeira, T. P. (2017) 'Peripheral urbanization: Autoconstruction, transversal logics, and politics in cities of the global south'. *Environment and Planning D: Society and Space*, 35(1): 3–20.

Cirolia, L. R. and Scheba, S. (2019) 'Towards a multi-scalar reading of informality in Delft, South Africa: Weaving the 'everyday' with wider structural tracings'. *Urban Studies, 56*(3): 594–611.

Coleman, N. (2020) 'Supplementary budget puts South Africa on the edge'. *Daily Maverick*, 7 July 2020. Available at https://www.dailymaverick.co.za/opinionista/2020-07-07-supplementary-budget-puts-south-africa-on-the-edge/#gsc.tab=0.

Farr, V. and Green, L. (2020) 'Amid escalating gang violence, the City of Cape Town wages war on the poor'. *Daily Maverick*, 8 July 2020. Available at https://www.dailymaverick.co.za/opinionista/2020-07-07-supplementary-budget-puts-south-africa-on-the-edge/#gsc.tab=0.

Fraser, N. (2014) 'Behind marx's hidden abode for an expanded conception of capitalism'. *New Left Review*, (86): 55–72.

García-Lamarca, M. and Kaika, M. (2016) '"Mortgaged lives": The biopolitics of debt and housing financialisation'. *Transactions of the Institute of British Geographers*, 41(3): 313–327.

Graham, S., & McFarlane, C. (Eds.). (2014). *Infrastructural lives: Urban infrastructure in context*. Routledge, London.

Hart, G. (2018) 'Relational comparison revisited: Marxist postcolonial geographies in practice'. *Progress in Human Geography*, 42(3): 371–394.

Jackson, S. (2015) 'Repair. Theorizing the contemporary: The infrastructure toolbox', *Cultural Anthropology*. Available at: https://culanth.org/fieldsights/repair

Levenson, Z. (2018) 'The road to TRAs is paved with good intentions: Dispossession through delivery in post-apartheid Cape Town'. *Urban Studies*, 55(14): 3218–3233.

Mafolo, K. (2020) 'Budget cuts slow City of Cape Town's ability to solve housing crisis'. *Daily Maverick*, 12 August 2020. Available at https://www.dailymaverick.co.za/article/2020-08-12-budget-cuts-slow-city-of-cape-towns-ability-to-solve-housing-crisis/.

Millington, N. (2019) 'Critical spatial practices of repair'. *Society and Space*. Available at http://societyandspace.org/2019/08/26/critical-spatial-practices-of-repair/.

Millington, N. and Scheba, S. (2021) 'Day zero and the infrastructures of climate change: Water governance, inequality, and infrastructural politics in cape town's water crisis'. *International Journal of Urban and Regional Research*, 45 (1): 116–132.

Montgomerie, J. and Tepe-Belfrage, D. (2017) 'Caring for debts: How the household economy exposes the limits of financialisation'. *Critical Sociology*, 43(4–5): 653–668.

Myers, G. (2011) *African Cities: Alternative Visions of Urban Theory and Practice*. Zed Books Ltd: London.

New Frame Editorial (2020a) 'Covid-19 brings rubber bullets and water for the poor'. *New Frame*, 10 April 2020. Available at https://www.newframe.com/covid-19-brings-rubber-bullets-and-water-for-the-poor/.

New Frame Editorial (2020b) 'Amid despair, grounds for new hope'. *New Frame*, 24 July 2020. Available at https://www.newframe.com/amid-despair-grounds-for-new-hope/.

Oldfield, S. and Greyling, S. (2015) 'Waiting for the state: A politics of housing in South Africa'. *Environment and Planning A*, 47(5): 1100–1112.

Ollman, B. (2003) *Dance of the Dialectic: Steps in Marx's Method*. University of Illinois Press: Illinois.

Pieterse, E. (2011) 'Grasping the unknowable: Coming to grips with African urbanisms'. *Social Dynamics*, 37(1): 5–23.

Sheppard, E., Leitner, H. and Maringanti, A. (2013) 'Provincializing global urbanism: A manifesto'. *Urban Geography*, 34(7): 893–900.

Shoki, W. (2020) 'The existing order of things'. *Africa Is a Country*, July 2020. Available at https://africasacountry.com/2020/07/the-existing-order-of-things.

Solnit, R. (2020) 'What coronavirus can teach us about hope'. *The Guardian*, 7 April 2020 Available at https://www.theguardian.com/world/2020/apr/07/what-coronavirus-can-teach-us-about-hope-rebecca-solnit.

Spaull et al. (2020) 'National income dynamics study – Coronavirus rapid mobile survey (Nids-cram) wave 1 synthesis report: Overview and findings'. Available at: https://cramsurvey.org/reports/

Webster, D., (2020) 'Counting covid-19's economic cost'. *New Frame*, 15 July 2020. Available at https://www.newframe.com/counting-covid-19s-economic-cost/.

21

GLOBAL SELF-URBANISM

Self-organisation amidst regulatory crisis and uneven urban citizenship

Francesco Chiodelli and Margherita Grazioli

Introduction: For a unitary conceptualisation of self-urbanism

In recent decades, the concept of self-organisation has attracted growing attention in the academic debates concerned with urban issues (de Bruijn and Gerrits, 2018). 'Self-governance', 'self-production', 'self-building' and 'self-help' are some of the labels used to indicate various practices that, despite their distinctive features, are all characterised by a prominent role of non-public (individual or collective) subjects and by a very limited role of public authorities in fields in which a public presence is usually considered necessary. All these urban variations of self-organisation can be included under the umbrella term of *self-urbanism*, which denotes practices and processes of making, shaping and living the urban space that are not prevalently and directly mediated or led by public institutions – that is, they are mostly endogenously produced (Moroni et al., 2020). Thus the generative and creative role of local social forces and actors emerges clearly. As such, self-urbanism must be read as one of the constitutive forces of current global urbanism, together with other factors and processes such as financialisation, climate change adaptation, ethnic and religious diversity and socio-spatial polarisation.

Self-urbanism is not a homogeneous phenomenon. On the contrary, it exhibits a high internal (descriptive and normative) differentiation, even more accentuated than in the case of several other processes that constitute global urbanism. As such, it has never been conceptualised in a unitary manner. In fact, different empirical forms of self-urbanism have generated and substantiated radically different theorisations in the field of urban studies – which, in some cases, have been developed along ideologically divergent interpretative lines, ranging from left-wing ideals of self-determination and emancipation to right-wing ones of the minimal state and individual self-regulation. This is the case of subaltern urbanism (Roy, 2011), urban enclavism and revanchism (Caldeira, 2000), radical urban autonomy (Vasudevan, 2015), neo-localism (Beito et al., 2002), to mention only some of them.

Despite this fragmented understanding of global self-urbanism, a common thread links the buildings occupied by housing rights movements in Rome to the gated communities for the super-rich in Johannesburg, passing through the informal settlements in the suburbs of Maputo, the unauthorised houses built by the middle class along the shores of the Mediterranean Sea, the retirement communities in the United States and the radical-environmentalist communities

DOI: 10.4324/9780429259593-24

born in Northern Europe in the wake of the alternative movements of the 1960s. It is the linking thread of a specific relationship of otherness (and sometimes even opposition) with respect to ordinary forms of public regulation, production and management of the urban.

The connection produced by this commonality among different empirical manifestations of self-urbanism around the globe thus entails the need to find an interpretative grid of these phenomena. This would not only shed light on the shared, deep roots of self-organised practices that seem radically different at first sight, but would also help to clarify their political, normative and epistemological implications.

For this purpose, this chapter will first analyse the birth of self-urbanism and identify its constitutive features. It will then argue that one of the main implications of self-urbanism is the rise of a new institutional fragmentation that overthrows the traditional boundaries drawn by public authorities (e.g. municipal borders) and fuels spatialised forms of unequal urban citizenship.

The rise of self-urbanism in the twentieth century

The generic theorisation of phenomena characterised by a significant degree of autonomy from the formal predicaments of public institutions is not new (for a historical overview, see Sørensen and Triantafillou, 2009). On the contrary, specific conceptualisations of self-organisation in the *urban sphere* began to emerge in much more recent times (e.g. in the 1960s in many European countries) as self-organised urban practices clashed with the stratification of comprehensive and detailed forms of public regulation, development and management of space and buildings implemented by public authorities, especially after the Second World War. Such comprehensive and detailed forms of public control spread first in the Western countries and then throughout most of the world (Hall, 2002 [1988]). As a result, nowadays the use and development of urban space almost worldwide are characterised, at least on paper, by the significant role of public authorities. This role takes material form in a complex assemblage of different layers of public regulation (e.g. property regimes, building codes, land-use regulations, rules for the use of public space, landscape codes, zoning plans, master plans, strategic visions). The processes of neo-liberalisation have only partially modified this situation so that nation-states and their local ramifications have retained control over some spheres (for instance, planning processes) while relying on the private market in other sectors (e.g. for the housing supply). Therefore, the outcome of neo-liberalisation cannot be read as a generalised deregulation, but rather as the development of a complex and contradictory framework in which intense forms of public withdrawal and privatisation in some sectors are accompanied by strict public regulation and control in others (Andersson and Moroni, 2014).

The complexity of self-urbanism in the housing sphere: A stratified definition

The practices that fall under the label of 'self-urbanism' are extremely heterogeneous in social composition, material forms, ontology, structure and scope. For our purpose of propounding a ground-conceptualisation of self-urbanism, it is, therefore, useful to identify the phenomenon's constitutive components. Our goal is not to propose static typologies or taxonomies but rather to provide coordinates that can make general sense of the complexity and diversification of such a fragmented (and territorialised) phenomenon. We focus on three main questions,[1] centring on one of the various spheres in which self-urbanism materialises – namely, housing.

Self-urbanism and internal coordination. This first analytical question concerns the level and type of coordination among the diverse individual actions composing self-urbanism. Despite being an important component of some cases of self-organisation, *intentional and collective agency* is not a constitutive feature of the phenomenon. In some instances, in fact, self-organised actions lack any collective intent and explicit central coordination, because they are based mainly on the individual intentions of actors (Rauws, 2016). Nevertheless, the sum of these individual(istic) acts of self-organisation can evolve into collective outcomes through a sort of spontaneous alignment, or self-synchronisation, which is not intentionally organised by any actor. In this regard, we may speak of *self-coordination* (Moroni et al., 2020). This applies, for example, in the case of informal settlements that develop through the incremental addition of housing units by single households. In other instances, by contrast, urbanites may deliberately join together in order to pursue a common goal or ambition so that their action is guided by some form of internal coordination (Rauws, 2016). In this second case, we may speak of *self-governance*. This is the case, for example, of common-interest housing developments (also known as 'homeowners' associations') that are privately governed residential communities in which the inhabitants directly manage several aspects of daily life that are handled by public authorities in 'ordinary portions' of the urban area (e.g. the provision of collective services, such as rubbish collection or street patrolling; Glasze et al., 2006; McKenzie, 2003).[2] Self-governance also applies to the squatting actions led by grassroots urban movements that re-appropriate and repurpose urban spaces and infrastructures in order to claim housing rights and the 'right to the city' (Squatting Europe Kollective 2014; Montagna and Grazioli, 2019).

Self-urbanism and the market. The second question concerns the relationship of self-urbanism with the market. In several cases, self-urbanism takes place outside (if not against) the circuit established by the (housing) market, such as when access to housing is obtained by occupying a public or private building. However, this unrelatedness with the market is not a necessary feature of housing self-urbanism. In fact, in many instances, ordinary practices of the real estate market are in place. If we refer again to the case of common-interest housing developments, here the access to dwelling units is ruled by traditional forms of real estate trade. Consider also the fact that non-ordinary (i.e. illegal/informal) forms of the real estate market can occur, as in the case of the housing black market in informal settlements.

Self-urbanism and public regulation. The third question concerns the relationship of self-urbanism with the public regulatory system. Self-urbanism can arise *within* or *outside* the public regulatory framework. In the former case, it maintains a formal/legal character. This happens, for example, when different rules of land and building use are respected, but the residents of the community (e.g. the members of a homeowners' association) decide to regulate certain aspects of the settlement's life (e.g. the provision of certain collective services or the conduct of decision-making processes) differently from the remainder of the urban area. Non-public rules and procedures are thus established by the inhabitants *in addition* to the public ones. Depending on the specificity of the self-organised practice, the degree of independence from public institutions may differ. Sometimes, broad spaces of autonomy from public control are established, as in the case of several homeowners' associations in the United States (Nelson, 2005) or radical autonomous communities, such as Christiania in Copenhagen (Coppola and Vanolo, 2015). In other cases, this autonomy is weaker, as in many cohousing communities[3] in Europe (Chiodelli and Baglione, 2014). This may be due to various factors: for instance, because the institutional and political conditions impede greater autonomy, or because there are no economic, organisational or human resources for residents to claim for greater independence.

If self-urbanism occurs mainly *outside* the public regulatory framework (i.e. in its violation), it acquires an informal/illegal nature. This happens when many (if not all) of the rules of ownership, use and development of land and buildings are breached. As a consequence, a sort of 'space of exception' within the publicly regulated urban body is constituted, and more or less structured forms of non-public control and regulation can be established *to replace* the public ones in force. Nevertheless, such self-urbanism practices do not develop in a space characterised by a complete lack of relationship with public institutions. In fact, there is always some nexus (partial, weak, mediated or indirect) with the latter (Chiodelli and Tzfadia, 2016), which, however, does not completely determine the development trajectory of these practices.[4] The expression 'informal housing and settlements' is often used to indicate these cases of self-organisation outside the public regulatory system. However, the expression should not be limited to denoting 'slums' in Africa, Asia or Latin America. It should also be used to refer to a variety of phenomena widespread in many Western countries, which include the following: the occupation for housing purposes of public and private vacant buildings (Grazioli, 2017), illegal Roma camps (Maestri, 2019), unauthorised housing construction, non-institutional refugee camps and other multifarious examples (Chiodelli et al., 2020).

The implications of global self-urbanism: The new institutional archipelago in the midst of the regulatory crisis

In light of the complexity of global self-urbanism described in the previous sections, it is clear that an imagery of marginality and exceptionality cannot be associated with that phenomenon. Self-urbanism includes many practices that are marginal neither in quantitative terms (e.g. informal settlements in many cities of Africa or Asia play a dominant role in the production of urban space), nor with reference to their social and public value (e.g. counter-cultural and political squatting influences urban politics significantly in many cities), nor with regard to the actors involved (e.g. many gated communities are inhabited by groups that are 'central' from an economic and political point of view). On the contrary, self-urbanism must be understood as a crucial component of the current urban age in different parts of the world. This is true not only from a factual point of view, but also from an epistemic one. In fact, regardless of their specific features, all (housing) self-urbanism practices allude to two pivotal, critical issues of current global urbanism.

The regulatory crisis. The first question concerns the current configurations of public regulation, development and management of urban space. The different variations of (housing) self-urbanism shed light upon the dissatisfaction and criticisms of growing segments of the urban population with regard to these forms of public control, which are unable to keep up with social innovation to effectively regulate and manage the growing complexity of urban phenomena or to respond to the demands, needs and aspirations formulated by different groups of urban dwellers and stakeholders. Self-organised urbanites counterpose such public failures with the concrete construction of alternative that are deeply rooted in situated local contexts. They range from extreme forms of elitism, in which a sort of 'secession of the successful' is claimed and public control is criticised for its excessive invasion of the private sphere, to forms of left-wing radicalism in which the disproportionate subjection of the public authorities to the private sphere (for example, to property rights) is condemned in the name of more accentuated, and inclusive, public intervention. Regardless of this polarisation, it is clear that increasing shares of the urban population are more and more critical of the contemporary public governance of urban space and that their agency builds (contradictory) fragments of alternatives. Against this backdrop, it is worth stressing that self-urbanism does not arise mainly in 'intentional voids' – that is, the spaces left intentionally by public authorities to favour the emergence of forms of self-organisation and social innovation. It springs above all from legislative loopholes or outdated rules, from ineffective and selective processes of policymaking, and from ambiguous and contradictory practices of street-level bureaucrats. These generate unintentional interstices (Brighenti, 2013) that are then crafted and contended by a myriad of urban actors as spaces of possibility in which to pursue their goals and agendas.

In short, self-urbanism alludes to the structural shortcomings of public intervention in the urban sphere – as it has been imagined from the beginning of the twentieth century and practiced until today – at its various levels (local, metropolitan, regional and national) and stresses the need to radically redesign the process by which the urban space is regulated, managed and developed.

The new institutional archipelago. The second question that arises from a ground-conceptualisation of self-urbanism concerns its institutional implications. Self-urbanism testifies that the current urban age is characterised not only by a deep socio-spatial fragmentation and unevenness, but also by a new *institutional differentiation* (and polarisation) *driven by non-public forces.* The current institutional geography of the urban, in fact, is no longer defined by the traditional boundaries (e.g. national and municipal borders) drawn by public authorities, which also established the limits and purview of their sovereignty, spatial control and regulative devices. This geography currently also arises – often only *de facto*, but sometimes also *de jure* – within and across these traditional public boundaries, which are increasingly traversed by complex regulatory jurisdictions that may emerge from self-organised, non-public processes. This determines new and complex institutional arrangements that are shaped by constant tensions between centrifugal forces (which aim to move further away from public control in the name of greater autonomy and secession) and centripetal forces (which aim to be incorporated into the 'ordinary' city). In essence, the spread of self-urbanism accentuates the transformation of the urban into an *institutional archipelago*, which is not only the result of governance/governmentality strategies and configurations of the public power, but also the outcome of self-organised actions by individuals or groups. As a result, new forms of unequal urban citizenship emerge within this institutional urban archipelago.

In the second half of the twentieth century, many countries saw the factual rescaling to the city level of many rights associated with citizenship. This happened, for instance, in the aftermath of the Second World War, with the expansion of social rights related to the rise of the welfare state (Marshall, 1964); a second, different stage of this rescaling process occurred with the crisis of Fordism in the 1970s–1980s and the subsequent entrepreneurial turn of many cities (Blokland et al., 2015). Such rescaling has led to the birth of a new layer (or kind) of citizenship besides the national one: 'urban citizenship' – or *citadineté*, to use the words of Henri Lefebvre (1972, 1990). The notion of urban citizenship implies that the fundamental rights of a person are no longer defined only by his/her membership in a nation, but are shaped *de facto* also by his/her belonging to a city – which, for instance, deeply influences the quality of the social rights that s/he can claim in the form of social services, such as school or hospitals (Chiodelli, 2013). Differently from national citizenship, *citadineté* is uneven and fragmented by its very nature, always in the making, in tension between pushes for its enlargement and pulls for its restriction (or selective recognition; Varsanyi, 2006; Yiftachel, 2015; Grazioli, 2017). There are various reasons for the uneven and unstable character of urban citizenship, such as the fact that many urban social rights are linked to localised public goods, which are impure by definition (that is, their externality exists as a spatial field of effects; Harvey, 2009) and are deeply moulded by urban policies and local politics. The institutional micro-geographies related to global self-urbanism are another key source of the uneven character of urban citizenship: they can generate different *spheres* of urban rights – and, as a consequence, different degrees of urban citizenship. In fact, the 'spatial autonomy' linked to many practices of self-urbanism often triggers processes of differentiation of the rights available for the subjects of the practice in question. This emerges, for instance, in relation to the case of homeowners' associations: by providing specific services only to their members (e.g. sport and recreational facilities, open spaces, private security), homeowners' associations grant them additional social rights, which are not available to the rest of the inhabitants of the 'ordinary' city (for instance, due to municipal budget constraints). Something similar (although for different reasons and through different mechanisms) happens in instances of political self-organisation, such as squatting for housing purposes. Here self-managed welfare provisions and public utilities replace those that are not provided (or overtly denied) by public action; they are devised by activists and squatters and then 'offered' to other marginalised urbanites as forms of 'commoning' (Linebaugh, 2008; Dadusc et al., 2019). In all these latter cases, self-urbanism reverses the constitution of citizenship as a device of differential inclusion (Mezzadra and Neilson, 2013) by *expanding it selectively*. By the same token, an opposite-directed process of *contraction* of urban citizenship can be fuelled by self-urbanism – for instance, when some basic social rights are denied in informal settlements under the pretext of their illegal status or when restrictive internal rules are enforced inside gated communities.

Conclusions: The epistemic value of investigating global self-urbanism

Diverse practices and processes of urban self-organisation have been at the centre of a thriving stream of academic research that, mainly over the past two decades, has stimulated significant and popular theorisations in the field of urban studies (such as, for instance, subaltern urbanism). One of the shortcomings of such research is its tendency to neglect the horizontal connection among different episodes of urban self-organisation around the world, which, however counterintuitive it may sound, must be read as different empirical manifestations of a single phenomenon that we called 'self-urbanism'.

As argued in this chapter, the specificities and roots of self-urbanism are to be found in the crisis of the regulatory mechanisms that have governed cities in the past century. One of the radical outcomes of such a crisis and the consequent spread of self-urbanism is the creation of a new institutional archipelago of non-public origin. This accentuates the fragmentation and unevenness typical of the notion of urban citizenship. The latter's political and normative implications can either stimulate policy innovations or contradictory governmental approaches, according to the situated contexts where these differential forms of self-urbanism unfold. On the one hand, self-urbanism can be a vehicle for the creation of alternative spaces and channels for claiming and regaining rights for marginalised groups (very much in line, for instance, with the repertoire of actions, demands and conceptualisations concerned with the right to the city, radical autonomy, subaltern urbanism). On the other hand, it can become an instrument that encourages the disengagement of public action from ruling the urban space and taking care of spatialised inequalities, thereby fostering the secession of the successful and the marginalisation of 'undesirable' urbanites.

In light of these elements, a ground-conceptualisation can help to grasp the epistemic value of self-urbanism as a key part of current processes of global urbanism, as well as being an analytical lens through which to observe its development trajectories and impact on broader questions pertaining to the segmentation of urban citizenship, spatialised inequality and urban justice(s).

Notes

1 While each of these questions is described in this section in dualistic terms for analytical purposes, the reality of self-organised practices is characterised by the simultaneous coexistence of fragments of all of them.

2 Retirement communities, gated communities and cohousing communities are specific kinds of homeowners' associations (Chiodelli, 2015). Today, homeowners' associations host almost 70 million people and represent around 80% of houses built in new subdivisions in the United States (Clarke and Freedman, 2019).
3 Cohousing is a form of intentional residential community characterised by communitarian services and facilities (e.g., kitchen and dining room) managed by the residents and for the residents, shared values among the inhabitants (e.g., environmental sustainability), high degree of participation of residents in the management of the daily life of the community and self-selection of the inhabitants. Cohousing was born in the 1960s in Northern Europe. It then spread mainly to Southern Europe and North America (Chiodelli and Baglione, 2014).
4 There are exceptions to this dualistic framework. For instance, there are self-organised housing practices that are conducted in a constant relationship and negotiation with public authorities. This is, for instance, the case of squatting for housing purposes as practised by housing rights movements in Rome (Grazioli and Caciagli, 2018).

References

Andersson D. E. and Moroni S. (Eds.) (2014) *Cities and Private Planning. Property Rights, Entrepreneurship and Transaction Costs*. Cheltenham, UK: Edward Elgar.

Beito D.T., Gordon P. and Tabarrok A. (Eds.) (2002) *The Voluntary City. Choice, Communities and Civil Society*. Ann Arbor: The University of Michigan Press.

Blokland, T., Hentschel, C., Holm, A., Lebuhn, H. and Margalit, T. (2015) 'Urban citizenship and right to the city: The fragmentation of claims'. *International Journal of Urban and Regional Research*, 39(4): 655–665.

Brenner N., Schmid C. (2015) 'Towards a new epistemology of the urban?' *City*, 19(2–3): 151–182.

Brighenti, A. M. (Ed.). (2013) *Urban Interstices: The Aesthetics and the Politics of the In-Between*. Farnham: Ashgate Publishing.

Caldeira, T. P. (2000) *City of Walls: Crime, Segregation, and Citizenship in São Paulo*. Berkeley: University of California Press.

Chiodelli, F. (2013) 'Planning and urban citizenship: Suggestions from the thoughts of Henri Lefebvre'. *Planning Perspectives*, 28(3): 487–494.

Chiodelli, F. (2015) 'What is really different between cohousing and gated communities?' *European Planning Studies*, 23(12):2566–2581.

Chiodelli, F, Baglione, V. (2014) 'Living together privately: For a cautious reading of cohousing'. *Urban Research & Practice*, 7(1): 20–34.

Chiodelli, F., Coppola, A., Belotti, E., Berruti, G., Clough Marinaro, I., Curci, F. and Zanfi, F. (2020), 'The production of informal space: A Critical Atlas of Housing Informalities in Italy between Public Institutions and Political Strategies'. *Progress in Planning*. https://doi.org/10.1016/j.progress.2020.100495.

Chiodelli, F. and Tzfadia, E. (2016) 'The multifaceted relation between formal institutions and the production of informal urban spaces: An editorial introduction, *Geography Research Forum*, 36: 1–14.

Clarke, W. and Freedman, M. (2019) 'The rise and effects of homeowners associations'. *Journal of Urban Economics*, 112(2019): 1–15.

Coppola, A. and Vanolo, A. (2015) 'Normalising autonomous spaces: Ongoing transformations in Christiania, Copenhagen'. *Urban Studies*, 52(6): 1152–1168.

Dadusc, D., Grazioli, M. and Martínez, M. A. (2019) 'Introduction: Citizenship as inhabitance? Migrant housing squats versus institutional accommodation'. *Citizenship Studies*, 23(6): 521–539.

de Bruijn, E. and Gerrits, L. (2018) 'Epistemic communities in urban self-organisation: A systematic review and assessment'. *Journal of Planning Literature*, 33(3): 310–328.

Glasze G., Webster C.J. and Frantz K. (Eds.) (2006) *Private Cities. Global and Local Perspectives*. London: Routledge.

Grazioli, M. (2017) 'From citizens to citadins: Rethinking right to the city inside housing squats in Rome, Italy, *Citizenship Studies*, 21(4): 393–408.

Grazioli, M. and Caciagli, C. (2018) 'Resisting to the neoliberal urban fabric: Housing rights movements and the re-appropriation of the 'right to the city' in Rome, Italy'. *Voluntas*, 29(4): 697–711.

Hall, P. (2002) *Cities of Tomorrow: An Intellectual History of Urban Planning and Design in the Twentieth Century* (3rd ed.). Malden, MA: Blackwell [or. ed. 1988]

Harvey, D. (2009) *Social Justice and the City* (rev. ed.). Athens and London: University of Georgia Press.

Harvey, D. (2012) *Rebel Cities. From the Right to the City to the Urban Revolution*. London/New York: Verso.

Lefebvre, H. (1972) *Espace et politique. Le droit à la ville II*. Paris: Anthropos.

Lefebvre, H. (1990) Ouverture. Du pacte social au contrat de citoyenneté. In Henri Lefebvre and Lefebvre, Henri, and Groupe de Navarrenx (Eds.) *Du contrat de citoyenneté*. Paris: Syllapse (pp. 17–37).

Linebaugh, P. (2008) *The Magna Carta Manifesto: Liberties and Commons for All*. Berkeley: University of California Press.

Maestri, G. (2019) *Temporary Camps, Enduring Segregation: The Contentious Politics of Roma and Migrant Housing*. London: Palgrave MacMillan.

Marshall, T.H. (1964) *Class, Citizenship and Social Development*. New York: Doubleday

McKenzie, E. (2003) 'Common-Interest Housing in the Communities of Tomorrow'. *Housing Policy Debate*, 14(1–2): 203–233.

Mezzadra, S. and Neilson, B. (2013) *Border as Method, or, the Multiplication of Labor*. Durham: Duke University Press.

Montagna, N. and Grazioli, M. (2019) 'Urban commons and freedom of movement: The housing struggles of recently arrived migrants in Rome'. *Citizenship Studies*, 23(6): 557–592.

Moroni, S., Rauws, W. and Cozzolino, S. (2020) 'Forms of self-organisation: Urban complexity and planning implications'. *Environment and Planning B*, 47(2): 195–202.

Nelson, R. H. (2005) *Private Neighborhoods*. Washington, DC: Urban Institute Press.

Pruijt, H. (2013) 'The logic of urban squatting'. *International Journal of Urban and Regional Research*, 37(1): 19–45.

Rauws, W. (2016) 'Civic initiatives in urban development: Self-governance versus self-organisation in planning practice'. *Town Planning Review*, 87(3): 339–361.

Roy, A. (2011) 'Slumdog cities: Rethinking subaltern urbanism'. *International Journal of Urban and Regional Research*, 35(2): 223–238.

Sørensen, E. and Triantafillou, P. (2009) The politics of self-governance: An introduction. In E. Sørensen and P. Triantafillou (Eds.) *The Politics of Self-Governance*. Furnham: Ashgate (pp. 2–22).

Squatting Europe Kollective (2014) *The Squatters' Movement in Europe. Commons and Autonomy as Alternatives to Capitalism*. London: Pluto Press.

Varsanyi, M. W. (2006) 'Interrogating 'urban citizenship' vis-à-vis undocumented migration'. *Citizenship Studies*, 10(2): 229–249.

Vasudevan, A. (2015) *Metropolitan Preoccupations: The Spatial Politics of Squatting in Berlin*. Hoboken, New Jersey: John Wiley & Sons.

Yiftachel, O. (2015) Epilogue—from 'Gray Space' to Equal 'Metrozenship'? Reflections on Urban Citizenship. *International Journal of Urban and Regional Research*, 39(4): 726–737.

PART IV

Governing global urbanisms

22

UNLOCKING POLITICAL POTENTIALITIES

Edgar Pieterse

Our training as critical scholars compels us to move from hard-earned theoretically informed diagnostics to critique, and along the way, we end up ensnared by certitude (Bergman, 2018). Cities are replete with opportunities for endless diagnosis and critique, especially as they suffer from acute neoliberal hangovers, but they are also ripe with possibility. In my reading, despite the dire state of mainstream politics, it is perhaps a time for cautious optimism about the prospects for system change at the metropolitan scale, no least in bare-knuckle cities across the Global South.

Can this line of exploration contribute to the broader concern of this volume whether something akin to global urbanism is conceivable? On the critical end of urban studies, there is an unavoidable debate between neo-Marxists with a post-structural bent to theorise emergent urbanisms through various disjunctive flows that operate at a global/planetary scale, on the one hand, and postcolonial theorists who insist on apprehending the city through a more situated, historicised and contingent perspective on the other (Derickson, 2015; both clusters disagree fundamentally with structuralist urban theories that draw on rational-choice perspectives or conventional Marxism). The clue that this debate may be overdrawn in terms of ontological differences can be discerned in the fact that many postcolonial urbanists draw on a rich and variegated critical theory oeuvre to make sense of the differential and undue power of capital whilst not overstating such power in terms of cultural specificities and endogenous generative capacities. In this sense, both theoretical approaches contribute substantively to the piecing together of strands of insight about similarities and differences across urban areas in all world regions, but both camps are fated to be partial and incomplete because by definition cities exceed the explanatory power of any one epistemological approach.

Due to my biographical becoming in the city, Cape Town to be precise, and urbanism writ large, I am drawn to a postcolonial reading that insists on tracing historical, spatial and affective clues to piece together highly specific expressions of urban (institutional) life and the assemblages of meaning-making it carries in tow. At the same time, due to the scale and overbearing weight of historical injury and social violence that overdetermine the parameters of emergence and spatial justice, the portal into deciphering the folds of the city must always be extremely concrete and in search of a theoretical language to account for specific practices (Bhan, 2019; Simone and Pieterse, 2017). The practice orientation also compels an obsession with unexpected possibilities

DOI: 10.4324/9780429259593-26

that arise when opposing, competing and subversive logics invariably rub up against each other in the machinic unfolding of everyday life (Amin and Thrift, 2002). This focus requires a search for political imaginations and experiments that can potentially catalyse systemic change towards the normative horizons of spatial justice, environmental restoration, beauty and psychic flourishing. In this moment, amidst the exposures of pre-existing fault lines of inequality rendered by the coronavirus disease 2019 (COVID-19) pandemic, I want to offer a sense of the kind of political reimaginations that become possible inside this approach to global urbanism.

The spatialisation of the sustainable development goals (SDGs) produces an understanding that urban transformation holds the key to the advancement of just, sustainable and radically integrative development (Pieterse et al., 2018). There is an indivisible connection between the realization of sustainable human settlements and changing patterns of production and consumption, which in turn depends on the proliferation of sustainable infrastructure and deep cultural change. These cultural materialisations are also preconditions for the achievement of climate policy goals (IPCC, 2018). Most importantly, a generous reading of the SDGs and allied development discourses recognises and accepts that social justice and environmental imperatives are indivisible and unattainable unless the logics of extractive capitalism are upended. These political/policy "achievements" are undoubtedly the outcome of sustained activism, critical scholarship, democratic innovations, millions of grassroots experiments in alternative forms of urbanity and so much more (Swilling et al., 2018). To be sure, all of this energy is also accompanied by incessant failure and dead ends and the tenacity to persevere—a politics that is more marked by setback than success.

In earlier reflections, I argued that these mainstream confluences allow us to imagine and project concrete landscapes of system change: various overlapping and interdependent urban operating systems (Pieterse, 2011). That typology refers to sustainable infrastructure, the inclusive economy and efficient spatial form, glued by processes of radical democratic political decision making. At the political heart of these operating systems live the decision making and regulatory force of the state enmeshed with the relative power of a plurality of private and civic actors. The operating system in this reading is both metaphor and an emergent assemblage. However, to identify specific priorities to design, mobilise for and effect system change, a spatial reading is called for. This imperative drives the argument to differentiate between two mutually constitutive scalar imperatives: localisation and regionalisation.

Localisation denotes radical infrastructure technology reforms that are potentially consistent with sustainable environmental flows (low carbon and efficient materials consumption), labour intensive and fundamentally affordable for households with low and variable incomes (Hyman and Pieterse, 2017). Affordability is connected to system design that relies on local labour, internment of value within a community or city and hybrid approaches grafted onto the makeshift practices of popular neighbourhoods (Brown and Stigge, 2017; Caldeira, 2017). Regionalisation refers to the long-term macro-infrastructural decisions to transition grids and entire infrastructural networks onto a low-carbon and resource-efficient footing in lockstep with regional ecosystems or green infrastructure. There is a burgeoning academic and consulting cottage industry on the substance and procedural aspects of these city-region infrastructural transitions (Ravetz, 2000). For example, the 1.5-degree report of the Intergovernmental Panel on Climate Change (IPCC) stipulates a detailed checklist for policy areas ripe for adaptation across four systems: energy, land and ecosystems, urban and infrastructure and industry (IPCC, 2018). To be sure, these processes are not apolitical and not immune to nefarious commodification interests and can easily slip into lazy and disingenuous "smart city" boosterism (Greenfield, 2017). The critical edges of these debates are focused on the deeper potential of these vital transitions to serve as the spearpoint for a larger economic transition into a more sustainable, just and inclusive global economic system (Swilling, 2019) coupled with alternative global governance systems.

The systemic notion of urban operating systems that feed into the focus on infrastructural and land-use transitions (to alter urban forms) typically runs into the constraints of our political imagination and, more deadly, our contemporary institutions in and for the city. Moreover, the cultures of politics, the dance of accommodation between entrenched vested interests and the relentless efforts at co-option from moneyed elites make it very difficult to see *how* radical alternatives can find meaningful traction. Recently, I reflected on the layered nature of these institutional dynamics as manifest in the African urban context (Pieterse, 2018). Building on that reading (not rehearsed here), I want to be more exploratory and speculative in the remainder of the chapter. Specifically, I want to offer a sketch of five generic actors, or caricatures, that have to be appreciated and set loose in specific urban settings if the potentialities of our moment are to be seized in full. The argument extends an older provocation on the essential role of epistemic communities in cities that crisscross diverse and opposing institutional settings and logics (Pieterse, 2006). Then, and in later iterations, I was interested in the art of forging deep collaborations around urgent transformative priorities—e.g. public transport, off-grid energy, universal access to sanitation—across the divides of social movements, municipal departments, parastatals/utilities, private businesses, think tanks and universities and the arts.[1]

In this chapter, I want to focus on five personas sketched as caricatures—the sentinel, artist, designer, hacker and subversive bureaucrat—that should ideally be enrolled in these kinds of epistemic communities in order to hold and activate the profound tensions that accompany our dissonant times (Simone and Pieterse, 2017). These caricatures are a rhetorical device to foreground the diverse, tension-filled sensibilities and skills that must accompany radical urban politics and play. The personas do not map neatly onto specific occupations, professions or actual people. Flesh-and-blood actors in the material world may fulfil an aspect of one or two of these personas, but as a composite, I am trying to provoke what creative agency might include in our over-scripted politics. At the very least, I am hopeful that it can alter strategic conversations about transformative urban strategies and tactics. In particular, I am interested in contributing to broader debates about how best to enrich and enlarge social imagination about radically inclusive and playful urban futures (Mulgan, 2020). My hope is that such an imagination should thrive within a global urbanism conceptual project.

Sentinel

In the daily rough and tumble of our dissonant times, it is hard to keep the faith. Amidst so much uncertainty about long-term employment security, access to social protection and the daunting impacts of technological change, the only political forces that seem to gain strength are on the edges of the democratic spectra (Fleurbaey, et al. 2018). Thoughtful and generous approaches cannot find resonance nor traction (Connolly, 2013). In the meantime, the storm clouds of intensifying climate change impacts, xenophobia, deepening inequality and cultural divides are darkening political and economic horizons, all heightened by the unfathomable impacts of COVID-19 and its afterlives. The causal drivers are paradoxically both obvious and opaque, making moral claims and judgement harder. This very dynamic prepares the ground for political identifications that are all about certitude, boundaries, exclusions and expulsions (Ferguson, 2009; Sassen, 2014). Commercial imperatives of social media economies with all of their cultural effects reinforce the most dangerous tendencies within these processes (Lanier, 2010). In such political atmospheres, it is impossible to discern how deep, transformative systemic change might come about and how the (humanist) promise of human rights, social justice, ecological regeneration, cosmopolitanism, aesthetic freedom and well-being might be advanced for the majority.

This is where sentinels step into the frame. These characters are contemporary griots who are able to make philosophical sense of our confusing times by offering a long view of history and ideas rooted in a language and sensibility for ethical judgement—phronesis. Their obsession is to generate concepts and theories that can conjure a narrative account of the now, thoroughly historicised and sometimes spatialised but always leaning into the future. Sentinels give us the conceptual and emotional stuffing to engage in anticipatory thinking because grounded interventions to bring light and possibility into the city cannot arise without a line of sight on the future. Most importantly, sentinels offer us a grammar and sensibility to stare imponderables in the face until a considered reaction appears, leading to proposition, strategy, intervention, learning, recalibration and an acceptance of radical incrementalism as unavoidable dispositions in the practice of insurgent urbanism. Much of the post-humanist, postcolonial work that remains open to emergent ontologies can be designated as reference points.

Artist

Artists dive headlong into indefinable questions or spheres of concern and deploy aesthetic registers to wrestle with the possibilities of sense-making, meaning, forms of exploration and expression that can only find respite once some kind of "artistic resolution" can be achieved. Such "resolution" is not necessarily fully intelligible or obvious to others but the artist, and even then, it might merely be at an affective level, just out of reach of cognitive sense-making (Massumi, 2002). However, throughout, the artist relies on what Gurney (2019) calls "artistic forms of thinking" to engage the animating problematic. For Gurney, artistic thinking references artistic practice and process. Artistic practices involve the deployment of various kinds of aesthetic craft and expertise to decipher meaning in-process and in situ (Gurney, 2019). It is only by working the canvass (or stage, composition sheet or open space), spreading the paint, retracing it until a certain form of clarity or resonance has been achieved through intent, trial, error and renewed intent that release eventually becomes possible, or just out of grasp, as is often the case with artistic enquiry. Immense labour and tenacity are called for alongside a profound fealty to beauty, craft and creation. Obsession is a useful trait. This reflects a pre-cognitive, affective, aesthetic form of knowing that is at home with uncertainty, not knowing in advance where an act may lead in order to navigate unlit corners of the imagination (Pieterse, 2019).

Artistic processes produce a form of knowledge and experience that helps one to learn how to inhabit complexity, uncertainty, paradox, pain, love, fear and many other waves of emotional intensity. Artistic thinking carries exploratory practice beyond the limitations of rational thought and didactic logics but can also offer valuable jump-off points for such discursive practices. Artistic thinking cannot be divorced from poetic and experimental forms of knowledge production that includes sketching, diagramming, mapping, play, pleasure-seeking, humour, irony and a willingness to erase and start again. Since the urban and global condition will only become more contradictory and prone to the unforeseen and unexpected, it is vital that epistemic communities are able to tap into the aesthetic and embrace the liberating force of accepting that one cannot always decipher an answer or solution and certainly not in advance. Some problems simply require inhabitation, immersion and struggle without guarantees (Simone and Pieterse, 2017). The humility, or provisionality, that stems from such acceptance is an essential ingredient to nurture strategic openness, generosity and an authentic culture of critical deliberation (Bergman,

2018). Moreover, artistic thinking undermines one-dimensional understandings of power and sociality that often accompanies old-school Marxian attachments to revolutionary militancy.

The sentinel helps one to understand the weight and significance of emergence in our times, but often their philosophical grammar remains elusive, imprecise, gestural and a source of creative confusion. Artists offer a non-representational approach to tap and excite our affective consciousness—a necessary complement to the work of systematic critical thought so that the overflowing potentiality of a given crisis or moment can be gleaned or simply sensed. In the absence of fellow travellers who are adept at artistic thinking and practice, at least we can remain mindful to take aesthetics seriously, to appreciate its power to ground and animate strategic repertoires. Aesthetics also hold the key to figuring out what cultural embedding might require in a given city, or neighbourhood or street.

Designer

If the sentinel can help to reframe a given challenge or threat by broadening our understanding of causal drivers and significance, the artist can enrich our repertoire of response and strategic action. Both of these personas bring greater complexity into the world. Both remind us to see, understand and inhabit paradox when wanting to act, injecting a necessary sense of reflection and maybe caution. This can be liberating when one is able to let go of the need for certainty, but it can also be confounding, especially when it is obvious that there is so much that is broken and unjust in the world. There is an undeniable tension between accepting complexity and the desire to act with intent and purposefulness. The tension stems from the concern that well-intentioned actions can have unintended consequences that could worsen a given situation. This is where the orientation and skill set of designers can be useful.

By definition, designers are always searching for and trying to effect simple but elegant solutions. Designers find ways to bound specific problems, even if they are complex and highly interdependent, and surrender to the professional and ethical imperative to act in response to a need—to configure a means of moving forward to address very concrete problems. Designers are uniquely skilled at being both deconstructionist and propositional in their outlook and disposition (Dunne and Raby, 2013).

In an urban setting overdetermined by inequality, unhealthy living conditions and limited economic opportunities, designers are invariably obsessed with picking a manageable problem to solve with an appreciation that four imperatives must be fulfilled: increased foundational security for those who are vulnerable, improved access for those excluded from routine urban opportunities and resources, affordable cost for accessing the service or opportunity and, always, quality. It can be substantiated in design terms by a further four qualifiers: usability of the technology proposed, functionality of use, aesthetic beauty and cultural resonance. Quality is not reducible to banal considerations about the materials used in a design. Rather, quality is always a contextual, culturally informed set of parameters for the design approach. These imperatives compel designers to work intimately with citizens and various institutional actors who want to pursue the right to the city and equity. It also demands of grassroots organisations and intermediary non-governmental organisations to expand their (learning and advocacy) repertoire to embrace design thinking because it invariably brings together spatial dynamics, financial parameters, operating procedures and, most importantly, ownership. A suggestive and open-ended activist designer workbook is *Making Massive Small Change* (Campbell, 2018).

Hacker

> To his (the hacker's) eye, all the works of mankind (sic)—and maybe even the very laws of nature—are contraptions, just like computers once were. Law, ethics, society, the economy: they're not sacred, they're not proper ways of life. They're codes. All systems of code can be disrupted, upgraded, improved without permission. Everything's hackable. People who realize this about our world are in the right. Lesser people are obscurantists, since they're forced to rely on instructions, regulations, patents, ethics, public opinion and similar un-hackerly crutches.
>
> (Sterling, 2015)

In order to take a contraption apart, one has to unlock its secrets. Thus in its purest form, computer hacking references the expertise of programmers who are able to penetrate the doors and intricate latches of software that houses vast systems of propriety data. Since governments and large corporations expend small fortunes to build these software systems, there is a delicious pleasure in unlocking the codes to these vaults or discovering backdoors that were never meant to be seen. To be sure, as Gabriella Coleman (2014) reminds us, there is both "craft and craftiness" in the lifeblood of the hacker. One part of the craft is learning the skills to be able to breach closed digital borders and another is programming fluency to not get lost in system minutia where the hacker has to roam but rather explore and identify more elegant ways of achieving a certain outcome. At his or her core, a hacker seeks to find ways to rewire code or circuits to achieve an improved or extended or alternative outcome than intended by the system designers. This requires a constant search for weaknesses and instabilities—points of entry and/or leverage.

It is easy to see how this sensibility of the computer hacker can be extended to other domains, digital and analogue as well. In fact, much of the survival and extension practices that support everyday life in popular neighbourhoods across the world display a clear hacker sensibility. Through the power of necessity, everything gets repurposed, reimagined, redeployed, reinvented and recirculated in novel circulations (Thieme, 2017). This is sometimes labelled as a DIY sensibility. This is accurate but also insufficient to capture the mobilisation of imagination and sociality—banal heroics.

Cities and urban politics are circumscribed by public bureaucracy encoded by multiple regulatory meshworks and reproduced through hierarchical forms of power that relish fragmentation and control. In democratic societies, these are overlaid by the regulatory and disciplinary logics and rituals of political parties that shape and feed off public bureaucracy. Most city governments have been operating on this basis for well over a century, generating administrative encrustations that harden through routine and time. (These entities are typically impervious to the (expensive) playbooks of strategy consultancy firms, even though they always play along until the right moment arrives to trip up the prolonged, expensive and progressively diluted effort.) In these settings, effective radical operators are able to understand the fundamental logics of bureaucratic power and recalcitrance and are able to deploy that knowledge to use the weight of administrative minutia against itself. This form of institutional hacking is indispensable for larger coalition-based struggles to transform the city. There is now a unique opportunity at the confluence of strident moves towards e-government and governance and higher public standards for transparency and accountability for hacker activism within and across public bureaucracies (Landry and Caust, 2017). Typically, opportunities can best be pursued in conjunction with specific claims and campaigns, whether it be for fairer taxation policies, better subsidies, more sustainable approaches to infrastructure provision, land reform and justice or improved access to the commons. The bureaucratic hacker is best placed to trace and expose how existing mechanisms within public regulatory systems can be utilised or redesigned to achieve alternative outcomes. However, such openings must then be converted into new political and institutional norms, and the best method to achieve that is regulatory activism, which brings me to the final caricature.

Subversive bureaucrat

Left critics enjoy a deeply ambivalent relationship with the state. On the one hand, they strongly favour an interventionist state that can lead from the front to advance social justice, and on the other, so much critical work rediscovers the complicity of the state with neoliberal intentions, and these critics cannot but be sceptical about whether the state can be trusted. This is obviously an oversimplification, but I am trying to expose a tension that others have written about more eloquently (Connolly, 2013; Ferguson, 2009; Stewart ,2007). This ambivalence discourages an intimate understanding of the workings of the (local) state and easily diminishes the quiet activism, heroics and staying power of progressives who believe you have to remain embedded if we are to achieve a time when the full radical potentialities of the state can be understood and deployed. In the last two decades, this kind of critical positioning has become incredibly hard as old-school Weberian hierarchical specialisation got blended with neoliberal managerialism (Landry and Caust, 2017; Mulgan, 2020). In other words, overly bureaucratic rules and power games got deployed to bolster a market logic of securing payment for services through a redefinition of citizens as customers and orienting limited public resources to customer segments with the highest purchasing power and greatest potential to enlarge expensive real estate. These local public finance approaches produced contemporary trends marked by rising inequality and too many households being cut off from essential basic services, economic opportunities and public space.

At the same time, as intimated in the opening of this chapter, we have seen a massive discursive shift over the last five years in particular where there is now a renewed recognition that neoliberalism has been disastrous for well-being, environmental health and economic inclusion (Fluerbaey et al., 2018). Furthermore, most acute urban development problems are profoundly interconnected and demand complex interdisciplinary responses, and the contemporary urban governments are singularly ill prepared to work in this fashion. The older literature on participatory governance has prepared the ground for forms of governance that involved diverse and multiple constituencies implicated by a given area of public concern. We have known for some time that participatory processes are indispensable for more effective outcomes, even if fraught with sticky dynamics that are unavoidable in multi-class processes. However, there is a chasm between the outdated command-and-control habits of public bureaucracies and the institutional imperatives of working across society, across specialisations and across scales. Fundamentally, this demands a shift away "from a 'no, because' to 'yes, if' culture — one that captures possibilities and potential", according to Landry and Caust (2017: 6).

When political work results in the necessary policy opening to adopt a "yes, if" disposition, subversive bureaucrats can truly come into their own. These actors are able to foster an alternative administrative and decision-making culture that over time will have to result in the adoption of new regulatory systems that operate on completely different assumptions to Weberian command and control or neoliberal risk aversion. The superpower of subversive bureaucrats is that they can understand and see the systems they are embedded in—the imperative of coalitions of change that weave actors across various pockets within the public sector and with allies in civil society, the academy and private business. Most domains of urban change—albeit thematic or place-based—cannot avoid messy processes of deliberation and action. Such processes require the design and construction of institutional scaffolds rooted in agonistic democracy and combined with the intentionality of co-produced plans, programmes and banal projects. Most importantly, all of these undertakings must be circumscribed by mechanisms to ensure transparency, measurement and learning. It is impossible to achieve nimble, adaptive forms of urban governance without built-in instruments that can induce reflexivity and accountability. Subversive bureaucrats are able to discern how best to deploy mainstream (or formal) policy narratives, everyday administrative talk and performance for these larger administrative and regulatory processes of system change (Kania et al., 2018).

In Conclusion

Due to the narrative and cultural nature of urban politics, it is not possible to effect deep systemic change without prioritizing two to three critical thematic areas that can be embedded in a compelling story about the "essence" of the city. There is an unmistakable art to the theatrics of bringing an issue into the public sphere and transforming it into a shared concern that can produce agonistic debate about what needs to be done. Once this crack exists, it then becomes imperative to forge a rich and affectively resonant strategy to move from diagnostic to proposition and action whilst practicing reflective learning alongside. The different caricatures invoked in this chapter all have a crucial role to play in these rhizomatic processes. It is beyond the scope of this chapter to offer examples and evidence, but I want to leave it in the safe hands of the readers' imaginations as they apply these possibilities to the context where they enact the city. That said, it is relevant to clarify that by invoking these caricatures, I am not trying to make a case for these actors to be deliberately sought out and enrolled. That would be far too mechanical a reading of the argument.

Instead, I am making a case for a much more nimble, experimental, strategically opportunistic, non-dogmatic and intelligent political sensibility that draws inspiration from diverse forms of knowledge to strengthen phronesis. By invoking these five caricatures, the argument is to foreground the range of perspectives, skills and dispositions that should ideally be called upon when the slow and difficult political work of urban invention and activation are pursued. Such a sensibility is consistent with a global urbanism project that allows for holding in tension circuits of power and exchange with variable spatial intensities, along with a fealty to uncover and theorise the intimate specificities of the everyday and its cultural layers. In this chapter, I have suggested that we are witnessing shifting political discourses about cities that open strategic landscapes ripe for systemic restructuring and reformatting. This intensifying dynamic provides opportunities of varying significance to imagine, instigate and test novel political repertoires. The five monikers—sentinel, artist, designer, hacker and subversive bureaucrat—are simply entry points to map much more variegated and imaginative forms of city politics whilst insisting that generosity has to live at the core of global urbanism, for we don't always know.

Note

1 During 2017–2018, I curated a deliberative process with 25 leaders across these sectors in Cape Town under the banner of a project called the Integration Syndicate. The idea was to forge an epistemic community out in the open and crystalise compelling provocations on how the city can be transformed over the next 20 years. The process and results are published in Pieterse et al., 2019.

References

Amin, A. and Thrift, N. (2002) *Cities. Rethinking the Urban*. Cambridge: Polity.

Bahn, G. (2019) 'Notes on a Southern urban practice'. *Environment and Urbanization*, 31(2): 639–654.

Bergman, C. (2018) 'The stifling air of rigid radicalism'. *New Inquiry*, 2 March, www.newinquiry.com.

Brown, H. and Stigge, B. (2017) *Infrastructural Ecologies. Alternative Development Models for Emerging Economies*. Cambridge, MA: The MIT Press.

Caldeira, T.P. (2017) 'Peripheral urbanization: Autoconstruction, transversal logics, and politics in cities of the global south'. *Environment and Planning D: Society and Space*, 35(1): 3–20.

Campbell, K. (2018) *Making Massive Small Change: Building the Urban Society We Want*. Hartford, VT: Chelsea Green Publishing.

Coleman, G. (2014) 'What is a hacker?' *Wired Magazine*, https://www.wired.com/2014/11/what-is-a-hacker/ [Accessed: 10 April 2020].

Connolly, W. E. (2013) *The Fragility of Things: Self-Organizing Processes, Neoliberal Fantasies, and Democratic Activism*. Durham, NC: Duke University Press.

Derickson, K. (2015) 'Urban geography 1: Locating urban theory in the "urban age"'. *Progress in Human Geography*, 39: 647–657.

Dunn, A. and Raby, F. (2013) *Speculative Everything. Design, Fiction and Social Dreaming*. Cabridge, MA: MIT Press.

Ferguson, J. (2009) 'The uses of neoliberalism'. *Antipode*, 41(1): 166–184.

Fleurbaey, M., with Bouin, O., Salles-Djeic, M., Kanbur, R., Nowotny, H., and Reis, E. (2018) *A Manifesto for Social Progress. Ideas for a Better Society*. Cambridge: Cambridge University Press.

Greenfield, A. (2017) *Radical Technologies: The Design of Everyday Life*. Brooklyn, NY: Verso.

Gurney, K. (2019) *The Mattering of African Contemporary Art: Value and Valuation from the Studio to the Collection*. PhD. Cape Town: University of Cape Town.

Hyman, K. and Pieterse, E. (2017) Infrastructure deficits and potential in African Cities, in Burdett, R. and Hall, S. (Eds.), *The SAGE Handbook of Urban Sociology: New Approaches to the Twenty-First Century City*. London: Sage Publishers (pp. 429–452).

IPCC. (2018) Summary for policymakers, in *Global Warming of 1.5°C. An IPCC Special Report on the Impacts of Global Warming of 1.5°C Above Pre-Industrial Levels and Related Global Greenhouse Gas Emission Pathways, in the Context of Strengthening the Global Response to the Threat of Climate Change, Sustainable Development, and Efforts to Eradicate Poverty*. Geneva: International Panel on Climate Change (pp. 1–30).

Kania, J., Kramer, M. and Senge, P. (2018) *The Waters of System Change*. Washington DC: FSG.

Landry, C. and Caust, M. (2017) *The Creative Bureaucracy & Its Radical Common Sense*. London: Comedia.

Lanier, J. (2010) *You Are Not a Gadget. A Manifesto*. New York: Vintage Books.

Massumi, B. (2002) *Parables for the Virtual: Movement, Affect, Sensation*. Durham, NC: Duke University Press.

Mulgan, G. (2020) *The Imaginary Crisis (and how we might quicken social and public imagination)*. Helsinki: Demos.

Pieterse, E. (2006) 'Building with ruins and dreams: Exploratory thoughts on realising integrated urban development through crises'. *Urban Studies*, 43(2): 285–304.

Pieterse, E. (2011) Building brave new worlds: Design and the second urban transition, in Smith, C. (Ed.) *Design for the Other 90%: Cities*. New York: Cooper-Hewitt National Design Museum (pp. 40–53).

Pieterse, E. (2018) 'The politics of governing African urban spaces'. *International Development Policy/Revue Internationale de Politique de Développement*, 10: 26–54. http://journals.openedition.org/poldev/2626.

Pieterse, E. (2019) Stitching fragments and Fractals, in Marx, G. (Ed.), *Ecstatic Archive*. Johannesburg: Goodman Gallery (pp. 29–35).

Pieterse, E., Parnell, S. and Haysom, G. (2018) 'African dreams: Locating urban infrastructure in the 2030 sustainable developmental agenda'. *Area Development & Policy*. doi:10.1080/23792949.2018.1428111.

Pieterse, E., Green, P., Knemeyer, B., Pulker, A. and Viviers, A. (Eds.) (2019) *The Integration Syndicate: Shifting Cape Town's Socio-Spatial Debate*. Cape Town: African Centre for Cities.

Ravetz, J (2000) *City Region 2020: Integrated Planning for a Sustainable Environment*. London: Earthscan.

Sassen, S. (2014) *Expulsions: Brutality and Complexity in the Global Economy*. Cambridge MA: Harvard University Press.

Simone, A. M. and Pieterse, E. (2017) *New Urban Worlds. Inhabiting Dissonant Times*. Cambridge: Polity Press.

Sterling, B. (2015) 'What is a hacker?' *Wired Magazine*. URL: https://www.wired.com/2014/11/what-is-a-hacker/ [Accessed: 10 April 2020].

Stewart, K. (2007) *Ordinary Affects*. Durham, NC: Duke University Press.

Swilling, M., Pieterse, E. and Hajer, M. (2018) Futuring, experimentation, and transformative urban politics, in Poli, R. (Ed.), *Handbook of Anticipation*. New York: Springer (pp. 1–28). https://doi.org/10.1007/978-3-319-31737-3_24-1.

Swilling, M. (2019) *The Age of Sustainability: Just Transitions in a Complex World*. New York: Routledge.
Thieme, T. (2017) 'The hustle economy: Informality, uncertainty, and the geographies of getting by'. *Progress in Human Geography*, 42(4) 529–548.

23

CLIMATE CHANGED URBANISM?

Harriet Bulkeley, Laura Tozer and Emma Lecavalier

Introduction

Over the past two decades, climate change has escaped the conventional spatial moorings of global environmental issues to become understood not only as a matter of the international arena but also one that has a particularly urban nature. Whilst other such challenges – from species loss to ocean pollution – remain resolutely global in the imaginaries they generate and in the form of their politics, climate change has come to be understood as fundamentally connected to global urbanism. In return, those seeking to articulate the ways in which cities are now necessarily global frequently point to their collective contributions to the changing atmosphere and to the levelling effect of its impacts. Urbanism is now global by virtue of its role in producing and inhabiting a climate changed world. As global urbanism comes to be deeply tied into questions of climate change and its future, in this chapter, we seek to explore how and why we have come to understand climate change in relation to the urban and at the same time to question the ways in which climate has come to change the nature of urbanism. We argue that the portrayal of climate change as a deeply urban issue has been the result of decades of effective political work to position cities as strategically (and economically) important in the governing of climate responses. At the same time, bringing climate into the city has begun to shift the terrain of the urban present and its futures. Yet the extent to which such shifts have been fundamental in reconstituting global urbanism is moot. We argue that evidence for transformative approaches to climate changed cities are as yet few and far between but suggest that they must form a cornerstone for the future of global urbanism.

Urbanising climate change

Cities have come to occupy a central place in the political imaginary of climate change. This relates both to their material entanglement with the global atmosphere and the political work that has taken place over the past three decades to position cities not only as part of the climate problem but also its solution (Bulkeley and Betsill, 2013). Whilst calculations vary and figures are heavily contested, it is thought that cities currently contribute between 60% and 70% of energy-related greenhouse gas emissions to the global atmosphere and that the vast majority of future

DOI: 10.4324/9780429259593-27

emissions of this kind will arise from the growth of cities in the non-OECD world (IEA, 2016). At the same time, cities are seen to lie in the path of the climate storm: large coastal populations and property assets are at risk from coastal flooding whilst others are seen to be vulnerable to the effects of extreme heat and water shortage (Revi et al., 2014). Whilst the discourse of climate change as a global commons problem in need of a collective international response has persisted, the powerful mobilisation of a narrative that has positioned climate change as a fundamentally urban problem by a host of cities and the transnational organisations that seek to represent them has questioned the naturalisation of the global climate problem as necessarily one of international politics. Coupled with images of climate-related urban disasters – from Hurricane Katrina to Hurricane Sandy, the Cape Town drought to the Parisian heatwave – the rendition of the globally urban nature of the climate change problem has now been widely accepted within both scientific and policy circles. If, for example, it took more than a decade for the Intergovernmental Panel on Climate Change to explicitly mention cities as a part of the climate challenge, by the time of its special report focusing on how to sustain a global temperature rise below 1.5 degrees, cities were identified "as one of four critical global systems that can accelerate and upscale climate action"(Bazaz et al., 2018: 6). Such has been the momentum behind the urbanisation of climate change that in 2018, the IPCC held a meeting dedicated to the urban contribution to the climate change challenge, with the objective of bringing together research, policy, and practitioner communities to carve out an agenda for the next decade of climate science (Bai et al., 2018). With a significant investment of both resources and political capital, CitiesIPCC was unique not only for its focus on the urban agenda and for bringing together diverse knowledge communities from beyond the academy but also as the first meeting ever held to establish the climate research agenda. Marking a significant step-change in the way in which climate change has come to be urbanised over the past three decades, it is a testament to the weight now placed on cities as central means through which climate responses can take place.

Indeed, it has been this framing of cities not only as fundamental to the climate problem but also as essential to its solution that has been critical to the urbanisation of climate change. Over three decades, the framing of the potential for an urban response to climate change has shifted. During the 1990s, the focus was on the actions that municipal authorities could undertake on a voluntary basis to demonstrate their commitment to a global cause – a frame within which cities were mostly regarded in terms of their public authorities and relevant action seen in terms of the kinds of action that could generate both momentum internationally for the cause and additional benefits in terms of financial savings locally. The landmark Kyoto Protocol, after a few false starts as US intransigence made its political future appear uncertain, ushered in a new decade of urbanising climate change as the issue came to be regarded as one of both political and economic strategic importance. The first few years of the new century saw a number of new organisations formed precisely in order to mobilise climate change as a matter of strategic urbanism, from the work of The Climate Group, established to generate interest amongst businesses and subnational governments in the benefits of climate action, to the C40, a transnational network established with the express purpose of bringing together global cities to form a 'climate leadership group' and the explicit engagement of a host of philanthropic bodies, development banks, and global organisations in promoting, supporting, and delivering an urbanised climate agenda.

Such manifestations of climate change as an urban issue were not simply urban in general but urbanised in particular ways that played into and through existing discourses about what it might mean to be a 'global' city on the one hand and along neoliberal lines on the other. For Marvin and Hodson, in an era of "resource constraints and climate change, national security, infrastructure 'protection' and economic competitiveness are being overlaid with concerns around energy security, constraints on water resources, the growth of diseases, increased flood risks and multiple aspects of demographic shifts" in such a manner that the reproduction of the (municipal) state came to be understood as a matter of ensuring 'urban ecological security' (Hodson and Marvin, 2009: 195). Consequently, they suggest, the positioning of global cities as an essential part of the solution to climate change can be read as part of advancing the interests of and futures for particular urban centres and specific communities within them, often at the expense of 'ordinary' cities. Likewise, in her work, Wakefield (Wakefield, 2018: 5) has shown how the mobilisation of discourses concerning the need for resilience in the face of climate change has generated an "imaginary of ubiquitous crisis", which in turn "is ushering in a new set of management techniques to further 'secure' the city and its populations" (see also Braun, 2014; Wakefield and Braun, 2014). Similarly, in their work on climate disaster preparedness in Cancun, Mexico, Manuel-Navarrete et al. (2011) found the persistence of development visions focused on mass tourism coupled with neoliberal discourses of adaptation and resilience served to further entrench urban inequalities. What these studies reveal is the variety of conditions under which climate action in cities can forge synergies between neoliberal economic thinking and global urbanism.

It is clear then that the urbanisation of climate change is neither politically neutral nor without ethical consequence. Positioning cities as climate heroes with added benefits – capable of saving the planet whilst also saving themselves – has served to do particular political work in rendering climate change a problem amenable to solutions which are compatible with neoliberal thinking about the nature of the state, economy, and individual. At the same time, focusing on some of these more prominent discourses is to neglect the multiplicity of ways in which climate has come to be framed as an urban problem with the potential for resolution. Climate change is being rendered urban through diverse actor configurations in ways that seek to generate a more progressive form of politics. Transition towns are one example of such action taking place through a transnational platform, where climate is made urban precisely in order to reconfigure global economic orders and to engender a sense of community, and there are quite literally thousands of other interventions taking place globally in which are also seeking alternative ways of advancing alternative urbanised climate agendas (Hoffmann, 2011; Bouteligier, 2013; Lee, 2013). In practice, urban experimentation forges dynamic connections across social, technical, and political systems, pluralising agency and shifting relationalities within and beyond the spatial bounds of the urban system. Consequently, this heterotopic world of urban climate change experiments helps us understand global urbanism by highlighting the many ways that interventions in cities influence interconnected systems with global reach, in turn usurping traditional spatial boundaries by dissolving distinctions between global systems and urban ones.

Climate changing urbanism

The growing urbanisation of climate change as an issue and a political arena over the past three decades has had consequences for cities – the concerns, visions, interventions, and struggles over climate change have come to change urbanism. In short, if cities have changed climate change, then climate change is also starting to change cities. It is difficult to navigate the global urban without tripping over the accoutrements of climate changed interventions, from green roofs, solar panels, and electric scooters to projects intended to provide community-based adaptation, climate-related interventions are woven into the urban fabric. Analysis of the kinds of activity taking place in cities also bears this out. In their assessment of how non-state actors can contribute to the goals of the Paris Agreement, Hsu et al. (2018: 8) find that by 2017, "7,378 [cities] from 133 countries, representing 16.9 percent of the global population" had declared some level of commitment to action. In the largest assessment of climate change action planning conducted to date, Reckien et al. (2018) surveyed the nature of climate action planning taking place in 885 urban areas in the European Union (EU)-28. They found that "approximately 66% of EU cities have a...[stand-alone] mitigation plan, 26% an adaptation plan, and 17% a joint adaptation and mitigation plan", though also, and critically, "about 33% lack any form of stand-alone local climate plan" with climate action, where it existed, linked into other agendas (Reckien et al., 2018: 208). Finally, climate action is not simply just a paper or plan-based exercise: Davidson et al. (2019: 2) report that the 96 cities that make up the C40 network (and which encompass 650 million people and 25% of global gross domestic product (GDP) have undertaken over 14,000 actions.

Global urbanism in a climate-changed world is materially different. The provision of water and electricity is changing form, with more or less intended outcomes for the future of urban infrastructures and service provision. In some cases, distributed and decentralised provision through rainwater collection and solar panels is having an empowering effect as cities and local communities gain control over access and costs (MacArthur, 2016; Burke and Stephens, 2017). Elsewhere, however, it may serve as a further means through which infrastructure provision comes to be splintered, as elites secure their own resources and abandon public utilities which can no longer afford to extend service provision to those who currently lack access without the profitable consumers who are going off-grid. Climate change is proving a force for the introduction of smart grids and digital technologies, including initiatives that have the intention of managing peak load and integrating renewable electricity into local grids. At the same time, such technologies are creating concerns about the increased surveillance of daily life and the potential effect of creating differential access to services between those who can and cannot afford to manage their peak demand (Bulkeley et al., 2016; Kitchin, 2017). New forms of urban nature are being deliberately introduced into cities in the form of parks and pollinator gardens, walls that grow vegetables and roofs for butterflies and drainage systems with reed filtration and the 'daylighting' of entombed urban rivers (Kabisch et al., 2016). Whilst often with laudable aims for enabling climate resilience alongside other sustainability goals, concerted efforts for urban greening are also being seen as a means through which new kinds of social and economic exclusion are being generated with evidence mounting of forms of 'green gentrification' (Anguelovski, Connolly and Brand, 2018).

Climate change is also serving as a means through which new forms of urban politics are being generated and existing political matters reconsidered. The last two decades have seen a growth of experimentation as a mode of urban climate politics. Experiments are "purposive interventions in which there is a more or less explicit attempt to innovate, learn or gain experience": urban climate change action has largely been characterised by these more informal, heterogenous and, on the whole, fragmented kinds of action taking place beyond the traditional channels of policymaking (Bulkeley and Castán Broto, 2013: 363; Hoffmann, 2011). Experimentation as a political practice is noteworthy for a number of reasons. First, it is facilitating new and unlikely partnerships across diverse political actors at multiple levels of governance. In a survey of climate experimentation in 100 cities worldwide, Bulkeley and Castán Broto found that whilst local governments led the majority of urban climate experiments (66% of experiments surveyed), non-public actors, such as non-governmental organisations (NGOs), community-based organisations, universities, and private-sector firms, were also actively involved in climate governance, comprising nearly a quarter of the experiments surveyed. Moreover, initiatives led by non-public actors were more commonly developed through partnerships, between international organisations and private-sector firms, NGOs and universities, community-based organisations, and regional governments. This suggests that the practice of climate experimentation is forging new relationships amongst non-state and non-public entities and establishing new political spaces beyond traditional policymaking arenas. However, as climate governance moves beyond the public sphere, questions of justice must be reworked in such a way as to reflect shifting responsibilities and rights to act, ensure the establishment or continuation of democratic processes, and recognise existing and emerging vulnerabilities to climate change and climate interventions at the local level (Bulkeley et al., 2014).

Equally fundamentally, bringing climate into the city is serving as a means through which urban futures are being reworked. Partly they are being reworked in material terms as current interventions for climate change are done with particular futures in mind, serving to prioritise some renditions of climate urbanism rather than others. Indeed, different ideas about what a carbon-neutral city will look like are leading to different socio-technical configurations for urban built environments (Tozer and Klenk, 2018). Instead of seeing the imagination of urban futures as a process of navigating amongst various material options with different politics (Stripple and Bulkeley, 2019), contestation and negotiation are often smothered by managerial discourses and consensual politics (Kenis and Lievens, 2016). But even more broadly, as climate has been brought into the city, it has reworked our relations to the future. In the grand plans of the twentieth century, urban futures were equated with optimism and progress, but modernist understandings of the future are being challenged by new ways of relating to a future that is already changed and may already be here. Climate changed urbanism is causing us to grapple with a new question about urban futures: What does it mean to have a future that is not limitless in its possibilities but already bounded by things foretold? Interventions seeking to reimagine and reconfigure global urbanism in a climate change light are heterotopic, generating spaces in which utopian plans and projects are enacted and contested, whilst all the time refracted through the realities of existing urban interests and inequalities (Edwards and Bulkeley, 2017).

Transforming urban futures?

Yet for all of the ways in which the mutuality of climate change and global urbanism have served to generate profound changes in both how we view the nature of global challenges and the configuration of global urbanism, in both a banal sense of the everyday experience of climate changed global urbanism and its political economies, it appears as if much is intact. Above all, global urbanism is insufficiently changed. Trajectories of global urbanism are likely to take us beyond a 2-degree rise in global average temperature above pre-industrial levels, and ways in which climate is rendered as an urban issue have not opened up space for the multiplicity of alternative sites and forms of political engagement through which action could be taken. What would it mean, then, for climate urbanism to be transformative of urban lives and futures? A decarbonised and resilient global urban system requires transforming energy systems, transportation, and consumption. These systemic changes will mean reducing material use, as well as the widespread integration of climate interventions into the urban milieu. But a transformative agenda for urban climate action takes into account that "interventions must be grounded in analysis of structural drivers, and differential capacities for change"(Romero-Lankao et al., 2018: 754) and that climate change is a matter of justice since it creates winners and losers both as impacts occur and as interventions to stop it unfold (Marino & Ribot, 2012). Rather than proceeding along neoliberal lines and through existing discourses about what it might mean to be a 'global' city, the urbanisation of climate change needs to take new political directions in order to enact this transformative agenda. A transformative agenda for urban climate action is about achieving decarbonised and resilient urban systems, as well as reconfiguring structures of power towards more just ends. This entails radical social transformation to address the underlying root causes of vulnerability (O'Brien, 2012; Pelling et al., 2015), as well as action to avoid exacerbating urban inequality whilst reducing greenhouse gas emissions and justly apportioning mitigation responsibilities (Rice, 2014; Fuller, 2017). Whilst many existing urban inequalities might appear unrelated to climate change at first glance, they can represent access points to address the climate crisis. Deepening and broadening material change in cities to address climate change means tackling issues such as differential access to services and infrastructure, for example. Moreover, opening up space for broader political action on climate change means recognising the ways in which urban imaginaries are limited by the influence of existing urban interests. The heterogeneity of current material conditions in urban areas and the ways in which it shapes different imaginaries of urban futures provides openings to understand not only how existing inequalities are contextually assembled but also to illuminate their fundamental relationship with climate urbanism. Enacting a transformative agenda for urban climate action will mean grappling with this question: What if transformative global urbanism means tackling some of the inequalities at the heart of the contemporary urban-climate condition, and what in turn would that mean for urban futures?

References

Anguelovski I., Connolly J., and Brand A. L. (2018) 'From landscapes of utopia to the margins of the green urban life: For whom is the new green city?' *City,* 22(3): 417–436. doi:10.1080/13604813.2018.1473126.

Bai, X. Dawson, R., Ürge-Vorsatz, D., Delgado, G., Salisu Barau, A., Dhakal, S., Dodman, D., Leonardsen, L., Masson-Delmotte, V., Roberts, D., Schultz, S. (2018) 'Six research priorities for cities and climate change'. *Nature,* 555: 23–25.

Bazaz, A. Bertoldi, P., Buckeridge, M., Cartwright, A., de Coninck, H., Engelbrecht, F., Jacob, D., Hourcade, J., Klaus, I., de Kleijne, K., Lwasa, S., Markgraf, C., Newman, P., Revi, A., Rogelk, J., Schultz, S., Shindell, D., Singh, C., Solecki, W., Steg, L., Waisman, H. (2018) *Summary for Urban Policymakers: What the IPCC Special Report on Global Warming of 1.5°C Means for Cities.* Cambridge: Cambridge University Press. Available at ipcc.ch/site/assets/uploads/sites/2/2018/12/SPM-for-cities.pdf.

Bouteligier, S. (2013) *Cities, Networks, and Global Environmental Governance: Spaces of Innovation, Places of Leadership.* New York: Routledge.

Braun, B. P. (2014) 'A new urban dispositif? Governing life in an age of climate change'. *Environment and Planning. D, Society & Space,* 32(1): 49–64.

Bulkeley, H., Edwards, G.A.S., Fuller, S. (2014) 'Contesting climate justice in the city: Examining politics and practice in urban climate change experiments'. *Global Environmental Change,* 25: 31–40. doi:10.1016/j gloenvcha.2014.01.009.

Bulkeley, H. and Betsill, M. M. (2013) 'Revisiting the urban politics of climate change'. *Environmental Politics,* 22(1): 136–154. doi:10.1080/09644016.2013.755797.

Bulkeley, H. and Castán Broto, V. (2013) 'Government by experiment? Global cities and the governing of climate change'. *Transactions of the Institute of British Geographers,* 38(3): 361–375. doi:10.1111/j.1475-5661.2012.00535.x.

Bulkeley, H., Powells, G., and Bell, S. (2016) 'Smart grids and the constitution of solar electricity conduct'. *Environment & Planning A,* 48(1): 7–23.

Burke, M. J. and Stephens, J. C. (2017) 'Energy democracy: Goals and policy instruments for sociotechnical transitions'. *Energy Research & Social Science,* 33: 35–48.

Davidson, K., Coenen, L., Acuto, M., Gleeson, B. (2019) 'Reconfiguring urban governance in an age of rising city networks: A research agenda'. *Urban Studies,* 1–16. doi:10.1177/0042098018816010.

Edwards, G. A. S. and Bulkeley, H. (2017) 'Heterotopia and the urban politics of climate change experimentation'. *Environment and Planning. D, Society & Space,* 36(2): 350–369.

Fuller, S. (2017) 'Configuring climate responsibility in the city: Carbon footprints and climate justice in Hong Kong'. *Area*, 37: 23–27.

Hodson, M., and Marvin, S. (2009) '"Urban ecological security": A new urban paradigm?' *International Journal of Urban and Regional Research*, 33(1): 193–215. doi:10.1111/j.1468-2427.2009.00832.x.

Hoffmann, M. J. (2011) *Climate governance at the crossroads: Experimenting with a global response after Kyoto*. Oxford: Oxford University Press.

Hsu, A., Widerberg, O., Weinfurter, A., Chan, S., Roelfsema, M., Lütkehermöller, K., Bakhtiari, F. (2018) Bridging the emissions gap – The role of non-state and subnational actors. In Christiansen, L., Bois von Kursk, O., Haselip, J.A. (Eds.), *The Emissions Gap Report 2018: A UN Environment Synthesis Report*. Nairobi: United Nations Environment Programme.

IEA (2016) *Energy Technology Perspectives 2016: Towards Sustainable Urban Energy Systems*. Paris, France: OECD Publishing.

Kabisch, N., Frantzeskaki, N., Pauleit, S., Naumann, S., Davis, M., Artmann, M., … Bonn, A. (2016) 'Nature-based solutions to climate change mitigation and adaptation in urban areas: Perspectives on indicators, knowledge gaps, barriers and opportunities for action'. *Ecology and Society*, *21*(2). https://doi.org/10.5751/ES-08373-210239.

Kenis, A. and Lievens, M. (2016) 'Imagining the carbon neutral city: The (post)politics of time and space'. *Environment & Planning A*, 49(8): 1762–1778.

Kitchin, R. (2017) Data-driven urbanism. In Kitchin, R., Lauriault, T.P. and McArdle, G. (Eds.), *Data and the City*. New York: Routledge, pp. 44–56.

Lee, T. (2013) 'Global cities and transnational climate change networks'. *Global Environmental Politics*, 13(1): 108–127.

MacArthur, J. L. (2016) 'Challenging public engagement: Participation, deliberation and power in renewable energy policy'. *Journal of Environmental Studies and Sciences*, *6*(3): 631–640.

Marino, E. and Ribot, J. (2012) 'Special issue introduction: Adding insult to injury: Climate change and the inequities of climate intervention'. *Global Environmental Change*, 22(2): 323–328. doi:https://doi.org/10.1016/j.gloenvcha.2012.03.001.

Manuel-Navarrete, D., Pelling, M., Redclift, M. (2011) 'Critical adaptation to hurricanes in the Mexican Caribbean: Development visions, governance structures, and coping strategies'. *Global Environmental Change*, 21(1): 249–258. doi:10.1016/j.gloenvcha.2010.09.009.

O'Brien, K. (2012) 'Global environmental change II: From adaptation to deliberate transformation'. *Progress in Human Geography*, 36(5): 667–676. doi:https://doi.org/10.1177/0309132511425767.

Pelling, M., O'Brien, K. and Matyas, D. (2015) 'Adaptation and transformation'. *Climatic Change*, 133(1): 113–127. doi:https://doi.org/10.1007/s10584-014-1303-0

Reckien, D., Flacke, J., Dawson, R. J., Heidrich, O., Olazabal, M., Foley, A., … Pietrapertosa, F. (2014) 'Climate change response in Europe: What's the reality? Analysis of adaptation and mitigation plans from 200 urban areas in 11 countries'. *Climatic Change*, 122(1): 331–340.

Reckien, D., Salvia, M., Heidrich, O., Church, J.M., Pietrapertosa, F., De Gregorira- Hurtado, S., D'Alonzo, V., Foley, A., Simoes, S.G., Lorencová, E.K., Orru, H., Orru, K., Wejs, A., Flacke, J., Olazabal, M., Geneletti, D., Feliu, E., Vasilie, S., Nador, C., Krook-Riekkola, A., Matosović, M., Fokaides, P.A., Ioannou, B.I., Flamos, A., Spyridaki, N., Balzan, M.V., Fülöp, O., Paspaldzhiev, I., Grafakos, S., Dawson, R. (2018) 'How are cities planning to respond to climate change? Assessment of local climate plans from 885 cities in the EU-28'. *Journal of Cleaner Production*, 191: 207-219.

Revi, A., Satterthwaite, D.E., Aragón-Durand, F., Corfee-Morlot, J., Kiunsi, R.B.R., Pelling, M., Roberts, D.C., Solecki, W., da Silva, J., Dodman, D., Maskrey, A., Pahwa Gajjar, S., Tuts, R. (2014) Urban areas. In Field, C. B., Barros, V.R., Dokken, D.J., Mach, K.J., Mastrandrea, M., Bilir, T.E., Chatterjee, M., Ebi, K.L., Estrada, Y.O., Genova, R.C., Girma, B., Kissel, E.S., Levy, A.N., MacCracken, S., Mastrandrea, P.R., White, L.L. (Eds.), *Climate Change 2014: Impacts, Adaptation, and Vulnerability. Part A: Global and Sectoral Aspects. Contribution of Working Group II to the Fifth Assessment Report of the Intergovernmental Panel on Climate Change*. Cambridge: Cambridge University Press, pp. 535–612.

Rice, J. L. (2014) 'An urban political ecology of climate change governance'. *Geography Compass*, 6: 381–394.

Romero-Lankao, P., Bulkeley, H., Pelling, M., Burch, S., Gordon, D. J., Gupta, J., … Munshi, D. (2018) 'Urban transformative potential in a changing climate'. *Nature Climate Change*, 8(9): 754–756.

Stripple, J. and Bulkeley, H. (2019) 'Towards a material politics of socio-technical transitions: Navigating decarbonisation pathways in Malmö'. *Political Geography*, 72: 52–63.

Tozer, L. and Klenk, N. (2018) 'Discourses of carbon neutrality and imaginaries of urban futures'. *Energy Research & Social Science*, 35: 174–181.

Wakefield, S. (2018) 'Infrastructures of liberal life: From modernity and progress to resilience and ruins'. *Geography Compass,* 12(7): e12377. doi:10.1111/gec3.12377.

Wakefield, S., and Braun, B. (2014) 'Guest editorial'. *Environment and Planning. D, Society & Space*, 32(1): 4–11.

24

THE GLOBAL URBAN CONDITION AND POLITICS OF THERMAL METABOLICS

The chilling prospect of killer heat

Simon Marvin

Introduction

Under conditions of climate change and continued urbanisation, one of the critical challenges of the twenty-first century will be to manage thermal metabolism in ways that sustain urban life during extreme heat events. Advances in microclimatic knowledge and expertise, emerging landscapes of novel thermal socio-technical infrastructures and new forms of digital weather monitoring and control are emblematic of this shift. Together, they reflect a new logic of re-engineering of the outside environment through fixed and permanent thermal infrastructures designed to protect against too much (or too little) heat with the aspiration of producing 'perfect weather' all year round. The drive is not simply to provide 'comfort and convenience' during heatwaves, but it becomes increasingly critical to secure urban life within safe limits through thermal modification. The chapter charts these developments and asks more critical questions about the social implications of thermal modification. Why has urban thermal metabolics become so important? What sorts of cities are investing in thermal modification? Who will be included or excluded from the thermal effect of these new capacities, and do they represent new forms of urban socio-spatial advantage and disadvantage that weakens protection for the less affluent? What are the wider socio-ecological implications of urban thermal engineering?

Overheating and urban thermal metabolism

Thermal management of the outdoor environment will become as critical to supporting urban life as air-conditioning has been to enabling urbanisation to transcend climatic limits by standardising indoor environments (Healy, 2008). There are three well-understood and researched sets of factors – that due to space we can only briefly consider here – whose complex intertwining can help account for the emergence of an intensified problematic of urban overheating (see Rikards, 2019 for a review).

The first is climate change that is already leading to significant temperature rises of 1.5°C that dramatically increase the intensity, frequency and duration of extreme thermal events. Critically, these changes will increase the scale of coverage and the variability of extreme heat events to affect many more urban areas. Life-threatening heatwaves are expected to escalate from being

DOI: 10.4324/9780429259593-28

exceptional to normal and regular events. The second is that the implications of these global and regional changes are amplified by urbanisation. The Urban Heat Island (UHI) is how urban climatologists understand the ways in which the urbanisation process itself – through removal of vegetation and introduction of hard surfaces – reshapes urban microclimates to the extent that the heat in urban areas is typically much higher than surrounding areas. There are also dramatic differences in temperatures within an urban area. This 'ratcheting up' effect of the material conditions of the urban environment creates much higher temperatures than would otherwise be the case – that can kill humans, animals, plants and result in technical failures of key equipment and essential infrastructure. The third factor is the structural differences in the ability of urban populations to cope with extreme heat (see Klinenberg, 2002). Poor-quality housing, absence of air-conditioning, lack of greenery and trees, low mobility and poor health all greatly exacerbate and intensify the destructive consequences of heat extremes on urban populations. Typically, low-income and black communities disproportionality suffer from the effects of heat events – they also have severe consequences for outdoor workers, the elderly, the ill and children who cannot easily cope with the effects of extreme heat.

This changing context means there is an urgent need to extend research interests into analysis of urban thermal metabolisms (see Capriotti and Romanowicz, 2013). The challenge is to construct an understanding of the material and social intertwining of both the urban environment *and* the circulations of heat. A urban thermal metabolic framework could be constituted through (i) knowledge and understanding of the physical heat inputs, sinks and outputs and its spatial and temporal variability in configuring the differential exposure of humans and materials to heat; (ii) the analysis of the social, material, ecological, technological and economic interrelationships that contribute to the production of particular thermal environments at multiple scales, and; (iii) unpacking the differing political contexts through which the thermal environment is constituted as both a problem and an object of intervention focussing in particular on the social and ecological inequalities that may be concealed (or revealed) by these responses. Whilst this short chapter cannot address all these issues, we explore the contours of an emerging urban thermal politics and in doing so highlight what is at stake for the future development of global urban studies.

An emerging urban thermal politics

International non-governmental organisations, climate experts and urban social movements are starting to examine how the new thermal landscape will reshape the future conduct and experience of everyday urban life. This begins to make visible the critical socio-material connections between global climate change, urban overheating and the thermal regulation of the human body. Critically, this work shows how these processes intensify existing patterns of urban inequality by creating extremely challenging and life-threatening thermal problems that are then re-intensified in the mid-century in three contexts:

First, 'Sustainable Energy for All' (SEA), in their 2018 report "Chilling Prospects", demonstrates the huge scale of the problem of extreme heat and the complex material and social transition required to provide sustainable cooling for all. This report identifies those countries and populations facing the biggest risk, the considerable economic costs of working hours lost, expected food losses and damaged medicines and vaccines from lack of cooling. A total of 1.1 billion people face cooling access risks, of whom 630 million are 'slum dwellers' in hotter climate urban areas with absent, intermittent or too expensive electricity. Another 2.3 billion people, primarily urban dwellers in China, India, Indonesia and Brazil in the 'affluent lower middle class', are about to be able to buy the least energy-efficient air conditioners – potentially increasing global energy demand an incredible 33 times by 2100. The SEA work outlines the need for both passive and active measures in existing and new cities and the development of new highly energy-efficient cooling options to provide sustainable cooling for all.

Second, the 'Union of Concerned Scientists' (UCU) in 2019 published a report titled "Killer Heat in the United States" that provides an analysis of the mid- and late century implications of continued climate change on the frequency, duration and location of heat extremes. If emissions stay on their current path, then extreme heat days will cause "large areas of the United States to become dangerously hot and would threaten the health, lives and livelihoods of millions of people" (2019: 1). Heat, it is argued, is likely to affect the lives of most Americans more than any other aspect of climate change. The UCU work on the spread and frequency of extreme conditions demonstrates that future heat extremes are quite literally 'off the charts' of the existing National Weather Service Heat Index. A third of US urban areas will experience an average of 30 days with a Heat Index of 105°F affecting over 90 million people. The implications will be unevenly felt – millions of outdoor workers will face elevated risks in construction, landscaping, military, police and postal sectors, and heat-related illness and deaths will rise. Because of the UHI, extreme heat is a threat to city residents, especially those of low-income status because they lack air-conditioning and live in the "hottest parts of cities" (2019: 25).

Finally, new urban coalitions are being formed in response to these heat extremes. A partnership of organisations has formed the 'Heat Relief Network' (HRN) to provide hydration stations, refuge locations and water donation sites across Phoenix, Arizona, to prevent heat-related deaths amongst vulnerable populations. In 2018, there were 181 heat-related deaths amongst the homeless, and three-quarters of these occurred in the 'outside' environment during July and August. By the end of the century, the climate in Phoenix will be similar to Kuwait today. Yet it is already too hot now – and new adaptations are being made to deal with extreme heat, as construction work is undertaken at night to protect workers and prevent damage to building materials, many outdoor workers start early and aim to finish at 2 p.m. at the latest and local parks are staying open late at night so that people can experience the 'outdoors' in the cooler night. In response to local demands, the HRN has installed a water misting network in a bus stop so that the homeless can get some relief from the intense heat.

What these critical interventions start to graphically illustrate is that extreme heat will not only make many urban areas uncomfortable but also drive many people indoors during the day; alter work patterns, especially for outdoor workers; reduce the effectiveness of infrastructure; and threaten life. Although urban coalitions are developing 'cooling strategies' for heat emergencies, the scale and scope of the socio-material response required to modulate the thermal environment are daunting.

Urban thermal infrastructures: Outside cooling and heating

In many global urban contexts, there are ongoing attempts to render a problematic, denuded or threatening outside environment manageable through new urban infrastructures of urban weather modification. Compared to the indoor environment, it is extremely challenging to temper and modify the outdoor environment. There are a wide range of different thermal adaptation techniques, including psychological adjustment to high heat, behavioural changes, environmental modifications and provision of technical infrastructures. The comparative analyses of the performance of these measures are clear that carefully configured packages using all of these measures are required. But the evidence is clear that technological infrastructures are the most effective in reducing or increasing temperatures and provide the greatest flexibility in 'targeting' thermal cooling and heating capacity. A new landscape of thermal infrastructures that intervene during disruptive weather events is beginning to emerge in a piecemeal and fragmented manner that seeks to modify the outdoor environment.

An increasing number of small- and medium-sized enterprises now provide outdoor cooling services in the United States and many other contexts Australia, China, Europe, the Middle East and Japan. For example, Koolfog based in California has been "enhancing outdoor environments" for more than 30 years (see www.koolfog.com). Koolfog's misting systems provide users with an "effective means to control their outdoor settings, allowing more consistency in patronage and happier customers", which is viewed as critical for the profitability of outdoor customer-focussed sectors, such as restaurants, resorts, retail centres and theme parks. Markets also include the military, firefighters and schools where they can ensure staff and pupil safety whilst "enhancing student performance" in heatwaves. Cooling systems can also be configured to remove heat from residential outdoor spaces in order to "extend the square footage of your house, the time you can spend outdoors". Additionally, Koolfog misting systems are targeted at plants and animals in greenhouses, dairy farms, equestrian centres, zoos and aviaries to provide improved health, increased productivity and decreased losses attributed to the "adverse effects of a 'less-than-ideal' environment".

Koolfog's misting systems are urban infrastructure systems that pump high-pressure purified water through special nozzles that create a very fine mist that cools the temperature by up to 35°F. This is a complex infrastructure that requires water, power, networks, treatment and control systems in order to construct a working socio-technical system. Figure 24.1 illustrates the main elements of the system and the five critical elements of the infrastructure. The misting infrastructure becomes a networked part of urban life, operating in a wide range of different contexts and at different scales from individual homes, networks of cafes and entire city blocks.

The second example of thermal infrastructure is focussed on the supply of outdoor heating in response to addressing a surplus of excess cold, snow or ice that disrupts urban life. The heating of pavements, squares and roads is utilised in Iceland, Norway, Italy, Japan and the United States, and it is also now being extended to northern China. Specialist companies also provide outdoor heating for cafes and restaurants, heated ramps, roads and distribution areas at airports, railway stations and even heat the drives of high-end residential homes. Many domains of urban life are

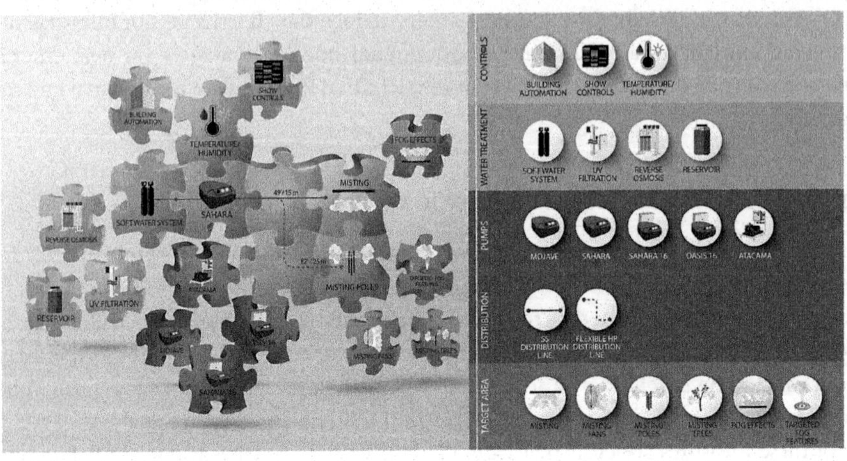

FIGURE 24.1 Main components of a misting system (source: Koolfog)

being subject to external heating either as a public infrastructure or private capacity in order to provide warmth and prevent the formation of or remove snow and ice more rapidly than would otherwise occur through 'natural' snowmelt.

The city of Holland in Michigan, United States, has the most extensive outside urban heating system in the world with nearly 5 km of heated pavements and roads (see https://www.holland.org/snow-free-holland). Holland has an architecturally significant historic core that was widely viewed as being at threat from a new indoor large-scale shopping complex constructed in 1988 with an 'all-year-round', temperature-controlled environment. In response, a local business leader provided funding for a small, heated road and pedestrian network that over time was extended to much of the downtown area. Waste heat from power generation is captured to heat water that is circulated through 190 miles of tubing laid underneath the roads and pavement, returning back to the Holland power plant – see Figure 24.2. The system pumps over 4,700 gallons of water per minute at 95°F and can melt approximately one inch of snow per hour. The system enables snow-free vehicular and pedestrian access all year round; it stretches the period in which customers can easily and safely use the downtown area and is credited with ensuring that the shops and restaurants can compete with the enclosed mall. It also avoids the need for salting, snow ploughing and snow disposal, and there are fewer personal and vehicle accidents due to slipping and sliding. Many urban authorities in North America, most recently Montreal, have been looking at the Holland system and the feasibility of installing the infrastructure in their downtown area.

The misting systems of Koolfog, the heating network of the City of Holland and the responses of many other companies and cities in the growing sector of outdoor weather modification are an infrastructure in the making. Selectively, the particular thermal qualities of the outdoor urban environment are being constructed as a problem which can then be rendered calculable and manageable through a new thermal infrastructure. The development of these infrastructures raises a set of critical issues about how they make, reinforce or create new contours of uneven access to thermal comfort and the conditions for ensuring the safe reproduction of life itself.

FIGURE 24.2 Urban snowmelt infrastructure (source: City of Holland)

Urban climate modification: Simulation and greenwash, comfort and convenience or essential climate security

Urban climate is now no longer being viewed as an atmospheric commons that is given, fixed by biophysical processes and immune to human shaping. We now have a context in which urban climates are not only subject to human modification through climate change and urbanisation but also being viewed as knowable, calculable and governable to achieve targeted priorities. Whilst at an early stage of development it is clear that a microclimate modification capacity is being developed to reconfigure thermal environments. Why does this matter to global urban studies?

First, these processes of thermal modification might be considered as a mode of climate greenwash that simply creates the impression of intervention through dramatic misting infrastructures without fundamentally altering the thermal metabolism of the city. Historically, when looking at the origins of the misting industry, many companies actually started in the entertainment sector – providing misting services as special effects for the film industry, theme parks and zoos to recreate tropical or exotic environments. The logic was not to actually change the weather but instead to create the impression of wonder, amazement, the thrill of being in a different location for users. The first is reputed to have been in 1970 at the exhibition in Osaka, Japan. Visitors noticed the cooling effect, and over time, many of the companies extended their offer to include misting for cooling purposes.

Second, cooling (and warming) is about enhancing human comfort and convenience. In this context, the point is that thermal infrastructure can provide the capacity to extend the use of outside spaces – even in extreme heatwaves and or the severe winter months. These forms of intervention provide a new thermal fix that can (within limits) transcend local weather conditions to create perfect all-year-round weather for enabling consumption activities outdoors. Clearly, the provision of a thermal infrastructure can provide an advantage when compared to unmodified thermal spaces. The providers of outside cooling services have constructed return on investment metrics for assessing the commercial value of the additional usage of outside space during heatwaves – for example, working with clients such as Starbucks to provide competitive metrics for intervening in weather modification. In new urban developments, specialist microclimatic consultants are developing packages of passive and active design measures to extend the usability of outside space. The most systemic recent example is Sidewalks Labs whose consultants have designed a package of digitally controlled measures to significantly increase the annual number of "thermally comfortable" daytime hours. It may be possible "to more than double the number of comfortable daylight hours each year for key spaces – from about 30% to 74%" (see https://rwdi.com/en_ca/knowledge/blog; Figure 24.3). Urban thermal infrastructures may constitute a new premium urban fix designed to provide comfort and convenience for select customers to continue comfortably using the outdoors in weather extremes.

The third potential logic is when cooling services become critical to sustaining human and non-human urban life and the performance of existing infrastructures as a logic of 'climate security'. There are a number of examples of powerful connections being made with the thermal metabolics of the body and the need for specialist cooling infrastructure. The Sydney City Council has opened air-conditioned public buildings during heatwaves for the vulnerable, but existing experience has shown that these are insufficient, and it has now recommended that "special-purpose refuges – hydration stations, air-conditioned refuges and emergency refuges" will be needed. A number of misting companies are providing cooling services to school playgrounds and sports areas justified by medical research showing that learning capacity deteriorates with heat extremes. Another area where companies have been extremely active is in providing

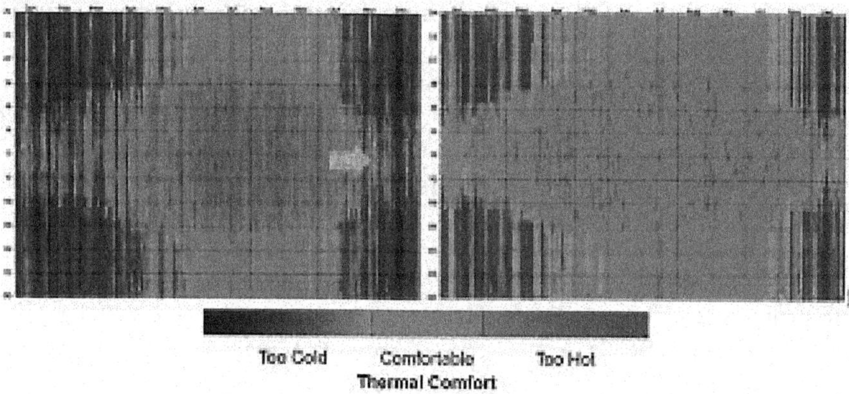

FIGURE 24.3 Extending usable outdoor daylight hours (source: RWDI)

portable air-conditioning units for outdoor dog kennels that protect pets from both heatwaves and intense cold. There has also been a significant growth in portable air-conditioning units to cool outdoor ATMs and vending kiosks that cannot operate effectively with the new extreme temperatures. In these cases, the ability to provide outside cooling is not solely for comfort and convenience; it becomes critical infrastructure for the maintenance of human and non-human life and to ensure the functioning of everyday infrastructures.

In the context of increasing climatic turbulence which will produce more frequent, longer duration and more intense heatwaves (and cooling) in urban areas, the strategic importance of active thermal infrastructure is likely to increase and potentially become normalised as an increasingly taken-for-granted solution. Whilst recognising the initial emergence and diversity of existing thermal interventions and the possibility of multiple and hybrid logics for the provision of infrastructure, there is clearly a subtle but fundamental shift happening in which these capacities are not solely about 'comfort and convenience' but becoming central to ensuring the maintenance and reproduction of survival of urban life. As such, we need to subject the emergence of urban thermal infrastructures to critical scrutiny to understand the extent to which these become uneven enclaves of cooling and heating in a denuded climatic landscape and the socio-spatial tensions and contradictions this produces. This raises critical challenges for how we conceptualise global urbanism that can rethink the categories, methods and geographies through which a reinvigorated politics of thermal metabolics can be understood.

Conclusion

The aim of this chapter has been to examine the global urban issues raised by an accelerated and intensified process of overheating. Research has already identified uneven social access to cooling and the social equity issues raised by exposure to extreme heat events. Future responses to overheating will clearly have major implications for social justice and for the maintenance of life itself as humans seek to adjust to a new world of thermal discomfort that exceeds the ability of existing techniques to provide artificial cooling. It is difficult to underestimate what is at stake here. Almost every aspect of life – the specifications of buildings, the operational norms of infrastructures and the reproduction of human and non-human life have been configured around a taken-for-granted and relatively stabilised thermal landscape that is now fundamentally changing.

The new thermal landscape will more frequently, intensely and protractedly exceed the safe, comfortable and reliable operating limits of existing networks, materials and life. Conventional responses to heatwaves are not reliable; for example, the effectiveness of air-conditioning systems falls dramatically in extreme heat, and the intense peak energy demands associated with heatwaves seriously threaten the reliability of energy grids. Whilst the existing focus has been on sustaining urban life by trying to provide thermal 'comfort and convenience' during heatwaves, the future challenge will be more serious as excess heat will not only kill the low income, ill and elderly but also threaten the lives of the prosperous, healthy and fit.

This intensified thermal politics needs to become a more central concern for urban research as it fundamentally reshapes the socio-material context of the future development of global urbanism.

There are three key research tasks for a critical global urban studies of thermal metabolics. The first task is to explore the interplay between an intensifying thermal metabolics and the complex geographies of uneven development. It is clear from the examples we have considered that the challenge of overheating cuts through existing topological models of urban/rural and first/third world and Global North/South. We need to understand how overheating and thermal engineering mutually reinforces existing, and produces revised, patterns of uneven development. Critical to this is understanding the interweaving of the differential spatialities and temporalities of excessive heat with existing processes of urbanisation and the uneven capacity to provide a systemic response.

Second, research is needed to build a more differentiated understanding of intersections between thermal metabolics and diverse and divergent urbanisation processes in different urban landscapes. For instance, how are unprecedented levels of densification made compatible with future forecasts of intense overheating, and how is infrastructure made to work and the survival of human life secured in such contexts? How do new thermal extremes reshape the metabolisms of human and non-human life in cities and in the extended operational resource landscapes that support urban economies. Third, research is needed to explore how thermal metabolics is addressed as a collective project in the highly diverse and variegated capacities of urban governance. Who or what will be the priority for public and private investment? Will economic development take precedence over wider public benefit given the pressure for new 'thermal fixes' of amenable climate contexts for continued growth? Will the infrastructures and systems necessary to guarantee comfort, convenience and even survival be delivered through privatised and liberalised providers in private enclaves or as a collective infrastructure? What new political struggles will emerge as thermal metabolics is intertwined into political struggles around access to basic infrastructures of water, food, energy and security?

References

Caprotti, F. and Romanowicz, J. (2013) 'Thermal eco-cities: Green building and urban thermal metabolism'. *IJURR*, 26 (6): 1949–1967.

Healy, S. (2008) 'Air-conditioning and the "homogenization" of people and built environments'. *Building Research & Information*, 36(4): 312–322.

Klinenberg, E. (2002) *Heat Wave: A Social Autopsy of Disaster in Chicago*. Chicago: University of Chicago Press.

Rikards, L. (2019) Making climates through the city, in Heijden, J., Bulkeley, H. and Certomà, C. *Urban Climate Politics: Agency and Empowerment*. Cambridge: Cambridge University Press (pp. 80–97).

Sustainable Energy for All (2018) Chilling prospects providing sustainable cooling for all. July 2018. Sustainble Energy for All: Vienna.

Union of Concerned Scientists (2019) Killer heat in the United States – climate choices and the future of dangerously hot days. July 2019. Cambridge, MA.

25

ON THE DEPLOYMENT OF SCIENTIFIC KNOWLEDGE FOR A NEW URBANISM OF THE ANTHROPOCENE

Vanesa Castán Broto

Introduction

According to Eugene McCann (2017: 312), urbanism "*connotes the widest sense of the urban, addressing not only urbanization and development but also ways of life that define urban areas in specific historical periods.*" The scale of unprecedented global environmental change, including climate change and mass extinctions at a planetary level, generates social responses that become expressed in a new urbanism of the Anthropocene. This new 'climate urbanism' is linked to the historical rise of climate change agendas that prioritize the urban as a site of action (Parnell, 2016). At the same time, climate urbanism underpins accelerated efforts to securitize urban areas both in terms of gaining protection from climate change impacts and reducing the effects of decarbonization actions on privileged lifestyles (Long and Rice, 2019). The global urbanism of our time is a climate one, motivated by our collective anxiety about our capacity to survive the global environmental crisis and grounded on our joint efforts to understand the Anthropocene.

The Anthropocene is also the age of global environmental assessments. These assessments examine the current state of knowledge about environmental problems and solutions on a planetary scale. They include various modalities of deliberative panels of international scientists, such as the Intergovernmental Panel on Climate Change (IPCC), United Nations Environmental Programme's International Resource Panel, the Intergovernmental Science-Policy Platform on Biodiversity and Ecosystem Services (IPBES), UN-Habitat's World Cities Report, and the Global Commission on Adaptation, among others. These assessments examine the organization and planning of human settlements, among other aspects of life in the Anthropocene. Global environmental assessments connect future visions and aspirations across multiple international contexts, supporting a scientific-led brand of global climate urbanism.

Global environmental assessments are tools to increase the scientific influence on policy decisions over time. Thus these assessments may shape how urban areas are seen and lived. For example, in 1983, the United Nations Secretary-General appointed Gro Harlem Brundtland as the head of the World Commission on Environment and Development, with the explicit purpose of building a shared global vision for all countries to pursue sustainable development. The report published in 1987 (Brundtland et al., 1987) inspired the 1992 Rio Declaration, a short but bold document that pointed toward an international consensus on the concept of sustainable

DOI: 10.4324/9780429259593-29

development. Sustainability then entered the mainstream, influencing social life from political debates to daily practices (Banister et al., 2019). Cities and urban areas became central in the implementation of sustainable development, despite growing critiques of the predominance of techno-economic fixes in urban sustainability agendas (Marvin and Guy, 1997; Hodson and Marvin, 2017).

Today's mission for international scientific panels is not only to deliver a consensual vision of a planetary plan for action in urban areas and elsewhere but also to provide new arguments for hope in the context of the impending catastrophe. Environmental activists like Greta Thunberg repeat the phrase "ask the scientists" as a mantra. However, neither assessments provide an ultimate objective truth nor are scientists infallible automata with the right answers at hand. The mantra "ask the scientists" is valid only insofar as global environmental assessments result from painstaking deliberation between people who have dedicated their lives and careers to understanding the subject matter. They are also developed with a sense of respect for policy makers, urban planners, infrastructure managers, business leaders, community organizers, activists, and citizens who put ideas in practice and who become the active agents of any form of global urbanism.

The influence of these assessments cannot be overstated. While they are all different in their conception and operation, some common characteristics explain their impact on shaping contemporary global urbanisms. Assessments act as institutions that validate what knowledge is relevant or not to address urgent problems. They make visible the connections between knowledge generation and environmental politics. In the long term, discourses advanced by international scientific panels become part of our daily vocabulary and repertoire of actions. In this chapter, I reflect on the experiences of working within international scientific assessments and expert groups and working to build new vocabularies for urban futures in the hope of influencing new climate urbanism models of the Anthropocene. Rather than providing ultimate truths, scientific panels at their best become open, deliberative forums for the generation of multiple alternatives and ideas that can be materialized in multiplex urbanisms. However, as the sections that follow explain, existing structures of knowledge production may limit their potential.

Building consensus in a heterogeneous, chaotic world

The literature on science and technology studies has linked the construction of particular imaginaries of the future to the actual performance of science as a form of consensus. Jasanoff (2015: 11) explains how "the imaginative work of varied social actors" enmesh science and technology "in performing and producing diverse versions of the collective good, at expanding scales of governance from communities to nation-states to the planet." The emphasis on consensus around complex environmental policies is not always actionable in concrete measures and plans for action (Gillard, 2016). Often, leadership is framed as consisting of daring activities to lead complex and sometimes not-so-popular green agendas. A critical barrier to deal with socio-ecological complexities in international scientific assessment is the limited dialogue between disciplines and the realization that most scholars will always use the frame of reference of their discipline as a starting point to engage in any form of scientific dialogue (Castán Broto et al., 2009). Since many cultural and personal boundaries have to be negotiated simultaneously in a short period, finding common disciplinary ground is essential.

Such dynamics disadvantage critical social science. For example, a critical social scientist whose analysis requires questioning deeply entrenched assumptions about knowledge production may find it challenging to reaffirm an opinion against a self-assured engineering scientist who automatically links a precise definition of the problem to the delivery of a solution on the ground. The critical social scientist may focus on problematizing the issue at hand. In contrast, the engineering scientist is already engaged in thinking about how to solve such a problem, accepting that finding a solution will require defining the problem in a precise (also read: narrow) way. Most scientists receive limited training in communicating across disciplines. In the end, only interpersonal relations may enable negotiating these painful divides (most often around a beer, which further excludes those who do not share that way of interaction).

If global environmental assessments are to play a role in shaping global urbanism, they must establish the relationship between citizens and resources. Urban infrastructure models define service provision within a set of socio-ecological relations. But what is urban infrastructure? I find it hard to imagine urban infrastructure as characterized by anything else than a profound heterogeneity that emerges from the assemblage of physical elements, alongside social relationships, expectations, and emotions. The focus on heterogeneous configurations of infrastructure requires understanding that multiple ways of governing and accessing infrastructure coexist in cities (Lawhon et al., 2018). However, such a perspective is not immediately suited to the construction of a space of intervention through infrastructure actions. Classical models of infrastructure, structured in a more or less centralized network, provide more persuasive answers routinely shared across the disciplinary spectrum. As a social scientist, I often find myself arguing for citizen and community-led alternatives for environmental management. These alternatives become immediately criticized for their lack of scale and replicability in the face of the promises of public and private provision of centralized infrastructure.

Many urban infrastructure models found in almost every report on sustainable urbanism are wholly inadequate to address the notions of urban infrastructure in actual cities, mostly characterized by incompleteness and heterogeneity (Lawhon et al., 2018). Most cities operate with multiple hybrid arrangements that planning models consistently overlook (Monstadt and Schramm, 2017). Heterogeneous, hybrid arrangements become visible alongside forms of spatial informality, but they are also present in urban areas thought of as 'formal' where change and experimentation are ongoing. There cannot be a consensual view of urban infrastructure, as it is built from complex arrangements of materials and social institutions and embedded in practices of everyday life. Static models of homogenous infrastructure are unable to explain how the urban environment is constructed through daily practices by urban dwellers in a myriad of overlapping infrastructure landscapes (Castán Broto, 2019). Nevertheless, models of infrastructure as organized in centralized networks pervade global environmental assessments.

I have so far been unable to communicate this understanding of infrastructure in any assessment. I struggle to describe it in a manner that makes my definitions workable in a solutions space. Nevertheless, the very complexity of urban infrastructure and the difficulties of translating this heterogeneous vision into a viable, solutions-oriented definition haunts many such reports. Taxonomical descriptions of urban infrastructure map the realm of intervention dividing it into ideas of green infrastructure, physical infrastructure, and social support infrastructures, which in the end become inadequate to address the challenges to the non-networked infrastructure on which the most disadvantaged people rely (circuits of fuel provision, shared support systems, informal structures of financial support and debt, and many other parts of the invisible city).

In conclusion, even considerable efforts to deliver diversity and plurality in sustainability assessments struggle to bring to the forefront visions of global urbanism as analytical models struggle to reflect the particular, the quotidian, the chaotic, and the random nature of urban life. These assessments seek to organize the world in neatly packaged spaces for action, which do not always match the realities of urban life. Areas of discussion, patience, and openness may enable engagement with temporary models of urbanism still open to negotiation. However, this may not be sufficient when confronting the entrenched hegemonic paradigms that support global environmental assessments.

The elusive policy-making audience

Delimitating the audience is key to global environmental assessments. Who will read the assessment and for what purpose? Scientists are one of the primary audiences for these assessment reports. They use the reports to frame and inform research. Assessment reports are routinely cited in the introduction of journal articles and have a powerful impact on helping define the relevance of research. In this context, authors can assume that the reports will be read with nuance by their colleagues and that the assessment will help to bridge debates across disciplines.

However, scientists are not necessarily considered the primary audience of global environmental assessments. Instead, global environmental assessments are thought to support deliberative processes in policy making (Kowarsch et al., 2016). In the changing context of environmental policy making,

> as assessments have become an established feature of the international environmental policy landscape, decision-makers, practitioners and scholars are demanding more explicit focus on analyzing the suitability of specific response options and policy pathways that range from technologies and behavioral change to policies, such as regulatory measures or market-based instruments. (Kowarsch and Jabbour, 2017)

In this new framework of solutions-oriented assessments, there is an increasing wariness about the extent to which such assessments can include out-of-the-box perspectives or alternative public policies, such as those which engage with radical, bottom-up experimentation in urban areas. Instead, scientists are called to consider the interest of policy makers and governments at the heart of the assessments in a context of constrained coordination capacities and spaces for dialogue (Carraro et al., 2015).

When assessments focus on cities and urban areas, they have to address the interests of policy makers, city managers, and planers in ways that restrict the framing of the report and define what is within or without its scope. For example, the terms of reference of a consultancy contract to deliver an urban-based assessment contained the following prescription:

> The primary audience of [this assessment] are policymakers. Another major audience for [this assessment] is the general public. It is thus essential that the discussion...should be empirical and evidence-based. Furthermore, it is essential that the chapter is written in a non-academic language, taking into account that many readers of [the assessment] will have English as their second language.

What policy makers are and what they want is a difficult question and one that requires situating those policy makers in a specific context of decision making. However, the image of the policy maker is a powerful one, and one that decidedly shapes the content of this type of assessment. The policy maker becomes a 'receiver' of the information. This image also informs judgments about what the report should include and exclude. While there is often an expressed aspiration that policy makers should define the problems for scientists to contend with, scientists or experts continue to identify issues in terms that fit their points of view regardless. However, the image of the policy maker as an audience alone shapes the justifications that scientists would offer for the framing of the assessment as something that is acceptable/useful/valuable for the imagined policy maker.

I experienced this recently when I tried to introduce an ecocentric (rather than an anthropocentric) perspective on environmental valuation in urban areas. As I presented an outline to a group of international experts, I argued for the consideration of ecocentric values with a perspective necessarily building on theories deriving from critiques in philosophy and critical social theory. The panel suggested that my proposal was too academic and out of the scope of the report. Despite the long history of ecocentric thought in environmental philosophy and politics, it remains an approach that eludes practical application (Hettinger and Throop, 1999). Moreover, ecocentric thought has been systematically excluded from political life, even among sectors promoting radical thinking about the environment (Kopnina, 2012). In the case of urban areas and climate urbanism, there is still a gap in terms of how far the ecological foundations of cities can be incorporated into assessments of global environmental futures beyond utilitarian perspectives on nature-based solutions and green infrastructure.

This experience is not uncommon. I have frequently observed that members of global environmental assessments put aside radical ideas as examples of elite academicism. Is this a common mechanism of exclusion of any form of radical thought? At the time of this anecdote, I was reading Reni Eddo-Lodge's popular best seller *Why I'm No Longer Talking to White People About Race*. The parallel reading made me realize how this style of argumentation is frequently deployed in other spheres of public life. Eddo-Lodge explains why she finds it useful to use the label "white feminism" to reflect on the negation of black feminists' critique of a particular brand of feminism defined in terms that respond to the needs of the white population (Eddo-Lodge, 2018). According to Eddo-Lodge, claims of elitism silence concerns that emerge from within marginalized and less-heard groups, such as black feminists in the context of British politics. Black feminists, for example, have embraced the notion of intersectionality as a means to advance politically both the agendas of feminism and anti-racism. However, as Eddo-Lodge explains, this has had complex consequences. She states,

> When black feminism started to push for an intersectional analysis in British feminism, the widespread response from feminists who were white was no one of support. Instead, they began to make the case that the word 'intersectional' was utter jargon – too difficult for anyone without a degree to understand – and therefore useless. (Eddo-Lodge, 2018: 160–161)

She then goes on to describe examples in which high-profile white feminists in the United Kingdom disregarded intersectionality as an esoteric, elitist discourse that cannot be used by girls or women (or policy makers!) for practical purposes. This dynamic is common and very dangerous. By signaling concepts and ideas as elitist, they are quarantined and remain undiscussed. New examples, empirical materials, and out-of-the-box ideas for possible responses lack development because the discussion gets stuck on the legitimacy of the debate alone. However, those who claim elitism are often those in better positions to argue the case. White feminists over black feminists, in the case of Eddo-Lodge's call for intersectional analysis. Engineers over critical social sciences in my case. If addressing this debate is difficult when we talk about minorities with a clear-cut claim to public life, such as black feminists in Britain, it is even more complicated when we are thinking about the political representation of those parts of the world that have no access to humans' representation of consciousness – e.g., advancing ecocentrism or ecological values or the values of future (not yet existing) generations.

The possibility for a more just, ecologically sound global urbanism of the Anthropocene requires the use of all the tools at our disposal, from complex theory to ready-made solutions. The sheer complexity of socio-ecological relations and their embeddedness in politics makes it impossible to ignore that climate urbanism requires sophisticated, theoretically rich critiques and that the charge of elitism may also silence the concerns of the most vulnerable and neglected in favor of an abstract, mythical figure of 'the policy maker.'

Toward recognition of multiple knowledge systems

Amid these conflicts, I find myself drawn to these global environmental assessments not only because of their importance but also because of their true potential to facilitate the kind of radical change to address the ecological crisis of the Anthropocene. Global environmental assessments have the potential to improve processes of deliberative governance and transform entrenched frames and paradigms. They are increasingly linked to multilateralism in international politics (Jabbour and Flachsland, 2017). Despite the criticisms, the Brundtland report remains an admired effort to deliver a policy discourse on sustainability (including urban sustainability) still relevant today after more than 30 years.

Members of current assessments, such as the IPCC (Ford et al., 2016) and IPBES (Díaz et al., 2015; Löfmarck and Lidskog, 2017; Tengö et al., 2017; Peterson et al., 2018), have made efforts to make the assessments open to a variety of ways of seeing the world. The incorporation of indigenous knowledge and local knowledge, for example, is increasingly a concern for both intergovernmental panels. However, the integration of indigenous and local knowledge into a scientific assessment means that such knowledge has to be validated in scientific terms. Indigenous knowledge views, however, cannot be subsumed or validated through scientific means (Gratani et al., 2011; Matsui, 2015). Validation may disrespect indigenous peoples' own identities and histories and overlooks the fact that, for indigenous peoples, the structural drivers of oppression and endangerment are associated with Western civilization and the knowledge structures that support it (Ford et al., 2016). Thus the question is not only bridging different disciplines and different theoretical and empirical orientations but also, most of all, bridging non-commensurable systems of knowledge. The strategy has to go beyond a mere question of integration, focusing on the recognition of the fundamental incompleteness of human expertise in any part of the world and the need for deliberative methods that enable the development of collective, solidary, transient agreements about making together a sustainable world. This is particularly relevant to understand the progressive potential of global urbanisms, through the recognition of the multiple ways of interpreting social and ecological relations and how they shape urban life.

I see three ways forward. First, to counteract the scientific and managerial consensus that tends toward the consolidation of uniform models of urbanism, we need to celebrate the defining characteristic of contemporary urbanism in a global context: its heterogeneity. Despite attempts to incorporate plural views, the style of scientific debate tends to privilege models of urbanism based on engineering assumptions about the world. How can we resist perspectives that reduce the range of interventions and the actors that can be included? Emotion-based or experiential arguments need to find a space in mainstream understandings of valid scientific knowledge.

Second, as actors within scientific assessments attempt to shape their messages to the needs and demands of policy makers (as argued by Carraro et al., 2015), we need to practice the art of constructing a diverse audience. Such art, of course, includes thinking of who are the experts/scientists authoring the report but also who are the various audiences that will receive it. Constructing the audience in this way means going beyond narrow imaginations of who are the policy makers and what do they want. Given the mandate of being policy relevant without being policy prescriptive (Blythe et al., 2018), how can these assessments shape meaningful debates about the potential for urban transformations?

Third, scientific assessments have the potential to drive progressive models of urbanism through challenging established paradigms of knowledge production, not only through the inclusion of multiple voices but also through questioning how forms of knowledge production appear to construct hegemonies that affect disadvantaged groups, in line with postcolonial thought. The deliberate attempts to include indigenous knowledge in the IPCC and IPBES are an example of the efforts within those reports to reflect upon the hegemonic dominance of scientific thought. These efforts so far are not sufficient, but the commitment to change is, in itself, an advance.

Overall, global environmental assessments have so far fallen short of challenging hegemonic practices of knowledge-making in the quest to address the root drivers of socio-ecological inequality. However, they have showcased the importance of deliberative approaches at the root of knowledge production. While knowledge-making processes are still shaped by the same unequal socio-political conditions that shape our society (racism, colonialism, patriarchy, and ableism), opening them to dialogue is a brave move and the only hope for a progressive global urbanism in the Anthropocene. But we need to make further efforts to ensure decolonizing, not just knowledge itself but also the fundamental processes whereby knowledge is produced.

References

Banister, D., Holden, E., Langhelle, O., Meadowcroft, J., Banister, D., Holden, E., Langhelle, O., Linnerud, K., and Gilpin, G. (Eds.). (2019) 'What next for sustainable development?' *What Next for Sustainable Development?: Our Common Future at Thirty*: 295.

Blythe J., Silver J., Evans L., et al. (2018) 'The dark side of transformation: Latent risks in contemporary sustainability discourse.' *Antipode*, 50: 1206–1223.

Brundtland G. H., Khalid, M., Agnelli, S., et al. (1987) '*Our common future*.' New York. WCED.

Carraro, C., Edenhofer, O., Flachsland, C., et al. (2015) 'The IPCC at a crossroads: Opportunities for reform.' *Science*, 350: 34–35.

Castán Broto, V. (2019) *Urban Energy Landscapes*. Cambridge: Cambridge University Press.

Castan Broto, V., Gislason, M and Ehlers, M-H. (2009) 'Practising interdisciplinarity in the interplay between disciplines: Experiences of established researchers. *Environmental Science & Policy*, 12: 922–933.

Díaz, S., Demissew, S., Carabias, J., et al. (2015) 'The IPBES conceptual framework — connecting nature and people'. *Current Opinion in Environmental Sustainability*, 14: 1–16.

Eddo-Lodge R. (2018) *Why I'm No Longer Talking to White People About Race*. Bloomsbury Publishing: London.

Ford, J. D., Cameron, L., Rubis, J., et al. (2016) 'Including indigenous knowledge and experience in IPCC assessment reports'. *Nature Climate Change*, 6: 349.

Gillard, R. (2016) 'Unravelling the United Kingdom's climate policy consensus: The power of ideas, discourse and institutions'. *Global Environmental Change*, 40: 26–36.

Gratani, M., Butler, J., Royee, F., et al. (2011) 'Is validation of indigenous ecological knowledge a disrespectful process? A case study of traditional fishing poisons and invasive fish management from the Wet Tropics'. *Australia. Ecology and Society*, 16: 25.

Hettinger, N. and Throop. B. (1999) 'Refocusing ecocentrism'. *Environmental Ethics*, 21: 3–21.

Hodson, M. and Marvin, S. (2017) 'Intensifying or transforming sustainable cities? Fragmented logics of urban environmentalism'. *Local Environment*, 22: 8–22.

Jabbour, J. and Flachsland, C. (2017) '40 years of global environmental assessments: A retrospective analysis'. *Environmental Science & Policy*, 77: 193–202.

Jasanoff, S. (2015) 'Future imperfect: Science, technology, and the imaginations of modernity'. *Dreamscapes of Modernity: Sociotechnical Imaginaries and the Fabrication of Power*, 1–33.

Kopnina, H. (2012) 'The Lorax complex: Deep ecology, ecocentrism and exclusion'. *Journal of Integrative Environmental Sciences*, 9: 235–254.
Kowarsch, M., Garard, J., Riousset, P., et al. (2016) 'Scientific assessments to facilitate deliberative policy learning'. *Palgrave Communications*, 2: 16092.
Kowarsch, M. and Jabbour, J. (2017) *Solution-Oriented Global Environmental Assessments: Opportunities and Challenges. Environmental Science & Policy*, 77: 187–192. https://www.sciencedirect.com/science/article/pii/S1462901117308353.
Lawhon, M., Nilsson, D., Silver, J., et al. (2018) 'Thinking through heterogeneous infrastructure configurations'. *Urban Studies*, 55: 720–732.
Löfmarck, E. and Lidskog, R. (2017) 'Bumping against the boundary: IPBES and the knowledge divide'. *Environmental Science & Policy*, 69: 22–28.
Long, J. and Rice, J. L. (2019) 'From sustainable urbanism to climate urbanism'. *Urban Studies*, 56: 992–1008.
Marvin, S. and Guy, S. (1997) 'Creating myths rather than sustainability: The transition fallacies of the new localism'. *Local Environment*, 2: 311–318.
Matsui, K. (2015) 'Problems of defining and validating traditional knowledge: A historical approach'. *The International Indigenous Policy Journal*, 6: 2.
McCann, E. (2017) 'Governing urbanism: Urban governance studies 1.0, 2.0 and beyond'. *Urban Studies*, 54: 312–326.
Monstadt, J. and Schramm, S. (2017) 'Toward the networked city? Translating technological ideals and planning models in water and sanitation systems in Dar es Salaam'. *International Journal of Urban and Regional Research*, 41: 104–125.
Parnell, S. (2016) 'Defining a global urban development agenda'. *World Development*, 78: 529–540.
Peterson, G. D., Harmáková, Z.V., Meacham, M., et al. (2018) 'Welcoming different perspectives in IPBES: Nature's contributions to people and Ecosystem services'. *Ecology and Society*, 23.
Tengö, M., Hill, R., Malmer, P., et al. (2017) 'Weaving knowledge systems in IPBES, CBD and beyond—lessons learned for sustainability'. *Current Opinion in Environmental Sustainability*, 26–27: 17–25.

26

GLOBAL CITIES AND THE BIOECONOMY OF HEALTH INNOVATION

Donald McNeill

Introduction

The relationship between cities and health has long been described, with the COVID-19 pandemic only the latest manifestation of how dramatic and unpredictable this can be. Achieving a 'universal' standard of public healthcare has long been a policy goal of global health agencies. The World Health Organization, most notably, has actively promoted a concept of healthy cities which is focussed on a teleology of continually improving urban health (World Health Organization, 2019). However, the relation between global urbanism and health is paradoxical and highly political. On the one hand, the act of living in global cities offers many urban dwellers a degree of access to the most advanced knowledge in terms of medical science diagnosis and treatment. Major cities sell a huge range of medicines, with chain retail outlets making them readily available. Pharmaceuticals are, then, part of the wider story of post-war market-making, but even more a specialist field than automobiles, air-conditioning systems, and clothing. And, due to the research and development costs involved, many drugs are 'born global'. They are expected to sell on a wide range of global and regional markets. On the other hand, urban life in megacities has connotations of growing pollution, excessive heat, and an advanced degree of alienation and mental health stress. Within this optic, cities actively structure social inequality, with the best health-care infrastructure gravitating towards wealthier neighbourhoods. This chapter works within this paradox and seeks to explore several approaches to understanding the nature of health and medicine in contemporary global society. Broadly speaking, it follows through the interest in the impact of neo-liberalism on urban governance (Brenner and Theodore 2002; Peck 2010) but is interested in the tension between the apparent widespread deregulation of markets with the highly regulated – and in many countries, still largely redistributive – model of universal health care through price-controlled or subsidised drugs.

To approach this topic, my focus is on the gap between public policy and global corporate firm strategy. As argued elsewhere (McNeill and Pollio, 2018), following corporations is a key in tracking global urbanism, as they contain the wherewithal to influence public policy at both national and inter-governmental levels through powerful legal and marketing divisions, well-resourced research and development units, and, often, through the employment of highly influential lobbyists. On the other hand, it is also important to consider the changing role and

DOI: 10.4324/9780429259593-30

approach to health policy of local and national governments and global institutions, such as the World Health Organization, in terms of how they frame appropriate medical responses and public health strategies.

The first section of this chapter discusses the relationship between global urbanism and public health, providing a schematic overview of some key urban theoretical approaches to health in global cities. The second part sketches the rise of the bioeconomy, the creation of a global medical economy based around pharmaceuticals and medical service expertise. It draws on work such as Aihwa Ong's *Fungible Life* (2016) to consider how the globalisation of health research and development, particularly around genetics, has a spatial logic in the active sites of knowledge production. It sketches the outlines of a moral urban economy around such innovation milieu and the relationships between venture capital and new health products.

Global urbanism and health

The growth in size of cities worldwide has brought with it a huge infrastructural challenge. While this is often framed in terms of the difficulties of clean water provision and transport infrastructure, we should also note the difficulties in providing an effective hospital and medical system infrastructure for the rapidly urbanising peripheries of large cities. Global urbanism as a phenomenon, I suggest, is thus often primarily a health governance challenge. To take some examples of recent scholarship, we can consider the following three areas of health which have received thoughtful elaborations in recent years by urban geographers and sociologists: air pollution, contagion, and mental health.

First, in terms of air pollution, Adey (2013) points out that it is

> the megacity's environment of interdependencies and connections which are both the source of its success as well as risks to their growth. Across megacities the world over, the acceleration of urban development moves hand in hand with the alteration of not only the urban surface but the aerial and atmospheric matter within which urban inhabitants must live. (p. 299)

There is now significant research on the negative air quality of many cities as a corollary of shifting patterns of manufacturing and growing rates of car ownership in densely populated megacities. As Adey continues, this creates a health divide as significant as other more visible modes of social segregation: "The separation of air and atmospheres is a simultaneous process of securitization and gentrification through the production of new urban natures" (Adey, 2013: 303)

Second, much of colonial and tropical medicine as a discipline has been concerned with the reduction of contagion and the containment of disease. In their study of SARS, Harris Ali and Keil (2006) argue, "The emergence of the global city network may have notable implications for the nature of the threats posed by infectious disease in the contemporary world in relation to both disease transmission and outbreak response" (p. 493). The periodic fears of the rise of a new pandemic, such as avian flu, had captured both public and intellectual interest in the mid-2000s, as discussed in Mike Davis's (2005) *The Monster at Our Door*, which offered a radical critique of the governmental and economic obstacles to disease control. The emergence of COVID-19 has been, for many commentators, an inevitability, with the speed of global viral transmission intensified by many global processes (Harris Ali and Keil, 2008).

Third, mental health is now identified as a core 'urban' branch of medicine which is at the heart of the global condition, linked particularly to the stresses posed by migration. For example, Lie and Rose (2017) highlight the negative mental health outcomes for many Chinese country-to-city migrants. To understand this, and theorise it, Fitzgerald et al. (2016a, 2016b) have suggested an attentiveness to the bieoconomies of urban experience, where they propose a series of ways of combining sociological and neuropsychiatric attention to the urban experience. In turn, they call for a renewed sociological – and I would add geographical – sensitivity to 'biological localities' and neurobiological streets and districts.

What this means is that global urbanist accounts of economic relations, gateway cities, and mobile expertise, to take some key preoccupations of recent urban theory, must be able to account for both individual genetic and physiological traits and histories, as well as chart a global capitalist health economy. This requires, as Whiteis (1998: 795) argues, recognition of "the relationship between patterns of unequal economic development and public health i.e. the pathogenic effects of corporatization, disinvestment, austerity and capital withdrawal".

This means that we are now caught between so-called diseases of affluence, usually chronic and non-communicable, such as diabetes, and diseases of poverty, often associated with infections caused by poor living conditions, as well as other key factors, such as poor nutrition, low vaccination, and poor access to preventative health care. These terms are, of course, relative and not always helpful.

There is still a view that public health impacts are the result of individual choices, ignoring the fact that social relations – not least intergenerational poverty – remain fundamental drivers of health outcomes (Whiteis: 1998). Recently, there has been growing research that poor health outcomes are physiologically intergenerational, and even epigenetic, where during the course of reproduction and birth, toxic environmental conditions are imprinted upon the genetic code of newborns. Sociologically born into cycles of poverty, individuals may also be genetically impacted, rendering them more prone to disease during their lifetimes. This is a controversial science, it should be noted, and when combined with new CRISPR technology, leads us to challenge new areas of how the reproductive rights of global populations are governed and the potential that some ethnic groups are subjected to intervention for wider social governance goals.

Relatedly, we can consider that the treatment of global health is hugely influenced by capitalist development histories. As Kirmayer and Pedersen (2014) note, there has been a shift from 'tropical medicine to global health', but one in which the focus was still largely on "the delivery of biomedical services modeled on urbanized Western medical practices, disease prevention, and therapeutics" (p. 761). This has, in turn, often focussed on the containment of potential pandemics through strong public health campaigns. The dominant approach to health remains wedded to an idealised Western model of health delivery:

> There are tensions between a public health approach, grounded mainly in biomedicine and current evidence-based practices (which are still largely produced in high-income countries), and a socially and culturally informed community-based approach that emphasizes the social determinants of mental health and the imperatives of listening to local priorities, strengthening community resources, and developing endogenous solutions. (Kirmayer and Pedersen, 2014: 760)

This is a legacy of colonial medicine which, genealogically, shapes the health systems of cities and nations.

In this context, it was significant that the World Health Organization's Commission on the Social Determinants of Health, established in 2005, made a strong shift away from "conventional epidemiological models" to a "social epidemiological approach, focusing on the contextual, supraindividual factors". However, as Kirmayer and Pedersen continue, "this shift in focus inevitably highlights two politically sensitive issues: first, the extraordinarily unequal social distribution of disease-related morbidity and mortality and second, the social causation of disease and disability" (Kirmayer and Pedersen, 2014: 765). This has included a narrowing and medicalisation of the causes of major poverty-driven diseases (p. 767).

The global urban health landscape is, then, something of a mosaic, which maps onto the contemporary spatial distribution of enclaved wealth amid large populations with either substandard or non-existent, but highly industrialised and massified, medicinal cultures. In the context of the Global North, this is characterised by the growing assertiveness of the 'wellness economy', which can be seen as a confluence of high value-added products, such as nutraceuticals, organic food products, social media–driven wellness gurus, and lifestyle facilities based around gyms and yoga studios. It is notable that many of these ventures are based on some kind of 'detox', which baldly implies a degree of consumer autonomy in controlling and responding to their quotidian range of toxifying behaviours. And many of these consumers are now requesting their own DNA profiles as a tentative step towards precision diagnosis, in turn creating a demand for biomedical products tailored to their needs.

The global bioeconomy

The other side of the global urban economy is the field of health innovation, which is strongly associated with biomedical and bioengineering venture capital investment. This sub-sector of the economy offers potentially lucrative returns to investors through patented drug discovery, covering a wide range of medicines and diagnostic intervention infrastructure, such as scanners and blood tests. Most governments have a strategic arm dedicated to the growth of their life sciences and biomedical industries, which are typified by long product development timelines, risky clinical trial procedures, and uncertain pricing futures. They persist with this because the role of public venture capital or tax incentives is seen as worthwhile given the potentially huge profits that can be made from successful drug development.

In her book *Fungible Life*, Aihwa Ong provides a case study of Biopolis, the Singaporean life sciences health precinct that she sees as being "at once embedded in the Asian tropics and densely connected to biomedical science sites around the world" (Ong, 2016: 1). Ong has a fascination with this because of its positioning in the dynamic science of genomics, now at a point where specific population aggregates might be monetised in a complex mix of national medical improvement, national economic growth, and global – 'cosmopolitan' – scientific advance. As Ong puts it,

> While building a research platform for novel problematization of and interventions into 'Asian' bodies, citizenship and well-being, Biopolis has also staked its ambitions in cosmopolitan science.... Some fitting questions posted by Biopolis may be how is scientific knowledge governed at a global scale, and what are the implications of a novel and distinctive 'Asian' model of knowledge production for understandings of life? (p. 5)

The importance of national health security became sharper after Singapore's unfortunate location as one of the epicentres of SARS. After this epidemic, Biopolis "was recast as a center for biodefense, and 'scientific entrepreneurialism' came to include this public responsibility to prepare for impending catastrophic events" (p. 10).

The intriguing point raised by Ong is the existence, or otherwise, of 'cosmopolitan science', the assumption that there is a singular global body to which scientific advances in medicine and surgery can be applied. While this is an appealing perspective, it has been challenged by the differentiated health outcomes at the population level of different ethnic genotypes and haplotypes.

The reduction in cost of DNA sequencing tests has driven a new genetic health economy. It is aligned with a growing scientific interest in precision medicine, which observes that the genetic code of any individual reacts very differently to particular medicinal treatments. This is a radical change in the direction of medical research. It is one which fragments medicinal development away from a global, mass medicine drug, likely purchased and subsidised for mass population prescription, to higher value but niche medicines probably targeted at advanced economies with well-established private health insurance regimes.

And so, this provides a dual rationale for government investment in biomedical research:

> By coding and valorizing genomic and ethnic variability, Biopolis scientists are better able to be competitive in the international arena of bioscience research and pharmaceutical investments while also continually re-emphasizing a reinvestment in the diverse 'races' that compose Singapore and Asia at large. (p. 13)

The Singapore case provides interesting insight into how global innovation, while often seen as part of a straightforward race to gain competitive advantage in the knowledge economy, is also an attempt to strengthen national sovereignty and the ability for societies to literally and metaphorically reproduce themselves. What is novel in the Singaporean case, and in Ong's analysis, is the urban spatiality of genetics, where urban populations can be easily accessed for large-scale clinical trials, and where teams of expertise can be assembled which may possess unique data sets and analytical capabilities.

Biopolis is, interestingly, cast in the mould of the suburban science park. It should be noted that this is against a trend of advocacy for public investment of downtown biomedical sites as part of master-planned 'innovation districts' where knowledge is, in the ideal world, co-produced, with proximity and co-located knowledge-generating outcomes which would otherwise not have occurred. This is an unproven claim, but it is an important element in a recent genre of urban writing. For example, a notable trend is the discourse of the 'return to the city', promoted in works such as Glaeser's *The Triumph of the City* (2011) and Florida's *The New Urban Crisis* (2017) regarding the apparent relocation of innovation from the suburban science park to the inner core of cities. Such districts are often anchored by health research and development, notably around large research hospitals, globally operative biotech and pharmaceutical firms, and privately held start-ups.

Regardless of the merits of various locational debates, there has been a compelling, and growing, political critique of the political economy of global pharmaceutical innovation, as seen in works such as Cooper's *Life as Surplus* (2008) and Sunder Rajan's *Biocapital* (2006) and *Pharmocracy* (2017). This critique focussed on, for example, the modes by which clinical trials – a fundament of the impact of global pharmaceuticals marketing on diverse populations – are introduced to vulnerable groups. It also includes the denial by the Indian Supreme Court of Novartis's attempt to 'evergreen' its anti-leukaemia drug, Glivec, instead allowing the production of cheaper generic versions by Indian manufacturers. Such cases show the importance that regulatory bodies play as part of a complex web of governance.

Global pharmaceutical firms are powerful international actors, active in lobbying, marketing, and shaping landscapes of health care on a global scale. The ways in which such pharmaceutical firms operate has been critically charted by numerous scholars, such as Goozner (2014) and Healy (2012). Firms have been actively criticised for both their operating mechanisms, such as the use of transfer pricing to avoid tax liabilities and their ability to exploit tax incentives for research and development. They have operated using widely differing standards of care through clinical trials and regulatory frameworks. While individual scientists often justifiably operate within a moral framework of relieving suffering and prolonging life, there have been numerous cases where the search for shareholder return has undermined any claims to social benefit. Recent examples include Theranos, the Silicon Valley blood testing start-up whose founders are now under federal investigation in the United States, or Purdue, who are accused of aggressively promoting the OxyContin opioid. It is useful to note that the family that owns the latter firm, the Sacklers, have been actively involved in cultural and educational philanthropy in leading global cities, and institutions which have received funding have faced a significant backlash. In China, the government charged GlaxoSmithKline with bribery and corruption in their attempts to influence doctors to adopt its drug products, with the firm having to pay a $500 million fine.

And so here we can see the contours of how two very different kinds of global urban scholarship begin to converge. Biomedicine and other related health 'growth' industries (such as medical tourism) are intrinsically woven into a globalised set of capitalist practices. Interestingly, although this can sometimes be attributed to neo-liberal patient market segmentation, these industries are also entangled with the uneven national markets in subsidised health provision. On a global stage, biomedicine, and even the ability to consume or administer medicine, is structured by prevailing religious practices, ethical beliefs, and national political systems. The standard global products supplied by major firms rarely arrive in epistemologically pure form in the bodies of recipients, whose consumption, use, and access to such drugs will be based on a textured mix of genetic predisposition and norms of cultural consumption (Hutch, 2013). To take one example, women in some patriarchal societies will have little autonomy in their ability to choose whether to access drugs at all.

To understand the implications for global urbanism, however, it is important to see health as being an infrastructure in the same way as transport or energy utilities. In times of crisis, as in the rapid development of supply chains and vaccination centres for COVID-19, the materiality of health – and especially drug – infrastructure becomes clear. Hospitals, a core element of the urban fabric of cities, and their networks of doctors and allied health centres, which are unevenly distributed through the suburbs of cities, are the local hubs of these global systems of provision. The ecosystem of research and development that we see in the innovation districts and science parks of the global innovation economy ultimately streams these products into the daily lives of urban dwellers, affecting their ability to participate in meaningful activities, as well as altering the mindset and creating new types of addictive behaviour.

Conclusion

If the global population is increasingly urbanised, medicalised, and vaccinated, then we can immediately see the importance of health infrastructure, production, and consumption, as being central to any discussion of global urbanism. As Ezzati et al. (2018) point out, "The concentration of knowledge and innovation, economic activity, healthcare, education, and other public services endows cities with the potential to deliver substantial improvements to the health and wellbeing of their residents and those of other parts of the country". As such, cities may have a leadership role in providing specialist services for non-urban residents who still make up almost half of the world's population. As well as direct clinical intervention, cities are likely to generate the training required to create a health workforce skilled in non-acute interventions.

However, the example of Singapore and its genetic research ambitions illuminates the geographical differentiation in the world's population and the limits of 'Western' medicine. It highlights the important need to understand the 'global' medicalisation of populations based on little underlying knowledge of personal, and regional, genetic characteristics. The role of global pharmaceutical firms, particularly those which have patented 'blockbuster' drugs, is a key element of the globalisation of health. National governments, particularly those which have committed to 'universal health care' (including recent entrants, such as Vietnam), play an important role as mass consumers of these medicines, which in turn feeds the market capitalisation of global pharmaceutical corporations. They have a carefully calculated understanding of risk. The wave of court cases that have, after many years, tracked Monsanto's influence on global pesticides or underscored the right to develop generic medicines, which in turn feeds into the drug consumption patterns of urban populations, is evidence of that.

This is the flip side of the biomedical innovation strategies that have long been a key element of the playbook of the global knowledge economy. It is not to say that these laboratory-driven economies are malign: there have been striking pharmaceutical and analytical advances that explain the phenomenal improvement in certain diseases – in certain places – over the past decades. However, it is important to recognise the global power structures, capital-driven, expertise-driven, that are shaping markets to provide a global standard in commodified health care for urban dwellers. In this context, the chapter proposes that there exists a global health/urbanism nexus driven by a complex set of neo-liberal urbanism, protected and regulated national health markets, and an array of research, production, and consumption sites of drugs, devices, and procedures.

References

Adey, P. (2013) Air/Atmospheres of the Megacity, *Theory, Culture & Society* 30 (7–8): 291–308.

Brenner, N. and N. Theodore (2002) 'Cities and geographies of "actually existing neoliberalism"'. *Antipode* 34(3): 349–379.

Cooper, M. (2008) *Life as Surplus: Biotechnology and Capitalism in the Neoliberal Era*. Seattle: University of Washington Press.

Davis, M. (2005) *The Monster at our Door: The Global Threat of Avian Flu*. New York: New Press.

Ezzati, M., C. J. Webster, Y. G. Doyle, G. Owusu, G. M. Leung (2018) 'Cities for global health'. *BMJ*, 363. https://www.bmj.com/content/363/bmj.k3794.

Fitzgerald, D., N. Rose and I. Singh (2016a) 'Living well in the neuropolis'. *The Sociological Review Monographs*, 64(1): 221–237.

Fitzgerald, D, N. Rose, I. Singh (2016b) 'Revitalizing sociology: Urban life and mental illness between history and the present'. *British Journal of Sociology*, 67(1) 138–160.

Florida, R. (2017) *The New Urban Crisis*. New York: Basic Books.

Glaeser, E. (2011) *The Triumph of the City*. New York: Penguin.

Goozner, M. (2014) *The $800 Million Pill: The Truth Behind the Cost of New Drugs*. Berkeley: University of California Press.

Harris Ali, S. and R. Keil (2006) 'Global cities and the spread of infectious disease: The case of severe acute respiratory syndrome (SARS) in Toronto, Canada'. *Urban Studies*, 43(3) 491–509.

Harris Ali, S. and R. Keil (Eds.) (2008) *Networked Disease: Emerging Infections in the Global City*. Oxford: Blackwell.

Healy, D. (2012) *Pharmageddon*. Berkeley: University of California Press.

Hutch, R. A. (2013) 'Health and healing: Spiritual, pharmaceutical, and mechanical medicine'. *Journal of Religion and Health*, 52(3): 955–965.

Kirmayer, L. J. and D. Pedersen (2014) 'Toward a new architecture for global mental health'. *Transcultural Psychiatry*, 51(6): 759–776.

Lie, J. and N. Rose (2017) 'Urban social exclusion and mental health of China's rural-urban migrants – a review and call for research'. *Health and Place* 48, 20-30.

McNeill, D. and A. Pollio (2018) 'Tracking the global urbanists', in M. Harrison and M. Hoyler (Eds.), *Doing Global Urban Research*. London: Sage, pp. 81–95.

Mittra, J. (2016) *The New Health Bioeconomy: R & D Policy and Innovation for the Twenty-First Century*. Basingstoke: Palgrave MacMillan.

Ong, A. (2016) *Fungible Life: Experiment in the Asian City of Life*. Durham, NC: Duke University Press.

Peck, J. (2010) *Constructions of Neoliberal Reason*. Oxford: Oxford University Press.

Sunder Rajan, K. (2006) *Biocapital: The Constitution of Postgenomic Life*. Durham, NC: Duke University Press.

Sunder Rajan, K. (2017) *Pharmocracy: Value, Politics and Knowledge in Global Biomedicine*. Durham, NC: Duke University Press.

Whiteis, D. G. (1998) 'Third world medicine in first world cities: capital accumulation, uneven development and public health'. *Social Science of Medicine*, 47(6): 795–808.

World Health Organization (2019) Healthy cities. January, 2020.

27

HACKING THE URBAN CODE

Notes on durational imagination in city-making

Swati Chattopadhyay

This chapter responds to two contemporary symptoms in urban studies. By themselves, they do not entail a crisis and would not invite comment if it were not for something more fundamental: the *durational imagination* that undergirds the understanding of urban materiality and the modes of envisioning cities in the new millennium.

The first is the recourse to the term 'hacking' to describe how grass-roots mobilisation might remake cities in the Global North in the thralls of corporate capital (Sassen, 2019). 'Open-source urbanism' is offered as a solution to closed elite systems, as if it is merely a technical logic that distinguishes the two or that the recommended strategy for mobilising popular knowledge systems is in any way new. Residents of scores of cities in the Global South hack the urban code on a daily basis and have done so for decades, if by hacking we simply mean the process of appropriating and potentially altering an established code (Bayat, 1997; Larkin, 2008; Holston, 2009; Simone, 2004; Hou, 2010; McFarlane, 2011; Chattopadhyay 2012; Rai, 2019). A liveable city is a hackable city.

The second is an allied emergent focus on ephemeral architecture and urbanism that attempts to grapple with social-structural changes in contemporary cities. This includes new forms of experiential retailing, street art and biennales, pop-up shops, temporary urban beaches and 'shared urbanism' that are understood to be plugging the gaps of global capital in the urban fabric of cities in the 'Global North' (Gwiazdzinski, 2014; Madanipour, 2017). This literature shares some conceptual similarities with studies of temporary urbanism in other parts of the world, such as the state-planned mega-religious event of Kumbh Mela in India (Mehrotra and Vera, 2015). The keen interest in sustainability and reversibility notwithstanding, scant attention is paid to the differential temporal horizons of state- or corporate-led interventions and popular movements. The terms used to describe short-term interventions or structures lasting for a few days to a few months or that are vulnerable to weather conditions—impermanent, temporary and ephemeral—are used interchangeably, as if they all obey the same temporal imagination and practice. This terminological laxity is not limited to urban studies. Historians who have studied temporal change as the defining condition of modern life have not sufficiently interrogated the socio-political import of different forms of short-lived interventions. Against a tendentious imagination of permanence is set everything that is presumed to be short-lived.

DOI: 10.4324/9780429259593-31

The problem of the 'global' in global urbanism is situated at the intersection of these two symptoms. First, scholars routinely speak of the 'global' by taking into account a handful of cities that function as centers of financial capital, mostly in the Global North, inattentive to the epistemic divide of North/South that their scholarship perpetuates. Second, there is a reluctance to acknowledge the theoretical import of scholarship on cities in the Global South, thus impoverishing any collective effort to tackle the problem of planetary urbanism. And third, the vast majority of scholars working on cities in the Global South and North are captive to city imaginations that assume permanence as a given and a desideratum. To understand urban processes and urban materiality that have some kind of global relevance, it is imperative that we closely attend to the durational imaginations in city-making: it is here that we might locate the *politics of surprise* that hacking the urban code entails.

What follows are some preliminary notes on the implications of the terms we use to discuss the temporal relations that constitute the urban environment. Here I am specifically concerned with how we understand the relation between agency and urban temporality.

Permanence and durability

The reiteration and reassertion of the Western classical connection between permanence and durability during the Renaissance provided a working language of sovereignty for the cities of emergent European nations. The striking formal logic of baroque plans with splendid vistas provided a legible diagram of (seemingly endless) territorial sovereignty, while the grand scale of the monuments and avenues demonstrating the capacity of the state to marshal resources and labour projected the military strength and permanence of the state (Mukerji, 1997). Such diagrams of spatial legibility generated an urban code that city residents were meant to learn and comply with: it set the parameters of movement, use and occupation, and it made the power of the state explicit and 'readable'. This urban code was transported globally with European colonialism and became a visible indicator of political strength and durability.

With the advent of industrial capitalism some 200 odd years ago, the relationship between permanence and durability was snapped. Capitalism's economy of scale demanded (i) rapid obsolescence of goods and spaces to create new opportunities for investment and (ii) unprecedented mobility of goods, services and peoples that created a situation of constant flux. And yet while 'all that is solid' melted into air, the value placed on permanence in urban planning lingered.

The invention of the principle of obsolescence in the early twentieth century strained the principles of architecture and urban planning. Building professionals were faced with a seeming conundrum as they negotiated between the structural durability of buildings and rapid change in building function based on the changing demands of the capitalist market (Abramson, 2016). What emerged is the defining characteristic of twentieth-century urbanism with its dual faith in a long-term spatial future and an uninterrupted flow of capital. The goal of sustaining a long-term economic/political future translated into a desire for permanence by building durable infrastructure, establishing zoning laws that guaranteed private investment and creating techniques of enumeration that sorted populations based on categorical distinctions of class, race, gender, ethnicity. Functionally, the durable infrastructure was meant to provide channels for the movements of goods and peoples and to locate and contain them. Such concrete immobile built environments that enabled capital fixity and social stability came with its own contradictions. The building professional's conundrum was a reflection of capitalist development itself that had to 'negotiate a knife-edge between preserving the exchange values of past capitalist investments in the built environment and destroying the value of these investments in order to open up fresh room for accumulation' (Harvey, 1989: 83). Thus some capitalist urban interventions were meant to last for a few years and some for decades, and a systematic evaluation of the longevity of the built environment, irrespective of the durability of materials, was undertaken to tackle this new turnover in the built environment (Abramson, 2016: 16–37). This pattern of obsolescence continues: residential buildings are planned to last a shorter duration than road infrastructure; office and commercial spaces face quick turnovers for reasons that have nothing to do with their structural integrity, and so on.

Twentieth-century architects and urban planners addressed the conundrum in various ways. Le Corbusier, who favored planning decisions tethered to long-term durability, made the well-known statement equating the lack of planning with revolution (Le Corbusier, 1989: 269). This was in direct contrast to the Futurists who called for every generation to build their own cities. The Futurists made such claims in the hope that 'no formal or linear habit can grow' in an architecture based on the fundaments of 'impermanence and transience' (Kern, 1983: 100). We are told that Antonio Sant'Elia objected to this call for impermanence because it seemed to defy the logic of architecture (Martin, 1968: 190). Reflecting on Sant'Elia's diffidence, Stephen Kern remarked, 'Imagine the pressure that would lead an architect to put his name to the first formal commitment to build buildings that would fall apart' (Kern, 1983: 100). Kern, like those whose works he was studying, failed to heed the important political distinction between durability and permanence. When in the mid-twentieth century Frei Otto, Yona Friedman, Archigram and others designed modular frame structures capable of accommodating expansion and change over time as 'liberating' solutions to the notions of permanence that defined standard building practices, they ran into another dead end. The more they championed the cause of flexibility, interchangeability and movement through modular extensions and plasticity of surface, the more rigid the infrastructure of their designs became. In responding to the self-revolutionising impetus of capital, these architects and urban planners were ineffectively tackling the fundamental connection between permanence and political culture: that claims of permanence/impermanence are inherently political. Permanence in the modern built environment has very little to do with the durability of materials; it has something to do with the perpetuation of social contracts. Permanence as durability survived in urban thought because it defined possibilities of exclusion and inclusion and set the terms of who belongs in a city (Chhabria, 2019: 86–88).

Rather than durability, permanence in the built environment is about the conferral of property rights in land that produces a distinctive urban order by creating and sustaining a land market that is conducive to capitalist accumulation. At some level, this might sound counter-intuitive. After all, duration/durability has been cited as the founding condition of property rights. Invoked by John Locke to envision a new political order, duration (explained as a distinction between perishable foodstuff as opposed to non-perishable goods such as metal) served as an analytic to make a case for money as a universal form of exchange that extends use value and enables the expansion of property (Locke, 1964). Primitives were defined by their short durational imagination, from which followed a whole set of attributes that negated their territorial rights and sovereignty.

Linked to the modern Western conception of property rights, permanence was used by European colonial and metropolitan powers against peoples whose modes of living did not require the construction of masonry structures. Inattentive to or ignorant of how temporality, property and territorial occupation were understood by indigenous peoples and tribal populations (Banerjee, 2006), the absence of durable structures became an excuse for dispossessing millions of people across the world.

What could be marked as durable in buildings, however, was neither simple nor clear. For a building to be considered durable, could a fired brick structure qualify if it had a thatch roof? Was brick construction durable, even if the bricks were not well made? What of well-built or even monumental construction of materials considered non-durable: unfired bricks and earth construction? Citing the case of colonial Bombay, Sheetal Chhabria points out that building materials went from 'unmarked' to 'marked' and organised in a fairly arbitrary matrix to legitimise the residence of certain populations and practices within the city and to exclude others (Chhabria, 88).

Planning and impermanence

Impermanence signals the absence of property-right protections, even if the structures are made of durable materials. In squatter settlements, construction of structures with materials considered durable by local authorities is a move towards legitimisation: to connect to the order of permanence that determines long-term spatial futures (Neuwirth, 2005). Similarly, when squatters build next to railway lines, they are not merely making their commute easier and cheaper but also locating themselves strategically in relation to infrastructure that is considered durable (Chatterjee, 2006). This facilitates the squatters' demand for legitimisation. The classic problem of sites and services planning, and the refusal of people to inhabit such planned facilities away from the city (and jobs and infrastructure), is that the potential residents recognise this as a strategy of unhinging populations from long-term spatial futures that cities promise.

Conversely, structures built for short-term events by state, municipal or corporate agencies of either durable or non-durable material assert legitimacy precisely because they are authorised. In contemporary cities as various as London, Glasgow, Brussels, Paris, Berlin, Singapore and scores of other locations around the world, civic agencies and municipalities have resorted to hosting art events, festivals, parades, food fairs on streets and various 'meet-your-neighbor days' to give people back their space in the urban fabric (Madanipour, 2017). Such attempts to *plan* occasions for community get-togethers are based on a recognition of loss of community and vibrancy of public space: beyond work hours, vast squares and ample streets sit empty and dead. They do not stimulate or simulate an image of active consumption. The fear of shrinking cities, à la Detroit, prompts attempts to allocate specific durations—summer evenings, lunchtimes, after-work hours which are identified and set apart—to host art events, and artists are permitted to alter the space for a few hours or perhaps a few days. The power to determine the duration and parameters of intervention rests with the authorities. Such short-lived events planned by the state or the municipality demand infrastructural organisation and a planning ethos that have a long tradition: state-sponsored games, *durbars*, world's fairs, expos, which in the long history of modernity are deeply connected to imperial regimes and imaginations. They bring together the ordering attributes of the planned environment and the anticipation of the temporary to set the stage for spatially and temporally bounded events where a well-choreographed narrative can be made to unfold. It is thus not a surprise that the principles of spatial organisation and design expounded in the temporary world's fairs and *durbars* were successfully transferred to planned cities in the early twentieth century. Daniel Burnham's Chicago plan and Edwin Lutyens and Herbert Baker's plan of New Delhi are two spectacular examples of that translation.

In contrast, bottom-up mobilisation of urban public space—festive or protest gatherings, graffiti and unauthorised occupation of public space for dwelling, selling goods and services—are considered threats because of the terms of their spatial and temporal unfolding. Although both state-sponsored events and public mobilisation may be temporary, the two types of events are different, not simply because of who organises them and what their architecture looks like but also because of the durational imagination on which they are premised and how this imagination impacts the event's spatial outcome. A more useful way to think about the distinction between two types of temporary interventions is that state-sanctioned events, such as expos and art festivals, recreate and reaffirm the authorised urban code of the city, whereas bottom-up mobilisations of urban space for living or protest hack the authorised urban code. In so doing, they hack not only the spatial architectonics but also the temporal imagination on which the code is based.

Hacking, as an unauthorised entry into a domain, involves three operations: determining a point of entry, deciphering the code, and temporary disabling of security systems while interpreting the code. Disabling the security system might entail changing the code permanently, which in turn would require a repair/replacement code. Changing the code alters the terms and conditions of access. Translating this process metaphor to urbanisation and city-making would mean recognising a vulnerable/amenable point of entry that would give a foothold to further appropriation of urban space.

In her discussion of the graffiti of the intifada, Julie Peteet pointed out that the hastily if 'painstakingly painted' wall writing in the occupied West Bank during the late 1980s and early 1990s was a product of particular spatio-temporal constraints of severe censorship and state violence. The wall writing registered dissent and transformed the wall into a site of political performance. Such wall writing was not only 'resistant' but also the 'mere appearance' of such texts 'gave rise to arenas of contest' in which the residents of West Bank became agents of a political process which was denied to them (Peteet, 1996: 40). The walls were transformed by graffiti and became interventions in public space—speaking to and forming audience communities by actualising the act of reading on the street. The graffiti took the negative codes of closure, delay and prohibition that define the boundary wall and turned them into an open and immediate invitation to read and share. The applied texts did not destroy the wall but changed its materiality by transforming a boundary wall into a canvas on which visions of the body politic that are not sanctioned by the state could be articulated. These comprised small acts, whose aggregate impact was immense.

Peteet noted that in order to understand the meaning of such graffiti, one needs to understand the clandestine, swift, dangerous conditions of production, as well as the 'optics' involved in the process of reading within a power structure that creates differential reading communities. The power of Palestinian graffiti derived from the risk involved in their appearance: it demonstrated to the occupying force its lack of territorial control. It changed the materiality of the wall and its claim to obdurate permanence by showing up its susceptibility to re-inscription and transformation and in so doing challenged the materiality—the projected material legitimacy—of the occupying state.

Political wall writing is, of course, only one means by which city streets are appropriated. The boundary wall as an edge, for example, facilitates the beginning of many other acts of habitation. This can be an unauthorised shop that uses the wall as its first move to set up space on a sidewalk. It may involve something as simple as inserting two hooks in the wall so that a light tarpaulin shed can be constructed to shelter shopkeepers, their wares and clients. In linking itself to the sturdy materiality of the wall, shopkeepers make their own claims to endurance and the right to city space. An intervention born out of scarcity of resources, such instances of vernacularisation constitute a process by which a formal urban code is appropriated and then recoded. This is the stuff of political vernacular in cities in much of the world. This is the 'global' of global urbanism.

The politics of surprise

The structural contradiction of capitalism—the necessity to invest in a durable infrastructure and the desire to rip off that infrastructure due to technological and financial needs—was mirrored in the thinking of critical theorists who viewed the goal of revolution as one of permanent change. In *Prison Notebooks*, Antonio Gramsci struggled with the tension between structural changes that are permanent or organic and conjunctural movements that are 'occasional, immediate, almost accidental' (Gramsci, 1991: 177). Since the goal of political transformation is to change the status quo permanently—that is, to wrest the state from the ruling classes, he bemoaned the 'failure' of subaltern groups to transcend their episodic and fragmentary history and achieve 'permanent victory'. He viewed these episodic, fragmentary challenges to the status quo as evidence of 'everyday experience illuminated by "common sense"'—i.e., the traditional popular conception of the world and hoped that such a fragmentary conception of the world could be educated and modernised to bring the subaltern classes into hegemony.

The temporality of the political is central to Gramsci's theorization, as it implicates agency. In his analysis, the victory of subaltern groups is short-lived because of a gap between their conception of the world and the modern conception of the political. They are subjects (agents) only contingently. The cultural world of the subaltern has to be remade to mobilise them as a 'class', a corrective response to their fragmentary history. He privileged 'high' culture as the source of fashioning a national-popular that by affording a language of modern collective consciousness would impart historical agency to the subaltern.

Despite the teleological basis of this conception of political agency, he found conjunctural movements highly suggestive: it is in the temporary organisation of forces that rudimentary elements of conscious leadership become visible, as does a capacity for critique. Citing Benedetto Croce's conception of 'passion as a moment of politics' and its inability to explain permanent political formations, he noted:

> Since it is impossible to conceive of a passion being organized permanently without its becoming rationality and deliberate reflection and hence no longer passion, the solution can only be found in the identification of politics and economics. Politics becomes permanent action and gives birth to permanent organizations precisely in so far as it identified itself with economics. But it is also different from it, which is why one may speak separately of economics and politics, and speak of 'political passion' as an immediate impulse to action which is born on the 'permanent and organic' terrain of economic life but which transcends it, bringing into play emotions and aspirations in whose incandescent atmosphere even calculations involving the individual human life itself obey different laws from those of individual profit. (139–140)

These moments of incandescence—these moments of passion during which the subaltern emerges as a popular force—thus became lessons in political agency.

Conjunctural moments are ephemeral, which is to say they combine the elements of short duration and unpredictability. Therein resides their power to surprise. This is precisely why the graffiti of the intifada were particularly problematic: the simultaneity of reading the texts on the street linked peoples and communities that had been fragmented by state power. This act of bringing together—the aggregate effect of small acts and small spaces—created platforms for political debate. The writing and reading took on the quality of a live performance that undermined hierarchical control and expanded the spatial parameters of the given order of public space. In hacking the urban code defined by the occupying state, wall writing *hacked time*: the ephemeral writing produced a conjunctural moment whose spatial boundaries were uncertain. And in so doing, it gave birth to a new kind of urban space that resisted appropriation by the state—indeed resisted representation.

The promise of the ephemeral resides in this capacity to transform while resisting representation. The ephemeral, intimately linked to the sensations of light, heat, texture, smell, interpersonal contact, plays on the virtues of small acts, of unpredictability and multisensoriality that harbor possibilities of spatial amplification. Like the smell of Proust's madeleines that occupy a larger place in the imagination than one is allotted in physical space, the urban ephemeral connects time, space and people in non-linear and surprising ways. It intrudes in the given order of urban space to show up the temporal limits of spatial legibility.

Conclusion: Reimagine temporality and scale

If global urbanism is not to be consigned to an empty (neo)liberal gesture of cosmopolitanism and treating cities of the Global South as if they are only good for thinking about cities in the Global North, we need to reconfigure the *temporal parameters* and the *scale* at which we imagine global urbanism. I conclude with five moves that are necessary for this reconfiguration: (1) attend to the theoretical distinctions between the ephemeral, temporary and impermanent; (2) think beyond permanence as a urban desideratum; (3) track the urban ephemeral as a process to recast urbanism as a politics of surprise; (4) reconsider the scale at which the politics of surprise occurs so that we may reimagine the scale of the global; and (5) reject the grandiosity of big plans and focus on the small scale, small acts and small spaces that constitute the urban vernacular. It is at this small scale that ordinary citizens across the world operate and that they can manipulate to imagine a new urban future.

Bibliography

Abramson, D. M. (2016) *Obsolescence*. Chicago: University of Chicago Press.

Banerjee, P. (2006) *The Politics of Time: Primitives and History Writing in a Colonial Society*. New Delhi: Oxford University Press.

Chhabria, S. (2019) *Making a Modern Slum: The Power of Capital in Colonial Bombay*. Seattle: Washington University Press.

Chatterjee, P. (2006) *The Politics of the Governed: Reflections on Popular Culture in Most of the World*. New York: Columbia University Press.

Bayat, A. (1997) 'Un-civil society: The politics of the "informal people"'. *Third World Quarterly* 18(1): 53–72.

Chattopadhyay, S. (2012) *Unlearning the City: Infrastructure in a New Optical Field*. Minneapolis: Minnesota University Press.

Gramsci, A. (1991) *Selections from the Prison Notebooks*, Quintin Hoare and Geoffrey Nowell Smith (Eds. and trans.). New York: International Publishers.

Gwiazdzinski, L. (2014) 'The malleable, adaptable metropolis: Toward a temporary and temporal urbanism'. *Stream*, 03. https://www.pca-stream.com/en/articles/the-malleable-adaptable-metropolis-toward-a-temporary-and-temporal-urbanism-8.

Harvey, D. (1989) *The Urban Experience*. Baltimore: Johns Hopkins University Press.

Holston, J. (2009) *Insurgent Citizenship: Disjunctions of Modernity and Democracy in Brazil*. Princeton: Princeton University Press.

Hou, J. (2010) *Guerilla Urbanism and the Remaking of Contemporary Cities*. New York: Taylor & Francis Group.

Kern, S. (1983) *The Culture of Time and Space, 1880–1918*. Cambridge, MA: Harvard University Press.

Larkin, B. (2008) *Signal and Noise: Media, Infrastructure and Urban Culture in Nigeria*. Durham: Duke University Press.

Le Corbusier (1989) *Towards a New Architecture*, Frederick Etchells (trans.). London: Butterworth Architecture.

Locke, J. (1964) *Two Treatises of Government*. P. Laslett (Ed.), Cambridge: Cambridge University Press.

Madanipour, A. (2017) *Cities in Time: Temporary Urbanism and the Future of the City*. London: Bloomsbury.

Martin, M. W. (1968) *Futurist Art and Theory 1909–1915*, Oxford: Clarendon Press.

McFarlane, C. (2011) *Learning the City: Knowledge and Translocal Assemblage*. London: Wiley Blackwell.

Mehrotra, R. and Vera, F. (Eds.) (2015) *Kumbh Mela: Mapping the Ephemeral Megacity*. Ostfildern, Germany: Hatje Cantz Verlag.

Mukerji, C. (1997) *Territorial Ambitions and the Gardens of Versailles*. Cambridge: Cambridge University Press.

Neuwirth, R. (2005) *Shadow Cities: A Billion Squatters, A New Urban World*. New York: Routledge.

Peteet, J. (1996) 'The writings on the walls: The graffiti of the intifada'. *Cultural Anthropology*, 11(2): 139–159.

Rai, A. (2019) *Jugaad Time: Ecologies of Everyday Hacking in India*. Durham: Duke University Press.

Sassen, S. (2019) Unsettling formal power systems, in S. Chattopadhyay and J. White (Eds.) *The Routledge Companion to Critical Approaches to Contemporary Architecture*. New York: Taylor & Francis Group, 363–372.

Simone, A. (2004) *For a City Yet to Come: Changing African Lives in Four African Cities*. Durham: Duke University Press.

28

GLOBAL URBANISM

Urban governance innovation in/for a world of cities

Pauline McGuirk

Introduction

Despite its currency, 'global urbanism' is a nebulous concept deployed in dissonant registers that offer diverse perspectives on the dynamics of a global urban age. 'Mainstream' global urbanism (Sheppard et al., 2015) suggests Global North urbanism, ordered by globalised neoliberalised capitalism and liberal democracy, as the universal essence and norm. In parallel, the planetary urbanisation thesis (Brenner and Schmid, 2015) posits a generalised global condition wherein cities' extensive demands on global flows and relations cast the planet as always and already in the grasp of capitalist urbanisation. Countering these unitary registers, postcolonial perspectives reject the telos of global capitalist logic shaping a singular (if variegated) global urbanism and suggest a world of multiple global urbanisms wherein multiple processes beyond neoliberalised capitalism shape cities differently in a 'differentiated, changing and globalising urban world' (Parnell and Robinson, 2018: 28).

In this chapter, I consider yet another invocation of global urbanism: one that echoes a postcolonial appreciation of the diversity of processes and relations defining a 'world of cities'. This is a form of global urbanism that frames maniform global challenges—climate change, resource consumption, economic expansion, inequality and poverty, biodiversity, migration, etc.—both as distilled in and governable through cities' diverse social, material and political capacities. Expressions of this framing can be found at multiple scales: in the urban turn in global policy evident in the United Nation's Sustainable Development Goals 11, in the proliferation of global intra-city networks (e.g. C40 Cities, ICLEI, The Things Network, Shareable) that catalyse cities' capacities to act collectively and in emergent governance configurations in multiple individual cities. Crucially, this conceptualisation of global urbanism as vector of global change and pragmatic site of intervention creates new expectations of urban governance as a means through which to innovate, tackle global challenges and foster prosperous and sustainable futures (Acuto et al., 2018).

Indeed, in the context of global urbanism, urban governance innovation is becoming 'the new normal' as cities experiment with novel arrangements, techniques and practices to deal with the complexity and dynamism of 'global urban' issues (Timeus & Gascó, 2018). Diverse globally referenced ecosystems of innovative urban governance are being shaped with the capacity to transform the parameters and practices of governance. In this chapter, writing from the Global North,

DOI: 10.4324/9780429259593-32

I sketch the ways in which emergent ecosystems are reworking urban governance in relation to global urbanism, recasting relations, mechanisms and operations. I then turn to the epistemological orientations useful to recognising innovative urban governance, drawing out the affordances of a postcolonial sensitivity, receptive to how such innovation is shaped across the 'relational multiplicities, diverse histories and dynamic connectivities' (Robinson and Roy, 2016: 180) of a 'world of cities' and attuned to understanding related progressive and potentially problematic possibilities. To conclude, I reflect briefly on the politics of urban governance innovation and the knowledge politics of related research in the context of global urbanism.

Innovating urban governance in/for a world of cities

Diverse suites of governance innovations are emerging as cities are increasingly expected to engender new capacities to find 'solutions' to global urbanism's manifold challenges. A 'mashup of new and diverse (urban governance) ecosystems' (Cohen et al. 2015, 12) is surfacing via novel arrangements, practices, institutional forms and relationships. Through these, governance capacity is being reworked, pluralised and distributed. Multiple/plural assemblages are being crafted that are reconfiguring the actants involved (actors, objects, technologies, organisations, representations, etc.) and producing new mechanisms and modes of governing. Some hint of their flavour can be read in the prefixes used to describe modalities of governance associated with nascent innovations – e.g., smart-, algorithmic-, experimental-, financialised-, co-produced urban governance. As to the geographies of these governance assemblages, they are being constituted across differentiated global topologies as cities are linked to policy mobilities, networks and connections. Even as these topologies are shaped evenly in politically and geographically differentiated ways that often favour Global North metropoles, the potential to reshape the geopolitics of governance is clear.

Urban governance innovation can be traced across three key dimensions. First is *relations*. Of course, government, business and community actors have long been imbricated in city governance. But these relationships are increasingly recomposed in a diversifying array of institutional mixes and scales that leverage new networks and capacities to govern urban challenges. For example, third-sector intermediaries, such as global philanthropies and charities (e.g. *NESTA*, the UK-based innovation foundation) and non-governmental organisations (e.g. the not-for-profit *Shareable*), are being embedded in urban governance in increasingly influential ways (Davidson et al., 2019). Some position themselves as boundary-spanning organisations, consciously leveraging governance capacity across diverse stakeholders and geographies (Acuto et al., 2018). *City Possible*, for instance, driven by the global corporation Mastercard, is proposing 'a new model for public-private partnership that brings together cities with a united private sector to address the most pressing urban challenges globally' (https://citypossible.com/). And urban government authorities are increasingly embedding themselves in international networks as they seek to develop collaborative capacity (such as C40 cities), provoking new tensions concerning local accountability and new patterns in the kinds of relationships that are disrupted or re-embedded (Davidson and Gleeson, 2018).

A second dimension is the proliferation of *mechanisms* aimed at nurturing innovation that builds operational and financial capacity. This includes 'experimental' mechanisms sourced from the business, technology or creative sectors: hackathons, living labs, sandboxes and other methods of crowd-based 'solution seeking' (Taylor-Buck & While, 2017). Indeed, the experimental itself can, indeed, be understood as a globally circulating trope that does political work in refashioning the scope of governance (Easterling, 2014). Multiple cities, for instance, have offered themselves as trial sites for automated vehicle experimentation willing to test both the technology's capacity to address urban mobility and energy challenges and the parameters of politically workable regulations: Melbourne's 2019 announcement of Uber 'flying taxi' trials in 2020 comes to mind. New financial mechanisms, such as social investment bonds, green capital and civic crowdfunding, are being invoked to drive municipal entrepreneurialism and, equally, to distribute responsibility for financing socially desirable investments. And new global benchmarking indices related to urban challenges, such as ARUP's *City Resilience Index*, are being mobilised locally and globally to hone city policies (and their integration into data science-informed modalities). These mechanisms are being layered across existing institutional orders, altering the social practice of governance, with the potential to shape new politics, possibilities and parameters for decision making and power dynamics in the context of global urbanism.

Finally, distinctive *operational transformations* are being fomented by the political necessity to maintain legitimacy as essential to the authority to govern. These include city authorities' consciously prototyping new ways of working, such as co-governance, co-design and co-production, across government, citizens, corporations, not for profits, philanthropies, transnational networks and more (e.g. the co-city protocols for collaborative, polycentric governance established across multiple Italian cities; Iaione, 2016). Meanwhile, alternative forms of citizen-led participation, long since practiced in Global South cities, are emerging to redress governance gaps (e.g. 'guerilla' governance, such as Toronto's Urban Repair Squad; Danenberg & Haas, 2018). And, of course, the affordances of smart technologies, open data, digital innovations and algorithmic governance capabilities are creating potentially paradigm-shifting operational transformations and institutional arrangements for city governance (Ashton et al., 2017).

The multidimensional innovation being cultivated by new expectations of urban governance in global urbanism warrants critical investigation of emergent politics, possibilities and geographies. All three dimensions of innovation entrain progressive and regressive potentialities. They produce new institutional settings and incorporate new actants, potentially invoking new forms of power (Clark, 2019). In so doing, they can produce new, or further entrench existing, geographies of inclusion and exclusion. They demand new skill sets and capabilities. They are reconstituting the multiscalar relations of cities as they bring, for example, global philanthropies, municipal financialisation, the 'smart city' phenomenon and sharing city movement into the urban governance 'mash-up'. As new technological capabilities, knowledge flows, financial instruments and data platforms that are 'partially or wholly outside the purview of government agencies' (Ashton et al., 2017: 1) are assembled together, governance innovations raise multiple questions requiring critical exploration about accountability and about the extent to which they reproduce or disrupt existing alignments of interests and intents.

Engaging in critical exploration requires a set of epistemological and methodological orientations sufficiently agile to capture the unruly ways in which global urbanism is reshaping the constituencies, topologies and diversely configured geographies of urban governance. Yet these orientations need also to remain attuned to the fact that, while urban governance capacity is reshaped in relation to global urbanism as a general condition, it is irreducibly situated in particular historical-geographical and political-economic contexts, reproducing multiple global urbanisms.

Researching urban governance innovation in/for global urbanism

Dispositif thinking—Emergent ecosystems of innovative urban governance, made up of diverse networks of actants, are being assembled in response to complex global challenges. This suggests a Foucauldian conception of governance viewed through the lens of a 'dispositif': 'a heterogenous set of discourses, practices, architectural forms, regulations, laws and knowledges connected together into an apparatus of government' that is loosely held together in the practice of governing (Braun, 2014: 49). Dispositif thinking displaces the city as a predetermined territorial object of enquiry. It suggests instead an approach focussed on the processes through which efforts to govern particular urban problems are conjured in different topologically configured contexts: the actants and practices gathered together to shape means of government and the multifarious claims and sources of authority that assert themselves in the process of governing across different urban domains: housing, economy, environment, energy, etc.

Methodologically, dispositif thinking lends itself to following process geographies wherever they lead, across 'the messy spaces within which the urban governs itself' (Rogers et al., 2014: 1558). It allows for the recognition of the emergence of multiple, contextually-shaped governance assemblages drawn together to address multiple articulations of urban challenges as they are expressed and problematised in diverse contexts. It provides a versatile optic, receptive to the way governance innovation invokes new topologies of actants and, in so doing, decentres and distributes authority across multiple overlapping sources. These operate simultaneously across different registers (including conventional registers such as control or discipline, alongside less conventionally recognised registers, such as persuasion, automation, nudging) and across differently configured scales and territories (the neighbourhood, precinct, city, interconnected systems of cities). They informalise governance capacity and generate autonomy from conventional hierarchies of government and new capacities to command support (see Bulkeley et al., 2018). State authority features in these topologies, of course, but it is over-determined by the multifarious political forms, practices and modes of action that come to the fore through innovations aimed at 'improving' particular urban 'problems' as they are differentially expressed and politicised in place (Magnusson, 2014). In denaturalising the tendency to start with 'the state', dispositif thinking can provide a powerful analytical lens for grasping the implications of urban governance innovation, which 'increasingly escape(s) the reach of, and is not directed towards, the geographically bounded state' (Boudreau, 2019: 6). It draws our attention to a multiplicity of (potentially dissonant) institutions, territories, techniques and technologies, claims and interests and how they come together in governing global challenges through the urban. Moreover, dispositif thinking recognises the multiple contextually specific inflections that shape both the problem to be governed and the assembled actants loosely connected in its governance.

The urban unbound—Geographers have understood urban governance as embroiled in the relational workings of processes and institutions beyond the city itself (McGuirk, 2004). The governance innovation spurred by global urbanism heightens the city's relationality as a node through which to assemble actants, policies and ideas to steer global attempts to (re)organise socio-political-ecological relations. Cities' worlding (Roy and Ong, 2011) in this way require an epistemological orientation that is more than multiscalar but topological: attuned to tracing the ways in which particular governance domains are simultaneously localised—produced in place—and globalised—stretched across space—with actants and knowledge flows from 'elsewhere' being drawn into local political relations (McCann, 2011) and local capacities being connected into 'worlded' settings (Parnell and Robinson, 2018) via, for example, boundary-spanning organisations (e.g. global philanthropies), global social movements (e.g. Transition Towns), social media campaigns (e.g. the #deleteuber campaign) and policy tourism (e.g. the Vancouver model).

So research on governance innovation in global urbanism calls for an approach that allows it to be understood both through the lens of local experience and via the web of connections that extend beyond the city and exceed the formal domain of 'the state', assembling proximate and distant actants into emergent urban governance formations (Allen and Cochrane, 2014). A postcolonial sensibility would also orient such topological reading across a wide array of contexts, eliding analytical distinctions between Global North and Global South and resisting interpreting connections in terms of assumed points of origin in the Global North being mirrored in Global South 'copies' (Robinson, 2016). Rather, all topological nodes would be placed into the same frame of analysis to trace the dispersed connections, processes and practices assembled in producing distinctive realisations of governance innovation.

Multiplicity—Different urban 'problems' for which innovative solutions are being sought—housing; mobilities; production; inequality; energy, waste and water; and food systems—are likely to produce distinctive actualisations of governance, operating through distinctive spatialities, temporalities and configurations of actants that materialise in specific urban contexts. Thus a multiplicity of domains of governance will coexist and overlap in any city, working across distinctive arrays of motivations and lines of determination. These will rub up against each other in ways that escape conception as unitary, state directed and implemented and that are irreducible to any singular or unifying logic. Think, for instance, of the contrasting governance aspirations, values, frames of reference (and, relatedly, distinctive topological geographies) evident in the Leeds City Lab, a public-private and voluntary sector collaborative initiative aimed at driving co-production to rework urban governance and address governance gaps in the United Kingdom's austerity context (Chatterton et al., 2018); and those informing urban energy governance in Sydney, which is being led through the top-end commercial building sector via the local adaptation of a collaborative local government/building-owner partnership model drawn from London; private-sector-led smart building energy management systems; and investor-driven sustainable building performance ratings and mandatory building standards (Carr et al., 2019).

Fluid categories—While research may start with a given city or a particular localised expression of an urban challenge, productive methodologies need to be sufficiently agile to follow the diverse and emergent configurations through which the 'will to improve' materialises. This is not merely a question of geography but of following how governance processes work. Particularly valuable here are orientations that resist *a priori* assumptions about the roles of the various actants and that lean away from pre-emptive categorisations ('state', 'market', 'civic/community', 'social', 'technical, 'local', 'global', etc.) often based in conceptions derived and universalised from specific contextual settings (Jaffe and Koster, 2019). They allow a conception of governing in which actants may play changing, multiple or ambiguous roles, not confined to predetermined categorising or routinised repertoires (see Boudreau, 2019), which is particularly productive given the plurality of practices and configurations associated with urban governance innovations and the growing propensity for promiscuously sourced experimental stances to test new modalities and registers of governance. They allow for more diffuse modes and practices of governance (from the algorithmic to the co-produced). And by shifting the analytical gaze from formal political actors to the process of governing, epistemological orientations that grasp this fluidity can expand our receptiveness to a broader range of governing practices and actants derived from a broader array of contexts. They set us about tracing how the elements of governing a given urban problem in specific situations are assembled in time and across space and how they come to take effect in the practice of governing.

Collectively, these four epistemological orientations are receptive to the proliferative nature of urban governance innovation; they pre-empt reading innovation in any given city as a localisation or derivation of a putative global urban norm. Instead, they position us to account for the unfolding of distinctive innovative governance configurations in specific places and moments, conditioned by uneven connections of people and places across space and scale and to globalised systems (Boudreau, 2019). They take urban governance in global urbanism as both globalised and irreducibly place-based, condensed across distributed geographies and differentiated topologies, whereby people, technologies, objects, ideas and policies are contingently arranged and fitted together across multiple governance domains (see Allen and Cochrane, 2014).

As a set of orientations, they are alive to diverse domains of governance innovation and to the profusion of actants whose roles and practices are not fixed and whose logics are not unitary. They are equipped to follow multiple connections that exceed specific urban jurisdictional boundaries and actualised as distinctive configurations between place-based and seemingly 'distant' actants across an array of contexts that may escape any neat articulation of 'Global North/ South. They are attuned to understanding the relations between urban governance innovations and a 'differentiated, changing and globalising urban world'.

Conclusion

In the context of multiple global urbanisms—not a generalised condition refracted in multiple contextualised renderings of a global norm but a 'world of cities' produced as distinctive or 'singular' realisations of the urban shaped by multiple globally connected processes (Robinson, 2016)—I conclude by reflecting on the politics of urban governance innovation and the knowledge politics of related research.

The emergent, globally referenced constellations of actants, institutional relationships, mechanisms and geographies that characterise governance innovation ecosystems raise substantial questions around accountability, the potential for elite capture and the re-authorisation of embedded interests and power hierarchies. Smart urban governance innovations, for instance, raise critical questions around their capacity to facilitate corporate dominance, market logics and 'smartmentalities' amongst citizens, irrespective of their capacities to improve the management of global urban challenges. It is essential then to maintain a critical research focus on how new hegemonies are shaped in global urbanism by the framing of cities and urban governance innovations as the conduits for managing diffuse global challenges.

Nonetheless, the spur that global urbanism provides to governance innovation also encapsulates progressive opportunities. With increased disaggregation of authority and engagement with diverse governance modalities comes the prospect of inscribing new aspirations, values and inclusionary practices into authoritative positions (Chatterton et al., 2018). The emergent mash-up of innovative urban governance ecosystems cannot be assumed to be united around a common rationality or a singular system of management or to be subject to delegated authority by the state (see Picker, 2019). Rather, as Braun (2014, 51) puts it, these ecosystems combine 'diverse elements into new heterogeneous formations…taking hold of new knowledges, technologies and practices that either did not previously exist or had not previously been appropriated as a means of administration'. With this comes the possibility of including new modalities of governance, expertise and logics that might radically expand governance repertoires: from the technological to the creative, the communal to the competitive, the everyday to the systematic. Critical global urbanism research needs to remain alert to these possibilities and, crucially, to their vulnerabilities as it works from vantage points that appreciate the inherent diversity that overdetermines different city governance contexts, where place-based iterations of urban challenges intersect with globally circulating governance practices and localised contestations (Beveridge and Koch, 2017).

Finally, in terms of knowledge politics, the epistemological orientations outlined in this chapter chime in significant ways with a postcolonial ethos and offer crucial reorientations to those of us writing and thinking from the Global North. They equip us to recognise more readily the resonances in governance practices across Global North/South contexts: for instance, between the characteristics of Global North urban governance 'innovation' and Global South 'informality' (see

Jaffe and Koster, 2019). This, in turn, encourages us to engage in earnest with scholarship from contexts habitually overlooked as dissimilar to 'the metropolis'. They caution us against deploying pre-given universalised categories rather than tracing practices. And they undercut interpretations of global conduits of innovative urban governance practices founded on imaginaries of (Global North) origin and (Global South) replication (Robinson, 2016). Practically, this de-hierarchised optic opens out the potential actants, mechanisms and practices that might productively be drawn into the urban governance playbook. This might rework the networks through which governance learning occurs, potentially providing new inspirations for innovations beyond the market-led and competitive, challenging their habitual exclusions. What might be learned, for instance, about governance to support the 'sharing city' ethos flourishing in Global North cities from the long-established coordination of sharing practices amongst shack dwellers and poor people's movements in Global South cities? Research attuned to multiple global urbanisms can be more fully receptive to the plurality of governance innovations being engaged globally and significantly expand the resources available to city governance, making visible 'another way in which to think about urban policy models as more-than-neoliberal or, indeed, as not necessarily neoliberal at all' (Bunnell, 2015: 1992), creating space for such alternatives to be legitimated and authorised.

The attention global urbanism focusses on urban governance innovation suggests an insistently critical stance on emergent hegemonies, uneven power relations and claims to authority that innovation may reproduce as much as disrupt. Equally, it suggests a generative, even hopeful, stance receptive to identifying more socially and environmentally just pathways for reformulating global urbanism.

References

Allen, J. and Cochrane, A. (2014) 'The urban unbound: London's politics and the 2012 Olympic Games'. *International Journal of Urban and Regional Research*, 38: 1609–1624.

Ashton, P., Weber, R. and Zook, M. (2017) 'The cloud, the crowd, and the city'. *Big Data & Society*, 4: 1–5.

Beveridge, R. and Koch, P. (2017) 'The post-political trap? Reflections on politics, agency and the city'. *Urban Studies*, 54: 31–43.

Boudreau, J. (2019) 'Informationalization of the state: Reflections from an urban world of translations'. *Journal of Urban and Regional Research*, 43: 597–604.

Braun, B. (2014) 'A new urban dispositif? Governing life in an age of climate change'. *Environment & Planning D*, 32: 49–64.

Brenner, N. and Schmid, C. (2015) 'Towards a new epistemology of the urban'. *City*, 19: 151–182.

Bulkeley, L.-A.A., McFarlane, C. and MacLeod G. (2018) 'Enhancing urban autonomy: Towards a new political project for cities'. *Urban Studies*, 55: 702–719.

Bunnell, T. (2015) 'Antecedent cities and inter-referencing effects: Learning from and extending beyond critiques of neoliberalisation'. *Urban Studies*, 52: 1983–2000.

Carr, C., McGuirk, P. M. and Dowling, R. (2019) 'Geographies of energy transition'. *Australian Geographer*, 50: 29–48.

Chatterton, P., Owen, P., Cutter, J, Symski, G. and Unsworth, R. (2018) 'Recasting urban governance through Leeds City Lab: Developing alternatives to neoliberal urban austerity in co-production laboratories'. *Journal of Urban and Regional Research*, 42: 426–443.

Clark, J. (2019) *Uneven Innovation: The Work of Smart Cities*. Columbia University Press: New York.

Cohen, B., Almirall, E. and Chesbrough, H. (2015) 'The city as a lab: Open innovation meets the collaborative economy'. *California Management Review*, 59: 5–13.

Danenberg, R. and Haas, T. (2018) New trends in bottom-up urbanism and governance, in Arefi M and Kickert C (Eds.), *The Palgrave Handbook of Bottom-up Urbanism*. Palgrave: London, pp. 113–129.

Davidson, K. and Gleeson, B. (2018) 'New socio-ecological imperatives for cities: Possibilities and dilemmas for Australian metropolitan governance'. *Urban Policy & Research*, 36: 230–441.

Davidson, K., Coenen, L., Acuto, M. and Gleeson, B. (2019) 'Reconfiguring urban governance in an age of rising city networks: A research agenda'. *Urban Studies*. doi 1177/0042098018816010.

Easterling, K. (2014) *Extrastatecraft*. Verso: London.

Iaione, C. (2016) 'The co-city'. *American Journal of Economy & Sociology*, 75: 415–455.

Jaffe, R. and Koster, M. (2019) 'The myth of formality in the Global North: Information-as-innovation in Dutch governance'. *Journal of Urban and Regional Research*, 43: 563–568.

Magnusson, W. (2014) 'The symbiosis of the urban and the political'. *International Journal of Urban and Regional Research*, 38: 1561–1575.

McCann, E. (2011) 'Urban policy mobilities and global circuits of knowledge: Toward a research agenda'. *Annals of the Association of American Geographers*, 101: 107–130.

McGuirk, P. M. (2004) 'State, strategy and scale in the competitive city: A neo-Gramscian analysis of the governance of 'global Sydney'. *Environment and Planning A*, 36: 1019–1043.

Parnell, S. and Robinson, J. (2018) 'The global urban: difference and complexity in urban studies and the science of cities', in S. Hall and R. Burdett (Eds.), *The Sage Handbook of the 21st Century City*. Sage: London, 13–31.

Picker, G. (2019) 'Sovereignty beyond the state: Exception and informality'. *International Journal of Urban and Regional Research*. doi:10.1111/1468-2427.12704.

Robinson, J. (2016) 'Reading cities through elsewhere: Comparative tactics for a more global urban studies'. *Progress in Human Geography*, 40: 2–29.

Robinson, J. and A. Roy (2016) 'Debate on global urbanism and the nature of urban theory'. *International Journal of Urban and Regional Research*, 40: 180–186.

Rogers, S., Barnett, C. and Cochrane, A. (2014) 'Where is urban politics?' *International Journal of Urban and Regional Research*, 40: 180–186.

Roy, A. and Ong, A. (2011) *Worlding Cities: Asian Experiments and the Art of Being Global*. Blackwell: Oxford.

Sheppard, E., Gidwani, V., Goldman, M., Leitner, H., Roy, A. and Maringati, A. (2015) 'Introduction: Urban revolutions in the age of global urbanism'. *Urban Studies*, 52: 1947–1961.

Taylor-Buck, N. and While, A. (2017) 'Competitive urbanism and the limits to smart city innovation'. *Urban Studies*, 54: 501–519.

Timeus, K. and Gascó, M. (2018) 'Increasing innovation capacity in city governments'. *Journal of Urban Affairs*, 40: 992–1008.

29

CORRIDOR URBANISM

Jonathan Silver

> This is not a concession, it's a giveaway of property belonging to the Greek people.
>
> Constantinos Tsourakis, a worker at Piraeus Port [1]

In May 2016, a controlling stake in the state-owned Port of Piraeus, Athens, was sold by the Greek Government in a deal for over €300 million. The sell-off was facilitated through Greece's privatisation agency as part of imposed conditions for a Eurozone 'structural adjustment' of the crisis-hit country. The 50-year lease to the Chinese COSCO Shipping Company is part of plans for this southern Mediterranean port to act as a European gateway for the Maritime Silk Road, the ocean route of the $1 trillion Belt and Road Initiative. Large- scale opposition to the sale of this site of global infrastructure was driven by the Union of Dockworkers, including through strike action, but was ultimately unsuccessful. The port's future is being reimagined as a high-tech enclave of global logistics and real estate development for the elite. Meanwhile, residents in surrounding neighbourhoods have been disconnected from the electricity network as part of a wave of attacks on basic human rights in the name of austerity and balancing the budget. As Petrova and Prodromidou (2019: 1382) documented across Greece, 'In the first 10 months of 2017, [the Public Power Corporation] cut the power supply of 240,000 consumers', leaving households without electricity and often left to resort to highly polluting toxic wood-burning fires. In Piraeus, the promise of infrastructure is configured around its potential to act as a global hub of exchange, mobility and circulation along new land- and ocean-based corridors. In nearby districts, everyday infrastructural life slowly breaks apart. With media attention focussed on asking, 'Could Chinese investment rescue ailing Greece?' (Morris, 2011), it failed to consider the experiences of workers and residents in Piraeus living through a 'splintered urbanism' (Graham and Marvin, 2001) and the contradictions that lay between restructuring global and local infrastructure. These networked dynamics suggest that inequality as techno-social differentiation is only intensifying in Athens as new corridor investment and social repression proceed apiece.

DOI: 10.4324/9780429259593-33

FIGURE 29.1 Port of Piraeus (source: Author)

> Foreign powers, particularly China, should immediately withdraw from their all projects in Balochistan.
>
> The Baloch Liberation Army[2]

In May 2019, an attack took place at the Pearl Continental Hotel in Gwadar, Balochistan, a small city set to become an intense hub of infrastructural investment for the new 2,700-km China–Pakistan Economic Corridor. The attack gained headlines around the world and signalled a period of intensified violent resistance by the Baloch Liberation Army. The $1.62 billion financing of a port expansion, as part of a 40-year lease, was part of a series of new projects, including delineating 2,292 acres for a special economic zone and promises of over $200 million for a new airport. This investment aims to turn Gwadar, in the words of Tariq Al-Shammar, into 'the next Dubai'.[3] It's designated as a critical infrastructural hub of the larger $52 billion China–Pakistan Economic Corridor, connecting the city of Kashgar, China, in the far-west province of Xinjiang/East Turkestan to the rapidly growing urban centres of Pakistan and outwards onto the Indian Ocean. The Baloch Liberation Army, an insurgent group that claimed responsibility for the hotel and other subsequent attacks, was reportedly motivated by an understanding of these investments as extractive infrastructure, designed to plunder the natural resources of the region. Further violent resistance to this infrastructural corridor is considered highly likely. Concurrently, many urban dwellers in Gwadar face an uncertain future due to the significant transformations of urban space required to enable plans for the China–Pakistan Economic Corridor to move forward. Transformations to Gwadar are mobilised in the name of progress, modernity and economic development, for both the city and the Pakistani nation. Fears are particularly acute for people in the central areas as the Federal Government plans to expropriate land, displace residents and demolish the old city under the colonial-era 1894 Land Acquisition Act. As a fisherman working in an industry that generates employment for up to 80% of the existing population lamented to Mishele Ijaz in media outlet Dawn[4], *'We knew we would be displaced.'* Like in Piraeus, promises of infrastructural investment, trade and opportunity may ring hollow for those navigating already-precarious livelihoods. The making of this infrastructural corridor has become a maelstrom of uncertainty and disruption for the urban majority.

These cases convey the intensity of extended infrastructure projects, which have come to be known as corridors and are constituted in and through very different urban contexts. If Athens conveys an unravelling crisis facing Global North cities in an era of austerity and decline, Gwadar's experience is of surging investments that will utterly rebuild this city, mirroring the rapid urbanisations of the South Asia region. Yet the transformations afoot in Athens and Gwadar are connected. Both cases highlight what Wiig and Silver (2019) describe as the shifting techno-territories of the world economy, China's geopolitical ambitions, the importance of cities across these new spatial articulations and the global urbanisms shaped by and shaping the corridor.

In this intervention, I suggest that infrastructural corridors interrupt our ways of knowing global urbanism, much like they are disrupting the lives of urban populations and the relational geographies of cities in a world economy. As the editors of this volume argue, and now with well-established claims of over 50% of the world population being urban (Soja, 2000) and an estimated urban population of 6.7 billion by 2050 (UN-Habitat, 2013), scholars need to continually assess assumptions about what we know and how we research global urbanism, which I understand through two distinct notions. Firstly, as a dynamic process of shifting urban geographies across the planet, associated city planning and social life, replete with justice and inequality, hope and despair. Secondly, it has become a well-established research epistemology of particular ways of understanding (i.e. scholarship and theory) what the urban means in the social sciences—that is, 'not a coherent body of theory or practice, but loose bundles of ideas and practices that travel across the world—increasingly through transnational networks' (Sheppard et al., 2013: 894). Corridors disrupt both these notions of global urbanism.

Corridors as a new techno-territorial phenomenon

I turn first to the ways that corridors are remaking global urbanisms through the significant techno-territorial restructuring of cities. This reordering of urban space through corridor investment prompts us to think anew about the unequal, differentiated experiences of urban life. Vital to such a task is recognising the spatial power of the proliferating number of large-scale infrastructure corridors that act as conduits for intensifying investment required for envisaged transnational trade routes. Perhaps best known through the Belt and Road Initiative but encompassing many investments, plans, visions, hopes and physical transformations, corridors are restructuring the extant topographies of global urbanism. Cutting across various territories, cities, countries and regions, these projects seek to materially integrate a new series of accumulation regimes out of historical conditions and geographical contexts (Grappi, 2018; Newhouse and Simone, 2017). Corridors are designed to move people, goods and information and include trade, transportation and logistics routes in addition to Information Communication and Technology networks, gas and oil pipelines and enclaves, such as the Port of Piraeus. Corridors are planned and operationalised with the intention of organising the underlying technological life of cities and hinterlands around particular (and often narrow) logics, calculations and geopolitical interests. The corridor is a strategic, infrastructural-led attempt to reorder territory and economy; however, this phenomenon remains little understood. With these investments demarcating futures incorporating hundreds of millions of urban dwellers, they require much closer attention.

The Global Infrastructure Connectivity Initiative (2017b) identified 30 newly established or in-planning corridors, alongside dozens of potential new projects. And yet the implications of this techno-territorial reordering on urban life have yet to be fully considered. This poses the question about how corridors are intersecting with the rhythms, circulations and experiences of global urbanism. A focus on the everyday ways in which these new modes of infrastructural life are produced, operated, experienced and navigated may provide one step towards making these processes visible. This generates imperatives for how research might theorise inter-relations between vast investments, everyday life and urban infrastructural politics from Athens to Gwadar. The rapid deployment of corridors across the planet constitutes an always assembling, contemporary and relational geography that demands careful comparative attention. We need to think anew about what global urbanism means in the twenty-first century from the vantage point of the corridor.

Urban researchers will need to get to grips with these new patterns of global urbanism being produced through the planning of and investment in corridors. The imperative, then, is to understand the estimated $94 trillion of finance into large-scale infrastructure projects planned up to 2040 (GICA, 2017) as a critical determinant in the geographies and experiences of global urbanism. The prioritisation of large-scale economic development projects and new high-tech enclaves may intensify unequal access to infrastructure already critical in shaping urban life (Chalfin, 2014). Corridors are not a new phenomenon. Yet as we shift to an urbanised planet and with an intensifying, rapidly converging urban crisis (whether social, environmental, pathogenic or economic), the proliferation of these initiatives occurs as advanced technology prompts reassessments of how the urbanisation process is planned, experienced and understood. Patterns of global trade and investment are also undergoing a historic restructuring in which the geographies of North Atlantic–dominated capitalism wane (Wallerstein, 1979) and new economic power centred across the Indian Ocean and the Eurasian landmass emerges (Kaplan, 2009), the scale and speed of new corridor plans are accelerating exponentially. The corridor posits a particular technological vision of the future in which connected urban regions become critical sites in the infrastructural-led reordering of the world economy (Kanai and Schindler, 2019); in effect, global urbanism is produced out of these multiple corridor-configured regional urbanisms. In urban space, this takes form primarily through new concentrated enclaves (Caprotti, 2014) and premium networked space (Graham, 2000). Responding to this intensifying techno-territorial phenomenon is both analytically and politically vital.

A new epistemology of global urbanism

Thinking the urban through the corridor requires new conceptual vocabularies because the corridor is neither city-centric nor planetary in its composition. Rather, these techno-territorial projects generate a relational urban geography across disparate parts of the world. Corridors complicate and blur distinctions, such as North/South, where the city begins and ends, relations between infrastructure and society, circulation and disconnection and new forms of urban geopolitics. Corridor urbanism can push scholars to experiment with shifting analytical approaches and making new dialogue with the city-centric (Burdett and Sudjic, 2007), planetary (Brenner and Schmid, 2014) or mega-region (Olds, 2001).

Critical to how urban studies has begun to understand corridors has been the ways in which infrastructure, as advanced technology, is deployed across urban space, drawing attention to the process of standardisation (Schindler and Marvin, 2018). Spaces of standardised technology are assumed to be globally replicable (Carse and Lewis, 2017) and repeatable systems of operation so that, for instance, container technology, security procedures, monitoring equipment, sensors, databases and computer systems in Athens and Gwadar are the same. Literature has emphasised the emergence of urban enclaves, which can be understood as the 'technological zones' (Barry, 2006) of new standardised, high-tech operations (Chalfin, 2010). Global functions and premium technology are often separated from their surroundings through securitisation, displacement and differentiated technology (Wiig, 2018). However, a corridor is never as standardised as designers, politicians, investors and planners intend or assume. Shifting beyond standardisation can help us to pay attention to the everyday life within and beyond these infrastructure projects, both in relation to and exceeding the logics of capitalist urbanisation. It is these messy, contingent dynamics through which the deployment of corridors encounters urban space that should be made central in analysis. Moving beyond the dominant narrative of standardisation involves a reading of the global urbanisms constituted through corridors. In recent years, a growing body of scholarship has investigated various aspects of the everyday life of infrastructure from gendered violence (Datta and Ahmed, 2020), to living with disruption (Ghanem 2018) and accessing water services (Truelove, 2019), prompting important advances in global urbanism from a technological perspective. This has included contributions to feminist (Truelove, 2011) and postcolonial (Lawhon et al., 2014) approaches to these networks that begin not with epistemological abstraction but rather with situated, textured accounts of everyday relations of infrastructure. At the foundation of such explanatory frameworks is a motivation to start with urbanism as a relational and shifting category (Peake, 2016) rather than something fixed and determined by broader processes of capitalist urbanisation (Derickson, 2015). This matters for understanding corridors because these infrastructural geographies remain dynamic, variegated, contested and amenable to being reshaped and reconfigured through calculative practices, technical adjustments, minor and major contestations and negotiation. Literature primarily focussed on urban regions in the Global South has long highlighted differentiated experiences of infrastructural life, showing how systems come to be experienced in highly unequal ways (Larkin, 2008; Monstadt and Schramm, 2017) across urban space and social groups (Simone, 2006). If we turn back to the disconnected households in Athens, or the potentially displaced inner-city residents in Gwadar, I would suggest that the promise of the corridor will not hold for everyone. Cities are being disrupted in profound ways in order to make operational these new techno-territorial articulations of the urban future.

A research agenda for corridor urbanism?

Infrastructural corridors disrupt both notions of global urbanism, the techno-territorial phenomenon and the research epistemology. I now want to think forward about four ways in which we might develop a research agenda that addresses these imperatives.

Firstly, we have to think anew about the corridor and understand the significance of these investments in generating patterns, flows, associations and constellations of the urban. Uncertainty reverberates across social worlds, environmental conditions degrade, extractions are intensified, circulations are accelerated and new technological configurations reprogramme the functioning of cities. Important here is recognising the corridor as neither city-centric nor planetary in its composition. Urban regions need to be delineated beyond bounded territorial units of comparison towards relational spaces within new corridor demarcations. Relational theorisation and associated methodologies required to investigate are essential in thinking about how inequality, crisis and urban technology might be configured, not in the city or across the planet but rather along the corridor. It prompts us to ask, how might we think about the Chinese–Pakistani Economic Corridor as a constituted space or territory of global urbanism? What kinds of tools, collaborations and practices might we need to undertake research on such vast urban constellations? Furthermore, attention to new corridor formations demand shifting away from the paradigmatic cities of urban studies to new centres of power, exchange and transformation (Robinson, 2006). What does it mean to understand global urbanism from the perspectives of the infrastructural processes and experiences of people in Gwadar or Kashgar rather than people in New York or London, or even Karachi or Beijing?

Secondly, focus needs to be on the displacement violence and pressures that are visited on communities in the actual physical deployment of corridors, as work has previously done on mega-projects and large-scale urban transformations. As scholars, we can help make visible the injustices being generated in the name of economic development and the promise of infrastructure. As I've previously written, the operationalisation of the Central Corridor in Kampala, connecting this Ugandan city to the Indian Ocean through Lake Victoria and the Port of Dar es Salaam, was predicated on the clearance of homes and businesses from lands surrounding the train track (Wiig and Silver, 2019). For urban populations such as those in central Gwadar, planned deployments of the infrastructures of the corridor generate uncertain futures, living with the threat and expectation of displacement, the disruption of livelihoods and households breaking apart. Mobilisation by communities can be supported through detailed and, if required, adversarial research practices that help to convey ways in which corridors are shattering everyday urban life and the impact on ad hoc social infrastructures that sustain social reproduction. We might ask, how do we understand urban displacement through a transnational trade route and the standardisation processes imposed on populations? Concurrently, what connects the infrastructural displacements experienced by urban dwellers in disparate cities, such as Athens and Gwadar?

Thirdly, as urban researchers, we might need to develop revised articulations of life inside and outside the zone of the corridor. Situated approaches can provide a textured understanding of how these infrastructures are transforming the urban experience in unequal ways and the new forms of associational life being configured in the making of corridors. Who gets access to the premium technologies of the corridor? Who is left to face more precarious lives when the corridor is operationalised? How do people subvert, exceed or navigate the capitalist logics imposed by the corridor? These questions can be enhanced through approaches that elucidate the everyday, variegated and diverse life of corridors experienced by populations. The purpose of this work may be to enter into conversation with existing theorisation and methodologies of global urbanism and in the process subvert top-down, epistemological abstractions of global infrastructure, economic development and standardisation.

Fourthly, research can open up the everyday ways in which corridor-driven infrastructural politics are beginning to crystallise, especially in relation to contestations and resistance. In Athens, the strike action by the Union of Dockworkers has shown one means through which collective mobilisation is occurring. In Gwadar, the developing insurgency by the Baloch Liberation Army suggests a more violent mode of resistance that, in its becoming, has brought intensive surveillance and repression from the state onto the local population. However, politics constituted through corridors do not stop at city boundaries; rather, they hold the potential to open new transnational solidarities and forms of resistance. In 2018, I interviewed Dolkun Isa, president of the World Uyghur Congress, about the role of the Belt and Road Initiative in 'pacifying' Xinjiang/East Turkestan and the ongoing, technological and infrastructural-led oppression of the Uyghur people (Chung, 2018). In Frankfurt, Dolkun thoughtfully explained to me that such corridors of movement, trade and circulation, notably the historic Silk Road, had always exceeded the narrow economic and military purposes of the powerful, instead bringing opportunities for the exchange of stories, cultures, struggles and solidarities. The corridor has the potential to close down, antagonise, repress and restructure urban life with new intensities and injustices. But as

Dolkun has suggested, this techno-territorial phenomenon might also open up new connections across disparate places and people that bring hope, resistance and articulations of a different type of corridor. As researchers, we must be attuned to and, where possible, participate in the forging of these new relations as a counterstrategy to the worst excesses of the deployment of corridors. To do so also involves considering whether another type of corridor is possible, an infrastructural reordering that puts the priorities of people and the planet over capitalism.

Acknowledgements

I would like to acknowledge and thank the Leverhulme Trust for funding the time to research and think about corridors, Leonie Newhouse, Omar Jabary Salamanca, AbdouMaliq Simone and Alan Wiig for important conversations in advancing these ideas and Colin and Michele for the invitation and editorial support.

Notes

1 Source: Reuters (2016) 'Workers protest as Greece sells Piraeus Port to China COSCO'.
2 Source: thedefensepost.com (2019) 'Pakistan: 5 killed in Gwadar hotel attack claimed by Baloch Liberation Army'
3 Source: Open Democracy (2017) 'Dubai and Gwadar: The silent economic war in the Gulf of Oman'
4 Source: Dawn (2018) 'How Gwadar's CPEC development might leave its people behind'

References

Barry, A. (2006) 'Technological zones'. *European Journal of Social Theory*, 9(2): 239–253.

Brenner, N. and Schmid, C. (2014) 'The "urban age" in question'. *International Journal of Urban and Regional Research*, 38(3): 731–755.

Burdett, R. and Sudjic, D. (2007) '*The endless city: An authoritative and visually rich survey of the contemporary city*'. Phaidon Press: New York.

Carse, A. and Lewis, J. A. (2017) 'Toward a political ecology of infrastructure standards: Or, how to think about ships, waterways, sediment, and communities together'. *Environment and Planning A: Economy and Space*, 49(1): 9–28.

Caprotti, F. (2014) 'Eco-urbanism and the eco-city, or, denying the right to the city?' *Antipode*, 46(5): 1285–1303.

Chalfin, B. (2010) *Neoliberal Frontiers: An Ethnography of Sovereignty in West Africa*. University of Chicago Press: Chicago

Chalfin, B. (2014) 'Public things, excremental politics, and the infrastructure of bare life in Ghana's city of Tema'. *American Ethnologist*, 41(1): 92–109.

Chung, C. P. (2018) 'China's Uyghur problem after the 2009 Urumqi riot: Repression, recompense, readiness, resistance'. *Journal of Policing, Intelligence and Counter Terrorism*, 13(2): 185–201.

Datta, A. and Ahmed, N. (2020) 'Intimate infrastructures: The rubrics of gendered safety and urban violence in Kerala, India'. *Geoforum*, 110: 67–76.

Derickson, K. D. (2015) 'Urban geography I: Locating urban theory in the "urban age"'. *Progress in Human Geography*, 39(5): 647–657.

Gellert, P. K. and Lynch, B. D. (2003) 'Mega-projects as displacements'. *International Social Science Journal*, 55(175): 15–25.

Ghanem, D. A. (2018) 'Energy, the city and everyday life: Living with power outages in post-war Lebanon'. *Energy Research & Social Science*, 36: 36–43.

Global Infrastructure Connectivity Initiative (GICA) (2017a) *G20 global infrastructure connectivity alliance: 2017 work plan*. Singapore: G20. Retrieved from https://www.bundesfinanzministerium.de/Content/DE/Downloads/G20-Dokumente/GICA-2017-work-plan.pdf?__blob=publicationFile&v=2.

Global Infrastructure Connectivity Initiative (GICA) (2017b) *Global infrastructure outlook*. Retrieved from https://www.oxfordeconomics.com/recent-releases/Global-Infrastructure-Outlook.

Graham, S. (2000) 'Constructing premium network spaces: Reflections on infrastructure networks and contemporary urban development'. *International Journal of Urban and Regional Research*, 24(1): 183–200.

Graham, S., & Marvin, S. (2001) *Splintering Urbanism: Networked Infrastructures, Technological Mobilities and the Urban Condition*. Psychology Press: London.

Grappi, G. (2018) Asia's era of infrastructure and the politics of corridors: Decoding the language of logistical governance. In Neilson, B., Rossiter, N., Samaddar, R. (Eds.), *Logistical Asia* (pp. 175–198). Palgrave Macmillan: Singapore.

Kanai, J. M. and Schindler, S. (2019) 'Peri-urban promises of connectivity: Linking project-led polycentrism to the infrastructure scramble'. *Environment and Planning A: Economy and Space*, 51(2): 302–322.

Kaplan, R. D. (2009) 'Center stage for the twenty-first century: Power plays in the Indian Ocean'. *Foreign Affairs*, 88 (2): 16–32.

Larkin, B. (2008) *Signal and Noise: Media, Infrastructure, and Urban Culture in Nigeria*. Duke University Press: Durham, NC.

Lawhon, M., Ernstson, H. and Silver, J. (2014) 'Provincializing urban political ecology: Towards a situated UPE through African urbanism'. *Antipode*, 46(2): 497–516.

McFarlane, C. and Rutherford, J. (2008) 'Political infrastructures: Governing and experiencing the fabric of the city'. *International Journal of Urban and Regional Research*, 32(2): 363–374.

Monstadt, J. and Schramm, S. (2017) 'Toward the networked city? Translating technological ideals and planning models in water and sanitation systems in Dar es Salaam'. *International Journal of Urban and Regional Research*, 41(1): 104–125.

Morris, C. (2011) 'Could Chinese investment rescue ailing Greece?' *BBC News*.

Newhouse, L. and Simone, A. (2017) 'An introduction: Inhabiting the corridor: Surging resource economies and urban life in East Asia'. *Cityscapes Special Issue: The Corridor*, 3–5.

Olds, K. (2001) *Globalization and Urban Change: Capital, Culture, and Pacific Rim Mega-Projects*. Oxford University Press: Oxford.

Peake, L. (2016) 'The twenty-first-century quest for feminism and the global urban'. *International Journal of Urban and Regional Research*, 40(1): 219–227.

Petrova, S., & Prodromidou, A. (2019). Everyday politics of austerity: Infrastructure and vulnerability in times of crisis. *Environment and Planning C: Politics and Space*, 37(8), 1380-1399.

Pieterse, E., Parnell, S. and Haysom, G. (2018) 'African dreams: Locating urban infrastructure in the 2030 sustainable developmental agenda'. *Area Development and Policy*, 3(2): 149–169.

Robinson, J. (2006) *Ordinary Cities: Between Modernity and Development*. Psychology Press: London.

Schindler, S. and Marvin, S. (2018) 'Constructing a universal logic of urban control? International standards for city data, management, and interoperability'. *City*, 22(2): 298–307.

Sheppard, E., Leitner, H. and Maringantli, A. (2013) 'Provincializing global urbanism: A manifesto, *Urban Geography*, 34(7): 893–900.

Simone, A. M. (2004) 'People as infrastructure: Intersecting fragments in Johannesburg'. *Public Culture*, 16(3): 407–429.

Simone, A. M. (2006) 'Pirate towns: Reworking social and symbolic infrastructures in Johannesburg and Douala'. *Urban Studies*, 43(2): 357–370.

Soja, E. W. (2000) *Postmetropolis: Critical studies of cities and regions* (No. 307.764 S6). Blackwell: Oxford.

Truelove, Y. (2011) '(Re-) Conceptualizing water inequality in Delhi, India through a feminist political ecology framework'. *Geoforum*, 42(2): 143–152.

Truelove, Y. (2019) 'Gray zones: The everyday practices and governance of water beyond the network'. *Annals of the American Association of Geographers, 109*(6): 1758–1774

UN-Habitat (2013) *State of the World's Cities 2012/2013: Prosperity of Cities*. Routledge: London.

Wiig, A. (2018) 'Secure the city, revitalize the zone: Smart urbanization in Camden, New Jersey'. *Environment and Planning C: Politics and Space*, 36(3): 403–422.

Wiig, A. and Silver, J. (2019) 'Turbulent presents, precarious futures: Urbanization and the deployment of global infrastructure'. *Regional Studies*, 53(6): 912–923.

Wallerstein, I. (1979) *The Capitalist World-Economy*. Cambridge University Press: Cambridge.

30

BEYOND-THE-NETWORK URBANISM

Everyday infrastructures in states of mutation

Yaffa Truelove

Introduction

One of the fundamental features of global urbanism and city life are the diverse infrastructures that "enable unprecedentedly large, dense, and heterogeneous groups of individuals to live together within contained spaces" (Rodgers, 2019: ix). Being mindful of the complex and heterogeneous urbanisms that exist across the globe, I take people's everyday struggles to access fractured urban infrastructures in the Global South as a starting point to examine the socio-material processes and practices that are both endemic to cities globally and reveal highly situated and particular types of urbanism on the ground that shape differing experiences of the city and urban condition. Taking urban infrastructure as an entryway into contemplating global urbanism, this chapter approaches global urbanism through a postcolonial and southern urbanism lens. This includes efforts to build a more robust scholarship that decenters EuroAmerican tendencies to universalize global urbanism based on theories, concepts, and empirical evidence that stem from a narrow set of cities in the Global North. As such, I argue that analyzing diverse, fractured and often insufficient infrastructures provides a critical window into thinking through the global urban experience that, in the words of Robinson and Roy (2016: 183), "might incorporate any city, and whose insights will be seen as provisional and revisable." I focus on the ways that diverse networks of water, waste, energy and roads in cities across the globe are tied to, and can help reveal, highly situated and differing symbolic meanings, institutions, resource flows, atmospheres, bodies and social power that shape the contemporary urban experience. Viewing these situated particulars provides an important tactic for expanding "new genres" of global urbanism (Parnell and Pieterse, 2016) and comparative inquiry of cities that draw from the regions, spaces, and everyday practices that have often been sidelined or absent from the wider canon of urban studies.

Rather than the early twentieth-century "modern infrastructure ideal," which promoted the universal extension of large-scale networked supplies of key urban resources, urban infrastructural access in the twenty-first century remains highly fractured (Graham and Marvin, 2002). In fact, according to Ahlers et al. (2014), in cities of the Global South as diverse as Jakarta, Lilongwe, and Delhi, anywhere from 40 to 70 percent of residents procure urban services, such as water, waste, and energy through informal and provisional modalities that go outside large-scale networks and

DOI: 10.4324/9780429259593-34

centralized delivery systems.[1] Thus in small, medium, and large metropoles around the globe, city dwellers often reside in a liminal infrastructural space. Everyday urbanism often consists of residents using their labor, capital, creative practices, and socio-political networks to procure and coproduce infrastructures of a growing urban majority, co-constructing cityscapes, such as those in Indian metropoles in which some estimates suggest only 4 percent of built structures mirror those of urban plans.

However, urban studies as a whole is only beginning to come to grips with these "heterogeneous infrastructural configurations" (Lawhon et al. 2018) and their rippling ramifications for urban life, politics, and unequal claims to the city. For example, an exciting new genre of urban scholarship is placing infrastructure at the center of analyses for the purposes of not only tracing material networks but also apprehending the social, political, and cultural life of cities through the complex relations that surround a city's socio-technical infrastructure (Larkin, 2013; Anand et al., 2018; McFarlane et al., 2017). Thus rather than urban infrastructure providing an apolitical or inert technology, infrastructure binds humans, natural resources, and material objects into complex socio-political relations in ways that powerfully shape sociality, politics, state power, and nation-building. However, to date, there has been a dominant scholarly focus on large-scale networks over and above the multiplicities of small, provisional systematizing networks that often constitute the "everyday" infrastructures of residential resource access in the Global South. The emergent qualities of these small and provisional integrative systems, which constitute the ordinary ways urban dwellers access water, sanitation, and energy, have significant implications for how we understand the differing urbanisms shaping the ways cities are unequally produced, inhabited, and governed in the everyday (Amin and Thrift, 2017). For example, how does "beyond-the-network" urbanism structure relations between bodies and things in ways that either foster, or hinder, urban inclusion, social differentiation, and/or social justice? As heterogeneous infrastructural configurations often exist outside the confines of official planning, how do such configurations shape political space and citizen's unequal experiences of everyday urban governance? Finally, given that urban residents' bodies, practices, and senses are deeply tied to local infrastructures, what might beyond-the-network urbanism disclose in terms of socio-spatial inequity at the bodily (rather than simply neighborhood or city) scale?

Drawing on an emerging collection of urban research on diverse infrastructures that extend beyond the grid, this chapter specifically explores how beyond-the-network urbanism structures urban space, social relations, politics, and bodies, shaping urban inequities and differing lived experiences of global urbanism. The chapter considers the ways beyond-the-network urbanism provides a unique vantage point for revealing a set of urban processes and practices that risk being overlooked in other ways of "seeing" the city and urbanization itself, including meta- and meso-level analyses of infrastructure and urban environmental change, received concepts of policy prescription, and some dominant theorizations from the broader canon of urban studies. In the following sections, I examine how beyond-the-network urbanism can be used as a heuristic device to help illuminate key dimensions of global urbanism, including: (1) urban fractures and fragments that reveal cities within cities, (2) everyday urban governance regimes "beyond the state," and (3) the bodily scale of contemporary global urban inequality.

Cities within a city: Urban fractures and fragments

In a significant number of cities of the Global South, access to centralized infrastructural systems of water, sewerage, and energy is tied to the formal or "legal" occupancy of land. Settlers and residents of informal settlements, who constitute a significant portion of the global urban population, often do not qualify for networked extensions or must jump through a complex series of hoops to prove their land residency by particular cut-off dates in order to connect to large-scale municipal networks (Björkman, 2015; Anand, 2017; Truelove, 2019). Congruently, austerity measures, privatization, and decaying, intermittent, and/or contaminated infrastructure (i.e., lead within water pipes) also result in a growing number of urbanites needing to go outside the network to procure vital services (Silver, 2019). Beyond-the-network urbanism or *the everyday practices and socio-material processes that constitute access to key urban services in spaces in which the network either fails to reach or is inadequate* is often a dominant mode by which a growing number of city dwellers experience the city's vital infrastructures.

Analyses of beyond-the-network urbanism bring attention to the scales and domains by which infrastructure is governed, accessed, claimed, altered, and mediated in highly heterogeneous ways. This first includes examining infrastructural fragmentation, consisting of both the ways infrastructural services are fractured, partial and unequal, as well as the accompanied fragmentation of "political and legal rights or economic opportunity" (McFarlane, 2018: 1008). Attention to infrastructural fragmentation provides an important vantage point to analyze differing ways of "seeing" and experiencing the city that emerge from the everyday relationship between people and the material environment.

Analyzing infrastructural fragments and fractures requires upsetting some of the dominant modes and scales by which theoreticians, practitioners, and even policy makers tend to approach the city. For example, there is a tendency to "see" and represent the city as bounded at the meso-scale within urban and development studies, bringing particular meta-narratives and urban-wide processes to the forefront while occluding others from view. Southern cities, for example, have been historically analyzed through the lens of underdevelopment and as "problem-ridden cities" (see Robinson, 2002; Roy, 2011) that lack basic infrastructures. However, problematizing southern cities through the "metonym of underdevelopment" (Roy 2011, p. 224) makes it more difficult to examine how southern cities (like their northern counterparts) experience a "splintering" (Graham and Marvin, 2002) of infrastructure that is tied to the production of both high-end and more impoverished spaces and infrastructures *in the same city*. The difference here is not reductive to one between the north (characterized as a place of more elite spaces and infrastructures) and the south (characterized as impoverished and underdeveloped) but rather between the differing logics and flows at work within both northern and southern cities that produce fractured space and fragmented infrastructural access. While Graham and Marvin (2002) argue that northern cities that once proclaimed universal access to infrastructures like water and sewerage in the twentieth century have later experienced a neoliberal splintering of space that produces fractured and unequal infrastructural access, postcolonial cities inherited a colonial legacy of fragmented and unequal infrastructural access that has persisted, and mutated, in contemporary times.

For example, Simone's (2004, 2014) work on the neighborhood scale within Johannesburg and Jakarta demonstrates how districts that often look the same on paper experience widely varying neighborhood networks, trajectories, and outcomes. These "cities within cities" emerge partially based on residents' differing social infrastructures that include networks of cooperation and knowledge sharing. In relation to fragmented water infrastructures, my own research within Delhi's informal settlements at the intra-neighborhood and intra-household scales shows how provisional infrastructures shape, and are coproduced through, gender/class/ethno-religious power relations. For example, struggles over restarting, accessing, and controlling a neighborhood tube well leveled gendered infrastructural burdens within households, empowering some groups (dominant caste residents) while marginalizing others (nondominant caste women and Muslim residents) (Truelove, 2019). Such studies reveal how differentiated infrastructural citizenship (Lemanski,

2019) comes into being not only through unequal rights from the state but also in tandem with complex material, gender, and ethno-religious power relations within neighborhoods and households. Thus attention to residents' diverse relations to the city's fractured materiality within neighborhood lanes and households provides an important pathway for formulating what McFarlane et al. (2017: 1394) call a "detailed analysis of the multiple forms of urbanism that emerge from the different spaces, contexts and presents that constitute a city."

Engaging with beyond-the-network urbanism, including the uncertainty and ongoing socio-material transformations that characterize provisional infrastructural modalities thus requires thinking through how the fractured materiality of the urban fabric is tied to new understandings, experiences, and articulations of the urban condition. For example, Lawhon et al. (2018) analyze how fragmented infrastructural configurations of the wastescape in Kampala can help mitigate configurations of risk and precarity, particularly for vulnerable populations of the urban poor. On the other hand, McFarlane and Silver (2017) show how fractured access to sanitation in Cape Town produces both alternative modes of "seeing" the city for marginalized populations, as well as new strategies and claims for linking unequal sanitation to the advancement of citizenship and rights. Understanding how changing and fragmented configurations of infrastructure unevenly affect differing social groups is thus intimately linked to our understanding of the politics of urban transformation and the competing logics by which differing groups and stakeholders imagine and structure city futures.

Beyond the network as beyond the state?

Provincializing infrastructure beyond large-scale socio-technical networks also reveals domains of political space and governance that exceed the state/nonstate binary and involve a wide range of political actors and institutions. These political assemblages frequently complicate both policy and conceptual approaches that posit a clearly delineated state/society, public/private, or formal/informal divide. At times, a given modality of infrastructure may start as formal (initiated or paid for by the state) but later be delegated to, or appropriated by, informal actors and/or embedded in informal processes or technologies. As Schwartz et al. (2015: 31) point out, in such cases, "informality and formality are interdependent" and constitute a "meshwork" of provisioning. Thus the "multiplicities of governance regimes" (Schindler, 2014) that accompany fragmented infrastructures rework and often blur state/nonstate and formal/informal dichotomies, requiring a rethinking of "actually existing" urban governance configurations rather than, in Amin and Thrift's (2018: 5) words, "forcing variegations at ground level into received categories of theory and discipline." Thus equally a challenge and an opportunity raised through beyond-the-network urbanism is to conceptualize and theorize infrastructural and political power that goes beyond neat bifurcations of the state and its outside: the hybrid and provisional assemblages of governance that, for a significant portion of urban dwellers, shape how urban power and politics are negotiated in the everyday.

Taking up this challenge, a new and growing literature documents the fluid, uncertain, and hybrid nature of fractured infrastructural configurations in which assemblages of state and nonstate actors, and formal and informal materials and practices remain in flux, constituting a "gray zone" (Truelove, 2019) of urban governance with uncertain and mutating effects on the ground (Björkman, 2015; Ahlers et al., 2014; Schwartz et al., 2015, Furlong and Kooy, 2017). Often, nonstate actors, such as electricians, plumbers, resident association members, tanker drivers, local leaders, strongmen, and nongovernmental organization workers take on state-like qualities or even state-related roles (see Björkman, 2015; Schwartz et al., 2015; Ghertner, 2017), at times helping to extend critical-to-life resources that have been made unavailable by the state to marginalized communities. At other times, the "infrastructural power" (Truelove, 2020) of these actors compounds the forms of unequal urban citizenship (Lemanski, 2019) and exclusion for particular social groups (such as informal vendors and strongmen who leverage control in neighborhoods through selling water at a price difficult for poor populations to afford). Such infrastructural governance practices reveal diffuse and decentered power relations, which often go under-theorized within scholarship that has a dominant focus on solely the state, private, or neoliberal infrastructural regimes, as well as within conventional policy prescriptions that view urban development as occurring through a linear movement from informal to formal (Ranganathan, 2018).

As hybrid actors and institutions go beyond being simply intermediaries of the state (Cornea, 2020), they play a significant role in not only regulating urban infrastructures and socio-material relations but also shaping political space and power in the city. Such everyday forms of governance help to reveal the highly diverse ways social and political power are experienced and mediated in the city beyond what Chatterjee (2004) terms "political society." In other words, as many residents calculate that approaching formal institutions and state agencies provides neither the most efficient nor direct way to build urban footholds and procure key resources, claims to infrastructure regularly extend beyond the political society of the urban poor petitioning the state through moral appeals and political mobilization. Instead, city dwellers participate in a much broader "politics of anticipation" (Simone, 2010) in which they flexibly anticipate and negotiate their circumstances, often with political actors at "the seam of the state" (Ghertner, 2017), bringing attention to the often improvised and experimental tactics of both residents and public authorities that shape social power and residents' abilities to maintain and build footholds in the city. Such practices produce differing effects for how the state (and its power) is imagined, negotiated, and, at times, circumvented, as certain types of informality thrive while others face punitive repercussions (Roy, 2011).

The body as infrastructure

Finally, beyond-the-network urbanism helps to reveal the complex ways the body is intimately connected to the "city's machinic materialities" (Lancione and McFarlane, 2016). This includes the everyday practices and labor of the body to access and maintain infrastructural networks and fragments, as well as urbanites' affective, aesthetic, symbolic, and subjective experience of these configurations. Attention to city dwellers' bodily practices and experiences of beyond-the-network urbanism can thus contribute an important vantage point to understanding patterns of urban inequality and transformation not often visible at the city, neighborhood, or even household level. For example, the bodily scale of beyond-the-network urbanism helps to reveal how particular intersectional inequalities (including gender, class, race, age, ability) are felt and experienced through the entanglement and co-constitution of infrastructures and the body, as well as how the latter helps in theorizing the former. Not only are particular bodies in the city shaped, marked, and affected by infrastructural configurations, but gendered, classed, and racialized bodies often become unequally enrolled in infrastructural networks themselves through broader structures of inequality: doing the labor, maintenance, and affective work that constitute "infra-making" (Lancione and McFarlane, 2016).

Thus examining the scale and domain of the body within fractured infrastructural networks brings attention to whose racialized/gendered bodies are mobilized in particular infrastructural practices and projects (Andueza et al., 2020; 6). For many urbanites, the labor of provisioning daily services, such as water, energy, and waste removal falls on specific (and intersectional) gendered, classed, and racialized bodies through wider urban discourses and social hierarchies. For example, informal urban water and waste networks are often considered part of gendered household labor through patriarchal discourses, consequently requiring disproportionate time and labor from poor (often lower-caste) women in order to service households (Agarwal, 2001; O'Reilly, 2006; Sultana, 2009). Here gendered bodies literally become part of physical forms of infrastructure (often substituting for absent garbage trucks or water pipes). In Fredericks's (2018) work on informal waste infrastructures in Dakar, Senegal, for example, waste infrastructure becomes "devolved" onto the body as poor women's bodies do the infrastructural labor of waste work within homes and streets. As women's bodies serve as waste infrastructure, such practices reify poor women's unequal gendered subjectivities as coproduced through infrastructures of discard. Here infrastructures serve to both (re-)make and become made through the labor of the particular bodies through intersectional social relations of gendered, classed, and racialized power.

The ways that bodies, affective experiences, and subjectivities become shaped by infrastructure (and vice versa) illuminates situated forms of infrastructural violence (Rogers and O'Neill, 2012) that increasingly characterizes global urbanism and the urban condition. The unequal consequences of fragmented infrastructures, including the labor and everyday practices to maintain and enable them to function, can deepen not only forms of material deprivation (through unpaid or low-paid hazardous infrastructural work) but also reify broader gender, class, and ethnoreligious discourses and hierarchies that fuel structural exclusions from rights to the city itself. For example, Alda-Vidal et al.'s (under review) work on the fragmented infrastructure of Lilongwe's water kiosks demonstrates the ways kiosks have both the potential to exacerbate infrastructural violence (disproportionately exposing women to night time risks when water is running at kiosks, as well as exploitative working conditions at the kiosk) and the promise to alleviate it (providing water to marginalized communities and paid labor opportunities to poor women). Such examples reveal the ways specific bodies in the city provide an often invisible infrastructure that keeps urban life ticking, remaking social and economic relations in the process.

Conclusion: Everyday urban infrastructures in mutation

This chapter argues that beyond-the-network urbanism structures the conditions and experiences of global urbanism for a growing urban majority, revealing (and shaping) critical dimensions of urban life and sociality. Not only do infrastructures beyond-the-network exist in contravention to urban plans, restructuring the urban fabric in new and constantly unfolding ways, but also these everyday provisional networks play a substantial role in shaping urban institutions, symbolic meanings, bodies, and social power relations that produce very different kinds of urbanisms on the ground. This chapter has sought to explore infrastructures beyond-the-network as a lens into viewing, and comparing, the differing forms of urbanism that arise across the world's cities from people's complex relationships to fragmented and provisional configurations. In particular, I have examined how beyond-the-network urbanism provides a vantage point for bringing visibility to (1) urban fractures and fragments that reveal cities within cities, (2) everyday urban governance regimes "beyond the state," and (3) the bodily scale of contemporary global urban inequality. Giving analytical visibility to these processes, scales, institutions, and practices requires scholars to view cities, and their complex socio-material networks, as always in flux, mutation, and transformation. As such, beyond-the-network urbanism can help reveal "new genres" (Pieterse and Parnell, 2016) of global urbanism and comparative inquiry that emerge from research on fragmented infrastructure and everyday practices in the Global South but have relevance for thinking through cities everywhere.

Furthermore, ever-mutating configurations of urban infrastructure provide an important research and analytical entry point for revealing the ways fractured modalities of access become a force for new forms of political space and power in the everyday city: the politics of not just making do but transforming urban conditions and remaking cities from the ground up. It is precisely the interconnections between urban-wide processes and diverse, and often unpredictable, configurations of local infrastructures and bodily experiences that help reveal and situate why particular groups in the city experience a spectrum of marginality or privilege through relations with the city's diverse infrastructures. Analyzing these ever-shifting diverse networks, and their potential to both alleviate urban inequities or exacerbate them, requires moving between scales, sites, and domains of analysis in order to connect the body to larger urban processes and policy shifts. As such, identifying interconnections between ecologies of materials, social relations, and resource flows at the local level can thus profoundly shape the politicization of infrastructure at the citywide scale – particularly movements to change exclusionary urban policies that curtail services to informal settlements or relegate hazardous infrastructural labor onto some bodies above others. In revealing these differing patterns and unequal experiences of urbanism in the everyday, there is a profound potential to identify, sharpen, and augment new forms of claims-making for more just and equitable contemporary cities.

Note

1 Such practices are not limited to the Global South, though they are more prevalent there (Schindler, 2017).

References

Agarwal, B.(2001) 'Participatory exclusions, community forestry, and gender: An analysis for South Asia and a conceptual framework'. *World Development*, 10: 1623–1648.

Ahlers, R., Cleaver, F., Rusca, M. and Schwartz, K. (2014) 'Informal space in the urban waterscape: Disaggregation and co-production of water services'. *Water Alternatives*, 7(1): 1–14.

Anand, N. (2017) *Hydraulic City: Water and the Infrastructures of Citizenship in Mumbai*. Duke University Press: Durham.

Andueza, L., Davies, A., Loftus, A. and Schling, H. (2020) 'Bodies as infrastructure'. *Environment and Planning E, onine first doi.org/10.1177/2514848620937231.*

Alda-Vidal, C., Browne, A. and Rusca, M. (under review) 'Gender relations and infrastructural labours at the water kiosks in Lilongwe, Malawi', in Truelove, Y. and Sabhlok, A. (Eds.), *Gendered Infrastructures: Dialectics in Form, Identity and Space*. West Virginia University Press: Morgantown.

Amin, A. and Thrift, N. (2017) *Seeing Like a City*. Polity Press: Cambridge.

Anand, N., Gupta, A. and Appel, H. (Eds.) (2018) *The Promise of Infrastructure*. Duke University Press: Durham.

Björkman, L. (2015) *Pipe Politics, Contested Waters: Embedded Infrastructures of Millennial Mumbai*. Duke University Press: Durham.

Chatterjee, P. (2004) *The Politics of the Governed: Reflections on Popular Politics in Most of the World*. Columbia University Press: New York City.

Cornea, N. (2020) 'Territorialising control in urban West Bengal: Social clubs and everyday governance in the spaces between state and party'. *Environment and Planning C: Politics and Space*, 38(2): 312–318.

Fredericks, R. (2018) *Garbage Citizenship: Vital Infrastructures of Labor in Dakar, Senegal*. Duke University Press: Durham.

Furlong, K. and Kooy, M. (2017) 'Worlding water supply: Thinking beyond the network in Jakarta'. *International Journal of Urban and Regional Research*, 41(6): 888–903.

Ghertner, D. A. (2017) 'When is the state? Topology, temporality, and the navigation of everyday state space in Delhi'. *Annals of the American Association of Geographers*, 107(3): 731–750.

Graham, S. and Marvin, S. (2002) *Splintering Urbanism: Networked Infrastructures, Technological Mobilities and the Urban Condition*. Routledge: New York.

Lancione, M. and McFarlane, C. (2016) 'Life at the urban margins: Sanitation infra-making and the potential of experimental comparison'. *Environment and Planning A: Economy and Space*, 48(12): 2402–2421.

Larkin, B. (2013) 'The politics and poetics of infrastructure'. *Annual Review of Anthropology*, 42: 327–343.

Lawhon, M., Nilsson, D., Silver, J., Ernstson, H. and Lwasa, S. (2018) 'Thinking through heterogeneous infrastructure configurations'. *Urban Studies*, 55(4): 720–732.

Lemanski, C. (Ed.) (2019) *Citizenship and Infrastructure: Practices and Identities of Citizens and the State*. Routledge: New York.

McFarlane, C. (2018) 'Fragment urbanism: Politics at the margins of the city'. *Environment and Planning D: Society and Space*, 36(6): 1007–1025.

McFarlane, C., Silver, J. and Truelove, Y. (2017) 'Cities within cities: Intra-urban comparison of infrastructure in Mumbai, Delhi and Cape Town'. *Urban Geography*, 38(9): 1393–1417.

McFarlane, C. and Silver, J. (2017) 'The poolitical city: "Seeing sanitation" and making the urban political in Cape Town'. *Antipode*, 49(1): 125–148.

O'Reilly, K. (2006) '"Traditional" women, "modern" water: Linking gender and commodification in Rajasthan, India'. *Geoforum*, 37(6): 958–972.

Parnell, S. and Pieterse, E. (2016) 'Translational global praxis: rethinking methods and modes of African urban research'. *International Journal of Urban and Regional Research, 40*(1), 236–246.

Peloso, M. and Morinville, C. (2014) '"Chasing for water": Everyday practices of water access in peri-urban Ashaiman, Ghana'. *Water Alternatives*, 7(1): 121–139.

Ranganathan, M. (2018) 'Rethinking urban water (in) formality' in Conca, K. and Weinthal, E. (Eds.), *The Oxford Handbook of Water Politics and Policy*. Oxford University Press: Oxford.

Robinson, J. (2002) 'Global and world cities: A view from off the map'. *International Journal of Urban and Regional Research*, 26(3): 531–554.

Robinson, J. and Roy, A. (2016) 'Debate on global urbanisms and the nature of urban theory'. *International Journal of Urban and Regional Research*, 40(1): 181–186.

Rodgers, D. (2019) 'Forward: Towards an infrastructural imaginary of the political' in Lemanski, C. (Ed.), *Citizenship and Infrastructure: Practices and Identities of Citizens and the State*.

Rodgers, D. and O'Neill, B. (2012) Infrastructural violence: Introduction to the special issue. *Ethnography*, 13(4): 401-412.

Roy, A. (2005) 'Urban informality: Toward an epistemology of planning'. *Journal of the American Planning Association*, 71(2): 147–158.

Roy, A. (2011) 'Slumdog cities: Rethinking subaltern urbanism'. *International Journal of Urban and Regional Research*, 35(2): 223–238.

Schindler, S. (2014) 'A New Delhi every day: Multiplicities of governance regimes in a transforming metropolis', *Urban Geography*, 35: 402–4419.

Schindler, S. (2017) 'Towards a paradigm of Southern urbanism'. *City*, 21(1): 47–64.

Schwartz, K., Tutusaus Luque, M., Rusca, M. and Ahlers, R. (2015) '(In) formality: The meshwork of water service provisioning'. *Wiley Interdisciplinary Reviews: Water*, 2(1): 31–36.

Silver, J. (2019) 'Decaying infrastructures in the post-industrial city: An urban political ecology of the US pipeline crisis'. *Environment and Planning E: Nature and Space*, online first doi:10.1177/2514848619890513.

Simone, A. (2004) 'People as infrastructure: Intersecting fragments in Johannesburg'. *Public Culture*, 16(3): 407–429.

Simone, A. (2010) *City Life from Jakarta to Dakar: Movements at the Crossroads*, Routledge: New York.

Simone, A. (2014) *Jakarta, Drawing the City Near*. University of Minnesota Press: Minneapolis, MN.

Sultana, F. (2009) 'Fluid lives: Subjectivities, gender and water in rural Bangladesh'. *Gender, Place and Culture*, 16(4): 427–444.

Truelove, Y. (2019) 'Gray zones: The everyday practices and governance of water beyond the network'. *Annals of the American Association of Geographers*, 109(6): 1758–1774.

Truelove, Y. (2020) 'Who is the state? Infrastructural power and everyday water governance in Delhi'. *Environment and Planning C: Politics and Space, online first* doi:10.1177/2399654419897922.

31

STILL CONSTRUCTION AND ALREADY RUIN

Mariana Cavalcanti

Thinking global urbanism entails engaging with at least two different levels or discussions of urban theory: first, the empirical fact of the transition to a predominantly urban planet and the concrete transformations this fact imposes upon the political economy of cities, logics of accumulation, structures of different markets, ways of life, aspirations and imaginations of the future. Second, the epistemological consequences of this shift to a predominantly urban world for urban theory and ideas of cities, the urban and the global – and, just as important, the politics of who gets to speak for who on what and where.

Since the urban transition is grounded in processes underway in the metropoles of the Global South, urban theory (as we know it "from the north") has come to be empirically challenged and analytically provincialized as one possible reading or perspective of what urban life can be (see Roy, 2009; Varley, 2013; Sheppard et al., 2015). The majority of the world lives in cities in which everyday life unfolds through spatialities and temporalities that are very distant from normative ideas of city life as prescribed by the concepts of city planning. And yet these normative ideas of urban theory "from the north" operate in and inform ideas and projects of everyday city life in the very cities and sites that challenge their universality. Thus planning normativities attain a particular type of reality as they make up the language through which urban life and form are thought and built. With this comes a certain future-bound imagination that lends meaning to and inaugurates disputes over ongoing transformations of urban space. The urban process can then be perceived not just as the production of space in itself but also as the production of particular ways of inhabiting time.

In this chapter, I engage with this problem by constructing an ethnographically minded genealogy of a recurring image used to describe Brazilian cities: that of the premature ruin or "still construction and already ruin". I trace this recurring image as it moves through different enunciations and subject positions over time: initially a metaphor to describe the specificity of cities of the New World by a French anthropologist in a context of national developmentism in the 1930s and '40s, later, and in the words of a songwriter, it becomes an allegory that spins vivid images of inequalities that helped describe what was already perceived as a new global order in the early 1990s. Finally, I trace its trajectory as an ethnographic motif that becomes exhausted but lends itself to a research agenda on the politics of urban infrastructure and city-making in Rio de Janeiro and hopefully elsewhere.

DOI: 10.4324/9780429259593-35

A metaphor of a monumental future

As science fiction narratives render evident the anxieties and future projections of the particular present in which they are written, travel narratives in many ways tell us more about the lifeworlds of their writers than about the places they narrate. This is the case even when these writers are unequivocally aware of the theoretical problems of Western claims to universality based on evolutionary ideas of civilization.

Anthropologist Claude Lévi-Strauss *Tristes Tropiques* is an account of his travels through Brazil in the 1930s. Originally published in France in 1955, the book is mostly dedicated to his encounters with indigenous populations in the country's hinterlands, but it spins a handful of incidental reflections about urban life in Brazil, taken also incidentally as a particular case of New World cities.

The first thing that struck Lévi-Strauss upon his arrival in South America was the spatial scale of American landscapes: "The streets remain streets, the mountains mountains, and the rivers rivers and yet one feels at a loss before them" (p. 83). This "monumental scheme of things", he writes, permeates one's (European? Aesthetic?) judgement:

> Those who call New York ugly, for instance, have simply failed to make the necessary change of registration. [...] The beauty of New York [...] results from the creation of a new kind of city: an artificial landscape in which the principles of urbanism no longer operate. (Lévi-Strauss, 1992, p. 83)

São Paulo, at the time, he claims, was also "thought ugly". But Lévi-Strauss preferred to call it "untamed". He adapted himself "to a system that has no backward dimension in time; and I enjoy having a different form of civilization to interpret".

The idea of beauty gets bundled up with images of good, proper or familiar city life – all notions materialized and made recognizable by their instantiation in so-called beautiful cities of the Old World. Incomprehension in such judgements, it seemed, ran both ways. The section titled "São Paulo" begins with the story of one of his "Brazilian girl-students" who "returned in tears from her first visit to France: whiteness and cleanness were the criteria by which she judged a city, and Paris, with its blackened buildings, had seemed to her filthy and repugnant" (p. 100). This incommensurability of ideas of urban beauty in the old and new worlds has to do with distinct experiences of producing space and inhabiting time.

The cities of the Old World contained a distinct monumental experience of a past-bound temporality; the seemingly endless landscapes of the New World presented a monumental experience of space – and of its future shaping into the space of the nation. This has an effect upon each of these "civilization's" ruins: whereas the Old World incorporated history into urban form as the result of temporal sedimentation, preserving and displaying its ruins with the aura bestowed upon them by notions of heritage, conservation and civilization, the New World, "with no backward dimension in time" instilled a new rhythm to the urban process that desacralized ruins (and pasts) by building over them. Consequently, these ruins were, paradoxically, "new":

> The older a European city is, the more highly we regard it; in America, every year brings with it an element of disgrace. For they are not merely newly built; they are built for renewal, and the sooner the better. (Lévi-Strauss, 1992, p. 100)

Lévi-Strauss rationalizes his discomfort by positing a distinction between the "cycle of evolution" of cities of the Old and New Worlds:

> It is not a case of new cities contrasted with old, but rather of cities whose cycle of evolution is very rapid as against others whose cycle of evolution is slow. Certain European cities are dying off slowly and peacefully; the cities of the New World have a perpetual high temperature, a chronic illness which prevents them, for all their everlasting youthfulness, from ever being entirely well: they pass from first youth to decrepitude with no intermediary stage (Lévi-Strauss, 1992, p. 100).

The production of ruins as a means of producing space and inhabiting time in modernizing Brazil was not without precedent. Decades before Lévi-Strauss, Euclides da Cunha, journalist and writer, author of *Os Sertões* (Rebellion in the Backlands) had already chronicled the ruined landscapes of the hinterlands of the state of São Paulo in the wake of "desert makers" who had ruined the natural landscape by setting up predatory coffee plantations before abandoning the area. The figure of the recent or premature ruin was something of a constant in Cunha's writings (Hardman, 1996). It also figured in his descriptions of abandoned cities in the Amazon region, a feat attributed to the "ruin builders" whose efforts to tame nature largely failed. This early focus on the producers of ruins (that perhaps entails a certain intimacy with "native" practices and dispositions) seems largely overlooked by the perspective of Europeans. Stephan Zweig, another European observer of São Paulo's growth in the 1940s, was also in awe of the urban process in Brazil. In his book *Brazil Country of the Future*, he remarks, "It is not the past or the present that make Sao Paulo such a fascinating city, but its growth and its becoming, the speed of its transformation, viewed in slow motion" (Zweig, 1941, p. 307).

Lévi-Strauss and Zweig write in a context of national developmentism, and they find themselves in Brazil and as observers as part of the nation-building and modernization efforts of the period. The idea of Brazil as the "country of the future" was constitutive of its nation-building discourses and practices. It also entailed the colonization of the vast hinterlands of the country. When *Tristes Tropiques* was first published in France in 1955, Brazil still had a glorious future ahead of itself. It was a nation about to execute the plan of building a modernist capital in its formerly deserted hinterlands in five years as part of a political project that proposed to develop the country at the rate of "50 years in 5". In this project, modernist architecture, rational planning, road building and industrialism were to do away with the weight of the past of underdevelopment. The monumental was in the future; the ruins in the present could be read as evidence and the material path to its inexorable making.

If travel narratives reveal something of the lifeworlds of their writers, then they are also addressed towards an imaginary "center" whose periphery they narrate. This center shifted in the postwar period, as American city growth and sprawl became a spectacle in its own right, and American planning and city-making produced and colonized a global imagination of urban life. Insofar as descriptions of city life are concerned, the meaningful split was no longer between the cities of the New and the Old Worlds (that themselves were still in ruins by the time *Tristes Tropiques* came out) but already between "north" and "south".

An allegory of the "global south" in the making

Looking at a Manhattan rebuilt many times over since Lévi-Strauss first set foot in the American continent, another French observer, Michel de Certeau, took up many of the themes of his predecessor. In the late 1970s, de Certeau described Manhattan as a "wave of verticals" (de Certeau and Mayal, 1998, p. 91). "Unlike Rome", he wrote,

> New York has never leaned he art of growing old by playing on all its pasts. Its present invents itself, from hour to hour, in the act of throwing away its previous accomplishments and challenging the future. A city composed of paroxysmal places in monumental reliefs.
>
> (de Certeau and Mayol, p. 91)

New York's artificial beauty for Lévi-Strauss becomes the privileged site for glimpsing the workings of power embodied in the practices of urbanism under capitalism for de Certeau.

Michel de Certeau's reading of urbanism and urbanization uttered from the top floor of the World Trade Center poses at least one problem to southern cities: it presupposes a rapidly changing but "tamed" city, to use Lévi-Strauss vocabulary. The figure of the ruin, of decay, of the uneven development of vacant spaces that Lévi-Strauss used to describe the New World is largely absent here: from the top of the World Trade Center, there is no time or space for abandonment or decay. The disciplinary project is institutionalized, naturalized and capillary: de Certeau's resisting walkers slip through the cracks of the grid of an ordered city – but the grid and the cracks are there to allow this slippage. The city charter may be always outdated because the city itself is always in the making, but it is equipped with infrastructure and its territory remains recognizable and congruent with hegemonic images of what a city should look like.

From the perspective of observers of the south, de Certeau's neat distinction between planners and users can always and easily be challenged. And yet his proposal to move from the concept city to the city of urban practices is still very much worth engaging. Down here, the city does not necessarily map onto its (formal) infrastructure or territory. There is always uncertainty as to the outcomes of any initiated construction in southern cities: buildings, infrastructures, roads and neighborhoods endure and change through the logics of informality, which entails all sorts of middlemen, mediators, usurpers, institutions and institutionalities whose power is as dispersed as it is fragmented in local arrangements or districts. Urbanism at the state's margins makes daily life a constant negotiation of uncertain if somewhat provisionally predictable outcomes. If, as Ananya Roy (2004) has suggested, urban informality is perceived as an organizing logic that connects different spaces, economies and markets, then our descriptions of urban life and the production of southern cities needs to bring to the fore other forces and actors of city-making. And this makes it necessary to look for middle perspectives, to grasp the myriad urbanisms that can be glimpsed from spaces located somewhere between the god-like view of planners and the myopic decisions of walkers of "tamed" streets.

In 1991, Brazilian songwriter Caetano Veloso released a song titled "Fora da Ordem" ("Out of Order"). Its lyrics begin with the image of the body of a *vapor* (low-ranking underage worker of the drug trade) found in the ruins of a school under construction.[1] The verses set a science fiction ambience, shedding light on details and textures of daily life as Veloso's gaze moves through piles of garbage in Salvador, the modernist ramps of Brasilia, open sewage in upscale neighborhoods of Rio de Janeiro and the streets of São Paulo. The boy's death could have taken place in any large late-twentieth-century Brazilian city, where "everything was still construction and is already ruin".

This verse is an explicit reference to *Tristes Tropiques*. "Still construction and already ruin" in Lévi-Strauss is a metaphor that describes the landscapes built out of the urbanistic "fever" of modern city-making in the New World, highlighting the future; Veloso turns it into an allegory of an urban world changing under globalization and neoliberalism, as seen from the (Brazilian) metropoles of the third world (before it was referred to as the "Global South") where the weight of the past and its endurance in the present cuts short imaginations of a better future. "Fora da Ordem" is about the demise of the "country of the future" temporality; its vivid and concrete images of the urban crisis express the premature failure of the promises of the Democratic order ushered in by the 1988 Constitution, passed after nearly three decades of a military dictatorship.

The god-like perspective is here countered not by a walker resisting power but by a "devil is in the detail" truth: the child is not killed in the street or in a public square (like thousands of Brazilians of color every year) but in a school under construction – in the key institution that was supposed to operate his inclusion to society. The violence in the foreground obfuscates the setting but the scene is banal and recognizable enough so that the ruin of the school under construction and the killing constitute the building blocks of a generalization that allows for a construction of the allegory of life conditions in a deeply unequal world. The building of democratic institutions lags behind the forces that shape daily life in the city. In Veloso's dystopic rendition of urban life at the end of the twentieth century, where the future has been cut short, people, infrastructure and the future share the same fate: "Boys and girls in the eye of the street/the pavement, the overpass/yelping at the moon/Nothing continues".

And yet there is still a claim to a totalizing perspective in Veloso's verses, a claim to a true point of view that grasps, if briefly, the totality of the city: "And the muzzle of the pistol/that the children bite/reflects all the colors of the city landscape/That is way more beautiful/and more intense/than in the postcard".

An ethnographic motif (undone)

In 2009, Rio de Janeiro won the bid to host the 2016 Summer Olympics. Back then, in the midst of a massive global economic crisis, Brazil showed signs of recovering from a long recession. Oil prices soared on global markets, flooding Rio de Janeiro with revenues of oil extraction along its coast. A massive refinery was under construction in the city's metropolitan region, with the promise of anchoring all sorts of growth in the "real economy". Inequality rates had been dropping for nearly a decade. An unprecedented political alignment between the municipal, state and federal levels of government created the conditions for what then seemed like a very successful city marketing campaign aimed at branding Rio de Janeiro as a world-class city suited to host global events, such as the Earth Summit, the 14th Catholic International World Youth Day and the final of the 2014 Soccer World Cup. The World Bank and Inter-American Development Bank offered infinite consultants and loans. The state stepped in as it sought to draw private and foreign investments through shady public-private partnerships, and Brazil's national development bank seemed as omnipresent a funder to grand development schemes as the multilateral agencies. The city hall announced that it would provide infrastructural upgrading to all of the city's favelas by 2020. The federal public housing program built massive projects in the peripheries that ended up constituting part of an eviction machine that dislocated more than 70,000 people. Massive investments in public transportation included the projected construction of four expressways equipped with bus rapid transit lanes and a new subway line. For a brief moment, it seemed, the "Brazilian giant would awaken", and the glorious future of past imaginings seemed once again feasible – or at least marketable. After all, even homicide rates in some areas had plummeted as the result of the Pacifying Police Units program that included the permanent police occupation of some of the city's favelas. Their military occupation was accompanied by monumental infrastructural works that included the construction of two massive gondola cable car systems built after the "Medellin Model" to cater to residents of the city's favelas.

The city turned into a constellation of overlapping construction sites. As the August 2016 deadline approached, however, many mega-projects got downsized, postponed, suspended or just altogether dropped, revealing their construction as "already ruin", even before the opening of the games. The Lava Jato investigation of the Federal Police initiated in 2014 brought to light a series of corruption schemes that involved the Brazilian oil company Petrobras but also construction companies and politicians involved in the making of the so-called Olympic city, stalling and delaying many ongoing works and projects. It precipitated and justified the 2016 coup d'état that removed President Dilma Rousseff from power and eventually led to the conviction of two governors of Rio de Janeiro involved in corruption schemes.

In early 2016, several Olympics-related infrastructures began to literally crumble, even before the opening of the games: a recently inaugurated bicycle path built to connect the established elite south zone to the booming west side where most Olympic events were to take place collapsed, killing two people. Two weeks later, one of the overpasses in the yet to be inaugurated expressway connecting downtown to the Olympic venues in the west zone exhibited cracks that cast doubt over the soundness of the entire newly built structure. The opening of the Athlete's Village was a monumental flop, as athletes were greeted by unfinished accommodations with leaking pipes, clogged toilets and failing electrical fixtures.

In the months leading up to the Olympics, "still construction, and already ruin" was a staple newspaper headline. Not as metaphor or allegory, but as an objective, commonsensical description of city-making in Rio in general. Its appeal lay not in the more intellectualistic dimensions of the image as metaphor or allegory that I have woven together analytically but in the fact that it resonated with how city residents received the announcements of public policies, social programs and development projects. The announced glorious future of the Olympic city was read in light of past frustrated promises in the everyday. "Still construction, already ruin" refers to the inhabited temporality of daily life sedimented by the repeated and frustrated promised futures of development, modernity, democracy: "nothing continues", as Veloso wrote.

Perhaps not so intuitive is the fact that its lived experience is productive of a present that bears its own futurity — however changing the contents of this future may be: if everything is "still construction, and already ruin", then the future remains somewhat open-ended, and agency may be up for grabs. It is not the future presented by the state, by public-private partnerships, the promises of neoliberal urban planning or multilateral agencies, nor is it dictated by the local drug lords, community leaders of all sorts, militia squads, grassroots activists or evangelical pastors. On the ground, these agents are very much aware of this, and they live their daily lives with an eye on changes in the structures of everyday life and the space it opens up for agency. "Construction" or "ruin" become foreign systems of classification, in a way, if or when the city is "still construction, and already ruin" in the everyday. The state/architect's blueprints and the "planning" temporality figure as incidental, if not marginal, elements in this horizon of possibilities.

One of the most constitute of Rio's "Olympic ruins" is the *Teleférico do Alemão*, a massive gondola cable car system installed in Rio's former industrial suburbs in the north of the city. The $140 million enterprise is visible from the runway of the international airport, and it was initially hyped in the media as the city's new postcard of the "pacified city". It was completed at the additional cost of the eviction of some 2,000 families. At the inauguration ceremony, Rio's mayor referred to the *Teleférico* as the "fantastic work of peace" achieved by what was then presented as the "permanent" police occupation of the complex by the pacification program. Inaugurated in 2011, at the height of Rio's boom, the *Teleferico's* operations were suspended just two months after the Olympics. Its massive decaying structure became exactly what favela activists and grassroots organizations had feared from the very beginning: an "Olympic" "white elephant".

The defunct "white elephant" spreads across five hilltops of the favela complex that it connects through 3.5 kilometers of cables that now sway dangerously in storm winds. Through these cables once glided 152 gondola cabins, each capable of holding eight to ten passengers.

The stations, according to the original project of the *Teleférico*, were supposed to host community centers, classrooms, government offices and many other promises that never came to term. After the shutdown of the *Teleférico* system, they were taken over by the remnants (or the ruins) of the now defunct Police Pacification Program. Very often, residents of Alemão denounce police officers' use of them for their own private purposes, including taking afternoon naps, storing all sorts of legal and illegal things and having parties. Their occupation remains a matter of contention.

On June 1, 2020, as the COVID-19 pandemic raged through Alemão, 28-year-old human rights activist, journalist and documentary maker Raull Santiago, who came to maturity under the construction of the massive *Teleférico* system, tweeted: "TELEFÉRICOS OF ALEMAO: Enormous structures with many rooms, that could be used by social projects or even now in the face of a global health pandemic, stalled since the Olympics, now the white elephant serves as a police base".

Born and raised in Alemão, Santiago looks around at the city from the ruins of one of the defunct *Teleférico* stations and sees neither construction nor ruin but a white elephant fit to be reappropriated and reincorporated as part of the city's infrastructure. While many others (foreign to Alemão) would dwell on tragic narratives denouncing the inevitable failure of an incorrigibly corrupt nation that would actually go through with an idiotic idea like building such a massive system of complex maintenance in one of the city's most conflicted areas, Santiago forgets the recent past because he is concerned with undoing the present everyday (the police appropriation of the space) and, tuned in to the latest change in the sociohistorical constellation (the pandemic), sets a political agenda for the immediate future.

Just a few paragraphs back, I suggested that "construction" and "ruin" may be foreign classifications to the natives of southern cities whose lives gravitate around city-making and everyday city politics. This is precisely because the "still construction, already ruin" temporality is a native understanding of the world's peripheries of city-making as a daily, embedded and embattled practice. It encapsulates how the temporality embodied experience of making/waiting for a city yet to come produce everyday life, history and perceptions of the future.

As such, this temporality is recognizable in many southern cities, and it has already been incorporated epistemologically into urban research – as the bourgeoning debates on global urbanism and the contemporary calls to build research agendas based on southern urbanisms make evident. For at least two decades now, urbanists and scholars of the city have devoted themselves to exploring the interstices of the formal/informal divide, underscoring the urgency of understanding the positive, productive and theoretical underpinnings of urban informality (Roy and AlSayyed, 2004; Fischer, 2014; Varley, 2013, to name only a few) and conceptualizing life at the margins of the state or of cities (Das and Poole, 2014; Lancione, 2016) that make up the improvised lives (Simone and Pieterse, 2018) of residents of cities yet to come (Simone, 2004). We now know much about the material and political infrastructures that make up the necropolitcs of the postcolony (Mbembe, 2003), and even the "small worlds" of those who attempt to foster "development" and instill order in unruly cities of the south (Roy, 2012). And we know that the figure of the ruin is itself productive of new analytical possibilities when viewed as part of the processes of state building and city-making in the Global South (see Gupta, 2018).

All of these themes are by now familiar to scholars of the urban. And yet challenging the discourse of formal urban planning does not solve the issue of what to do with it as a "native" utterance (voiced by many different "types" of "natives") that operates in the real cities we observe. In this chapter, I attempted to contribute to these debates by stepping back in time to explore a recurring image in descriptions of Brazilian cities in order to capture a dimension of urbanization that allowed me to move through different subject positions that also highlight the historicity of who gets to read and narrate the city to what audience. I believe these narratives are revealing of the historical dimension of the making of everyday life in cities of the south. In this process, and through Raull Santiago's tweet, the image/metaphor/allegory/ethnographic description ends up dissolving itself in its very endurance, in the juxtaposition of the rubble of past projects and future expectations of city dwellers.

One of the challenges of thinking global urbanism has to do with the understanding that we are all natives in this predominantly urban world in which some of us are (still?) trying to do ethnography. Bringing the "still construction, already ruin" temporality to bear upon theorizations of urban studies is an attempt to move away from the narratives of failure that stem from the future- and object-bound temporality of European modernity. Incorporating this temporality into urban theory as an everyday experience of city residents of the Global South entails moving away from the anthropology of the city to an anthropology of city-making, or from an urban anthropology to an anthropology of urbanism or, to be more precise, an ethnography of the many urbanisms that make up daily life in cities. The epistemological difference is that while discussions on whether we conduct ethnographies of cities or in cities, the (real or imagined) "city" remains stable as a normative image; when we focus on city-making we attend to the fact that the city is not a given or a goal but always in the making and to bring this "making" in as the privileged object of inquiry. This subtle shift helps to recast normative ideas of city-making as one among many native utterances about cities and urban futures.

Note

1 Vapor barato/Um mero serviçal/Do narcotráfico/Foi encontrado na ruína/De uma escola em construção…//Aqui tudo parece/Que era ainda construção/E já é ruína/Tudo é menino, menina/ No olho da rua/O asfalto, a ponte, o viaduto/Ganindo prá lua/Nada continua.
E o cano da pistola/Que as crianças mordem/Reflete todas as cores/Da paisagem da cidade/Que é muito mais bonita/E muito mais intensa/Do que no cartão postal.

References

Das, V. and Poole, D. (2004) *Anthropology in the Margins of the state. Santa Fe*: School of American Research Press.

De Certeau, M. and Mayol, P. (1998) *The Practice of Everyday Life: Living and Cooking* (Vol. 2). *Unversity of Minnesota Press.*

Fischer, B. (2014). 'A century in the Present tense'. In Fischer, B. McCann, B. and Auyero, J. (Eds.), *Cities from scratch: Poverty and informality in urban Latin America.* Durham: Duke University Press, pp. 9–67.

Gupta, A. (2018). 'The future in ruins: Thoughts on the temporality of infrastructure'. In Anand, N., Gupta, A. And Appel, H. (Eds.), *The Promise of Infrastructure.* Durham and London: Duke University Press, p. 6279.

Hardman, F. (1996) 'Brutalidade antiga: Sobre história e ruína em Euclides'. *Estudos Avançados, 10*(26): 293–310.

Lancione, M. (Ed.) (2016) *Rethinking Life at the Margins: The Assemblage of Contexts, Subjects, and Politics.* London: Routledge.

Lévi-Strauss, C. (1992) *Tristes Tropiques. 1955* (trans. John and Doreen Weightman). London: Penguin Books.

Mbembe, J-A. (2003) 'Necropolitics'. *Public Culture*, 15(1) 11–40.

Sheppard, E., Leitner, H. and Maringanti, A. (2013) 'Provincializing global urbanism: A manifesto'. *Urban Geography, 34*(7): 893–900.

Roy, A. and AlSayyad, N. (2004) 'Urban informality: Transnational perspectives from the Middle East, Latin America, and South Asia'. In Roy, A., and Alsayyad, N (Eds.), *Urban Informality: Transnational perspectives from the Middle East, Latin America and South Asia. Lanham and* Oxford: Lexington Books, pp. 7–30.

Roy, A. (2005) 'Urban informality: Toward an epistemology of planning'. *Journal of the American Planning Association*, *71*(2): 147–158.
Roy, A. (2009). The 21st-century metropolis: New geographies of theory. *Regional Studies*, *43*(6): 819–830.
Roy, A. (2012) 'Ethnographic circulations: Space-time relations in the worlds of poverty management'. *Environment and Planning A*, 44(1): 31–41.
Simone, A., and Pieterse, E. (2018) *New Urban Worlds: Inhabiting Dissonant Times*. John Wiley & Sons.
Simone, A. (2004) *For the City Yet to Come: Changing African Life in Four Cities*. Durham: Duke University Press.
Varley, A. (2013) 'Postcolonialising informality?' *Environment and Planning D: Society and Space*, *31*(1): 4–22.
Zweig, S. (1941) '*Brasil, país do futuro*'. Rio de Janeiro: Guanabara.

32

THE MIGRATION OF SPACES

Monumental urbanism beyond materiality

Morten Nielsen

> For everything that accords with the values of what we call 'civilization', its cities and monumental architecture, its social classes and elaborate lifeways, its incredible technologies, mathematics and self-expression in the control and knowledge of writing and speech, amounts to an overdetermination of the containment of sense by itself.
>
> (Wagner, 2001:30; italics added)

In the now classic "The City in History", Lewis Mumford describes the monumental architecture of the citadels of ancient cities (1961). As an expression of the ruler's power, which almost equalled that of the 'mighty god', the purpose of monumental architecture exhibited through 'costly building materials and all of the resources of art' was twofold: To 'produce respectful terror' and allow residents to partake in the divine personality that was manifested in the institution of kingship (op. cit.: 64–70). With the urban monumental art of the citadels, cities could be scaled to awe the beholder and thereby codify a collective imagination in terms of an eternal cosmological order. As Mumford describes,

> Here art came in to establish and re-enforce, with an effect beyond that of mere words, all that the new order had brought in to alter the dimensions of the older, purely agricultural regime: above all, *the power of the disciplined imagination itself to translate the possible into the actual*, and to enlarge the humble habits of everyday life into structures of magnificence.
>
> (op. cit.: 68; italics added)

In other words, the monumental architecture of the citadels of ancient cities served not merely to articulate a deep sense of eternity; it also triggered new imaginary potentials – a transformative social and ontological force that offered itself to the beholder in and through the materiality of the thing but which was not necessarily contained by it (Bergson, 2005: 9–10). To be sure, monumental architecture could be considered a material medium for producing docile subjects, but it was as an engine of imaginary transformation that its true powers resided (see Nielsen and Pedersen, 2015).

DOI: 10.4324/9780429259593-36

In what follows, I want to use Mumford's discussion of monumental architecture as an experimental heuristic for thinking about global urbanism. My particular focus for this series of reflections will be the multistranded and heterogeneous implications of large-scale speculative urbanization projects in the Global South. During recent decades, there has been a massive upsurge of ambitious urban development projects being implemented throughout the Global South allegedly with the aim of converting existing cities into 'world cities' geared towards integration within a global economy that operates outside territorial state-defined restrictions (Houston, 2002; Shatkin, 2011; Watson, 2013). Many of these urbanization projects appear to have been launched as property investment strategies, which aim to benefit elite groups of financial and political entrepreneurs while doing little or nothing for the growing urban populations living under fragile conditions in informal settlements without access to basic services let alone civic rights (Goldman, 2011). In cities, such as Dar-es-Salaam (Tanzania), Kinshasa (Democratic Republic of the Congo), Maputo (Mozambique), Ho Chi Minh City (Vietnam), Bangalore (India) and Singapore, massive urban development projects have been planned and some built with the explicit (or implicit) objective to generate financial growth from sequestered and privileged zones while disregarding the demographics of the broader socio-economic contexts. Often at the cities' peripheries where informal urban sprawl is most intense, land is being re-parcelled and legally redesigned in order to increase its financial value as expansion zones for new urbanization projects (Pitcher and Moorman, 2015). Not unlike the situation in many Euro-American cities, such recent urbanization projects have been driven by a desire to build new and detached urban satellites and enclaves rather than rehabilitating the existing built environment (Murray, 2015a). In doing so, financial investors allying with local city builders and state officials deliberately try to circumvent the mundane chaos of existing and poorly functioning city administrations so that the fortunate few can retreat to sequestered urban spaces with efficient infrastructure, up-to-date services and a comprehensive system of security based on surveillance and restricted access. That is at least the ideal. In reality, however, many speculative urban development projects have failed, thus causing wider economic and social disruptions with devastating consequences for local urban residents (Marcinkoski, 2015). And even in those instances where speculative urbanization projects do result in the building of segregated urban enclaves for the privileged few, it is rarely possible to maintain continuous detachment from the surrounding society. In Euro-America, the making of sequestered urban spaces has been described as a socio-economic effect of what Davis has called an endemic 'ecology of fear' (1998). By contrast, in many sub-Saharan African cities, for instance, the spatial lines of separation that isolate the affluent few from surrounding urban spaces follow a much wider and less coordinated meshwork of social divisions and political fissures and with deeper social, cultural and economic underpinnings (Nielsen et al., 2020; Pieterse, 2011).

But the lack of physical markers to distinguish privileged spaces from the wider social infrastructure of "incessantly flexible, mobile, and provisional intersections of residents that operate without clearly delineated notions of how the city is to be inhabited and used" (Simone, 2004: 407) does not mean that the materiality of the (non-physical) distinctions does not do its work. It simply needs to be understood differently. According to Ong (2011: 14), different cities may become connected through the circulation of 'urban models', which can be defined as particularly desirable global technologies and aesthetics that are disembedded 'from (their) hometown and adopted in other sites'. In many cities, physical spaces, built forms and sets of urban practices are reimagined and explored anew through the activation of desires and aspirations that are associated with urban milieu elsewhere. So, for instance, in Kinshasa, billboards all across the city make promises of an urban future that is no longer mirrored on a version of Belgian colonial modernity but, rather, captures 'the aura of Dubai and other hot spots of the new urban Global South' (de Boeck 2011: 274). Here solidarity with and referencing to another place comes to affect the pace of certain spheres of Kinshasa without necessarily involving the exact blueprint implementation of a 'Dubai model'. What is at stake in such instances, rather, is the production of an urban sensitivity that infuses the city with new imaginary potentials through the gathering of outside elements, what Ong defines as 'worlding' (2011: 11). Although heterogeneous and always contested, worlding practices conjure up new worlds across existing urban milieu through the juxtaposition of spaces, aesthetics and ideas from distinct localities.

Hence I will venture to suggest that if there is anything inherently global about contemporary cities, it is this. With the circulation of ideas and ideals of what constitutes desirable urban worlds, disparate connections are forged between different cities, which carry concrete visions of new urban futures and thereby also a sense of experimental world-making that transcends the confines of specific local sites. To be sure, this does not imply that urbanism anywhere can be considered a reflection of a self-replicating universal form, such as speculative global capitalism (Roy, 2016). Rather than having a singular universal causality as its engine of genericity, we may argue that each city and every singular urban process produce their own abstractions, which are always more than the concrete spaces and specific aesthetic and social imageries from which they derive their directionality and force (Nielsen and Simone, 2016).

Global urbanism is, therefore, never equal to whatever the cities of the world may be taken to be. The 'dis-individuated' abstraction (Galloway, 2014) does not correlate with the particulars of the city in relation to which it has become an unmarked and generic universal. The urban abstraction is essentially a 'stranger' (Laruelle, 2011) to the concrete specificity of the city and that is why its legibility becomes problematic. Conventionally, legibility emerges from the fixation of things (Taussig, 1993). As people, concepts and spaces are conquered and momentarily stabilized, scales are invoked that offer a universal readability of those very same phenomena that were vital in the making of the former as a 'conjuring of a dramatic performance' (Tsing, 2000: 119). Any perspective of things requires a precise 'spatial dimensionality' (ibid.) by which to gauge the articulations of socio-material life and, from there, act proficiently in and on the world. But we also know that scales are not just neutral frames for somehow viewing the world objectively (Wastell, 2001). While we may continue to insist that scales are set up prior to measuring the effects of actions, it is, in fact, by contextualizing, scaling, spacing that the idea of the scale itself is achieved (Corsín Jiménez, 2005; Latour, 2005). To be sure, we would never know what actions to gauge for their effects without the distinctions we impose upon them through the use of scales, and so the instruments we operationalize when measuring things end up also producing their particular qualities (Wagner, 1981).

The question is, then, what the implications might be for our understanding of the dynamics and modulations of global urbanism when it occurs through the workings of 'dis-individuated' abstractions? What happens, in other words, if we remove something from the constraints of legible scales and proportions – that is, if we move beyond the acts of immediate measuring and proportioning that seem to offer the most available form of readability of the city? What kinds of perspectives might be accessible if the contrasts, differentiations and distinctions that mark the deep physicality of speculative urbanization projects in the Global South were momentarily removed, and we had to leave the cities to their own devices? Would that even be possible? Here I think we need to return to Mumford's reflections on the relationship between 'the disciplined imagination' and monumental architecture. For is it not precisely the work of the 'disciplined imagination' that allows the materiality of large-scale – but still not built! – monumental urbanization projects to significantly affect and orient certain registers of social life in cities in the Global South?

Staying with Mumford's discussion of monumental architecture, we might speculate that its imaginative potentials derived not only from the 'respectful terror' that was evoked from the divine-like aesthetics of the physical buildings and monuments (1961: 65). Rather, the monumentality of the buildings triggered a transformative urge to lift out the materiality of the citadels of ancient cities certain visions of the world that arose from it but which it could not contain. The monumentality of urbanism that I am focussing on here has, therefore, less to do with its particular physical proportions than its considerable capacities for producing ideas and ideals of desirable cities that might circulate across different urban milieu. Obviously, it is not only very large speculative urban development projects which seem to have wedged into their physical structures a transformative capacity for producing portable visions of the world. But there is no doubt that the intensity by which desires and aspirations associated with such projects migrate across different urban terrains has a particularly pronounced effect on the pace of cities in many parts of the world and especially so in the Global South. Although impossible to capture by any legible scale, the proportions of these phenomena are, indeed, monumental.

The Department of Urban Planning (DPU) at the Maputo Municipality in Mozambique has never operated as a well-oiled, efficient administrative machine. If anything, it is a slow-moving, somewhat dysfunctional but fairly stable vehicle, which lacks a workforce, financial resources and political attention (Andersen et al., 2015). Architect Anselmo Goveia has been with the DPU for more than 20 years and is now head of a small team that works exclusively on private construction projects near the coastal sections of the new ring road that was opened in 2015. While being constantly overloaded with work tasks, Goveia is excited about the prospects for the city and urban planning in particular, which new large-scale urban construction projects seem to carry. In the spring of 2016, I sat down with Goveia to talk about his views on the many planned but still unrealized urban development projects in and around the Maputo city centre.

> Well, this is much better than before! We are so few technicians working here, you know. We can't deal with all of the city and the Mayor knows that!.... So, we need condominiums (*condomínios*) like the ones they have in South Africa. Do you know how they build Waterfall City? The government hands over a piece of land to the investor and then they take care of business. That is what we want here!

Waterfall City, which Goveia explicitly refers to, is an impressive new property development located halfway between Johannesburg and Pretoria. With a total cost exceeding R45 billion, it is the 'most ambitious city-building project ever undertaken in Africa, let alone South Africa' (Murray, 2015b: 505). Similar to many other master-planned private cities that have been constructed in or near urban areas throughout the Global South during recent decades, Waterfall City involves a fundamental reconfiguration of the relationship between public administration and private investors where the latter increasingly come to perform the functions that are conventionally expected of municipal and state authorities. For all practical purposes, the mandate to govern the new city has been outsourced to private actors, who are obviously interested first and foremost in financial gains rather than social cohesion and wide-ranging urban integration (Murray; 2017; see also Acuto, 2010; Kanna, 2011).

It is significant to note that, at the time of our conversation, the construction of Waterfall City had not been completed yet, and so Goveia was reacting to the design of an urbanization project whose full realization was stencilled onto a future canvas, the texture of which was still relatively unknown. Still, the lack of physical coordinates did not prevent Goveia and many other officials with him from using the projected urbanization projects as an imaginary conduit for recalibrating their understanding of what the city and urban planning might be. Indeed, whereas officials at the Maputo Municipality and elsewhere had generally considered urban planning as the main administrative strategy for providing a minimum of spatial regularity in the city as a whole, it now seemed that urban planning was in the process of being morphed into a transactional object, which could be negotiated and outsourced to private actors who would manage enclaved slices of the city (Easterling, 2016).

Consider, for comparison, the situation in Bangalore, India, where a project of 'world-city making' promoted by various parastatal agencies has given rise to a new art of urban government guided by neoliberal speculation (Goldman, 2011). Guided by an increasingly global 'master plan' agenda, the aspiration for 'world-city making' has generated widespread anxieties and social tensions, not least caused by accompanying processes of dispossession and mass displacement of poor urbanites living in areas destined for mega-city projects. But the introduction of global discourses on world-city projects has also produced imageries of alternative future scenarios, which already affect the socio-political dynamics of the city. In his discussion on 'speculative urbanism' in Bangalore (op. cit.: 559–560), Goldman describes a deal-making event at a beach resort in Goa where urban entrepreneurs from Bangalore and other potential Indian 'world cities' got together to discuss urban development. Seemingly inspired by the event, a city commissioner from the city of Jaipur took the podium to express his excitement:

> Thanks to these…meetings, I have begun to see my home city differently. When I look out from Jaipur's main railway station, I can see makeshift huts with women cleaning dishes and children playing and grazing their animals… But why couldn't we build right along the station a line of nice hotels, corporate centres and shopping malls? Now I can imagine that Jaipur too can become a world city that can generate jobs and money…. From this view, our cities are full of untapped value and potential, making them a very exciting place to be.

While the two contexts are seemingly too diverse to warrant comparison, it is, nevertheless, interesting to note how they both reflect a certain imaginative vibrancy prompted by projected large-scale urbanization projects. In both instances, the aesthetic monumentality of the future city does seem to 'awe and overpower the beholder' (Mumford 1961: 65) – not because it exists in a physical form but precisely because it does not. By not being captured by its physical form, monumental urbanism can roam freely. Whereas most cities remain where they are, monumental urbanism migrates.

In *Extrastatecraft* (Easterling, 2016), Easterling describes how a model of the economic free zone administered outside of recognized regulations has become an item of global export, which circulates between different localities 'beyond the reach of state jurisdiction' (op. cit.: 16). As a form of 'extrastatecraft', however, it does involve the activities of state agents, who seek to profit from the opportunities that offer themselves with the free zone. According to Easterling, the economic free zone has become 'the most popular mode for the contemporary global city, offering a "clean slate" and a "one-stop" entry into the economy of a foreign country" (ibid.). Similar to the free zone, which operates almost like a sovereign space irrespective of its locality, monumental urbanism conjures widespread desires for global city-making. But it does so in a slightly different way. In a sense, monumental urbanism only exists through its effects, as a series of speculative lines that are being drawn across existing city spaces with the force of future imageries. No urban models are being exported here other than as a longing for something that might not have existed in the first place. And it is this imaginary vibrancy that invests certain registers of urban life in the Global South with an almost staccato-like rhythm and urgency (Nielsen, 2016; see also Lefebvre, 1995).

If monumental urbanism cannot be properly gauged by way of legible scales, its social, economic and political effects can, nevertheless, be determined with relative ease. Take Maputo, for example, where a partially virtual building frenzy has fundamentally reconfigured the configuration of the urban landscape. Already in 2012, there were more than 80 gated compounds in and around the city centre (Costa, 2014) and only within the last five years, 15 new complexes have been built along the picturesque Costa do Sol road bordering the coastline (Nielsen and Jenkins, 2020). And, still, very few if any of these ongoing building projects manifest the proportions of the urban development projects, whose monumental aesthetics they also reference. But as urban land is reimagined and surveyed and building projects are initiated without ever reaching their expected outcome, the imaginary infrastructure of the city also changes. Although this can be considered a 'worlding' process, which conjures new urban scenarios through the juxtaposition of elements from distinct localities (Ong, 2011), it is one which gradually reduces the imagery of the city to an agglomeration of detached enclaves.

Even when city planners fail to realize their grand visions of building global metropolises, the social and material fabric of the cities continue to reverberate with their desires and aspirations. In this regard, monumental urbanism is nothing but the process of contraction and expansion of the imaginary space that moves with these desires and aspirations across different physical localities and which only ceases to invest the cities with its unique qualities if and when it can be contained by a physical form. Monumental architecture returns to the specificity of the city in the form of a 'stranger' (Laruelle, 2011), whose unmarked and dis-individuated qualities are inserted in local urban domains by way of the difference they constitute in relation to what already exists. If the distance between urban abstraction and local specificity is reduced over time, the former can no longer maintain its aprioricial function in relation to speculative urban development. It is no wonder, then, that large-scale urbanization projects activate the 'disciplined imagination' in an attempt to translate the possible into the actual (Mumford, 1961: 68). But it seems that a unique characteristic of speculative urban development is that this translation occurs precisely the other way around to what might be expected. The social and economic scaffolding of Maputo's urban territories cannot carry Goveia's lofty aspirations of transferring the model for Waterfall City from South Africa to Mozambique. And Goveia probably knows that. But the imagery of an autonomous zone one step removed from the chaos of everyday hassle has made its way to the city where it invests an interwoven cluster of spaces and relations, fissures and tensions with similar or at least analogue rhythms. In that sense, monumental urbanism operates backwards, as it were. The drive towards large-scale urbanization projects recursively acts on itself by progressively refusing to be contained by its own (physical) model (Wagner, 2001). The translation of the possible into the actual, in other words, occurs by relativizing the desire for large-scale speculative urbanization projects through a series of self-replicating transformations of the monumental urbanism that probably never existed in the first place.

Considered as a modulation of monumental architecture, global urbanism is a mirror image that is constantly changing. It is always 'off', as it were, in relation to the cities, whose material aspirations, social desires and infrastructural imageries it is allegedly capturing. In cities as different as Maputo and Bangalore, speculative urban development is invested with the qualities of a series of aprioricial abstractions that never seem to fully fit the specificity of these urban environments. And it is precisely because of this lacking adequacy that the pace of the cities can be constantly increased, not in order to reach eventual alignment between abstraction and specificity but, rather, to allow for the differential movements to play themselves out in an ever-more expansive speculative space (Simone, 2020).

References

Acuto, M. (2010) 'High-rise Dubai urban entrepreneurialism and the technology of symbolic power'. *Cities*, 27(4): 272–284.

Andersen, J. E., Jenkins, P. and Nielsen, M. (2015) 'Who plans the African city? A case study of Maputo: Part 1 – the structural context'. *International Development Planning Review*, 37(3): 331–352.

Bergson, H. (2005) *Matter and Memory*. New York: Zone Books.

Corsín Jiménez, A. (2005) 'Changing scales and the scales of change. Ethnography and political economy in Antofagasta, Chile'. *Critique of Anthropology*, 25(2): 157–176.

Costa, A. B. d. (2014). *Condominiums and exclusive neighbourhoods in Angola and Mozambique* (Unpublished manuscript). ISCTE-IUL.

Davis, M. (1998) *Ecology of Fear: Los Angeles and the Imagination of Disaster*. New York: Metropolitan Books-Henry Holt.

De Boeck, F. (2011). 'Inhabiting ocular ground: Kinshasa's future in the light of Congo's spectral urban politics'. *Cultural Anthropology* 26(2): 263–286.

Easterling, K. (2016) *Extrastatecraft: The Power of Infrastructure Space*. London & New York: Verso.

Houston, C. (2002). 'Corporate cities – urban gateways or gated communities against the city? The case of Lippo, Jakarta', in T. Bunnell, L. Drummond and K. C. Ho (Eds.), *Critical Reflections on Cities in Southeast Asia*. Singapore: Times Academic, pp. 143–164.

Kanna, A. (2011) *The City as Corporation*. Minneapolis: University of Minnesota Press.

Latour, B. (2005) *Reassembling the Social. An Introduction to Actor-Network-Theory*. Oxford: Oxford University Press.

Galloway, A. R. (2014) *Laruelle. Against the Digital*. Minneapolis: University of Minnesota Press.

Laruelle, F. (2011) 'The Generic as Predicate and Constant: Non-Philosophy and Materialism', in Bryant, L., Srnicek, N. and Harman, G. (Eds.).

Lefebvre, H. (1995). 'Elements of rhythmanalysis', in E. Kofman and E. Lebas (Eds.), *Writings on Cities*. Oxford: Blackwell, pp. 219–240.

Marcinkoski, C. (2015). *The City That Never Was*. New York: Princeton Architectural Press.

Mumford, L. (1961) *The City in History. Its Origins, Its Transformations and Its Prospects*. New York: Harcourt Brace Jovanovich Inc.

Murray, M. J. (2015a) 'City Doubles': Re-Urbanism in Africa', in Miraftab, F., Wilson, D. and Salo, K. (Eds.), *Cities and Inequalities in a Global and Neoliberal World*. New York: Routledge, pp. 92–109.

Murray, M. J. (2015b) 'Waterfall City (Johannesburg): Privatized urbanism in extremis'. *Environment and Planning A*, 47(3): 503–520.

Murray, M. J. (2017) *The Urbanism of Exception. The Dynamics of Global City Building in the Twenty-First Century*. Cambridge: Cambridge University Press.

Nielsen, M. (2016) 'Urban times : Temporal topographies and non-scalable cities'. *Ethnos*, 82(3): 393–405.

Nielsen, M. and Jenkins, P. (2020). 'Insurgent aspirations? Weak middle class utopias in Maputo, *Mozambique*'. *Critical African Studies*. https://doi.org/10.1080/21681392.2020.1743190.

Nielsen, M. and Pedersen, M. A. (2015) 'Infrastructural involutions. The imaginative efficacy of collapsed Chinese futures in Mozambique and Mongolia', in Rapport, N. and Harris, M. (Eds.), *The Imagination. A Universal Process of Knowledge?* Aldershot: Ashgate, pp. 237–262.

Nielsen, M. and Simone, A. (2016) 'The generic city: Examples fra Jakarta, Indonesia, and Maputo, Mozambique', in P. Harvey, C. Bruun Jensen and A. Morita (Eds.), *Infrastructures and Social Complexity: A Routledge Companion*. New York: Routledge, pp. 128–140.

Ong, A. (2011). 'Worlding cities, or the art of being global', in A. Roy and A. Ong, (Eds.), *Worlding Cities. Asian Experiments and the Art of Being Global*. Malden, MA: Wiley-Blackwell, pp. 1–26.

Pieterse, E. (2011) 'Grasping the unknowable: Coming to grips with African urbanisms'. *Social Dynamics*, 37(1): 5–23.

Pitcher, A. and Moorman, M. (2015) 'City building in post-conflict, post-socialist Luanda: Burying the past with phantasmagorias of the future (Luanda, Angola)', in Pieterse, E. and Edjabe, N. (Eds.), *African Cities Reader III: Land, Property and Value*. Vlaeberg: African Centre for Cities and Chimurenga Magazine, pp. 123–135.

Shatkin, G. (2011). Planning privatopolis: representation and contestation in the development of urban integrated mega-projects in: A. Roy and A. Ong, (Eds.) *Worlding cities: Asian experiments and the art of being global*. New York: Wiley Blackwell, pp. 77–97.

Simone, A. (2004) 'People as infrastructure: Intersecting fragments in Johannesburg'. *Public Culture*, 16(3): 407–429.

Simone, A. (2020). '(Non)Urban Humans: Questions for a Research Agenda (the Work the Urban Could Do)'. *International Journal of Urban and Regional Research*, 44(4): 755–767.

Taussig, M. (1993) *Mimesis and Alterity. A Particular History of the Senses*. New York: Routledge.

Tsing, A. L. (2000) 'Inside the economy of appearances'. *Public Culture*, 12(1): 115–144.

Wagner, R. (1981) *The Invention of Culture*. Chicago: The University of Chicago Press.

Wagner, R. (2001) *An Anthropology of the Subject. Holographic Worldview in New Guinea and Its Meaning and Significance for the World of Anthropology*. Berkeley: University of California Press.

Wastell, S. (2001) 'Presuming scale, making diversity. on the mischiefs of measurement and the global: Local metonym in theories of law and culture'. *Critique of Anthropology*, 21(2): 185–210.

Watson, V. (2013). 'African urban fantasies: dreams or nightmares?' *Environment and Urbanization*: 215–231.

33

LAND AS SITUATED SPATIO-HISTORIES

A dialogue with global urbanism

Wing-Shing Tang and Solomon Benjamin

Questioning a singular global narrative

We question if global urbanism, even when conceived relationally as a bridge across contexts, can be a viable operative concept posing a singular explanatory logic? This particular theoretical anxiety has emerged over the past decade from international relations countering ideas of a dominant 'Westphalian world' (Kayaoglu, 2010) and corresponding ideas of 'southern urbanism' (Roy, 2009). We seek power-laden spatial frames reflecting ahierarchical relationships to reject those that are sequential or temporal and thus project a binary between the 'north and south'. For instance our two cases, Kowloon and East Delhi, which illustrate the 'complexity of situated histories' rather than adhering to a method exploring diversity via a 'Chinese and Indian comparison' framed under a tacit assumption of both being influenced by global capital. Our approach invokes the *tongbian* philosophy (Tang, 2019; Tian, 2005) and considers the complexly embedded, sedimented and different logics concerning the occupancy of land (Benjamin 2017).

Complicating land is central, as it remains at the nexus of varied forces with relational specificities across time-space contexts. Invoking Li (2014), land cannot be rolled up and taken away but also evolves to be a deeply material space whose worn textures reflect hauntings of earlier occupancies with diverse logics, where an intensity of bodies who have sat and slept on, used it as a site of prayer, to sell stuff from and when set on a pole, protect against a harsh sun or heavy rain. Land tenures move beyond cartographic imprisoning to reflect accumulative practices, memories, values and meanings and thus are central to our framing of 'situated histories'. In this operative frame, we look beyond the visible to dig deep into the ground to unravel diverse materiality and relationalism developed from the deep soil up: layer by layer, temporally and in spatial relation with many other events that open different logics and forces of transformation, both locally and far distant apart. Such centring of land as an analytic rejects its relegation as implicit in both postcolonial and planetary urbanization thinking, one of the 'lower elements' constituting diversity under an overarching frame of global capital.

While we recognize the importance of the philosophy of internal relations in itself, the *tongbian* philosophy is central to emphasize spatial complementaries and problematize contradictions in land transformation. In doing so, it overcomes one of the shortcomings of dialectics informed by the philosophy of internal relations (Ollman, 2015): a determinism effected by ideas of the

DOI: 10.4324/9780429259593-37

global expansion of capital from 'the north to the south' *as a unilateral trajectory into neoliberalism.* Mumbai posed as an emblematic 'southern financial centre', Delhi as a national capital of policy directives and Hong Kong and Shanghai as financial outposts now incorporated into the global circuits; they all reflect a postcolonial contemporary within a unilateral trajectory into neoliberalism. In this the logic, land transformation gets clouded under clear and easy narratives of 'global gentrification' and urban renewal (Tang, 2017).

For instance, the framing of 'southern urbans' poses urbans as a diversity of 'landing sites' constituted around 'local' political economies, under capital's planetary condition. However, *tongbian* philosophy connects to the complexity of land tenure beyond such implied colonial cartography. As a way to differentiate between diversity and 'maximal difference' via spatial analysis, between symbolic spaces versus substantive ones, it allows us to consider colonial and postcolonial reads of the urban. For instance, colonial urban history reveals how land tenurial under colonial administrations remained unsettled and politicized sites: in Bombay (Chhabria, 2018); urban sites across Canada, such as 'Turtle Island'; and Palestine (Bhandar, 2016); Australia (Verran, 1998); Niger, among other sites in Africa (Lund, 1998); and Indonesia (Li, 2014). Land tenurial practices preceded and persisted through colonialization (albeit in a changed form in contemporary cities). In bridging simplistic binaries of 'use and exchange value', they disrupt assumptions of trajectory from common tenures to the absolute individual titles. However, much of the postcolonial writings miss such complexity: set into 'passivity' in a neat trajectory from a 'mystical commons' to disciplined territorial formations of neoliberal 'speculative urbanism', tenure lies on a ladder that has been mobilized by followers of Desoto to propagate the Torrens titling responsible for ethnic cleansing in Australia. Land turned into singular meaning remains not only a problematic analytical frame (speculative urbanism, frontier urbanism, gentrification, southern urbanism and planetary urbanization) but also risks reinforcing a real politics of violence and erasure. This implicates 'policy-oriented' elite researchers who naively frame narratives of 'citizenship rights' linked to access to water, corruption of proper policy by touts, brokers to routinely conflate these as 'land mafia' with criminal groups. In effect, this reinforces powerful land developers by de-legitimizing and criminalizing competing claimants to land and infrastructure.

A political shift implies moving beyond just diversity: Mere identifying and labelling more modes of ownership, tenure and usage and, accordingly, elaborating the usually neglected ones, such as those in 'the south', is only the first step. Once this is done, the political space around an interrogation of ownership, tenure-fixes and usage diachronically enables us to make sense of its present from its past and connected rather than independent of others. Thus it is equally imperative to comprehend synchronically the relationships between one and another. The diachronic and synchronic relationships to pose understandings of *'maximal difference'* resulting from varied forces interacting in mutually embedded ways. Such sharpened analysis is used when we move into the postcolonial analysis of land tenurial practices. Here criticality via the *tongbian* philosophy is possible via nuanced work. For instance, Bhandar's (2016) analytic of planning in Palestine usefully mobilizes law in recombinant ways that rework linear views of land transformation. The attention to societal dynamics is emphasized by legal pluralism theorists (Benda-Beckmann et al. 2006; Webber, 2009). Such complexity concerning different histories of land tenures and occupancies is discernible in three realms: (1) 'everyday' bureaucratic politics that include spaces of planning departments, such as in a Taiwanese municipality (Bernstein, 2008); (2) judicial spaces mediated varied property-territorial logic, as witnessed in Taiwan's urban sacred ground (Lin and Tsai, 2014) in India (Balachandran, 2018) and in Toronto spaces of planning regulations (Valverde, 2008); c) land as situated history gets 'peopled' where tenures, claims and meanings shape occupancies to be 'layered over time', as reflected in Simone's (2015) mobilization of 'indeterminate' space where ambiguity opens up such potentiality, where land in these conceptions remains inherently 'unsettled' and capable of reflecting but also generating new power structures.

Tongbian *philosophy*

We find the *tongbian* philosophy useful to address such complexity and centrality to the urban. Our two sites in Hong Kong's Kowloon located in Sham Shui Po (SSP) and in East Delhi's Viswas Nagar (VN) show assumptions of land being 'settled' into inalienable private property. Such a move, framed to be inevitable under global capitalism, is not just factually incorrect (Verran, 1998: 246) but, importantly, misses out contemporary archives where the ambiguity land points to forms of power realms constituting complex state spaces (Ho, 2001; Benjamin, 2019). Such terrains, reflective of land's tenurial complexity, are necessarily 'unsettled' and remain embedded in varied political and social spheres (Hann, 1998). The complexity of a heterogeneous state underpins the spatial story of land revealed analytically via a *tongbian* approach. Considering such complexity requires systematic uprooting the common parlance according to *yin/yang*-informed *tongbian* philosophy (Ling, 2014; Tang, 2019; Tang, 2020; Tian, 2005). To be brief, the latter argues that everything consists of two mutually embedded, opposite poles. "Each pole of the pair grows from the other and needs the other as partner. They are in ceaseless interaction, implying the process of becoming, continuity and change" (Tang, 2019: 464). This understanding has ontological, epistemological and methodological implications for any understanding of land development.

This philosophy argues that no one has an ontological privilege over the other, as one is always defined by the ceaseless interaction with the other. In the West, there is ontological privilege to inalienable individual rights. In terms of private land property in the ownership model, one is entitled to the alienable right to possess all kinds of property rights (including surface, productive, development, pecuniary, restrictive and disposal). Analyses about land development around the world have then assigned the ontological privilege to property title, ownership, possession and exchange (and, as an extension, further assuming this to form the basis of economy). Besides, there are, within the ownership model, many discrete opposites, such as landlord and tenant, exchange and use value, legal and illegal, formal and informal. In contrast, the *tongbian* philosophy argues the converse: one's ontology comes from one's impact on the other. It underscores "a simultaneity of oneness and multiplicity, difference and commonality, continuity and change" (Ling, 2014: 43). It is inappropriate to forget about the other and the interaction between the two polar opposites; there is the missing other besides what is prescribed in the ownership model.

According to the *tongbian* epistemology, knowledge is developed within the context. Since the Enlightenment, emphasis has been on rational conceptualization. The modern concepts of property rely on a bureaucratic and abstract representation of fictitious ownership interests in an administrative archive. As a result, there is the widespread application of the spatial sciences of cartography and geography in land management and the planning of urban space and the design of the urban environment. These techniques reduce knowledge of land development to the perspective of the surface and in the form of commodity and mask the structural depth of social reality. The land question is usually simplified as an economic question of supply and demand, ignoring the moral, social and political questions. In contrast, *tongbian* assumes no essential features, favouring how events are situationally correlated. Knowledge about land development, then, requires placing the latter within a web of multiple correlations of relations in a down-to-earth, concrete context.

Tongbian philosophy, informed by the interaction of the two polarities of *yin* and *yang*, denotes a nuanced methodology for research on land development. The famous *yin/yang* diagram, with each half of the whole always retaining within it an element of the other, implies that, besides the absence of the binary within the duality, there are always *YIN*-in-*yang* and *YANG*-in-*yin*, constantly approaching each other and becoming the other. Thus it is difficult to perform analysis in compartmentalized boxes of time and space, as well as of past, present and future. Instead, any informed methodology must elaborate the mutual embeddedness between the two polar opposites in time-space and past-present-future. Since mutual embeddedness is not restricted within one *yin/yang* whole but also others, becoming takes place not only diachronically but also synchronically (Tang, 2020).

We can argue that a huge change is in *tongbian in spatial complexity rather than dialectics in spatial contradictions*. Land, from the *tongbian* approach, turns out to be both conceptually useful and an empirical/methodological entry point: as a nexus, it is influenced and acted upon via other forces but also via its accumulated meanings, which form a significant generative realm on its own terms. We see an analytic based on a spatial story of practices as a productive frame to look at such complexity. The way social groups and institutions engage with land and via it allows us to explore nuances of how we think about the state as mutually embedded spaces and to bridge understandings of 'the urban' and capital.

SSP, Kowloon, Hong Kong

Hong Kong as a setting for 'signature' architecture – high-density development with renowned skyscrapers in the city centre – may reinforce a narrative of modernity, but this has, in fact, masked its situated land histories (Tang, 2014, 2017). Chinese customary land practices have continued to exist since the British colonized Hong Kong Island in 1842. The 'perpetual hereditary tenancy' system divided arable land into two landholders: the topsoil (*dipi*) and the subsoil (*digu*) refer to the separation of the right to use from that to own. Owners of topsoil were free to inherit, mortgage, transfer and dispose of it, irrespective of any change in the subsoil ownership. Ownership could also be collective by lineage, clan, etc. Land transaction was possible in the form of signing red deeds (with the county magistrate's consent to officiate the transaction) or white deeds (involving only the buyer and seller). This system had refused to put tenants and landlords into straitjackets of dualistic categories. In contrast, land practices in the West at that time emphasized private, individual land ownership and the legal status of title, not deed. As colonization proceeded apace, the land system had become further complicated due to formal protests from the Chinese government, the establishment of the Land Court in 1898 and the promulgation of the Extension of Laws Ordinance of 1900 to the new territories.

Unlike others around the world, the Hong Kong government has always been under the command of the British government at a distance while living with the Chinese within, as well as outside, the city boundary. As a result, it has been the predominant arbitrator of development in two ways. First, it has been concerned more with socio-economic order – alienable property rights – than with liberty – the inalienable individual rights. Second, the requirement of financial independence imposed by London even at the embryonic stage of colonialism has rendered land-related revenues the basic concern of the city. It is then the logic of the government, not capital *per se*, that counts. The hegemonic government regulates the fictitious capital from monetary capital, productive capital to commodity capital and modifies the prevailing practices once its hegemony is challenged. For the other characteristic, the dominance of property development since early colonization cannot be downgraded as secondary to industrial capital accumulation, more so as the more recent switching of capital from the latter. Since Hong Kong was colonized in stages, 1842, 1869 and 1898, the legitimacy of the government and the complexity of land ownership, entitlement and usage have been constantly challenged and defended over time. These challenges have serious spatial implications, as the stages of colonization have basically restricted the government's legitimate deployment of land resources to two sides of the harbour until the 1960s.

This brief account of the situated histories of land has highlighted that there are so many opposite forces in constant interaction, such as tenants/landlords, techniques/laws, building/agricultural land, crown/private land, government intervention/capital accumulation. There is no better district within Hong Kong than SSP on the Kowloon Peninsular to illustrate these complexities. Situated just north of Boundary Street (the boundary stipulated under the Convention of Peking in 1860), SSP, originally occupied by rural villages, is so distinguished a place in Hong Kong's spatio-temporality (Smart and Tang, 2014). For a long while since 1860, SSP was the place where the aforementioned forces mutually interacted. On the ground, literally, being part of a county in the dynastic agrarian empire, the land was still stipulated by the Chinese customary land practices. In the air, literally too, one could sense the atmosphere of an 'urban' centre established within the spatiality of the British empire on the other side of Boundary Street, where leasehold was widely practiced (Tang, 2014). The juxtaposition of two mutually embedded systems rendered SSP an extremely dynamic and fluid place of negotiation and resistance. For almost half a century since 1898, SSP was the testing ground for the government to implement the colonial administration, including land. This is exactly where the situated land histories are best illustrated.

VN East Delhi

East Delhi's VN in 1995 had evolved into Asia's largest electrical cables and conductors manufacturing cluster, supplying to Nepal, Myanmar, Bangladesh and even Pakistan (Benjamin, 2005). As a settlement characterized by home-based factories, its origins lay in agricultural land subdivided into housing. This is common not just in South Asia but also in large parts of Latin America built around a politics of settlers lobbying for the upgrading of basic infrastructure and services. Particular to VN, but also Delhi's 60 other industrial clusters, has been their economy, premised on evolved tenurial forms: real estate surpluses fund machinery, while land subdivisions and building new floors house incoming migrants: both skilled managers, foreman and a range of workers. This economy real estate nexus remains materialized in the complexity of leases as a property complex: common law's presumptive claims (unlike absolute titles) blur the line between owners and tenants, sub-tenants. As evolving industrial production drew in agricultural surpluses from North-West India, it, importantly, built powerful political alliances across party lines concerning social connections of long-standing and recent residents. Municipal elections were driven with voter turnout of over 80% of worker-dominated 'vote banks' demanding basic infrastructure. Shaping more than 60% of Delhi's territorial arrangement around 'unauthorized colonies' generates huge local government revenues; a key space of intense negotiation is that of liaison agents, touts, municipal councilors driving what elites fearfully term 'vote bank' or 'land mafia politics'. Land tenures inter-mixed both economy and political constituency of mostly small firms, while real estate funded inroads into Delhi's complex municipal political space, including its water and electrical supply agency, and higher, to engage circuits at the national parliament. Infrastructural improvements could be customized: engaging the corridors of the electricity supply agency brought in new electrical transformers to the denser industrial areas but also to counter master plan 'single zoning'. Importantly, none of this was 'informal'. From the middle 1970s, upgrading procedures have been written into the *porous bureaucracy*'s administrative procedures (ibid: 249). By the mid-1990s, the embedded politics of such specialized home-based clusters totaling 260,000 firms and a huge revenue base pushed the electricity authority to call for planning reforms concerning regularization. This was and continues to be deeply threatening to South Delhi elite and corporate groups who benefitted from master planning. They banded together with the English media and international donors to portray areas like VN as sites of land-use violations, encroachments and pollution. Around this logic, the Supreme Court in 1995 ordered factory sealing and shifting industry out of the national capital. The huge protests by factory owners, foremen and workers followed sealing Delhi closed for an entire week. Existing power realms remained re-configured till the late-1990s when a 'under the radar' pushback by constituencies across East, North, and West Delhi through middle-level administrative and judicial bureaucracies shifted the arena back to the Delhi administration's more accessible bureaucracy.

In the *tongbian* frame, attentiveness to complexity and continuity in histories and genealogies of tenurial arrangements engage lower and middle juridical and bureaucratic spaces in a frame, whose logics pose ambiguity and opacity to subvert elite groups, including elitist researchers. Similarly, rather than reading planning as oppositional politics, 'mixed land use' reflects simultaneously spaces within and outside master planning; finally, the mutual embeddedness of tenure and economy shows capital subverted within and beyond the commodity process.

Situated histories as difference

Our emphasis on this analytic, albeit briefly due to space, is not pedantic or essentializing when exploring SSP and VN as complex entanglements of situated histories engaging real estate with bureaucracy and planning as power to reflect elite power politics and shape it in material and immaterial ways. While land in the City of London or Wall Street of New York is different from that in, say, Hong Kong or Mumbai, contrasting dualistic positions cloud analytics: the latter as variegated extensions of the former or the Global North in general, or, conversely, Hong Kong and Mumbai represent alien autonomous parts of the Global South. Instead, in a *tongbian* world, constituents remain in constant interaction. Mumbai and Hong Kong would also see connections outside of those with New York and London. Moreover, we cannot see Hong Kong, Bombay, Delhi as being homogenous, as territorial formations within 'a city', emerging out of heterogeneous logics, built on varied histories of land. Importantly, these are not a move from their Central Business District (CBD) to 'the urban periphery' following a spatio-temporal logic but under complex connections within contested institutional politics. Rather, we explore spaces transformed via negotiations, appropriation reflected in substantive historical research across different contexts and focused on such complexity in colonial administrations and their extensions to contemporary times. It is no surprise that scholarship on contemporary politics' sacred territories in Northern Canada or Australia opens meanings to land via legal bureaucracies understood as practiced spaces in resonance with India or Taiwan. Land imbibed with situated histories pose particular spatialities of connections and ways to think about difference.

Bibliography

Balachandran, A. (2018) Petition town: Law, custom, and urban space in early colonial South India in iterations of law, in A. Balachandran, R. Pant and Raman B. (Eds.), *Legal Histories from India*. Oxford University Press: New Delhi, pp. 147–167.

Benda-Beckmann, F. and K. von Benda-Beckmann and M. G. Wiber (Eds.) (2006) *The Property of Properties, Introduction in the Changing Properties of Property*. Oxford: Berghahn Books.

Benjamin, S. (2005) 'Touts, pirates, and the ghosts,' in *Bare Act Sarai Reader*. Sarai–CSDS, New Delhi, pp. 242–254.

Benjamin, S. (2017) 'South Canara urbanism's diverse logics', in D. Zérah (Ed.), *Subaltern Urbanization in India: An Introduction to Ordinary Towns' Dynamics*. Springer: New Delhi, pp. 199–233.

Benjamin, S. (2019). Occupancy Urbanism. In A. M. Orum (Ed.), *The Wiley Blackwell Encyclopedia of Urban and Regional Studies* (Vol. 3, no page numbers). Malden; Oxford; Chichester, West Sussex: John Wiley & Sons Ltd. https://doi.org/10.1002/9781118568446.eurs0226

Bernstein, A. (2008) 'The social life of regulation in Taipei City Hall: The role of legality in the administrative bureaucracy'. *Law & Social Inquiry*, 33(4): 925–954.

Bhandar, B. (2016) 'Possession, occupation and registration: Recombinant ownership in the settler colony'. *Settler Colonial Studies*, 6(2): 119–132.

Chhabria, S. (2018) 'The aboriginal alibi: Governing dispossession in colonial Bombay'. *Comparative Studies in Society and History*, 60(4): 1096–1126.

Hann, C. M. (1998) 'The embeddedness of property', in C. M. Hann (Ed.), *Property Relations: Renewing the Anthropological Tradition*. Cambridge University Press: Cambridge, pp. 1–47.

Ho, P. (2001) 'Who owns China's land? Policies, property rights and deliberate institutional ambiguity'. *The China Quarterly*, 166: 394–421.

Kayaoglu, T. (2010) 'Westphalian Eurocentrism in international relations theory'. *International Studies Review*, 12(2): 193–217.

Li, T. M. (2014) 'What is land? Assembling a resource for global investment'. *Transaction NS*, 39: 589–602.

Lin, D. and P. Tsai (2014) Customary law and modern society: Land disputes of Jisigongye (祭祀公業, ancestral worship property) in Taiwan, in H. Fu'. and J. Gillespie (Eds.), *Resolving Land in East Asia: Exploring the Limits of Law*. Cambridge University Press, Cambridge, pp. 331–355.

Ling, L.H.M. (2014) *The Dao of World Politics: Towards a Pot-Westphalian, Worldist International Relations*. Routledge: Abingdon.

Lund, C. (1998) 'Struggles for land and political power on the politicization of land tenure and disputes in Niger'. *Journal of Legal Pluralism and Unofficial Law*, 30(40): 1–22. doi: 10.1080/07329113.1998.10756496

Ollman, B. (2015) 'Marxism and the philosophy of internal relations; or, How to replace the mysterious "paradox" with "contradictions" that can be studied and resolved'. *Capital & Class*, 39(1): 7–23.

Roy, A. (2009) 'The 21st-century metropolis: New geographies of theory'. *Regional Studies*, 43: 819–830.

Simone, A. (2015) 'What you see is not always what you know Struggles against re-containment and the capacities to remake urban life in Jakarta's majority world'. *South East Asia Research*, 23(2): 227–244.

Smart, A. and W.-S. Tang (2014) On the threshold of urban Hong Kong: liminal territoriality in New Kowloon, in A. C. Dawson, L. Zanotti and I. Vaccaro (Eds.), *Negotiating Territoriality: Spatial Dialogue between State and Tradition*. Routledge: Abingdon, 230–248.

Tang, W.-S. (2014) Where Lefebvre meets the East: Urbanization in Hong Kong, in L. Stanek, C. Schmid and Á. Moravánszky (Eds), *Urban Revolution Now. Henri Lefebvre in Social Research and Architecture*. Ashgate: Surrey, 71–91.

Tang, W.-S. (2017) 'Beyond gentrification: Hegemonic redevelopment in Hong Kong'. *International Journal of Urban and Regional Research*, 41: 487–499.

Tang, W.-S. (2019) 'Town-country relations in China: Back to basics'. *Eurasian Geography and Economics*, 60: 455–485.

Tang, W.-S. (2020) 'Land development in Hong Kong'. *Equilibra*, 24: 250–257. doi:10.1406/98116.

Tian, C. (2005) *Chinese Dialectics: From Yijing to Marxism*. Lexington: Lanham.

Valverde, M. (2008) 'The ethic of diversity: Local law and the negotiation of urban norms'. *Law & Social Inquiry*, 33(4): 895–923.

Webber, J. (2009) 'The grammar of customary law'. *McGill Law Journal*, 54(4): 579–626.

PART V

Contesting global urbanism

34

WOMEN ORGANISING, ADVOCACY AND INDIAN CITIES IN BETWEEN INFORMAL DWELLING AND INFORMAL ECONOMIES

An interview with SEWA's Renana Jhabvala

Renana Jhabvala is a community and social worker based in Ahmedabad, where she has carried out extensive work with women working in the informal economy, tackling issues concerning trade union organising and poverty. She is known for her work with the Self-Employed Women's Association (SEWA) and her writings focussed on informality, gender and power dynamics. She has won several awards for her activism, including the prestigious Padma Shri, awarded to her by the Government of India in 1990.

MICHELE LANCIONE: Both Colin and I were honoured to meet you at a British Academy and National Institute of Urban Affairs' event on 'Governing the Plural City', which took place in Delhi in December 2018. Inspired by your intervention, we would like to take this opportunity to hear more about how you have been engaging Indian cities, gender and politics in your work. Would you mind introducing yourself to our readers, perhaps saying something around your activism with SEWA?

RENANA JHABVALA: I have been associated with SEWA, full time, for over forty years. SEWA (Self Employed Women's Association) is primarily a trade union of women in the informal economy, with a total membership of around 1.7 million women, of whom just above one-third comes from urban areas. SEWA was founded in 1972, in the State of Gujarat, but today, SEWA is present in 14 states of India. Just after a few years from our foundation, we realised that just fighting for rights was not enough. And so we focussed on helping our members to set up what we call developmental organisations, which mainly take the form of cooperatives, such as a cooperative bank, and then production cooperatives, like weavers, farmers cooperatives and so on. As time went on, we realised that only being a cooperative may not be enough for our members, so we also helped them to form companies. So our approach combined cooperatives, companies and advocacy for rights and development together.

COLIN MCFARLANE: Could you just say a little bit about how you individually became part of the movement? What was it that drove you to become involved SEWA?

RENANA: I'm from Gujarat, but I grew up in Delhi, where I did my bachelor's degree in mathematics. Since I was doing well, I got a scholarship to go to Harvard, in the US, for a master.

DOI: 10.4324/9780429259593-39

When I was approaching the end of it, I decided to move into a more practical subject, so I applied to do a PhD in economics, and I got into Yale University with a good scholarship. But before starting the PhD, I came back to India, for a visit, where I was really struck by the poverty I was seeing in front of me. So I decided I had to do something. I took a year's leave of absence from Yale, and I asked my parents if they knew of any women's right-based organisations. And they found that SEWA, which was just a very small organisation at the time. I think there were only but five people there. So I came to meet Ela Bhatt, who was the founder of SEWA, and I said I'd like to work with them for a year. And she said, *Okay!* So, I worked for a year, and then I took a second year's leave of absence, and then I took a third year's leave of absence.... And then I never went back.

MICHELE: That is a wonderful journey! Could you tell us a little bit more around what did not make you go back to the USA and in particular around the challenges that your constituency face in Indian cities? Earlier you refer to matters of formality and informality, and we would also be interested in hearing more around those.

RENANA: I did not return to the US as I was fully involved with SEWA and found the work important and challenging. You may have heard of WIEGO, Women in Informal Employment Globalizing and Organizing, I am one of the founders. They'd done a study for us some years ago and found that eighty per cent of worker in urban India are what one would consider "informal". This doesn't necessarily mean all of them are poor, but we would say at least fifty per cent, sixty per cent of those are. In Indian cities, big industries have gone out of the urban core, leaving behind very small manufacturing, small workshops and home-based work – which are sectors with a huge degree of informality. Among these, the largest number of women workers are what we would call "home-based workers": women who use their home to make something for to sell out there, in the economy. Of these, there are two types. First, there are the "piece-rated workers", women who work for a supply chain. Sometimes those supply chains may be international, linked to really big brands, but more often they're either national or just local brands. The second group comprise women who are sort of self-employed, working their own little workshops and then sell their products in the market themselves. The distinction is not fixed: often a piece-rate worker can also become self-employed, and vice versa. But what does not change is that the largest number of women in urban area are doing some sort of home-based work. Official statistics are not capturing the full extent of home-based workers. And the same goes for other women working in the informal economy, including domestic workers, street vendors or women selling their labour and services, like carrying loads on their head and so on.

COLIN: I was going to ask, Renana, one of the questions that people sometime have in relation to India and domestic workers is that domestic workers sometimes live in informal settlements, but sometimes the latter does not necessarily take the form of what some would call a "slum". The housing conditions of informal workers vary greatly. Domestic workers often live in poor-quality housing near their place of work, but that housing is not technically "informal". It may be a formal housing addition, a subdivided house or something like that. Have you found that is the case?

RENANA: I prefer to use the notion of informal settlements since when you think of a "slum", you think of a dirty place, with drainage run coming out, disease and the like. But as you say, the picture is nuanced. Informal settlements take many different forms. In our extensive work around the topic, we found that informal settlements show different progression towards formality. So you have settlements which are so informal that they have no piece of paper that tells anybody that they have any right to that place. Then you have informal settlements which go a couple of steps better, and they've lived there for a little while and maybe they have an electricity bill or maybe they have a ration card which shows that they have lived there for a while. Then you go a little better and you find that places got authorised by the city government or the state government, but it's not really authorised because it's still agriculture land, so you can't sell it. This is because most of these documents are not freehold, but only leasehold agreements. This is fine when someone moves there, but if then that person want to move out, things get complicated because they can't technically sell land with a leasehold document. And so you have the phenomenon of illegal sales and a further complication of the "informality" picture.

MICHELE: From that point of view – that of extensive urban areas with complicated histories of entitlements and legal issues – what is for you the biggest challenge? On the basis of your experience and of your own organising, what do you think that would be, for these cities and their people, the biggest challenge they will have to face in the next fifteen years to come?

RENANA: Let me focus on three things. Number one is infrastructure: drinking water, drainage, sewerage, sanitation.... The problem is that cities do not have the capacity to provide enough of these basic infrastructures. Their first choice is always for toward the authorised, formalised, better-off settlements. Then the second challenge is solid waste management, because we are producing huge amounts of solid waste in Indian cities. I don't have the figures with me now, but it's just adding and adding and adding and then we have the landfills, and the landfills are filled up and becoming dangerous. I think sewerage and solid waste management, those are the two major, major challenges for Indian cities, which, of course, become much worse for informal settlements.

RENANA: The third challenge is around urban land. Cities are facing major challenges because the Indian laws concerning land are extremely complex, and the agencies managing those are overlapping, creating confusion. Moreover, our land records are very poor, so everything remains informal. Delhi is a particularly bad example at that. In Delhi, sixty percent to seventy percent of ownership is still informal, and these are not just the poor, but also real estate people who buy and sell land that might not be technically for building. Communities, even middle-class ones, come up in places that are not authorised for construction. This is a major challenge for the present and future of our cities. How can someone improve their housing conditions if they don't have formal titles? How can you get access to finance? These are issues that affect many urbanites in India, not only the poor.

COLIN: I think one of the things that we're really interested in is how the story and the challenges you just illustrated to us might or might not relate to "global urbanism".

RENANA: To me "global urbanism" is both about informal settlement and the informal economy. I want to go back to the latter for a moment. If you take WIEGO's figures again, the informal economy is very large everywhere, not only in India. It varies from fifty percent in South-East Asia to something like eighty percent in South Asia and many parts of Africa. Informal workers are not really recognised as part of the economy, as it is the case, for instance, with street vendors, who are always removed because they're "illegal", or home-based workers, who are not supposed to work in their home. In this context, the interesting part is that when city planners think about the city, when they plan, they rarely think of the economic in any extensive sense. They think about beauty; they think about residences; they think about how to make the city more liveable but not how to make the city more suitable to *every* economic level. So, they don't think about growth at all when they do city planning. They don't think about employment. They think about living as somehow detached from the live and felt realities of 'work'.

RENANA: In relation to the 'global urbanism' part, I can say only that in the informal economy, the conditions are really more or less the same everywhere, and the same goes for city planning. And so, therefore, the kind of problems that arise for the poor are the same. In a sense, I'm not going to be able to link it with global capital, but I feel that the way that both planning and development of cities is happening all of the world is the same. Everybody seems to aspire to the same things. In India, everybody wants to become Shanghai: and those models of developments don't take into account in any way the life that goes on into the informal economy or informal settlement.

MICHELE: Thank you for this very powerful line, Renana. Do you think that, despite the similarities across the processes making up urbanism globally, there is something else, specific to your context, that you wish to highlight? Perhaps something related to the ways in which you do your organising to respond against these global trends. You have written extensively about the way women can be empowered through membership-based organisations, so perhaps you could expand on the specific challenges of doing that in the context of an homogenising global urbanism?

RENANA: First of all, I really feel that the most effective way of being included and pushing back, against these forces of exclusion, is through organising and through making your voice heard, especially at the level of policymaking. Let me give you two examples. The first relates to the whole issue of informal settlements. Now there is some space for people living in those settlements to have their voice heard, and that is because of the democratically elected municipal bodies. The idea is that if we can get the informal settlement people to organise and form themselves into what we call a community-based organisation, we are able to influence the local municipal counsellor. So, if there's no drainage in their area, they can go to the municipal counsellor; they can do sit-ins outside her house. This enables the possibility of resistance.

RENANA: The other thing that has happened is related to the large-scale demolition that used to take place in India, without providing alternative housing. Many politicians – both at the municipal and state level – have lost their seat because they didn't do anything about large-scale demolition in their area. Now [thanks for the direct link between a specific area and a democratically elected municipal body] things are changing. Of course, the people who don't want informal settlements, pushed by the real estate developers and other powers, have their voice much stronger. But we can still push back, through community-based organisations. In some places, we have helped these communities to come together at the city level as federations to influence planning.

On the other hand, it's very difficult to organise informal economy workers because for most of them, there is no space of representation. This is because most of them don't have legal title to their dwelling land or their workspaces: their settlements are not recognised; their work isn't either. We have to find other ways. The most successful organising we've had, especially with street vendors, has been through the courts under the right to livelihood.

It was about fighting for a right to business and trade, which in turn helped them to get some recognition and some spaces. The global is important here. In the early days, we used to organise street vendors at the local levels. But then after India became more global, which was in the nineties, the voice of street vendors disappeared under the power of investment from big companies, which reduce the visibility of those perceived as small, local issues. So, in response to that, we helped to form a national alliance of street vendors to tackle the issue more systematically. The outcome was a national policy for street vendors in 2004 and a law protecting street vendors in 2014.

MICHELE: At this point, do you want perhaps to tell us a little bit more about what you are going to do next, based on your successes and maybe also on some failures?

RENANA: Maybe two things. The first relates to the big challenges brought forward by housing finance for the informal sector. On the one hand, if you have informal income, then banks will not give you a loan. On the other, even if you have formal document for work, but you live in an informal settlement [so you can't have legal rights to the land], you will not get a loan anyway. To tackle these, four years ago, we have set up a housing finance company. Our aim is to give loans to 'informal workers', as well as to change the law in relation to the kind of documents that one needs to show to access finance. That's one side of it. The other side is related to the ability of informal workers and settlers to have their voice heard. Playing at the global level is important in this respect. So, at the international level, we were one of the founders of the international network (WIEGO), which created a strong link between research, activism and advocacy.

COLIN: Perhaps as a final question, would you mind going back to the future of the cities you know most, Ahmedabad, Delhi.... What sort of cities are we likely to see, say, by 2030? And what is in your view the role of the kind of groups you're working with in the city twenty years from now, thirty years from now?

RENANA: With Delhi, things are extremely complicated because there are so many authorities. Ahmedabad is in a different position because the municipal corporation has a lot of power. I feel positive about integration of informal settlements into the city structures; we see this happening more and more. The informal settlements issue, I think, will get integrated in some ways. But it's the informal economy which is getting a lot of pushback, leading towards growing unemployment and a very, very high level of inequality of incomes in urban areas. This provides for a really scary mix. At the moment, India doesn't have that much urban crime like other parts of the world, and that's because people do have places to work and live. But if you take the informal economy away, without integrating it with something else, at a time in which people's aspirations have grown, this will really lead to very negative situations. I'm not an optimist about integration of the informal economy into the city.

35

FROM A NEAPOLITAN PERSPECTIVE, REACHING OUT BEYOND PREVAILING CULTURAL MODELS*

An interview Emma Ferulano

Emma is a community organiser, activist and grass-roots researcher based in Naples, in the Italian region of Campania. She is a co-founder of several associations operating at a grass-roots level in the didactic, intercultural and social fields in Scampia, one of Naples' most notorious neighbourhoods. Inspired by principles of self-organisation and independence from politics, regardless of party, over the last two decades, Emma's work has involved the establishment of important projects, such as the association of social promotion *chi rom e…chi no* (www.chiromechino.it), *Arrevuoto-Teatro e Pedagogia*[1] (www.arrevuoto.org); *La Kumpania*, a social enterprise offering intercultural gastronomic menus and employing both Roma and non-Roma women (www.lakumpania.it) and *Chikù Gastronomia Cultura Tempo libero* (www.chiku.it), based in Scampia and the first Italo/Roma kitchen and cultural space within the Italian territory. These projects apply years of experience to the task of fostering employability, which is crucial in the context of a chronically poor, peripheral and young neighbourhood. *Chikù* is a polyfunctional space: an intercultural kitchen that brings together discourses of economic self-sufficiency and dignified employment for a prevailingly female audience, organising cultural and social events open to the neighbourhood and to the city at large, as well as hands-on didactic workshops for children, adolescents and families.

MICHELE LANCIONE AND COLIN MCFARLANE: If you have to consider Southern European cities, and specifically Naples, from a "global" point of view, what comes to mind?

EMMA FERULANO: Naples is the largest city in southern Italy, a point of convergence for hundreds of thousands of people living in the province and region. It's the seat of important government institutions, as well as home to century-old universities and international scientific research centres and boasts a vast cultural heritage. Its strong sense of identity derives from its beauty and charm but also from traditional stereotyped and old-fashioned elements often reinforced by a certain "Neapolitan rhetoric" that pervades the everyday narrative. A perfect example of the latter is the inauguration of an international sports event called Universiadi 2019, where pizza, Mount Vesuvius, the siren Parthenope, Pulcinella, soccer and the song "Funiculì Funiculà" were all protagonists in what can be described as a global media circus.

* Special thanks to Lucy Lennon for the great and careful work of proofreading of the text.

DOI: 10.4324/9780429259593-40

In addition, Naples often suffers from being labelled as a "city of crime", where the presence of the Camorra and the level of danger involved were until not too long ago perceived as dangerously high. I'd like to take the opportunity to briefly focus on the extensive public debate that followed the launch in 2006 of Roberto Saviano's novel *Gomorra*, subsequently adapted to film in 2007 by director Matteo Garrone and then serialised in 2014 on SkyTV, all of which contributed to illustrate both the current role and operations of the Camorra not only to the general public in the south but in northern Italy and abroad, beyond the confines of a restricted academic audience. It was a great cultural operation that laid bare the objective relations between institutional structures and the private sector with the Camorra and its illicit acts – in keeping with a notion of world "globalisation of the mafia", to use an expression by the Sicilian writer Leonardo Sciascia. The debate that ensued from this cultural and mediatic operation split Naples into two positions: the first highlighted the importance of understanding the strong ties between legal and illegal power structures by reviewing the complex and nuanced spheres of influence held by the Camorra (and more generally the mafia) in the south and the consequences on their development – with a view to intervening on socio/cultural/economic levels, while the second was critical of the way in which a cinematographic and attractive depiction of the Camorra, its protagonists and criminal activities would reflect on Naples and its image. It also found the adoption of negative "idols" counter-educational, particularly for the younger generations in search of role models.

I believe the second attitude prevailed. As a result, today Naples is a city more concerned with offering a "clean" image of itself as a European city than attempting to resolve structural problems and dysfunctions resulting from a fatal combination between institutional shortcomings and weaknesses, political will and Camorra interests within the territory. Naples is full of contradictions, a southern city struggling to improve its quality of life but without satisfactory results, despite being the recipient, along with other southern cities, of special and substantial EU funding from 2000 to 2020. European parameters appear very distant from realities on the ground, with its official administration characterised by a high percentage of public-sector jobs in both central and local government, as well as an inordinate number of public officials in permanent positions but often lacking skills, goals and resources. This is coupled with a haphazard management of day-to-day issues and essential services (managing public spaces, acts of civil disobedience, street cleaning and garbage disposal) and haphazard policymaking. It's sad to observe the government's failure to implement the recommendations of the National Integration Strategy for RSC[2], as set out by the EU, in this case on a national and not just regional level. Unfortunately, public bodies in Naples and in other Southern European cities tend – or pretend – to act like their Northern European counterparts, affecting a modern, functional and law-abiding stance. But this is only a masquerade, a façade, which leaves the gates wide open to the very worst and unsolvable contradictions. The challenge facing these cities is to conjugate past and future, where positive elements of "innovation" are respectful of tradition, but where tradition can avoid the shortfalls of a stereotyping that feeds into phony tourist consumerism or worse still, where it justifies inaction.

ML AND CMF Can you give us some concrete examples of challenges and opportunities faced by the ones living in, with and through Naples's everyday life?

EMMA There are numerous examples illustrating how everyday life is far from "easy" for urban dwellers in Naples and in other Southern European cities. For instance, in Naples, public transport, which is one of the most extensive and advanced networks in Italy, suffers from bad administration and rivalling political interests; the garbage crisis is periodical and never-ending, a result of bad governance and an absence of civic public training; public health care is increasingly inaccessible thanks to budget cuts and extensive process of privatisation, despite an excellence in research. Public spaces are not easily or always accessible, even with thousands of people flocking to the streets in their leisure time, especially as life in the *"vicolo"*[3] in the city centre is a natural expansion of cramped houses and green spaces often closed for "security reasons". The city's outskirts lack adequate play areas for children, with cars ruling the streets and pavements, despite the fact that Naples had the youngest demographics for the entire country. Gender discrimination is also a big issue, with gender inequality a deep-seated and central feature for the greater part of southern Italian culture. There is also a palpable discrimination in local government policy between the urban centre and the hinterland, which reinforces a cultural discrimination ingrained in the population. Chronic unemployment, a parallel illegal workforce, a youth in crisis, emigration, exploitation, lack of adequate accessible housing, refusal to recognise rights to citizenship, all these issues are as stratified as the secular layers that characterise the city. The outcome is far from a modern scenario.

Naples is a city characterised by a coexisting level of social tensions and classes that barely connect, despite the cliché of a warm-hearted and united population, reflected in the adage "*simm e napule paisà*".[4] The working classes are class-bound within a socio-economic stratification, with few realistic possibilities of upward mobility and emancipation, often struggling for survival within a parallel system of an informal or illicit economy. For their part, the middle classes (including intellectuals and academics) are more generally trapped within a positive self-narration of *"napoletanità"*, far removed from the city's objective day-to-day issues and with strong prejudices against the poor and vulnerable components of the population. Attitudes that both condemn the poor for their poverty, at times with aggressive overtones, but can also be described as "do-gooder" paternalism, which sees the populace as an expression of the Neapolitan character. Both are discriminatory and illustrate a will (both conscious and unconscious) to maintain the status quo, impeding emancipation and a hope for a better life.

At the same time, Naples is also a vibrant, politically charged and multicultural city, with long-standing communities established in neighbourhoods – from Sri Lanka, Cape Verde, Eastern Europe, Africa, China, as well as Roma from ex-Yugoslavia, Romania and Bulgaria – that for decades now have been sharing territories, business, markets and social interaction. There are also sets of renowned and innovative antagonist movements cutting across the entire city, engaging in several areas (that include housing, public spaces, garbage disposal, citizenship rights, migrants, education, grass-roots cultural and artistic productions). These movements work from below to connect practices and theories, and despite facing major difficulties, in my view, they also express the best aspects of the city, with its obstinate determination to "awaken" people and trigger deep social change.

ML AND CMF We know about your impactful and inspiring work in Scampia. Can you give our readers a sense of the "global" challenges of Naples, starting from the specificity of Scampia and other interstices?

EMMA My work takes me both to Scampia, on the outskirts of Naples, and to the greater metropolitan area, in other words the vast northern province that includes Giugliano di Napoli, Caivano, Acerra, Afragola, Casoria – the so-called *terra dei fuochi* (fire-lands) and the province of Caserta with Casal di Principe, Mondragone and Castelvolturno (the set for *Gomorra*). These areas, where I spend the better part of my time, and the centre of Naples, where I live, provide me with a vantage point and dual perspective from which I will try to illustrate some aspects of the current situation, which can be described as in transition. The centre of Naples has been overwhelmed by a process of touristification and gentrification, a transformation of the city that caters to tourists and their money, with all its consequences. The city has become a kind of vast commercial centre, organised around buying and consuming (food is probably the main product), with higher rental costs, all this in a city that had traditionally been cheaper than central or northern Italy, resulting in real estate speculation and a lowering in the quality of food and other goods on offer. Inner-city, working-class areas have become less dangerous, and people can now move safely around neighbourhoods such as the Quartieri Spagnoli, Sanità or Forcella – historically very poor and somewhat seedy – seeking to get on the tourist bandwagon. But both tourists and the local population experience the same discomforts from a dysfunctional public service – emblematic is the case of the chronic malfunction of the Circumvesuviana train line, catering to thousands of commuters from the provinces to the city centre, as well as to thousands of tourists travelling from Naples to Pompei, Ercolano and Sorrento on the Amalfi Coast – areas listed as UNESCO World heritage sites – and the same goes for access to public spaces – for instance, it's almost impossible to find a public beach given the encroaching privatisation of these spaces. At times, the city's troubled soul with its unresolved tensions rears its ugly head and you find yourself witnessing a street murder in plain daylight or when adolescents give violent expression to their malaise at being excluded from the mainstream (the so-called baby gangs is one latest example of this, resulting in an increasing criminalisation of the working class's youngest members). It's a little ironic that these very aspects of "real life" allow you to live "authentic experiences" in Naples, probably the reason why hundreds of foreigners, mainly students and artists, flock to Naples to visit the city, then opt to stay or somehow include Naples in their lifestyle.

ML AND CMF Naples is also a space with considerable social and political experimentation "from below", so to speak, isn't it?

EMMA Yes! Another feature for which Naples is famous is with areas in the city where groups of people of every age have tried to utilise public spaces that have been abandoned over time by institutions, setting a trend that is almost unique for Italy. "FREE SPACES", self-organised communities in the centre of Naples have become a point of reference for thousands of people – including in other cities and abroad – that locate free and open spaces to occupy for cultural activities, with workshops, meetings, political discussions, places for people to meet and spend time together without needing to pay for the privilege, places that operate outside the commercial circuits. The current administration's merit is in having understood how the city as a whole can benefit from this kind of action and acknowledge the need for a juridical solution to legitimise the squats and protect them from eviction or repression. But are these solutions temporary? What happens if the administration changes? Suburban peripheries and provinces follow a different trajectory, with a different level of development, economic growth and systems of production. Away from the city centre and tourist hotspots, these are unfamiliar areas where on the one hand criminal systems can thrive and prosper while public administrations attempt to invest in infrastructures and construction. In Scampia, we can observe a slow improvement in housing, with new constructions replacing none too salubrious social housing, such as the notorious Vele complex, finally providing alternate accommodation to Neapolitan families who have spent decades waiting to be transferred, with all the social cost that that has entailed – poverty, insecurity, mental and physical illness and delinquency. Re-housing is an ongoing process[5] but in the meantime rates of unemployment, poverty, truancy and social exclusion are still too high. When we began working in this neighbourhood, 15 years ago, Scampia was the largest drug market in the world, with strong ties to a globalised Mafia and with an army of ordinary citizens living in poverty and isolation, lured into employment by the drugs market. The poorer classes and youth were the most affected by this situation. Today, following a major institutional crackdown with hundreds of people in prison – the percentage of inmates in Naples is very high and regards Italians, compared to the rest of Italy where often most inmates are foreigners – criminal organisations and the drug market have changed, but still survive. In the meantime, welfare and job policies are still inadequate to support families. On the positive side, school networks, efficient public health centres, alongside social and cultural intervention, are now guaranteed thanks in particular to figures – teachers, educators, school officials, doctors, social workers, associations, priests, volunteers, public managers, artists, local residents, citizens and activists – who have succeeded in establishing the largest informal network of active citizenship for the entire city.

ML AND CMF Do you think that the challenges you have just highlighted for Naples apply also to other Southern European, or Italian, cities?

EMMA Looking further afield, we could say that areas in southern Italy share many features and ostensibly the same fate, albeit for different reasons. From more urbanised and industrialised areas to rural and agricultural ones, the eternal conflict between growth and progress (see Pasolini) and the right to a dignified life (often denied or trampled on) still remains actual – even more so following the global pandemic – where the economic abyss between the north and south in the world becomes more and more visible, as is the need to change systems of production and social reproduction in ecological and human terms. Just think of the examples of contemporary slavery with the migrant fieldworkers from Puglia to Calabria, to the industrial areas – Taranto is emblematic – where the health of citizens and workers is constantly imperilled; think of the areas exploited and ravaged in the name of profit: fracking, major constructions, waste dumping.... In each one of these cases, which are merely examples, the only serious opposition that has emerged is from the same territories and inhabitants, as they painstakingly try to join forces and increase levels of awareness for reclaiming dignity and rights, for themselves and for future generations. The road to equality and social justice is still long and laborious, and in the absence of a serious and committed policy, it is all too easy to slip back to how it was.

ML AND CMF Can you tell us something about the work you do with Roma people and more generally about racism in the urban context you are familiar with?

EMMA The presence of Roma communities – mostly from ex- Yugoslavia and Romania – is another link shared by peripheries and cities in Southern Europe and a very complex crucial issue that illustrates both the difficulties hampering good housing policies and the limits of a political vision.

Within this context of major renewal for both the city and its outskirts, Roma people are not included, both for issues of citizenship and rights and as a result of political and cultural discrimination. This is a wider scenario that also embraces the rest of southern Italy, and Italy more generally, in a disquieting, regressive wave, legitimised by the recent declarations and directives of the ex-minister of the interior.[6] Also, in a "open" city such as Naples, the Roma question is a very hard political one, making it extremely complicated for any authentic dialogue leading to concrete results. After more than a decade where forward-thinking visions and practices concerning housing for Roma people seemed almost feasible, after numerous cycles of local, national and international roundtables backed by important centres and networks working against Roma discrimination, after multiple attempts to solicit EU intervention for changing local conditions, often with the active involvement of Roma people – something that is not always possible thanks also to a patronising attitude adopted by certain non-Roma NGOs and forms of institutional blackmail – after all these years and despite all these efforts, the results are overwhelmingly negative, with little or no positive outcomes, and with a high risk of evictions – whether forced or "voluntary" – of Roma communities, each with their individual trajectory and destination. Just prior to lockdown resulting from the COVID-19 pandemic, there was the formal acknowledgement by a department within the prime minister's office, called UNAR,[7] of the failure to apply National Strategy for RSC communities, proof of the inability by the department to adopt a strong political stance contrasting the clearly racist tones of our national government. The only work being done at a national level seems to be a self-defensive resistance to daily attacks at an institutional level, a sort of *"si salvi chi può"*.[8] Roma communities are moving elsewhere. Traditional activists are increasingly fatigued and left with few illusions, sporadically attempting to solicit public awareness and look for support from wider networks. But existing European networks are far away, while international observatories such as Amnesty International, UNHCR and OECD are very concerned by the Italian situation, particularly with regards to racism against the Roma community, but despite their many recommendations, they lack the power to act concretely towards changing the course of local and national politics. The one good news is the increasing number of young Roma people and activists, a good reason not to give up the fight.

ML AND CMF What does it mean to organise resistance, and how do you think that this relates to other methods adopted elsewhere?

EMMA Naples is a city where informal spaces of speech, reflections and actions are widespread. Due to the lack of a strong public administration and services, it's very frequent for citizens, groups and communities to intervene directly in public issues at stake and in political life, to improve social justice and quality of life. Inside the "movement", there is respect, solidarity and, probably, the will to "move" together, but it's not always easy to create unified city networks, to exit from a certain localism, and concretely it's not easy to create bridges between centre, periphery, province, etc. Over the years, symbolic, social and informal interventions, such as "Carnevale Sociale" or "Mediterraneo Antirazzista", were established as periodic appointments involving the whole city and its neighbourhoods, also attracting an increasing number of people from international networks. In fact, it's possible to talk about a "Naples model", organised around some keywords (public space, social justice, housing rights, welcome refugees city, open arms/open port…).

Personally, in keeping with my idea of resistance and "influence" on political life, I have tried in different ways to create connections between people, groups, spaces, cities and province, revolving around several issues and using different instruments, at times quite down to earth – Roma issues and field research, whether independent or commissioned; didactic and theatre workshops through *Arrevuoto* or other educational methods and measures; publicising "our model" of social enterprise with Roma and non-Roma women; also through very concrete and popular actions, such as caterings; over a decade's worth of participating in and contributing to the local paper "Napoli Monitor", especially on Roma issues; working on creating more connections with universities and the academic world in general. This has all been possible thanks to the support of the association "*chi rom e…chi no*", as well as a strong belief in my work and the pleasure I derive from it, creating projects with people whom I consider my travel companions, along with a trust and hope in the possibility of creating a better world.

Gender issues are something that we put into practice before even addressing these issues from a theoretical point of view because we are an association made up primarily of women, and since 2009, we have developed our activity with "La Kumpania" around a feminine universe, both Roma and non-Roma, "discovering" and fighting against the double and triple discriminations towards women, fostering a will to improve emancipation and, in some case, liberation. We adopt a soft approach, such as ethnic feminism, and in this way are able to include children, as well as the community as a whole, in our action and reflection, through dialogue and with practical examples of how to improve quality of life through work, relationships and by emerging from isolation and adopting a different life model. Of course, any initial dialogue is between us women, where we learn to know each other and to respect our different lifestyles and choices.

Personally, I have always perceived myself as a kind of "foreigner", caught between two worlds, probably the result of a very matriarchal family and avant-garde grandmother and mother, each with very independent lifestyles and no (material) dependence on men, open relations that in the context of southern Italy makes for quite a rare female role. It's not always easy, given the prejudice faced at times, as well as the difficulties and a certain solitude in my private life – even and my father are sensitive to and share my principles and practices – resulting from my complete dedication to public and political life and my valuing freedom from any kind of oppression. But I'm sure that in our narrative, the combination between strong characters and differences, at times in conflict, has been the right key for producing positive changes and improving journeys of emancipation paths.

In general, and particularly in suburban areas and those affected by major structural construction projects, it's interesting to find examples of resistance, ecological and grass-roots, scenarios in which an attempt is being made to create new communities starting from specific battles against environmental devastation affecting people's day-to-day life. And it's all the more interesting to fitness that often it's women who are at the helm of these battles for life, even if at times we're talking of embryonic role models at this stage, with minimal political awareness, but they are closely linked to the eco-feminist paradigm and enjoy a greater visibility in the southern part of the world, where they can radically influence more dominant growth models.

ML AND CMF Lastly, from the vantage point of your own practices in Naples, what do you think should be done and by whom in order to tackle the challenges facing Southern European cities?

EMMA Creating networks between cities and experiences; from around the world improving local community processes but keeping them open to external realities; improving connections between activism, visions and methodologies; giving more focus to and caring more for the younger generations, local populations, migrants and where necessary changing narratives; keeping up with the times. It could also be useful to pause and take stock, study and analyse models, successes and failures resulting from social and cultural interventions over the years in an attempt to construct a critical future, hopefully different from today's culturally dominant models.

With all probability, this is one of the keys to opening up southern Italian peripheries to the world, and set them under a different light, organising discussion groups with public institutions, though not always with positive or indeed any result, fostering a climate of coexistence and possible cooperation between people.

Training is paramount: self-training, drawing comparisons with success stories in other southern parts of the world, renewing didactic and methodological approaches for practising urban resistance, both environmental and social, as well as for movements of cooperative education, and in the relations between public space and citizens, with a focus on children (in this time of COVID-19 we directly participated in the creation of interesting informal movements and networks linking educators, parents, activists and citizens trying to affirm the rights of girls and boys to retake, live and rehabilitate whatever public spaces and nature can still be found within the city) and with a clear vision and direction as regards rights for women, migrants and youth.

Notes

1 *Arrevuoto* is a theatrical pedagogical project launched in 2005 in Scampia from a collaboration between Teatro Stabile of Naples Mercadante, Teatro delle Albe of Ravenna, Napoli Teatro Festival, the association "chi rom e…chi no", along with numerous artists, educators, musicians and actors. *Arrevuoto* proposes a theatrical-didactic workshop applying a methodology that creates connections between the city and its difficulties – in terms of social, economic, cultural opportunities – to overcome isolation and incommunicability. For the last 14 years, more than 150 Roma, migrants, Italian youth between the ages of 10 and 25 have been involved in a six-month-long path of theatrical workshops, supported and followed by a professional team, an activity that concludes with the performance of a collective show.

2 The so-called Regioni Obiettivo Convergenza, Campani, Calabria, Puglia, Basilicata, Sicilia, Molise, Sardegna

3 Very typical small street of Naples.

4 An extremely typical expression in Neapolitan dialect meaning "we are all from Naples, patriot!", in a proud and jingoistic attitude that affects unity and brotherhood.

5 For updates and additional information regarding recent demolition of one of the Vele, see GCR (2020).

6 On 15 July 2019, the Italian minister for the interior Matteo Salvini issued a directive to all Italian prefectures ordering signalling of all Roma settlements – especially informal ones – to identify and resolve all precarious situations in terms of hygiene, documents, etc. In essence, and it's clearly indicated in the directive, this authorises forced eviction of settlements without the need to provide alternate solutions, something which in recent years has become standard Italian practice, going against both EU recommendations and juridical condemnation.

7 Ufficio Nazionale Antidiscriminazioni Razziali, a national government agency for fighting discrimination against RSC communities.

8 "Every man for himself".

References

GCR (2020) 'Fall of Gomorrah': Demolition of Naples' infamous housing estate begins'. *Global Construction Review*. https://www.globalconstructionreview.com/news/fall-gomorrah-demolition-naples-infamous-housing-e/.

36

URBAN STRUGGLES AND THEORISING FROM EASTERN EUROPEAN CITIES

A collective interview with Ana Vilenica, Ioana Florea, Veda Popovici and Zsuzsi Pósfai

Ana, Ioana, Veda and Zsuzsi are scholars and activists based in Belgrade, Bucharest and Budapest. They kindly agreed to perform this interview collectively, working together to bring to the fore the powerful text that follows. Ana is active in urban movements in Serbia, as well as within the European Action Coalition for the Right to Housing and the City (EAC). She is also one of the editors of the *Radical Housing Journal* and the *Interface: A Journal for and about Social Movements*. Ioana is a researcher at the University of Gothenburg focusing on urban policies, urban power structures and related inequalities, as well as a housing activist based in Bucharest, Romania. Veda is an activist and engaged theorist based in Bucharest too, where she has been involved in resisting evictions, organising occupations and building community around collective space. Veda currently works as the facilitator of the EAC. Both Ioana and Veda are involved as militant researchers and organisers with The Common Front for Housing Rights and with the national housing justice confederation BLOCK for Housing, active in several cities and villages in Romania. Last but not least, Zsuzsi is a member of Rákóczi Collective, a group working on establishing rental-based housing cooperatives in Hungary. She is also a founding member of Periféria Policy and Research Center, an independent, critically engaged organisation working on issues of spatial justice and housing based in Budapest. Her work revolves around political economy, with an emphasis on understanding macro-scale dependencies, with a thematic focus on housing.

MICHELE LANCIONE AND COLIN MCFARLANE: If you have to think at Eastern European cities from a "global' point of view, what comes to mind? How can you position Eastern European cities within global urban geographies of exchanges, material and immaterial flows, and the likes?

ANA VILENCIA, IOANA FLOREA, VEDA POPOVIC AND ZSUZSI PÓSFAI: The historical position of Eastern Europe within global material and immaterial flows has been a semi-peripheral one, depending on and serving the Western core, while continuously trying to maintain its position as a less exploited territory than the Global South and the further East (Arrighi, 1994). Eastern Europe was largely rural at the beginning of the twentieth century and is still recording around 45% rural population today. Thus, historically, it has represented an extraction site for raw materials and agricultural products and, in the last three decades, for cheap

DOI: 10.4324/9780429259593-41

labour with "European quality"; this is considered by national and international decision makers and lobbyists as its main competitive advantage today.

The dependent relations towards the West came with a long series of disadvantageous commercial treaties (Ban, 2014), with accumulating cycles of extraction through financial relations (Raviv, 2008) and with a deepening debt relation both on the scale of the state and of individual households (Becker et al., 2015). With Western Europe struggling to tackle its own economic crises since the 1970s, the peripheries of Europe became an important "buffer zone" – mainly for investing surplus capital and for reducing costs of production by relocating certain production facilities to these countries. Thus we can broadly say that the West exported its economic crisis effects to the East – strong examples of which could be experienced in the years following the 2008 crisis (idem). In the meantime, generations of local elites have gazed towards the West with an aspirational ethos and with limited bargaining power when attempting to strengthen their (local/regional/global) positions. Since the nineteenth century, the dependency also came with a fast circulation of capital, technologies, political forms and discourses. This also deepened with the economic and political project of the European Union. Thus it is particularly interesting to understand dependent relations within Europe – which on a global scale is often seen as a "black box" of "advanced" economic development.

In parallel, the semi-peripheral dependency came with discourses of European belonging (Popovici and Pop, 2016) marked by what decolonial scholars termed "nested orientalism" (Bakić-Hayden, 1995): in dominant discourses, Eastern Europe has never been "actually Europe"; it always had to prove its Europeanness; it had to hide or reject its Oriental past as connected to the history of the Ottoman Empire or the Eastern Bloc; it had to prove its "return to Europe" after the 1945–1989 "deviation towards" or "kidnapping by" communist regimes (Florea, 2015). The main paradigm used in the last 30 years to describe the social/political phenomena in the region has been about "catching up" with the West, when in fact the uneven and dependent development of Southern and Eastern Europe has been a prerequisite for the economic advancement of core European countries (Becker et al., 2015).

In this broader process, cities have been crucial arenas of capitalist investment, and especially so during periods of financialisation, such as we are experiencing since the 1970s. Furthermore, the "catching up" or "transition" narrative always accompanied local bourgeois projects. Their rise ensured and was ensured by the growing affluence of their cities inside the region (Ban, 2014). But this growth was based on a competitive developmental model and always fueled by resources extracted from the hinterland, producing a nested hierarchy of cores and peripheries on different spatial scales (Hadjimichalis, 1983). Thus it meant a sharp hierarchy between cities, a deep inequality between capital cities and other urban localities, and an even deeper urban-rural divide. This internal polarisation is a required element of economic growth on the macro-scale (which can be seen as an apparent catching-up process) on the semi-peripheries (Hürtgen, 2015). As in other (semi-)peripheral regions of the world, growing Eastern European (EE) cities are dependent on extracting rural resources (from raw materials to cheap wage labour) and on attracting foreign investments, with subsidies, tax cuts and narratives of (catching up) Europeanness. In this developmental model, EE cities are assigned a role of performing Westerness, always proving their "progress", "development", "technological advancement", "democracy", etc. The most successful cities (usually the capital cities) are connected to the West and Western capital but always in asymmetric relations which facilitate resource extraction.

ML AND CMF: The world is increasingly urbanised, and it will be even more so in the next decade. This brings very specific opportunities and challenges according to local histories and current socio-economic, cultural and demographic dynamics. In this context, what is "global urbanism" as seen and experienced from Eastern Europe?

ANA, IOANA, VEDA AND ZSUZSI: We consider that the specific opportunities and challenges for EE cities in a globalised and urbanised world stem from the overlapping of their (semi-) peripheral position, their history of Western dependency, the bourgeois developmental projects to which they were associated, the 1945–1989 modernisation projects through which they were (re)built or abandoned, the post-1989 trade and credit agreements that the different EE states signed. We think there are varieties in their current situations, also according to (1) their relationship with the EU and their EU accession process, (2) their slightly different privatisation paths after 1989, (3) the post-1989 and post-crisis (uneven) urban development strategies at national and local levels. We think the varieties of Eastern European urban experiences are helpful in tackling the "global urban".

First, we should note that there are differences between cities in the EU internal and external peripheries: despite the fact that cheap labour, raw materials and financial extraction have similar mechanisms, in the internal peripheries (EU member states), cities were required to be (and were made) more predictable, open and stable. Here, massive Western investments came a bit earlier and were also part of these countries' process of EU accession – during which they were required to undergo liberalisation in various sectors of their economy. As a result, EE cities in EU member states have become important locations for attracting Western foreign direct investment and have also become arenas for credit-led financial extraction. Also, because of their EU membership, these EE cities could access structural development funds, which had an important impact on how urban restructuring has been done in the past years.

Secondly, as privatisation policies in the 1990s and austerity measures after 2008 were implemented under the pressure of international credit institutions and international corporate bodies (World Bank, International Monetary Fund, EU, international commercial banks, American Chamber of Commerce, the German Development Agency GIZ, etc.), they were quite similar across the region. Their social impact generally consists of the rise in unemployment and in-work poverty, wide-scale deindustrialisation and decay of industrial cities (Chelcea, 2008), strong work migration to the West, huge rise in housing costs, urban homelessness and housing precarity, the deregulation of real estate developments and household credits (Florea and Dumitriu, 2018) – processes which strongly mark the physical and social structure of our cities.

Specific to the region is the unsustainable structure of household debt, which was especially manifest in the years following the 2008 crisis, when many households went into bankruptcy. Beyond credit-based indebtedness, high levels of utility arrears are also prevalent in the region (Bródy and Pósfai, 2020). These forms of indebtedness can push people into migration – abroad, but also to urban outskirts, to smaller towns or even to rural areas where housing is more accessible. Also specific to the region is a high homeownership rate – around 90% of the housing units, which is higher than in any other region of Europe.[1] This specificity has its origin in the wide right-to-buy privatisation policies in the early 1990s and the property restitution policies[2] in the last decades – resulting in the sell-off of the vast majority of previously publicly owned housing – and in the strong policy support for private homeownership after the largest waves of privatisation. Indeed, support for homeownership was pushed as the main housing strategy of the EE countries, especially by the World Bank and EU, but it has also been a clear political priority of national governments – as a way to ensure the retrenchment of the social state and the creation and deregulation of the real estate and mortgage markets (Vincze, 2017). Similar to other semi-peripheries, global corporate landlords are preparing to expand in EE cities, feeding on household debt and dispossession (Mikuš, 2019). A particular form this takes are international debt collector companies expanding their presence in EE countries, buying both collateralised and non-collateralised debt, capitalising on defaulting household debt originating in the pre-crisis credit boom.

In these processes, middle-size cities in EE face special challenges. They illustrate our third point here about uneven urban development strategies at national and local levels, generating specific conditions and variations among cities. As in other semi-peripheries, middle-sized cities are marked by a permanent tension between growing and shrinking, testifying to the moving frontline between capital investment and household strategies for social reproduction (Gagyi and Vigvari, 2018; Pósfai, 2018). If these cities strive for economic prosperity, they find themselves in harsh competition to attract global capital by any means. The growing ones are directly dependent on larger international investments (such as manufacturing plants for German car companies), struggling to have special industrial zones, dedicated to niche industrial sectors. Other middle-size cities use cultural strategies to attract capital, framing "urban renewal" projects as their catching-up magic stories (Oancă, 2018). When the tourism industry came to European peripheries, it came with the "European Capital of Culture" festive imaginaries (Gazeta de artă politică, 2016), but with unstable job contracts, huge public resources put into public-private partnerships for private profit, Airbnb tax evasion, social and ethnic cleansing of the touristic urban areas.

ML AND CMF: Continuing on the challenges that Eastern European cities are facing, can you take one or two examples and unpack those for our international readership? We would like to hear about the way you conceptualise those issues from the ground of your own practice.

ANA, IOANA, VEDA AND ZSUZSI: Continuous challenges for Eastern European cities appear at the intersection of international, national and local scales: as global capital comes to extract resources from the region, national and local governments often respond with deregulations and anti-social policies, facilitating further housing financialisation.

Urban regeneration projects have been often serving as Trojan horses for these processes, resulting in the displacement of the impoverished. Such projects are usually facilitated by municipalities and the state, undertaken in partnership with foreign capital or through EU projects. Their optimist urban visions and seductive catching-up imaginaries promise growth, increase of employment, salvation from recession while facilitating the flow of public money into private hands, and processes of social cleansing.

One of the most indicative recent examples in the region has been the Belgrade Waterfront Project mega-development in Serbia. This project, being granted "national importance" status, has been planned through a public-private partnership (PPP) with the Eagle Hills company from the United Arab Emirates. Its status enabled the legislative "fast tracking", avoiding and violating planning and construction requirements. The PPP binding contract revealed that, for the Belgrade Waterfront Project, the Serbian government granted land rights to a newly formed company and agreed to invest a huge public budget in its preparatory work, including the displacement of the main railway and bus station. Before the initial phase of the construction work, more than 200 families were evicted, some of them remaining homeless and placed in temporary accommodation. These issues have not been left unanswered: a new urban movement, Don't let Belgrade d(r)own, arose in the city, opposing the project and the deregulations around it.

As we said before, housing has been an important battlefield in East European cities. In many of them, property restitutions to interwar owners and their heirs have been a key process in the capitalist restoration and development in ex 'state socialist' countries. Since the early 1990s, in Romania's larger cities, restitutions have been a major reason for mass evictions, as new owners raise rents to unaffordable levels, sell their properties (with tenants inside) to speculators or "cleanse" the poorer tenants from centrally located properties to boost real estate prices. Tens of thousands of families have been evicted by restitution profiteers only in Romania (Vişan and FCDL, 2019).[3] EE states, under their unequivocal pro-capital anti-social(ist) urban regimes (Vilenica, 2019), have failed to provide adequate public housing to those evicted, leaving hundreds of thousands in improper housing, informal housing and homelessness.

In the last years, household debt has been the main articulation of processes of housing financialisation in Eastern European countries, directly connected to the broader context of economic dependency described previously. Through this, households become sources for financial extraction on an international scale. For a small unpaid debt, households can be evicted without any alternative housing being provided, and with the only roof over their head sold in auctions to repay the debt. In Serbia, with the law on debt enforcement (Vilenica and Pantović, 2019) and the introduction of public-private debt enforcers that profit from default cases, evictions became an everyday reality. In Hungary, the turning point in debt-related evictions came in 2016, when the unfolding new wave of mortgage lending made it necessary to "clear banks' portfolios" – although evictions related to utility debt had been common even before. In parallel, the state has destroyed social infrastructure that could protect people from homelessness. Social housing has been residualised and turned into the leftovers of previous 'state socialist' housing systems (Pósfai and Jelinek, 2019; Vilenica, 2019). Potential users of social housing have been divided into vulnerable groups that compete for housing crumbs.

To further facilitate the process of financialisation and to further limit the power of what they portray as "needy citizens", EE governments have been progressively introducing legal measures in order to criminalise poor households and housing movements. For instance, in May 2019, Serbia introduced new amendments to the criminal law and the law on debt enforcement in order to criminalise the obstruction of bailiffs during eviction actions. Hungary was the first country in the world to criminalise homelessness through its Constitution in autumn 2018. Since 2011, Romania has changed its civil code, enabling and speeding up evictions.

In parallel, through the new "identity urbanism" based on cultural strategies for development, in the process of establishing and urbanising neoliberalism, anti-fascist and socialist histories have been manipulated and emptied out of their political context (Florea, 2015; Vilenica, 2019). New populist and nationalist narratives have been promoted as "cultural heritage", together with narratives of European belonging/becoming, in order to create continuities with/legitimations of the pre-war capitalist past and nationalist futures. The touristification process has been creating new types of extractive territories based on romanticising the peripheral Other. At the same time, touristification brought in global industries connected to digital technologies, such as Airbnb, Uber, video surveillance, all framed in catching-up modernisation imaginaries, all designed for extracting and consuming our cities.

ML AND CMF: Since you are all part of international networks or struggles, we would like to hear your reflections in terms of ways of thinking that arise from the space that you occupy and spaces you know of. So, what does it mean to organise resistance in the contemporary EE city, and how do you think that this relates to other modalities to be found elsewhere?

ANA, IOANA, VEDA AND ZSUZSI: Resistance movements in Eastern Europe found themselves after 1989 in the particular historical moment of "postsocialism", a paradigm dominated by neoliberal imperatives and producing specific adverse conditions. The post-Cold War victory of capitalist liberal ideology broadly meant the structural, material and symbolic attack on socialist, communist and generally left-oriented ideologies, socialities and political organising. Histories of resistance – to either state authoritarianism or capitalist values and materialities – that were anchored in socialist times were either erased or forgotten (Udvarhelyi, 2014; Gagyi, 2015). This normative narrative has been negating the rich history of resistance and has put EE political subjects in an infant position of needing to be educated on democratic values and resistance organising. Against this backdrop, fostering agency, asserting dignity and reclaiming one's own collective historical past become crucial for regional left social movements. Delegitimising the socialist and communist world – not only its state structures but also its social behaviors and mentalities, everyday life and popular values – has become the dominant ideology through which local elites ascend to power and become recognisable to Western transnational structures. Additionally, it enables policing and soft-to-hard repression of all broad-left movements that articulate class analysis and anti-capitalist visions.

In spite of such adversity, progressive radical left politics have always been present. For instance, in Bosnia and Herzegovina, Croatia, Bulgaria, Slovenia and Macedonia, occupations, strikes, protest and plenums have become more visible in the last decade and are the indisputable sign of new "radical politics" based on direct democracy that is seriously undermining neoliberal ideological consensus. Parallel to this however, one can notice within urban social movements the rise of a particular liberal direction appropriating left vocabularies of the commons and social justice. Often, this direction takes the form of urban liberal movements with a green thumb but with little to no class or anti-racist analysis (Vilenica, 2017). Aligned with middle-class interests and resting on the political subject of the concerned citizen (Florea et al., 2018), such agendas may oppose projects of urban development or renewal only as long as they threaten middle-class ideals and lifestyles: the reduction of green space, increase of traffic or the destruction of architectural heritage, petty corruption in allocating building permits. Such articulations mask the violence of evictions and displacement of communities; moreover, in their discourse, anti-capitalist sentiments are silenced or at best marginalised while anti-racist positions are dismissed to the category of identity politics.

Another condition to consider is that members of regional social movements anchored in the urban environment face everyday precarity. Mass privatisations and loss of social and labour rights are affecting social movements both within the horizon of their struggle and the everyday survival strategies of their constituencies. Financial precarity, loss of labour, social and health rights mark differences between EE social movements and their Western counterparts. Reconciling life and political work is, thus, a challenge with local particularities often translating in everyday life as great obstacles to developing political work. The processes of privatisation and liberalisation in the past decades has meant a loss of resources for building up resistance movements – both in terms of historical knowledge of social organising and of material resources. To these losses, the infrastructure of regional social movements is marked by NGO-isation and a dependency for funding from Western structures. This dynamic has deep consequences for the autonomy and continuity of our movements.

In spite of this, new radical grass-root movements and infrastructures of care have been emerging. In Serbia, for instance, the anti-eviction housing movement The Roof has been progressively growing in membership and public influence since May 2017. Its organising is based on direct actions and case-based work. The movement has a clear anti-capitalist politics, and its strong media presence has contributed to the change of discourse regarding primacy and inviolability of trustee-debtor relationships by putting the focus on a home as an inalienable right.

In Hungary, Serbia and Croatia, after the 2008 crisis, forex mortgage debtors have organised themselves to demand a re-evaluation of the legitimacy of the credit contracts that pushed them into bankruptcy, and achieved important political responses (Mikuš, 2019; Florea et al., 2018). A housing movement called The City Is for All has been organising in Hungary since 2009, first mobilising people living in homelessness (in shelters and in self-built shacks) and then pushing against evictions and mobilising social housing tenants. Their strategies range from direct support to individuals, to demonstrative actions and occupations. In the past decade, this group was important in putting housing on the political agenda in Hungary.

In addition, solidarity housing care infrastructure has been emerging in the midst of the migrant crisis at the borders between Serbia, Bosnia-Herzegovina, Macedonia, Hungary and Croatia. International activist groups have been working together with migrants in order to build infrastructure for domesticity in extreme conditions (Vilenica and Stojić, 2019). This housing "mobile commons" (Trimikliniotis et al., 2016) has been emerging in forests and occupied abandoned structures, such as factories, storage places and farmhouses.

In order to counteract the vulnerability of individual movements in the region, we think it is crucial to form international networks which give tangible support to groups working on the ground. Most of us are or were previously linked to the European Action Coalition for the Right to Housing and to the City – which is a European network aiming to avoid the reproduction of core-periphery inequalities within its organising structures. Furthermore, we also work to create specifically Eastern European networks, such as MOBA Housing Network, aiming to establish new housing cooperatives in the region.

ML AND CMF: Thinking about some of the things that you just enumerated and about the future of the EE city, what can cities do to tackle these issues in EE? What should be done and by whom?

ANA, IOANA, VEDA AND ZSUZSI: Aspirational Western models dominating the visions of regional liberal movements reproduce and enhance an archetype of the activist as a Western, urban, white figure. Strategies of resistance developed outside of such models, not only the ones based on left-oriented ideologies but also the ones coming from modes of living and resisting outside of conventional categories of "the citizen" must be brought to the fore in our movements. Such models are especially linked to racialised histories of citizenship in Eastern Europe, testified to by experiences of Roma communities (Kóczé et al., 2019).

While both research and organising have shown us that processes of housing injustice rest heavily on anti-Roma structural racism, building our movements with an anti-racist basis has revealed the diversity of the resistance tactics employed by Roma communities. Challenging categories of citizenship and property while critically addressing issues of access, assimilation and inclusion, resistance tactics used by Roma communities forge often more radical understandings of how evictions happen, of the history of the property regime or the racist politics of urban renewal (Lancione, 2017; Vincze et al., 2018). While strong anti-racist positions consolidate the radicality of both the analysis and the strategies employed, building solidarities within neighborhoods or with Roma rights organisations enables housing movements to contribute to a broader regional and global anti-fascist resistance.

Looking more closely at the particularities of the EE city, we see that it is in the midst of becoming more and more segregated. While continuous processes of gentrification, urban "renewal" and displacement are rapidly changing our cities, three decades ago, they retained lower levels of ethnic segregation compared to other regions. Whole neighborhoods and still some city centres contain an assemblage of social classes and backgrounds sometimes marked by tensions with each other. This experience, coupled with an anti-racist position, constitutes a resource for our movements towards developing radical class and anti-racist solidarity.

As our struggles proceed, we continuously create the material and discursive possibilities of emergence and development for our movements. For this, we often look for further resources in the fields of art and academia. Housing movements in Eastern Europe have developed innovative tactics by using theatre or video documentary to not only represent experiences but also deepen community cohesion and solidarity. Used frequently as agitation and pedagogical tools, such tactics are embedded within the strategies of movements (Vişan and FCDL, 2019). Moreover, the leftist international academic and art fields support the EE movements' need for autonomy by serving two functions: as a way through which movement members are able to reconcile wage work and political work, and as a financial redistribution channel.

In Romania, the fields of art and academia can be an arena for recuperating, excavating and resurrecting histories of organising and resisting, silenced by anti-communist ideology. This has become a crucial imperative for local movements and their need for autonomy, agency and epistemological dignity. Although long-time allies of social movements, both fields of art and academia can also be territories of co-optation or dissipation. This is why we constantly need to pose questions concerning issues of depoliticisation or compromise in the context of strategies anchored in artistic or academic methodologies.

Broadening the tactics of our movements also implies continuously developing our alliances and communication. The most challenging aspect of this consists of the ongoing debate regarding working with state institutions. In Romania and Serbia, the housing movements still try to push the state to act in the public interest, while in Hungary, this has been hopeless in the last decade. In articulating this, differences must be kept in mind in terms of municipal and governmental administrations, regime changes, national, regional and EU level differences. Nonetheless, housing movements in EE consider an antagonising position to be key in both challenging capitalist, racist complicities of the state and resisting co-optation and assimilation of grass-roots movements in the state apparatus. Alliances beyond state structures remain crucial in broadening and strengthening our movements. Some of the allies considered by housing groups in Romania, Hungary and Serbia consist of labour unions and autonomous agricultural movements. Tackling these intersectionalities opens up our analysis to a more in-depth look at class issues as they are related to the dismantlement, precarisation and cultural suppression of both the urban and rural (agricultural) working class. Key issues such as social housing, ghettoisation, social reproduction costs, critique of property and the further development of mutual aid networks and vocabularies of the commons can be more widely addressed in such new alliances.

ML AND CMF: Back to more conceptual issues. How do you conceive the future of global urbanism worldwide? What kind of alliances are needed, in which domains, by whom and why (if any)?

ANA, IOANA, VEDA AND ZSUZSI: One of the questions that we have been asking is how can we think of global regions in different timelines. Underlying such questions is the certainty that breaking from the linear narrative temporality of Western modernity – a temporality that places the West as the most advanced present – needs to be part of any EE struggle. Much of our concerns are shared across other non-Western geographies and epistemologies too (de Sousa Santos, 2017).

Instead of considering the Western city as the enlightened democratic future of the neoliberal utopia that we need to "catch up" with, we consider it a dystopian future threatening our livelihoods. Reframing such temporality empowers us to see "transition" not only as a neoliberal ideological intervention but a narrative that we can intervene into. Thus transition stops being an unchangeable sequence of events that has to necessarily lead to the absorption of our cities and socialities in the neoliberal conditions of Western urbanities. Peripheries are not "catching up" but advancing from their own historical experiences towards still unknown futures that are yet to come. To imagine a transition that would get us out of this paradigm of "catching up" with a dystopian future, we need to redefine Eastern Europe as a territory with histories of redistribution and resistance.

In the case of Yugoslavia, if we look back, we can find ideas and movements to learn from, such as the concepts of societal property, workers self-management and the Non-Aligned Movement of the pre-1989 regimes. These constructs and practices generated important urban politics and territorial-administrative social organising in a dynamic and contradictory process. Societal property was the concept and practice of "a property-without-proprietor", a novelty in the dark history of the ever-present world domination of private property ideology. The dynamic process of "urban self-management" as a form of local territorial and political organisation has been envisioned, but never fully put in practice (Rakita, 2014). On a broader scale, the non-aligned movement tried to build an alternative to the binary (East/West) division of the world and had a significant role in supporting the anti-colonial liberation struggles of countries in the Global South; however, Yugoslav leaders failed to recognise systemic international racism and their own privileged position in such a divided society (Subotić and Vučetić, 2019).

Today, the pressures coming from the neo-colonial dystopian West have triggered a more and more consistent call for action. This call is materialising as a resistance to new enclosures, dispossessions and anti-social measures, as well as a process to build new social institutions that will protect us from further extraction. Besides direct action, advocacy and protest movements, cooperative movements have been emerging at the local level. In order to make these structures stronger, there is a need to go beyond financial dependency on external structures and create our own financial mechanisms (interconnected at international scale). All of this builds on our need for an intervention into the temporal developmental narrative of the neoliberal neo-colonial transition: socialist histories of alternatives to capitalism are reclaimed so as to build continuity with postsocialist resistance movements. The movements we are part of have detached themselves from narratives of Western democratisation and development, and see the future as an international anti-capitalist struggle that fights the dystopian present of the Western city.

This contribution was made open access with the support of The Swedish Research Council FORMAS (Grant No. 2016-00258).

Notes

1 But also a very high overcrowding rate, a very low percentage of public and social housing (generally less than 5%), comparable to other (semi-)peripheries in the world.
2 In the late '40s and early '50s, the EE regimes nationalised multiple-flat houses, villas and properties of landlords with more than three to four housing units, redistributing them to the large strata in need of housing. This process happened especially in cities, where housing needs were growing and where housing was concentrated in bigger property portfolios due to previous bourgeois developmental projects. After 1989, in some Central and Eastern European (CEE) countries, such as Romania, East Germany, Poland, ex-Yugoslav countries, the pre-nationalisation owners and their heirs reclaimed these properties – a process of (re)privatisation through restitution/ retrocession.
3 In countries where housing was not restituted to previous owners, but privatised to sitting tenants in the early 1990s (which was the case of Hungary, for instance), the main reasons for displacements and evictions are different.

References

Arrighi, G. (1994) *The Long Twentieth Century: Money, Power, and the Origins of Our Times.* London and New York: Verso.

Bakić-Hayden, M. (1995) 'Nesting Orientalisms: The case of former Yugoslavia'. *Slavic Review*, 54(4): 917–931.

Ban, C. (2014) *Dependență și dezvoltare. Economia politică a capitalismului românesc.* Cluj-Napoca: Tact Publishing.

Becker, J., Jager, J. and Weissenbacher, R. (2015) Uneven and dependent development in Europe – the crisis and its implications. In Jager, J, Springler, E. (Eds.), *Assymmetric Crisis in Europe and Possible Futures.* London and New York: Routledge.

Bródy, L. and Pósfai, Z. (2020) Household debt on the peripheries of Europe: New constellations since 2008. Periféria Working Papers, nr.3.

Chelcea, L. (2008) 'Bucureştiul postindustrial. Memorie, dezindustrializare şi regenerare urbană'. Iaşi: Polirom Publishing.

Florea, I. (2015) The ups and downs of a symbolic city: The architectural protection movement in Bucharest. In K. Jacobsson (Ed.), *Urban grassroots movements in Central and Eastern Europe.* Farnham and Burlington: Ashgate/Routledge.

Florea, I. and Dumitriu, M. (2018) 'Transformations of housing provision in Romania: Organizations of subtle violence'. *LeftEast*, https://www.criticatac.ro/lefteast/transformations-of-housing-provision-in-romania-organizations-of-subtle-violence/ (online only, accessed on 15 January 2021).

Florea I., Gagyi, A. and Jacobsson, K. (2018) 'A field of contention: Evidence from housing struggles in Bucharest and Budapest. *Voluntas*, 29: 712–724.

Gagyi, A. (2015) 'Why don't East European movements address inequalities the Way Western European movements do? A review essay on the availability of movement-relevant research'. *Interface: A Journal for and About Social Movements*, 7(2): 15–26.

Gagyi A. and Vigvari, A. (2018) 'Informal practices in housing financialisation: The transformation of an allotment garden in Hungary'. *Critical Housing Analysis*, 5(2): 46–55.

Gazeta de artă politică (2016) 'Capitale Europene ale Culturii'. *Contextul politicilor economice și culturale ale UE*, issue 15.

Hadjimichalis, C. (1983) Regional crisis: The state and regional social movements in Southern Europe. In: Ostrom, K., Seers, D. (Eds.), *The Crises of the European Regions.* London and Basingstoke: Palgrave Macmillan, 127–147.

Hürtgen, S. (2015) Das Konzept der strukturellen Heterogenität und die Analyse fragmentierter Wachstumsgesellschaften in Europa. Working Paper der DFG-KollegforscherInnengruppe Postwachstumsgesellschaften, Nr. 2/2015, Jena.

Kóczé, A., Zentai, V., Jovanović, J., Vincze, E. (Eds.) (2019) *The Romani Women's Movement : Struggles and Debates in Central and Eastern Europe.* London and New York: Routledge.

Lancione, M. (2017) 'Revitalising the uncanny: Challenging inertia in the struggle against forced evictions'. *Environment and Planning D: Society and Space*, 35(6).

Mikuš, M. (2019) 'Contesting household debt in Croatia: the double movement of financialization and the fetishism of money in Eastern European peripheries'. *Dialectical Anthropology*, 43(3): 295–315.

Oancă, A. (2018) *Bidding Wars: Enactments of Expertise and Emotional Labour in the Spanish Competition for the ECoC 2016 Title.* Budapest: Central European University.

Popovici, V. and Pop, O. (2016) 'From over here, in the periphery: A decolonial method for Romanian cultural and political discourses'. *LeftEast*, https://www.criticatac.ro/lefteast/from-over-here-in-the-periphery-a-decolonial-method-for-romania/ (online only, accessed on 15 January 2021).

Popovici, V. (2020) 'Residences, restitutions and resistance: A radical housing movement's understanding of post-socialist property redistribution'. *City*, 24(1–2): 97–111.

Pósfai, Z. (2018) Reproducing uneven development on the Hungarian housing market. PhD dissertation, University of Szeged. http://doktori.bibl.u-szeged.hu/9753/1/p%C3%B3sfai%20zs_phd%20final_2018.04.05..pdf.

Pósfai, Z. and Jelinek, C. (2019) Reproducing socio-spatial unevenness through the institutional logic of dual housing policies in Hungary. In T. Lang and F. Görmar (Eds.), *Regional and Local Development in Times of Polarisation.* Singapore: Palgrave.

Rakita, M. (2014) Teritorijalno-politički oblici organizacije sistema lokalne samouprave i prostorno-fizički aspekti društvenog razvoja u samoupravnom socijalizmu. In M. Stojčić and D. Olujić (Eds.), *Nasleđe jugoslovenskog sociajalizma.* Belgrade: Forum za primenjenu istoriju.

Raviv, O. (2008) 'Chasing the dragon east: Exploring the frontiers ofWestern European finance'. *Contemporary Politics*, 14(3): 297–314.

de Sousa Santos, B. (2017) A new vision of Europe: Learning from the south. In: G. K. Bhambra and J. Narayan (Eds.), European Cosmopolitanism. Colonial Histories and Postcolonial Societies. London and New York: Routledge. 172–184.

Subotić, J. and Vučetić, S. (2019) 'Performing solidarity: Whiteness and status-seeking in the non-aligned world'. *Journal of International Relations and Development*, 22: 722–743.

Trimikliniotis, N., Parsanoglou, N. and Tsianos, V. (2016) 'Mobile commons and/in precarious spaces: Mapping migrant struggles and social resistance. *Critical Sociology*, 42(7–8): 1035–1049.

Udvarhelyi, É. T. (2014) *(In)justice on the Streets (Answers to Hungary's Housing Crisis).* Budapest: Napvilág Kiadó.

Vilenica, A. (2017) 'Contemporary housing activism in Serbia: Provisional mapping. *Interface: A Journal for and About Social Movements*, 9(1): 424–447.

Vilenica, A. (2019) 'Contradictions and antagonisms in (anti-) social(ist) housing in Serbia'. *ACME: An International Journal for Critical Geographies*, 18(6): 1261–1282.

Vilenica, A. and Stojić, M. (2019) 'Enforcing and disrupting circular movement in an EU borderscape: Housingscaping in Serbia'. *Citizenship Studies*, 23(6): 540–558.

Vilenica, A. and Pantović, N. (2019) 'On the frontlines of Serbia's struggle for housing justice'. *Roar Magazine*, https://roarmag.org/essays/serbia-housing-justice-struggle/ (online only, accessed on 15 January 2021).

Vincze, E. (2017) 'The ideology of economic liberalism and the politics of housing in Romania'. *Studia Ubb Europaea*, LXII(3): 29–54.

Vincze, E., Petrovici, N., Ra ţ, C., Picker, G. (Eds.) (2018). *Racialized Labour in Romania: Spaces of Marginality at the Periphery of Global Capitalism.* Cham: Palgrave Macmillan.

Vişan, N., FCDL (2019) *Jurnalul din Vulturilor 50. Povestea Unei lupte Pentru Dreptate Locativă.* Cluj and Bucharest: IDEA and Hecate.

Voicu, S. (2017) *In the Middle of the Night. Anti-Communism, Anti-Corruption and National Pride. EuroNomade*, http://www.euronomade.info/?p=8805 (online only, accessed on 15 January 2021).

37

PLANNING, COMMUNITY SPACES AND YOUTH URBAN FUTURES

From Accra, in conversation with Victoria Okoye and Yussif Larry Aminu

Yussif Larry Aminu is the founder of Spread-Out Initiative (SOI), a non-governmental organisation (NGO) set up to support children and youth from diverse backgrounds in Nima, a neighbourhood in Accra, to realize their potential and attain their dreams of a better future. He is also a Pan-African and a philanthropist who believes in preserving cultures and people's histories. Victoria Ogoegbunam Okoye is a Nigerian-American doctoral researcher based at the University of Sheffield who worked in Accra for several years on community and youth projects. Victoria and Larry connected in Accra and are doing collaborative research in Nima. They are interested in how to bring people's art, culture, sociality, and history into design rather than imposing western philosophy or ideology.[1]

The interview examines the intersections of spatial planning, community living spaces and the challenges and opportunities of young urbanites in and beyond Accra, and what all of that might mean for how we understand – and change – global urbanism. Global urbanism in this interview emerges partly in the form of exclusive and capital-focussed urban development projects, but it also emerges as a politics of culture and identity in which multi-cultural and multi-generational neighbourhoods are both the ground for more inclusive global urbanisms and the challenge for urban planning now and into an urbanising future.

MICHELE LANCIONE AND COLIN MCFARLANE: If you have to think at the ways in which African cities, and in particular Accra, are "made global", what comes into mind? Can you give us an example of the lived and felt ways through which global flows (investments, ideas, projects, and so on) intersect with the everyday life of people in Accra?

YUSSIF LARRY AMINU: Our leaders are bringing these so-called developmental projects, and I see it as a form of colonialism. The moment there's gentrification going on in the city trying to destroy some buildings and bring some foreign or western buildings into our community, I see it as directly replacing our culture. Gentrification is a violation of people and our community culture. You can see the new mosque that they [Turkish government] built at Kawukudi. Although the mosque is beautiful, it's going to fail. Me, who believes in my history and culture, I don't see any beauty in it because it doesn't fit in my culture – it's Turkey's

DOI: 10.4324/9780429259593-42

culture, and they give it to us. I prefer the old mosque. They can even modernize our old mosque; that's better than taking an architectural plan from somewhere else and replacing it.

I see gentrification as violation of culture and identity and even religion of the indigenous people who are staying in the community. I prefer you show my child or my daughter the building that I was born in; the building that my mom, my grandparents were born in; the mosque they prayed in. It's better than showing these western buildings. I see it as a form of colonialism. It's better for our leaders, those in development, to train our indigenous architects to go by what we have; it's better than trying to infuse or bring this western planning into our cities. I don't believe in that. So, as an African, as a Pan-African, I'm telling you, Africa has everything. We have everything: if you look at the history of Egypt, Africans have Egypt. So why are we trying to copy something foreign? Why don't we stick to what we have?

VICTORIA OKOYE: Larry makes an important point: these international companies, in partnership with the national and city government, are asserting their own designs on the city. This produces gentrification that not only dispossesses urban residents; it also means loss of culture.

There is also another way that we've discussed Accra as a "global" city. As part of our collaborative research together in Nima, we interviewed Nat Nuno Amarteifio, who is former Accra mayor (1994–1998), as well as an architect and urban historian. He reminded us that Accra has always been shaped by global encounters, including multiple different groups of Africans coming from so many different locations, captured Africans taken involuntarily to the Americas and Europe, some of whom have also returned as Diasporans. Amarteifio pushed us to see Accra, from its very beginnings, as a site of in-migration and migration. This enables us to think differently about the contemporary city and groups, such as Nima residents, who are framed as migrants. He spoke of the Ga people, who migrated to Accra in the thirteenth or fourteenth century from somewhere around Benin City in Nigeria. There is the Alata, who were procured by the Europeans from Allada, a regional slave market of the Dahomey Kingdom in present-day Benin, and brought to Accra to build the James Fort. They intermarried with and became Ga. We know about the Tabon people, repatriated freed persons who left Brazil to return back to the continent, who also became part of Ga society.

Accra is also a site from which many Africans were taken as part of the slave trade, sent to the Americas and the Caribbean. He told us about the Yoruba women traders who were the original market vendors for Makola Market, the largest public market in Accra. We also can't forget the Muslims who came from Zaria, Sokoto, and Lagos, the Hausa people, the Fula people, Northerners, all of these people who came and settled in Accra and established communities that have become neighbourhoods: Nima, Madina, New Town, for example. Accra has always been, from its very beginnings, a very global city, and not just in terms of encounters with Europeans, which often dominates our imaginations and narratives of globality. Accra is an African city in a very wide sense, where migrations have directly taken from and contributed to the city as we know it. The city's genealogies emerge from so many places and move forward to so many places. We are learning, and we are reminding ourselves that Accra has always been a global space, always been a place where there have been exchanges, flows, not just of goods like gold, ivory, palm oil, but also of people.

ML-CMF: The world is increasingly urbanised, and it will be even more so in the next decade. This brings very specific opportunities and challenges according to local histories and current socio-economic, cultural, and demographic dynamics. Since you work extensively with the youth, what are the biggest *present* challenges for urban development and young people in Accra and, more generally, in urban Africa?

VICTORIA: Both in our research collaboration in Nima and in other opportunities where I have had the chance to work with youth to document and reflect creatively on their experiences of their neighbourhood spaces, what is revealed for me is the immense imagination and vision that young people have. From their positions as youth, they have come to understand power dynamics, access, and opportunity in their urban environments, and they have intimate knowledge of urban challenges. They also have wonderfully critical ideas about how to intervene in space and how to make improvements.

Yet the opportunities for their visions to be included in the making of the city – the recognition of their deep insights into their own communities – is missing from city planning, design, policy, and management approaches. Fundamentally for me, this is the challenge: Among those living in the city, who has the opportunity to contribute to and to inform the city vision? The opportunity to contribute and work with city authorities is so dramatically narrow, and exclusive, and reinforced by an emphasis on technical knowledge, best practices, and assumptions about what constitutes an appropriate and "modern" future. In a context like Accra with a large youth population, this creates a situation where those who will inherit the city are further and further distanced from the opportunity to shape the very spaces they live in. Not only that, but this is the mode of engagement/disengagement that youth are learning about how to make a city and about who the city is for (and not for).

LARRY: In my experience, I take Nima and Jamestown, Fadama and Ashongman as familiar places that I know. Youth have a serious problem of not having spaces because authorities are focussing on how to build structures, car parks, and other things. But these authorities always forget to include youth recreational centres in their planning, and they also forget about learning centres, like libraries, for the youth.

So, if you look at the Airport City [high-end commercial business complex], it's now developed; it's a city within a city. You see tall buildings over there, but it only serves the purpose of the workers in that building. Not even a space for relaxation – if you're tired from work, and you want to relax; you can't get it. This is the problem we face in Accra. The urban town planners now, they focus mainly on how to build a street and structures and highways. So, you will see a beautiful house, but it doesn't even have a car park. In Ashongman Estates where I'm staying now, you see very nice houses like in the commercials, but you don't even see a car park. They have to demolish space for a car park. So, these are the challenges we are facing in Accra now. You can go to a nice space now, but there is no space for hospitals, for cemeteries, no spaces for car parks. These are the changes they forget. There is no space for recreational centres for the kids. They don't put all these things in their plan – they only think about the beautiful structures, how to make tall buildings, but there are the things they forget to include.

In my experience working with kids, this is what I want to advocate: anytime town planners are developing for the community, they have to include recreation centres, learning centres, places for adults to relax, to read, to rest for a while and continue. All these things are missing now in urban town planning.

VICTORIA: Larry's framing of the lack of space for youth in Nima (and other neighbourhoods) is very illustrative of the general challenge for the city of Accra, and other cities like Lagos, with respect to their youth populations: There is lack of social, educational, especially non-consumptive spaces for youth, where they can exist and know that they belong in the city. In many instances in these cities, the spaces for youth are the leftover spaces or the spaces yet to be taken over by urban "development": the open fields, sidewalk pavements and alleys, the space under overpasses, semi-constructed buildings. There is a challenge in both the lack of physical spaces for youth activities, as well as the lack of space for youth to participate in the planning of the cities that they will inherit.

ML-CMF: What you are saying about this lack of "space", of a certain kind of community-based space, it's insightful. It signals a certain kind of urban logic at play. Can you unpack this problem further, from the ground of the neighbourhood you work from, Nima?

LARRY: As our research taught us, the lack of space is the main problem in Nima and communities like Nima. Planners didn't include all these spaces for the community, and now people navigate and occupy spaces. Some even use their power to occupy a space, whether political power or chieftaincy power [influence through traditional authorities]. The problem is the space; we don't have enough space. No places for markets, for children's recreation centres, libraries, and these kinds of things. The little space we have for children to play around or organise community activities, the vendors also occupy the space because the market is also too small for them. So, the little space we have in the community, people occupy the space for selling, building, to sleep. These are the challenges that I see in Nima that I think we can deal with, and we need to find positive solutions. Pedestrians' lives are in danger; they have to walk in the street, but there is no space – some people park their cars there, motorcyclists are riding roughly in the street, all of these are contributing to serious problems and dangers. People just use power to occupy space, and it poses a big challenge.

VICTORIA: You are identifying all of these different issues that exist in Nima and impact the urban spaces. I think these are challenges that we see, not just as in Nima but also places like Nima. We know other African cities that are facing very similar challenges.

LARRY: Like Lagos.

VICTORIA: Many academics and government authorities are continually saying the problem is a lack of planning, lack of planning, but you have also mentioned that there is a local way that people operate in Nima and that any planning should respect the culture, the religion, etc. My fear is that with the vision the "modern" government has for the city, if government comes and plans in Nima, this would mean losing important elements of culture.

LARRY: I agree.

ML-CMF: It seems as if you are talking about the need for a more dialogical way of planning and managing urban spaces in Nima, and beyond. What would be the role of local communities in that?

VICTORIA: I believe that we should not be surprised when people don't follow the government plan, because the plan is also not in line with people's needs. I believe the design should be informed by how people actually move, live, socialize, work. I mean, why did someone locate the borla [refuse] container in that spot? Was there some sort of group agreement and negotiation when it comes to locating the container? Maybe somebody had even been using that space for something before – what now? So, I imagine group discussions, negotiations, and agreements. You actually bring people together to have conversation, maybe even arguments, but you build agreement. That's the education. If everybody agrees to what works, then maybe it is more likely that people will respect it. Then if people don't respect, other people can even call them out. I agree it's not only about the planning, but it's about how do you bring the community in – not just in terms of the plan but the enacting of the plan because the community residents are also planners, right? What we're seeing now is everyone operating according to their own individual plans because they have to. This is the way I'm seeing it, based on what you're saying: How to move from government planning to more community planning. Not just a plan by the government, but a shared plan by the people.

LARRY: That's what I'm talking about, the way we did [our collaborative] research. The research has brought us a lot of things we didn't know. If the planners involve communities like we did in the research, if you involve the community, small number of the community who are also influential, they can inform the larger public, and the community also sees it as their property. They see somebody from outside is coming to write about our community, and work with the kids, and the kids also wrote their stories. These are the things that are the challenges. If the government or town planners involve the community, they will see that Nima people cherish mosques more than the rooms they play in because of their religious motivation. The same thing for the churches – people respect their church more than their house, so you can't just come and spoil their church. So, town planners must involve the community in their planning. If they can include the culture, religion, community, I think that's best. If you want to build a 100-story building, then you can let the community tell you what they want. Then if you do it that way, you won't have a problem. If you come and do what you want, that's where there will be argument and problems. If you involve the community and you want a borla container to be at the gutter or Nima bridge, they will tell you where they want it to be. If you follow what they are telling you, you won't have a problem. If you don't follow, then you will have a problem; they will also adjust to get what they want. It comes down to education – there must be conversation, negotiation, even argument.

ML-CMF: There are some big challenges to be tackled at the intersection of community's ideas and needs, foreign-fetched planning models, and investments.

LARRY: Yes. The challenge the urban planners are now facing with the communities is to include the culture and religion and beliefs of that community, so it will balance the development and the people in the community. If you look at Ghana example, private companies bring their own plans, government takes their time to look at the plan and to see the benefit of the plan for the community, then government will just take the money and let that company do what they want. Our leaders are self-centred; they aren't human-centred; they don't care what is happening in our community. If a foreigner comes and wants space, they will just go and sell our space to them. The development is ok, but it also affects the community. If government can also look at the types of plans and development projects that they do to also involve the community, it will be nicer, rather than focussing on money. For now, they take money, then later on it brings a problem. Like the mosques at Kanda Highway I mentioned before; it looks very nice, but because we weren't part of the decision making, it worries us. It's development all right, but the development does not serve the purposes of the community.

ML-CMF: You are known for your work with Spread-Out Initiative (SOI), an educational and training programme for youth living in Nima, one of the poorest areas of Accra. Why have you decided to work there? What do you aim to achieve and what have you already achieved?

LARRY: So, one of the reasons why I decided to operate in Nima is that I was born and raised in Nima. When I grew up, I saw a lot of opportunities that I missed when I was young, so I don't want these kids to miss these opportunities. This is why I created SOI, to be a platform where each and every kid and even youth and adults can learn, interact, excel, share their knowledge with the world. Nobody can change the community if we residents will not stand on our own feet to start initiatives like SOI. If not, then forget it, the community will not progress. There is a need for we who were born and raised in the community to start visionary initiatives that can bring people from outside – researchers, investors – to support the community.

VICTORIA: Larry, you talk about opportunities that you missed when you were growing up that you don't want these kids to miss. What were these opportunities?

LARRY: One example is now we are in the information age; the whole world is vying for information. So, if you are well informed, you are ok – everyone wants to work with you because you are current; you are well informed. When I was young, I didn't have this opportunity. Things like doing art, I didn't have it. Things like interacting with volunteers from across the world, I didn't have these kinds of opportunities. That's why SOI is important so that these kids can come and interact with different kids in their own community, to learn different cultures, different languages around them. I didn't have these opportunities, but these are things that as a community, I don't want them to miss. SOI works with more than fifty students in Nima, and these are good students; they can pick each other as friends; they can learn from each other; they learn about each other's tribes, their religions.

When I was young, Nima was for Muslims; we didn't know there are Christians, churches, more than fifty-something languages in our community. We didn't know tribes across Africa are resident in Nima. So, when I was doing social interventions, we learned, interacted, shared, and I learned we have everything. What I was searching for somewhere else, I learned we have it in our own communities. This is the opportunity. Because they have the opportunity to do for ourselves. The community has everything it takes to do for themselves, when they are aware of the culture, religion, and opportunities around them. They can know all these things with a platform like SOI. When we are together, we interact with each other; we begin to know each other – our strengths, weaknesses, and qualities. This is the one reason why I created SOI.

The second reason we created SOI is to bring Blacks in the Diaspora to come and work with their brothers and sisters in Africa. At least we need to have direct contact with our brothers and sisters in the Diaspora to learn and share opportunities, so both communities benefit from each other. This is another reason why we created SOI, and we accept Diasporan volunteers to teach our kids, and then they also learn from our kids. Here in Nima, it's like you're being with a family to relearn what they denied you over there. These are the things we are also creating here.

Also, the last thing is about the researchers and the community. It's part of our strong mission to make sure researchers work along with the community. The researcher will come in as an expert and the community will also serve as an information system so that in the end, the community will learn from the researcher, and the researcher also learns from the community. In the end, we all share the findings so the community will see what's happening in their community through the researcher. You remember, we did something like this. The researcher will not come to the community, do something, and just go. The researcher and the community have to work together.

M L-CMF: On the basis of the experience you just narrated, Larry, or your research experience, Victoria, how do you navigate the nexus of urban activism, NGO work, local governance, state government and supra-national politics? How is, in other words, your work affected by its broader political context, and are you able to affect the latter in any way?

LARRY: Yeah, it's difficult. I don't believe in politics, and I don't believe in politicians; that's my stand. With my work as an NGO, I am very, very careful. I make the community aware of the way I operate, I make it clear that my NGO is non-governmental, nonpolitical, and nonreligious. These things [political and religious affiliations] are not part of our activities, and they will never be part of our activities. But it's difficult as an NGO to operate in Nima because politics has taken over the minds of eighty percent of inhabitants here. If you are doing a special intervention and they [the community] know you from a political party before, then it will be difficult for you to navigate. But because they don't know me – I am not affiliating with any political party – whenever I have the opportunity to talk, it gives me power to navigate the community, and I'm not afraid to approach anyone. I'm not involved with any politics or political party. I'm not influencing any political party. I'm not going in favour or against any party.

We want social justice to prevail, so we don't care which religious group or political party that our activities will affect. We are representing the whole of society. My work is to engage the whole society, the whole community, not a political party or religious group. Some NGOs, I don't want to speak for them, but they are affiliated with a political party, so when their party is in power, they operate, but when their party is not in power, they don't operate. This is one of the challenges they face. I am ready to work with government institutions that are independent to achieve my purpose, but politics mixing isn't part of my work. This is the strategy I used to be able to run SOI up to this level.

VICTORIA: As part of our collaborative research with SOI, Larry and I interviewed a city planning authority to understand from a local planning perspective how authorities perceive Nima and their plans for the neighbourhood's future. It was a rich interview, and the official expressed a strong interest in our research project. At the end of our activities, we invited him to the office, and the SOI students we'd been working with presented their materials to him – their photographs, their maps, their stories, their insights into the neighbourhood's challenges and possible social, infrastructural, and other material interventions. The planning official said that he was wowed by their ideas, that in many instances, they were identifying solutions that the planning authority was itself trying to figure out. So, I think this was a small, but hopefully important, moment that can lead to further opportunities in which the city authority engages these young people as experts in their own neighbourhood. I think this also shows the ways in which it's important that our research goes beyond an intellectual curiosity to working to establish new relationships and creating spaces where change, no matter how small, might become possible.

ML-CMF: How should we conceive the *future* of cities in Africa, with particular reference to their youngest inhabitants?

VICTORIA: Young people are the largest demographic in African cities. In the next two or three decades, it will be these young people who are designing, planning, and otherwise defining African cities. The challenge then becomes today, what opportunities are we providing for these young people? In terms of learning about and taking pride in their own cultures, learning how to think and work critically? I think here is where we see the critical role of

Spread-Out Initiative and other youth-focused NGOs, in training, educating, and equipping young people to become leaders in their own community so that we can have better cities tomorrow. We have already inherited histories and countries in which we must work at urban and national and international levels across differences; these differences can bring us together or drive us apart.

This isn't a particularly African challenge, but African actors in cities must determine, hopefully, African solutions. I think in the same way that these young people will need to work together – across their ages, genders, ethnic identities, religious affiliations, socio-economic classes, languages, etc. – SOI is creating this space and encouraging this way of working and learning, through art, through technical training. I think it's particularly significant that SOI operates outside of the formal educational institution and curriculum. SOI and similar NGOs are teaching young people ways of coming together and thinking, working, and imagining together, critically and creatively. It's now through my engagement with SOI that I am imagining this as a radically possible future for African cities and the essential role of our next generation in that future.

LARRY: The future of Accra for young people is bright; it's positive. What's happening now, it's their time, not our time – our time has passed. What's happening now is their time; they're going to be familiar with developmental projects now. Accra has everything that it takes in terms of development, technology, Internet, everything. In Accra, the future is bright, compared to other African cities.

Cities are developing very quickly. If you look at the population, it's always growing, becoming large and larger. The problem too is our cities are choked – people don't have places to sleep. My worry is in this rapid urban development, town planners shouldn't forget to include the culture, the religion of the community, these are the things that if the urban planners include in their planning, it's best – you cannot develop a community and at the same time erase their culture and beliefs. If the urban planners can consider all these things in their design of urban cities, it's better. The developers should not forget the children and their recreational activities; they need space to learn; these are the social priorities, and they should include those into urban planning. Social priorities should be part of the plan – those things the youth need. This is the future I see. For example, the way we brought in Baba [city planning authority] to listen to the ideas of the young ones so he can also implement it in their planning. Now they know the young ones have something to contribute.

Note

1 We want to thank Victoria for her invaluable help in bringing this interview to fruition.

38

A COUNTER-DOMINANT GLOBAL URBANISM?

Experiments from Lebanon

Mona Harb

As I write this chapter, in addition to the global pandemic, Lebanon is undergoing multiple economic and financial crises that will very likely terminate its long-praised, political-economic model of consociationalism. The weeks following October 17, 2019, when thousands took to the streets, starting off what many called a "revolution" seem quite far away. At the time, a large and diverse public loudly mobilized and organized against the oligarchic rule of warlords and business elites dominating political-economic life since the end of the civil war in 1990. How does Lebanon's uprisings matter? This small country on the East Mediterranean, squeezed between occupied Palestine and Syria, is a remarkable laboratory where urban tropes and paradigms are swiftly debunked as its notorious disorder and chaos expose the futile boundaries of dominant binaries (e.g., state/society, formal/informal, structure/agency). Reflections on the city in the global age and on global urbanism are thus enriched by exploring the dynamics of urbanism and its contention in Lebanese cities. This chapter seeks to show how uprisings in Lebanon are substantively informed by the work of urban activists and scholars who have been nurturing for decades a much-needed political imaginary that actively contributes to sustaining mobilization and struggle against hegemonic modes of spatial production.

Numerous protests have marked the political history of Lebanon, and the postwar period has been ripe with mobilizations, several of which were anti-sectarian (Kingston, 2013; Clark and Zahar, 2015). Yet the ongoing uprisings of October 17 have been drastically different.[1] First and foremost, they are decentralized and not merely concentrated in Beirut as they were during the #YouStink mobilization of 2015 (Aburish, 2015; Abi-Yaghi et al., 2017). This time around, protestors have been active in major and intermediary cities throughout the country. Second, the uprisings cut across sect, class, age, gender, and sexuality, with a significant role for women and LGBTQ—especially in Beirut. Third, the uprisings managed to symbolically shake political sectarianism and its leaders. They forced the resignation of a government, the postponement of a parliamentary session, in addition to engaging in numerous forms of direct action, including striking, the blocking of highways and main streets—paralyzing work and public life numerous times, and the occupation of downtown Beirut's and Tripoli's main squares. Protestors also publicly shamed the ruling elite, through chants, slogans, and art, and defamed politicians who were patronizing restaurants and hotels, or attending concerts, chanting against their corruption, forcing them to leave these places. Fourth, and

DOI: 10.4324/9780429259593-43

more interestingly for this chapter, the uprisings are profoundly urban in both dimensions identified by Nicholls and Uistermark (2017). On one hand, their density and concentration in key cities enable networking and access to relational opportunities, which support diversity and possibilities for emotional interactions, yielding affect and passion, as well as supplying symbolic meanings to contention, thus allowing their sustenance over time. On the other hand, the uprisings' actors are strategically selecting and utilizing a range of urban spaces in their actions, from squares to streets, to roundabouts, highways, and specific sites, changing their locations and routes, depending on actors, issues, frames, stakes, and times. Fifth, albeit still fuzzy, the uprisings incorporate disparate (and desperate) yearnings for an effective "state" (*dawleh* in Arabic) that would secure rights and be accountable, that would protect the public, and that would provide all its dwellers with dignified livelihoods.

Many of these yearnings incorporate an urban component: they reject the ways by which cities and localities have been urbanized and managed, and urban rights squandered.[2] Protestors' demands include rights to affordable housing, public transportation, and basic services, as well as free access to public open spaces and the coast. They also denounce, directly and indirectly, how the dominant value of land has become a mere commodity to be exchanged. Their yearnings and claims stem from their awareness that alternatives to the dominant, extractive mode of urbanism do exist. Indeed, urbanists across Lebanon have been imagining and elaborating these alternatives in their studies, projects, and other initiatives as I discuss later in this chapter. Many of them are among the protestors and have been disseminating for a while their visions for another type of urbanism, one that is inclusive and ecological and where the core value of land is grounded in the shared commons.

Lebanon's uprisings have much in common with other contestations sweeping cities across the globe, from Santiago to Khartoum, to Hong Kong: they decry systems of political and urban rule that are unequal, extractive, and predatory and make loud demands for social and environmental justice. In this chapter, taking Lebanon as a laboratory, I show how protests incorporate urban initiatives and experiments with political imaginaries that include modalities for a counter-dominant global urbanism, one which challenges the growth-led global urbanism that structures many cities and localities across the world. Indeed, the study of Lebanon's uprisings over the past decade informs well how urban movements intersect dynamically with political mobilization, and how they can contribute to shape the much-needed political imaginaries and "real utopias" (Wright, 2010) of alternative urban futures.

The chapter starts with a brief discussion of how utopian horizons can productively contribute to the consolidation of oppositional politics. I then present the experiences of urban activists in Lebanon who have been insistently crafting inclusive and ecological modalities of city-making and urbanism, enabling the deployment of political imaginaries that have informed advocacy and protests over the past decade. The chapter closes with reflections on how urban activists and academics are experimenting with a varied set of tools—ranging from visual representation and mapping, to legal and archival research, to intellectual tools of inquiry, to planning and design tools that are able to delay, halt, and/or challenge the growth-led dominant model of urbanism and its associated systemic inequalities. Through these tools, their experiments forecast how the political horizon of social and environmental justice can shape the future production of space and thus sketch the modalities of an alternative kind of city-making: a counter-dominant global urbanism crafted within the interstices of activism and academia.

Utopias and imaginaries

In a context of shrinking horizons for oppositional politics, Hage (2015) asks what alternative politics could herald the possibility of social change, of another political and economic order, against the permanent state of impasse that we seem to be stuck in? Wright provides interesting cues as he advocates for "a vital belief in a utopian ideal [which] may be necessary to motivate people to leave the journey of the status quo" (2009: 4). He argues for the necessity of "'real utopias,' utopian ideals grounded in the real potentials of humanity, utopian destinations that have accessible waystations, utopian designs of institutions that can inform our practical tasks of navigating a world of imperfect conditions for social change" (2009: 4). More concretely, Espiner (2018) conceptualizes an open-ended form of utopia grounded in feminist thought and combined with strands drawn from anarchism and radical environmentalism. Like Wright, she notes how the dominant discourse that dismisses utopianism as irrelevant (given there is no alternative to capitalism) is misleading. For her, the real question is whose vision of utopia is dominant: currently, utopia is fueled by capitalist norms and discourses, by inequality, patriarchy, settler-colonialism, and anthropocentricism. Espiner calls for an alternative utopianism, led by an ethics of feminism, inclusion, ecology, and equity. She reminds us that utopianism is already present in our everyday realities and "pertains to any feelings of hope, and thinking of collectiveness and communalism". Utopias exist in and through micro interventions and actions of progressive activists, feminists, anarchists, radical environmentalists, in addition to the less visible tinkering of people within bureaucracies, state agencies, private enterprises, and homes. In other words, utopias are already out there, but they are often fleeting and imperfect and difficult to maintain.

The discussion of utopias echoes well with the pleas made for expanding the political imaginaries of another kind of city-making, aptly expressed by Kaika and Swyngedouw's who call upon "extend[ing] the intellectual imaginary [towards] new politicized avenues for producing a new common urbanity" (2014). Indeed, as Lefebvre (1974) argued almost half a century ago, political change cannot be conceived of without transforming processes of spatial production. Such attempts at transformations towards other kinds of urbanities and city-making are already being experimented with, similarly to the utopias that already exist all around us, albeit fleetingly and in fragments: Pieterse and Simone call them "radical experiments" and underscore how they can shape "a tactical politics of possibility" (2017: 124, 159). I also locate such attempts and experiments within the works of urban activists in Lebanon to which I now turn.

Lebanon's yearnings for another urbanism

Despite (or because of) the hegemonic apparatus of political sectarianism, mobilization and advocacy in Lebanon have been active across many spheres, including the urban, since the early 1990s. One of the key elements that enabled the consolidation of urban activism was the establishment of several graduate programs in urban studies and planning in public and private universities (Verdeil, 2019). In several, the curricula included readings in critical urban studies and case studies from the Global South that advanced ideas and practices of social and environmental justice, as well as inclusive urban governance and livelihoods. At the American University of Beirut (AUB) where I have been teaching since 2002, my colleagues and I were committed to teaching these values, and training students to become critical scholars and/or reflective practitioners. Four of us (Mona Fawaz, Ahmad Gharbieh, Howayda Al-Harithy, and I) were also actively engaged in trying to influence urban policies and, as of 2006, established several initiatives seeking to make the dominant practices of growth-led urbanism more inclusive and integrated. We engaged in matters related to postwar reconstruction (Al-Harithy, 2010), urban security (Fawaz et al., 2011), public space (Fawaz et al., 2015), urban and municipal governance (Harb, 2010; Harb and Atallah, 2015), urban social movements (Harb, 2019), and displacement (Fawaz et al., 2018), among others. Several of our programs' students graduated and actively engaged in urban advocacy. One helped with the establishment of the AUB's Civic Centre for Engagement and Community Service, which he now directs. Another was instrumental in leading the community-based action research that led to the reconstruction plan for the Nahr al-Bared Palestinian refugee camp (razed to the ground by the Lebanese Army in 2007) and is now coordinating a remarkable urban initiative that seeks to protect the social value of lands in Saida (*Lil Madina*, For the City). Two others have been developing noteworthy urban activism methodologies that are now crystallized in an urban action platform named "Public Works". Several joined political movements and feminist collectives.

Among other urban advocacy efforts, I discuss here four initiatives that were quite successful.[3] First, the "Stop the Highway" campaign succeeded in 2014 in preventing a highway from being built through Mar Mikhail, a neighborhood rich with urban heritage, greenery, and socio-spatial practices. Activists surveyed and mapped the area, conducted archival and legal research that showed the highway site had been expropriated decades ago, and thus could be used as a neighborhood park instead of an infrastructural route. They consulted a transportation engineer who conducted a traffic study that clearly revealed how the planned highway would not diminish traffic, as municipal agents claimed. Armed with this informative dossier, they lobbied the municipality to conduct an Environmental Impact Assessment study, which confirmed the project's negative impacts and ultimately did not give the mayor much choice but to halt it. The second story concerns the reopening of the city's largest public park, Horch Beirut, in 2015—a park that the municipality had kept closed off for decades using an array of unconvincing reasons ranging from the incivility of people to security threats. The Nahnoo nongovernmental organization had been lobbying the mayor for many years, to no avail, until the newly appointed city governor decided he wanted to score points against the mayor and opened the park. Today, Beirut's dwellers are able to enjoy Horch Beirut, which is still managed by Nahnoo volunteers. The third story is that of the "Campaign for the protection Dalieh" which managed to prevent in 2015 the construction of a private resort on the beautiful rocky coastal area of Dalieh el-Raouche, that is freely used by hundreds of people during weekends and holidays. Activists organized a series of protests, media campaigns, a legal inquiry (that demonstrated how the site was historically a commons and was privatized under shady circumstances), and an international design competition that provided alternative ecological interventions for the site.[4] Dalieh is still not protected, but activists garnered enough media and public attention to make the site owners (who are the family members of the assassinated billionaire prime minister Rafic Hariri) cautiously weigh any development decision, as it will raise major contestations. The fourth story is that of Save the Bisri Valley—an ongoing campaign that is fighting the building of a dam funded by the World Bank. The campaign was able to pressure the Bank to suspend its funding for the project in September 2020, after a concerted set of strategies ranging from campaigning to lobbying and coalition building to mobilizing and direct action.

All four activist initiatives were not only successful in preventing projects threatening the public domain but also in advancing other imaginaries of that public domain: imaginaries of vibrant, diverse, and inclusive built and natural environments. These imaginaries were envisioned and materialized in the discourses, representations, maps, and drawings of the campaigns, perhaps most notably in the speculative visions of Dalieh's design ideas competition, which outlined scenic perspectives of a protected seacoast celebrating nature and culture.

A key milestone in urban activism efforts is embodied in the municipal campaign of Beirut Madinati (Beirut, My City, or BM) in 2016, to which Mona Fawaz and I actively participated, in addition to a number of AUB's urban graduates as well as colleagues from other urban studies and planning programs. Although the campaign did not lead to an electoral win, it was experienced as a victory by many:[5] it scored 32,000 votes, some 30% of the total number of voters. If the elections had followed a proportional representation law, BM candidates would have won seven seats on the municipal council. But the electoral law in Lebanon is one of gerrymandering and does not allow city residents to vote for their representatives.[6] Yet BM's results demonstrate how the accumulation of instances of mobilizations throughout the postwar decades, even if fragmented, discontinuous, and transient, have fostered learning and "interstitial openings" that enabled "seeds to grow into potentially disruptive mobilizations" (Nicholls and Uistermark, 2017). While limited in time and space, urban activism in Beirut, epitomized by Beirut Madinati, embarrassed, held accountable, and constrained hegemonic urbanism and perhaps most importantly allowed the imagining of an alternate urban politics and of utopianism.[7] Beirut Madinati's urban planners and designers reimagined the city, drafted a detailed urban agenda for the municipality, and visualized how the city's neighborhoods could transform, if its candidates won the elections. They produced three-dimensional urban sketches showcasing before/after sceneries of Beirut's imagined future public spaces, streets, markets, and coastline—underscoring a plurality of other kinds of socio-spatial practices, premised on the social value of land, a shared commons, inclusive governance, affordable housing, liveability, and political ecology, as well as on play and sensations. Similarly to Dalieh's competition's representations, these sketches provided a tangible and concrete imaginary that enabled people to envision other urban realities and futures—"real utopias".

A counter-dominant global urbanism?

As Lebanon illustrates well, the dominant model of growth-led global urbanism is not sustainable. Other modalities of city-making are imperative and need to be incorporated in the struggles and movements seeking political and economic change. These modalities need not be grand plans and projects, and they probably have little potential of being productive if they are. Instead, as small, disconnected experiments and initiatives, they can connect and coalesce at key political junctures and thus have more impact, as the experience of Beirut Madinati showed. Such experiments are required to rehearse other social lives, other political ecologies, even other economies, and an alternate city life, grounded in solidarities, and in the commons, outside the reign of dominant global urbanism. They allow the imagining of "real utopias" and thus are prefigurative of a counter-dominant global urbanism, grounded in the political horizon of social and environmental justice, which urban activists advocate as their ultimate cause. Extensive collective work is needed to challenge hegemonic processes of spatial production and sustain urban experiments in order to continue nurturing "real utopias"—even more so in times of multiple crises. While local councils, mutual aid initiatives, unionizing, and other forms of political organizing structure are its typical constituents, the academy can also have an important contribution in anchoring and advancing such praxis.

Two years ago, in 2018, my three AUB colleagues and I established the Beirut Urban Lab at the university: we conceived it as a collaborative research space to explore urbanism and city-making from our position in the Global South, using intellectual tools of inquiry, archives, and legal research, as well as mapping and visualizations.[8] The Lab provides us with a platform through which we bring together our own efforts and we engage with the like-minded urban scholars, practitioners, and activists I presented earlier. Collectively, we seek to experiment with urban tools that can narrate and represent utopias in ways that enable the production of counter-dominant political and urban imaginaries. Inspired by other academics and activists around the world who have been investing in such efforts to try to advance social and environmental justice in cities and localities,[9] we see ourselves as part and parcel of a local, regional, and global ecosystem of political and urban change that has been diligently thinking of and experimenting with other ways of city-making. We thus seek to further connect our localized struggles to other international ones and are invested in exploring means to amplify our quest, physically and virtually, through debates, conferences, workshops, exhibitions, and substantive collaborations.

This ecosystem of urban advocacy efforts and initiatives seeks to challenge the modalities of spatial production and of city-making that serves the pro-growth interests of the few ruling elites. It tinkers with a range of ingredients that help us rethink the city and its urbanism in the global urban age of systemic inequalities and ecological disasters. In times of crises, experiments with spatial production and city-making adapt to respond to urgent needs, such as securing access to housing and food, often leading to the emergence of solidarity initiatives that feed into the crafting of political imaginaries.[10] During times of protests and uprisings, ingredients get materialized in the built environment in more-or-less violent and transient ways, through the appropriation of vacant lots and buildings, the occupation of dilapidated structures, the squatting and reclaiming of private property, the graffiti on security walls and other fences. Indeed, moments of crises and uprisings allow "real utopias" to be experienced and enacted in squares, streets, markets, and other sites. They provide opportunities for protestors and urban activists to share their narratives and representations, their critical mapping and visualizations through which some sites are reenvisioned and reimagined as places for prefigurative rehearsals and transactions. If we do not want to "end up confused, rudderless, and cynical" (Kelley, 2002: xii) in this global urban age, urban activists inside and outside the academy need to continue experimenting with meaningful modalities of city-making that provide the urban ingredients and political horizons that consolidate praxis, direct action and collective organizing, and maintain nurturing imaginaries and utopias, against all odds.

Notes

1 After a long pause due to COVID-19, the uprisings in many Lebanese cities resumed in June 2020, though with less numerous crowds and more acute police repression.

2 Such yearnings are saliently present in protestors' chants and various activists' social media posts, especially those of feminists.' A Facebook post of a leading queer Lebanese feminist (dated 20 November 2019) summarizes well this quest for a common urbanity: "The revolution showed us… what it means to love, enjoy, and to be happy in our city. We saw the difference in the form and feeling of the city before 17 October and after. From the beginning of the revolution, people felt their power and ownership of the city; life, light and movement (and recycling!) returned to the city center, and street vendors selling corn returned to the [seaside] corniche, and art shows are taking place in front of Grand Theatre, art and graffiti are back on the walls and on the floor, fishermen are back to the Saint-George's bay, and I now like walking leisurely to downtown and spend time there, without consuming. I even love the holes in the asphalt caused by burned tires, even if I have allergies, but I feel am reclaiming the streets through them. The revolution gave us erased and forbidden parts of our memories, our movement, our feelings, and of our city. The revolution gave us back ourselves and our cities".

3 This discussion is based on interviews I conducted during November 2015 in Beirut, on participant observation within these groups and informal conversations with their activists, in addition to a desk review of their social media campaigns, their websites and reports, which I have been conducting since 2013. For more, see Harb (2016).

4 For more information about the Dalieh campaign, check Saksouk-Sasso (2015).

5 For more information about BM, check www.beirutmadinati.com, as well as its Facebook, Twitter, and Instagram accounts. To get a sense of the scale of BM's campaign, note that its crowdfunding efforts gathered close to $200,000 in less than three weeks. BM's municipal list included 12 men and 12 women, none of whom came from the political establishment—a first in the political history of elections in Lebanon. It faced the list sponsored by Saad Hariri, the son of late prime minister Rafic Hariri, led by Jamal Itani, the director of Solidere—the real estate company that epitomized the pro-growth, neoliberal reconstruction of Beirut—and grouped all representatives of sectarian political parties.

6 For a good overview of the electoral law in Lebanon and its gerrymandering features that enable political sectarianism to reproduce, check Salloukh (2006).
7 As a former BM member, I am well aware of the movement's many limitations and do not seek to romanticize its experience here. For a critical review of the campaign, see Cambanis (2017).
8 More details about the AUB Beirut Urban Lab are available on its website: www.beiruturbanlab.com.
9 Examples include 596 Acres, www.596acres.org, which "champions resident stewardship of land" in New York City; CLUSTER, www.clustercairo.org, which maps informal practices in Cairo to encourage alternative urban intervention strategies; and Visualizing Palestine, www.visualizingpalestine.org, which seeks to advance rights-based narratives.
10 Experiments are also taking place through social media platforms leading up to many support initiatives. On WhatsApp, groups are forming and sharing job opportunities. On Facebook, the LibanTROC platform listed more than 54,800 users in June 2020 and manages daily dozens of support requests—namely, through information exchange and barters, which are mostly solved within a day or two. Its founders have managed to gather enough resources to rent out a house in a periphery of Beirut, Choueifat, and operate it as a community center providing bedding and food for people in need, while helping them identify job opportunities.

Bibliography

Abi Yaghi, M. N., Catusse M. and Younes M.(2017) From *isqat an-nizam at-ta'ifi* to the Garbage Crisis movement: Political identities and antisectarian movements, in Di Peri, R. and Meier, D. (Eds.) *Lebanon Facing the Arab Uprisings*. Palgrave-Macmillan: London, 73–91.

Aburish, Z. (2015) 'Garbage politics'. *Middle East Report*, 277.

Al-Harithy, H. (Ed.) (2010) *Lessons in Postwar Reconstruction: Case Studies from Lebanon in the Aftermath of the 2006 War*. Routledge: London

Cambanis, T. (2017) *People Power and its Limits. Lessons from Lebanon's Anti-Sectarian Movement*. New York and Washington DC: The Century Foundation.

Clark, J. A. and Zahar, M. J. (2015). 'Critical junctures and missed opportunities: The case of Lebanon's Cedar Revolution'. *Ethnopolitics*, 14(1): 1–18.

Espiner, S. (2018) 'Daring alternatives: Social theory and open-ended utopianisms'. *New Zealand Sociology*, 33(1): 5–27.

Fawaz, M., Gharbieh A., Harb M. and Salamé D. (Eds.) (2018) *Refugees as City-Makers*. AUB-IFI: Beirut.

Fawaz, M., Gharbieh A. and Salamé D. (Eds.) (2015) *Practices of the Public in Beirut*. AUB-IFI: Beirut.

Fawaz, M., Gharbieh A., Harb M. and Bekdache N. (Eds.) (2011) *Beirut wa sukkaniha: al-kharita al-amniyya [Beirut and Its Citizens: The Security Map]*. AUB: Beirut.

Hage, G. (2015) *Alter-politics: Critical Anthropology and the Radical Political Imagination*. Melbourne University Press: Melbourne.

Harb, M. (2016) Cities and political change: How urban activists in Beirut bred an urban social movement. *POWER2YOUTH Working Paper*, No 20, IAI: Rome.

Harb, M. (2010) *Le Hezbollah à Beyrouth (1985–2005): de la banlieue à la ville*. IFPO-Karthala: Paris.

Harb, M. (2018) 'New forms of youth activism in contested cities: The case of Beirut'. *International Spectator*, 53:2, 74–93.

Harb, M. (Ed.) (2019) *Activism in Lebanon: An Overview of Political, Feminist, LGBTQ and Urban-Rights Movements*. AUB-Asfari Institute: Beirut.

Harb, M. and Atallah S. (Eds.) (2015). *Local Governments and Public Goods: Assessing Decentralization Experiences in the Arab World*. LCPS: Beirut.

Kaika, M. and Swyngedouw, E. (2014). 'Radical urban political-ecological imaginaries. Planetary urbanization and politicized nature'. *Eurozine*, 14 March.

Kelley, R.D.G. (2002). *Freedom Dreams. The Black Radical Imagination*. Beacon Press: Boston.

Kingston, P.W.T. (2013) *Reproducing Sectarianism: Advocacy Networks and the Politics of Civil Society in Postwar Lebanon*. SUNY Press: New York.

Lefebvre, H. (1974) *La production de l'espace*. Economica: Paris.

Nicholls, W. and Uistermark, J. (2017) *Cities and Social Movements: Immigrant Rights Activism in the United States, France and the Netherlands*. Wiley Blackwell: London.

Pieterse, H. and Simone, A. M. (2017) *New Urban Worlds. Inhabiting Dissonant Times.* Polity: London.

Saksouk-Sasso, A. (2015) 'Making spaces for communal sovereignty: The story of Beirut's Dalieh'. *Arab Studies Journal*, 22(1): 296–319.

Salloukh, B. F. (2006) 'The limits of electoral engineering in divided societies: Elections in postwar Lebanon'. *Canadian Journal of Political Science*, 39(3): 635–655.

Verdeil, E. (2019) *L'émergence des études urbaines au Liban. Engagements critiques locaux et mondialisation des pratiques académiques.* Cities Are Back in Town Working Papers, no. 2. Sciences Po: Paris.

Wright, E. O. (2010) *Envisioning real utopias.* Verso: London.

39

LIVING IN THE CITY BEYOND HOUSING

Urbanism of the commons

Belen Desmaison

Why does the vast majority of humankind now prefer to live in cities? Harvey defines the city as "the perpetual production of the commons" (2012: 67). For him, the metropolis constitutes "a vast common produced by the collective labor expended on and in the city" (ibid.: 78). Monbiot defines *the commons* as "a particular resource; a particular community that manages that resource; and the rules and negotiations the community develops to manage it" (2019). Harvey adds to that definition two important characteristics: the social relation built around managing resources is both unstable and malleable and – when referring to urban commons – 'resource' refers to the "actually existing or yet-to-be-created social and/or physical environment deemed crucial [for the community's] life and livelihood" (2012: 73).

Building from the definition of urbanism as the study of the physical needs of urban societies, I would like to highlight the importance of those needs and activities that are necessarily communal and cannot be reproduced and managed at the individual or small-scale local level and which sustain dwelling in the city. With the rise of capitalism and the accompanying commodification of space at a global scale, urbanism (particularly in its practical dimension) has prioritised the market value of both land and housing. This text defends the need to revalue the social and communal values of the urban, pushing a necessary shift in global urbanism towards a twenty-first-century urbanism of the commons and of shared public space and infrastructure. To illustrate the argument, I will explore examples brought from literature and from practices observed in Latin America, particularly in Peru, that portray alternative ways of understanding the commons which propel transformative practices through the co-production and communal management of social infrastructures.

The most important *value* of a city does not come from investments in private properties but from the communal socio-spatial resources it offers. The study of the commons, of redistribution and of care, requires expanding the notion of dwelling in the city by researching and building theories on urban everyday experiences outside the household and the physical resources needed to sustain those experiences. Equally important, urbanism can give rise to the collaborative design of practices and strategies to transform both the social and the physical dimensions of the urban towards more socially just and environmentally sustainable cities. For this collaboration to occur, urbanism entails greater participation of everyday citizens and civil society and a reconfiguration

DOI: 10.4324/9780429259593-44

of current relationships and governance practices between the state (in its multiple levels), citizens and academia in producing and sustaining the commons. Global urbanism in the twenty-first century can be, as Muxi puts it, a tool towards fairer distribution and of care, by being public, communal, open, consensual and collaborative (Liba, 2020). An urbanism that listens to both the territory and the people and one that not only theorises and studies the urban but that also provides novel methodological approximations to transform existing realities.

Neoliberalisation urbanism: Individualisation limits dwelling and social mobilisation

Currently, urbanisation processes are dominated by one political-economic trend influencing urbanisation processes across the planet: neoliberalisation and three of its most fundamental and interrelated consequences on urban centres: privatisation, commodification and individualism. In this scenario, urban theory and state-led policies in Latin America have largely focused on the residential dimension of the urban, explored in housing policies, land markets and self-built neighbourhoods (Gilbert, 2014). Following global trends, finance took over housing provision in Latin America, seeing citizens as consumers in capital markets. This approach to housing, which focuses on the provision of either a built unit or a land plot for a single family, is divorced from other dimensions of the urban, including the commons, offering a simplistic and mechanistic view of what dwelling in a city really means.

Neoliberalisation is understood by Roy and AlSayyad as "an ideology and practice advocating privatization and austerity" and which leads to "an intense commodification of informal land and housing markets" (2004: 30). The austerity described is a retreat from the state and the promotion of greater 'freedom' of the market and civil society. This autonomy, Gilbert argues, has served to "fill the gap left by the demise of the fledgling welfare state" (2004: 114) in a time when "real estate has become more profitable and important than industry" (Madden and Marcuse, 2016: 47). The predominance of real estate as a source of capital accumulation has led to a conflict between "housing as a lived, social space and housing as an instrument for profit making" (ibid.: 19).

In the Global South, the process of commodification of dwelling and its divorce from other dimensions of inhabiting the city is more predominant in the residential land market than in housing, mainly due to the inability of the state and the market to produce affordable housing for the poorest citizens (Fernández-Maldonado, 2014). This situation is accentuated by regularisation policies that provide property titles to (often illegal) land occupations (Calderón, 2016). For instance, in Peru, public land is occupied by neighbourhood associations to obtain property titles. Once land occupation is recognised, the owners of those sites gain the right to access basic urban infrastructure (water, sanitation, electricity). Thus, more than a site-and-services scheme, the twenty-first-century land market in Peru works as a site-*then*-services process.

Breaking through fragmentation and the local scale

Ábramo (2012) describes *informal human settlements*[1] in Latin America as a result from an intersection between unmet housing needs, speculation and the "promise of social mobility, as they become the material embodiments of notions of progress" (Caldeira, 2016: 6). On the latter, Peruvian *barriadas* have become a form of investment for future rentability for citizens or informal urbanisers that occupy land without necessarily residing there (Figure 39.1). They operate under the premise that the value of that land will increase with the future arrival of basic services. This results in the loss of the social and cultural significance of *barriadas* as they become sites

FIGURE 39.1 Recently occupied land with empty shacks, showing squatters do not reside there. Writing on the image on the right reads, *Si Vive* (Someone lives here) (source: Belen Desmaison, 2019)

for individual profit (Calderón, 2016: 24) where large-scale mafias occupy land as a speculative activity. Hence, both the housing and the land markets have in common that their fundamental function for dwelling and residency is obscured by their real estate value, even if this value is yet a future promise. The speculative and profit-seeking dynamics observed on a broader, global scale in phantom urbanism practices[2] are emulated on a smaller, local scale through land trafficking that unveils an interdependency between political leaders and the urban poor.

Public policies for *barriadas* are divorced from other dimensions of the urban, prioritising legalisation of individual properties while omitting broader upgrading programmes, plans, and strategies that focus on the urban commons and shared spaces. Exalting the real estate value of housing and land at the individual household level weakens the development of social capital and mobilisation. Social mobilisation was (and still is) geared towards obtaining land titles and basic services and, once those demands are met, communal organisation and mobilisation is greatly debilitated (Hordijk, 2010). Furthermore, in Latin America, the emphasis on land titling and recognition at the individual level results from a "political strategy consisting in protection [and legal recognition of] the invasion in exchange for poor people's support" (Castells, 1983: 191). Thus the absence of policies and programmes for broader neighbourhood upgrading that reinforces the precariousness of a sustained social mobilisation is partly due to a lack of political will as they are considered to be less efficient methods to gain popular approval in comparison to the provision of land titles.

Additionally, in large metropolises like Lima, social organisation is becoming increasingly fragmented into smaller units as the remaining unurbanised land is more difficult to access due to the natural topographic conditions of the hillsides and because of the different timings in which these associations occupied land (Barreda and Ramírez Corzo, 2004). This is evidenced in how different but geographically close units lack familiarity with each other and often compete between themselves in their struggle to obtain legal recognition and basic urban services. In this context, the unsustainability of current individualisation practices in city-making becomes evident, pushing towards a reconfiguration of global urbanism, which currently reinforces patterns of multidimensional fragmentation of the urban. Can a shift towards the commons decrease this incremental fragmentation[3]?

Towards an urbanism of the commons

During the second half of the twentieth century, Peru's *barriadas* became the focus of two theoretical schools that reverberated across the globe: the importance of self-build housing as a process for the development of social capital and mobility (Turner, 1976) and the regularisation of land occupation as a means for the poorest sectors of the population to enter the real estate market as property titles signified access to economic capital (de Soto, 1986). They both portray informality – of housing on one side and economic on the other – not as a problem but as a solution produced by citizens themselves against the failure of the state and the market in meeting their demands. Informal urbanisation processes that took place mainly in Lima during the '60s, '70s and '80s nowadays show greater levels of urban consolidation and access to services (Hordijk, 2010). These sectors of the metropolis have become its middle class and exemplify the symbolic image of the self-made, hard-working citizen (Caldeira, 2016). However, as previously mentioned, twenty-first-century urban expansion is happening in harder to reach areas, greatly limiting the capacity of recent land occupations to consolidate and to procure access to services and infrastructure (Ábramo, 2012).

Both Turner's and de Soto's vision glorify the entrepreneurship, perseverance and capacity at the individual household level or, at the upmost, at the local neighbourhood level. Both authors emphasise the dwelling unit and neighbourhood associations as platforms for social and economic development. Turner's theory on self-built housing and how it better adapts to people's needs and promotes greater capacity than already-made, state-provided housing sheds little light into how these benefits could be articulated into a broader vision of the urban commons at a district or metropolitan level. This gap is found not only in terms of the spatial design and management of public spaces and services but also in terms of socio-political governance and other aspects of the urban everyday experience that extend beyond the individual household. As Harvey comments, "the question of how to manage the commons at large as opposed to small and local scales (…) tends to be evaded" (2012: 69). He further elaborates on how the common resources "required for urban populations to function effectively, such as water provision, transportation, sewage disposal, and open space for recreation, have to be provided at the metropolitan, regional scale" (ibid.: 80) Thus the challenge for twenty-first-century urbanism is how to overcome the limitations of producing and managing the commons at larger metropolitan, national and even global scales, as well as on how to sustain social organisation beyond the procurement of land titles and basic services. Cities are more than the sum of several 'local' commons. The resources to be managed require their governance to be approached in a systemic, interconnected manner so that the responsibilities and benefits of managing those resources are distributed in a way that is just and equitable.

In land occupations in the peripheries of Lima, neighbourhood associations organise themselves in *faenas* (workdays) for the construction and maintenance of public commons such as stairs, roads, football courts and retention walls (Hordijk, 2010: 6). The COVID-19 pandemic brought light to the potential to self-organise towards mutual support and care, as shown by the spontaneous appearance of community kitchens in Lima and other Latin American cities. Despite illustrating the potential of community union and solidarity, these interventions remain at the micro-scale partly due to the prevalence of fragmented and small-scale neighbourhood associations with limited negotiating power and resources. Additionally, these projects are not integrated with wider city-scale planning and increase risk exposure by giving access to previously unreachable highly vulnerable soil (Espinoza and Fort, 2017). Thus local processes of production of communal spaces are not currently capable of tackling the absence of quality urban services and spaces that promote capacity-building for civil society. Although the activities carried out help to strengthen social organisation at the local level, they offer a limited vision of what the urban has to offer and on how to promote a broader and sustained communal mobilisation.

Hence, there is a well-studied trajectory of solidarity and communal work which represents the base for the collective management of the commons. The challenge for global urbanism lays in the articulation and integration of these efforts into wider and more complex territorial management and in the procurement of more horizontal deliberation platforms that sustain social organisation and participation over time. Thus, our understanding of the commons needs to be expanded. They can range from the neighbourhood scale as in, for example, community centres that offer food and care facilities at the local level, to a more representative and democratic management of broader geographical resources, such as valleys and streams, or city-wide systems, such as public transport and health. Moreover, the understanding of the urban commons as sites of social encounter and exchange can be widened to address and integrate economic and productive activities that take place in shared spaces, better adapting to the commons of self-built urbanisations that predominate across cities of the Global South (Habitat III, 2016).

One possibility to amplify and sustain the effort towards a governance of the commons could be to promote socio-political structures that strengthen and build upon already-existing communal organisations, capacities and economic-productive activities which are, in turn, integrated and articulated with urban design schemes that provide spaces for these interactions to occur (Figure 39.2). Thus how the commons are implemented and managed will require a reconfiguration which allows more direct participation and interconnectedness of a multiplicity of agents in decision-making processes. The current dichotomy of state and the market managing the urban can be reconfigured and amplified to include civil society and academia.

One of the main challenges for global urbanism in the twenty-first century lays in the reconfiguration of governance structures to be able to arrive at pertinent solutions and proposals at the different interconnected scales of the urban, particularly pertaining to communal spaces and infrastructure of encounter and exchange. While academia can provide studies to render visible ongoing and potential practices and platforms, as well as design new approaches and strategies to govern and implement systems of communal spaces, new coalitions can be formed between citizens and government units of different scales to put them into practice. Lastly, it becomes necessary to transform current policies as to protect the commons from appropriation and destruction from private interests (Harvey, 2012: 80). On this, there is a decades-long tradition in Latin America on participatory action research (Freire, 1970; Fals Borda and Rahman, 1991) that represents a solid base from which the management of broader and more complex commons at the city-wide and territorial scales can be re-envisioned, prioritising the voices and needs of citizens over market interests.

The vital spatial and material aspects of the commons to be emphasised and worked upon in urbanism at scales beyond the local are the networks of spaces, infrastructure and services which are necessarily provided and managed in a communal manner by an array of urban agents. In short, the spaces that highlight that inhabiting the city actually occurs outside the household unit and those spaces which have been left behind by neoliberalisation urbanism.

Klinenberg defines *social infrastructure* as the physical places and organisations that "shape the way people interact" and that "determine whether social capital develops" (2018: 5). The author, however, tends to focus on the social and cultural spaces and buildings, while seeing basic urban services or hard infrastructure as "designed to promote efficient circulation of people or vital resources" (ibid.: 20). This vision enhances social atomisation by not recognising these services as platforms of possible encounter and interaction if designed and implemented to serve those purposes, as well as how the design quality of these spaces and services also impacts the potential to develop social capital and capacities. For example, the design of disaster risk reduction infrastructure that incorporates the provision of social spaces as in the case of contention walls and tree plantations geared towards reducing soil erosion which simultaneously function as lineal parks, community gardens and public terraces.

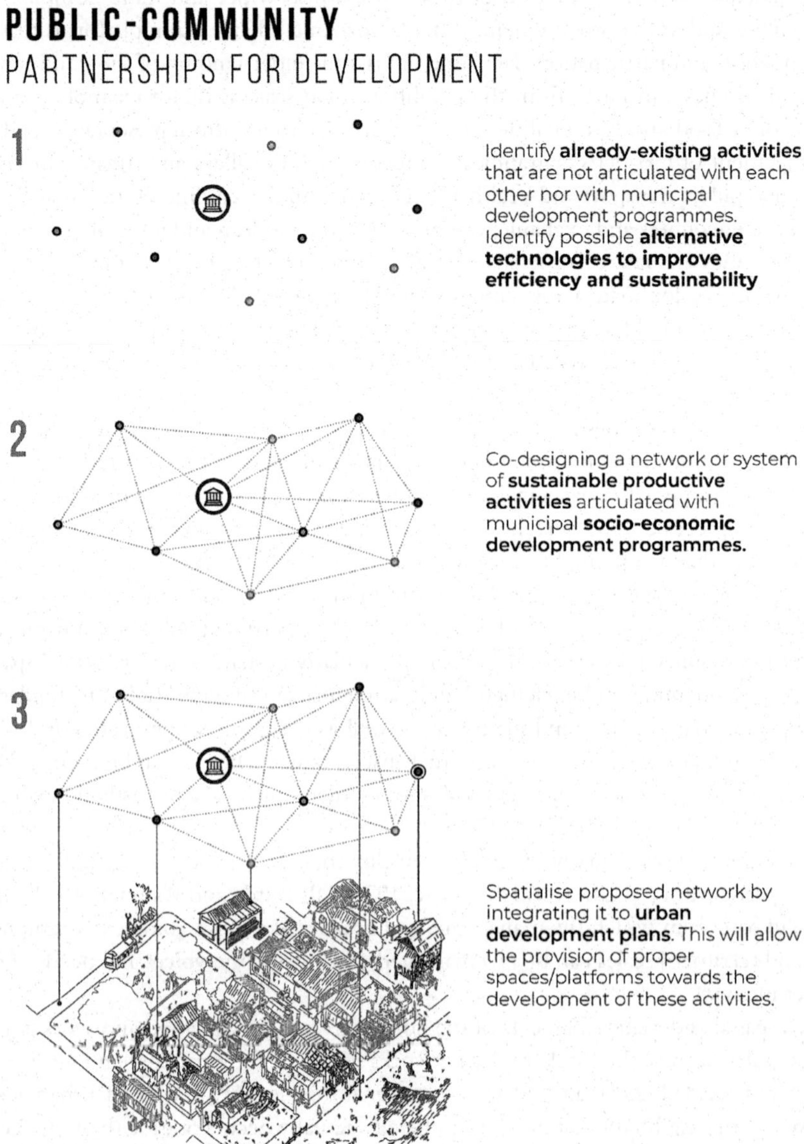

FIGURE 39.2 Recognising already-existing, economic-productive activities, articulating them with each other and with socio-economic development programmes and policies and integrating this network with urban design (source: Desmaison, 2019: 265)

Another body of work explores the linkages and interdependency between the social and physical – the soft and the hardware dimensions of the urban – as one is needed for the flourishing of the other (Graham and McFarlane, 2014). Amin and Thrift examine not only the social dimension of urban infrastructure but also its direct influence on local and economic networks embedded in urban flows (2017). An emphasis on the economic-productive aspect of the urban commons can enhance their reproduction, sustainability, and maintenance, while broadening the scale of operation beyond currently atomised social organisations. The emphasis on economic-productive activities does not disdain cultural or social aspects; rather, it seeks to expand current understandings of social infrastructure, as spaces where these activities occur are also sites of exchange and for building a sense of community. For instance, community centres can be managed by neighbourhood associations, promoting access to income to those that manage them and providing free time to users by reducing the time allocated for individual caring responsibilities, while community gardens can improve access to food and reduce expenses. The opportunity to design, implement and manage networks of interconnected communal socio- environmental infrastructures emerges. In an abundance of theories and debates about infrastructure, the challenge for global urbanism becomes bridging the gap between theory and practice, and between academia, policymakers and citizens, by delineating pathways to operationalise theory and linking it with practice through concrete transformations of the built environment and how it is governed and managed.

Following Turner's emphasis on the processes of design and implementation as fundamental for capacity building, the shared construction of the urban commons (understood as a socio-spatial process) could serve as a platform for a renewed dialogue and consolidation of trust between citizens and their democratic representatives. Under this premise, a renewed understanding of urbanism arises, promoting the co-construction – not of the 'usual' suspects of communal design (libraries, schools, cultural centres) in an isolated manner – but of strategies for strengthening already-existing networks of daily activities and livelihoods (Figure 39.3) and the links between these everyday activities (citizens), development programmes and policies (local governments) and urban infrastructure networks (the national government). These strategies seek to not only study and build theory around urban everyday experiences but also to present urbanism as a field in which practitioners directly participate in the processes of co-production of the urban commons and its processes while taking into consideration environmental sustainability and social resilience (Figure 39.4).

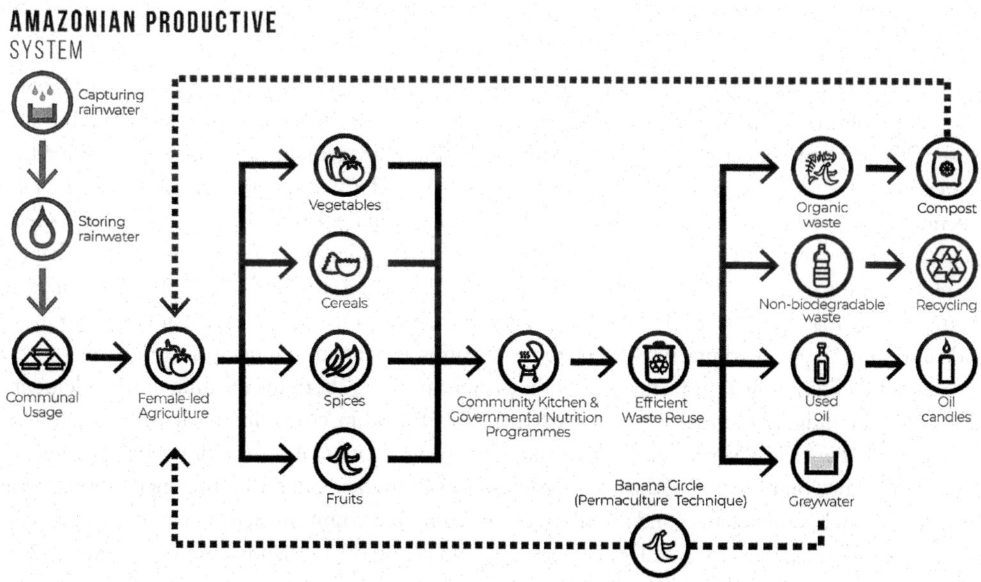

FIGURE 39.3 System of economic-productive activities and renewable energies to supplement these activities (Source: produced by undergraduate students of the Faculty of Architecture and Urbanism of Pontifical Catholic University of Peru. Teachers: Desmaison, B., Espinoza, K. and Văsquez, U., 2017)

FIGURE 39.4 Co-construction of a communal laundry and showers. Students and researchers from Pontifical Catholic University of Peru and the National University of the Peruvian Amazonia, volunteers, the community of Santo Tomás, the Municipality of Maynas and the Municipality of San Juan Bautista (source: Jorge Soria, 2019)

This text seeks to highlight social infrastructures as emerging forces in the discourses and practices of global urbanism rather than being invisible elements operating at its margins (McFarlane and Silver, 2017). Their collective design, implementation and management offer opportunities for the materialisation of novel social pacts that encourage, solidify and sustain communal organisation while opening new possibilities for more collaborative and horizontal relationships between civil society, public representatives and academia. I build upon the call to understand how everyday life is shaped by global power relations and forces by pushing forward the need to actively and collectively engage in the transformation of social infrastructures and their governance as practices that challenge individualised and fragmented practices currently dominating urbanisation processes worldwide.

Sassen highlights how transformations within a city can bring about the creation of new elements that could reconfigure global governance systems and citizenship, especially in times in which "the growing numbers and diversity of the disadvantaged in these cities takes on a distinctive 'presence'" (2014: 267). The consolidation of this growing 'presence', brought about through strengthening global networks of successful practices and traditions of managing the commons across cities and urban scenarios, provides platforms for a collective reimagination of the urban (Sendra and Fitzpatrick, 2020). Global urbanism, both as a theoretical and practical endeavour, is capable of interconnecting a multiplicity of actors and processes operating across geographies, to learn from each other and build translocal alliances to mobilise, share common struggles and to claim their rights to 'the cities', pushing towards a reconfiguration of governance structures of the commons at the metropolitan, national and international levels and actively engaging in their production and maintenance.

Acknowledgements

This chapter was written with the support of the research project KNOW: Knowledge in Action for Urban Equality, funded by the Economic and Social Research Council under the Global Challenges Research Fund (GCRF), project number: ES/P011225/1. It also portrays results of the research project CASA [Ciudades Auto-Sotenibles Amazónicas], funded by the International Development Research Centre, the Climate and Development Knowledge Network and Fundación Futuro Latino Americano (first phase) and by the Pontifical Catholic University of Peru (PUCP) (second phase).

Notes

1 The difficulties of naming these settlements cannot be fully analysed and developed in this article. For the limitations of the dichotomy 'formal/informal', see Gilbert (2004) and Roy and AlSayyad (2004). This chapter refers to those settlements that were occupied before basic urban services (electricity, water and sanitation) were available through the construction of self-built housing, albeit recognising the limitations of that short definition.

2 'Phantom urbanism' describes the rapid appearance across the globe of large-scale housing projects that remain mostly unoccupied because units were bought by wealthy people who rarely reside there or because their prices are too high to be purchased by poorer sectors of the population (de Graaf, 2018).

3 Fragmentation of the social as seen by weaker and smaller neighbourhood associations, a fragmentation of policies which focus on housing and land units divorced from other urban and common spaces and a fragmentation of the territory itself into increasingly smaller neighbourhood units that are not connected to each other.

References

Ábramo, P. (2012) 'La ciudad com-fusa: Mercado y producción de la estructura urbana en las grandes metrópolis latinoamericanas'. *EURE*, 381(114): 35–69.

Amin, A. and Thirft, N. (2017) *Seeing Like a City*. Cambridge: Polity Press.

Barreda, J. and Ramírez Corzo, D. (2004) Lima: Consolidación y expansión de una ciudad popular, in DESCO (Ed.), *Las ciudades en el Perú*, pp. 199–218. [Online Document] Available at http://www.desco.org.pe/ (Accessed on 19 April 2020).

Caldeira, T. (2016) 'Peripheral urbanization: Autoconstruction, transversal logics, and politics in cities of the global south'. *Environment and Planning D: Society and Space*, 35(1): 3–20. doi:10.1177/0263775816658479.

Calderón, J. (2016) *La Ciudad Ilegal: Lima en el Siglo XX*. Lima: Punto Cardinal.

Castells, M. (1983) *The City and the Grassroots: A Cross-Cultural Theory of Urban Social Movements*. London: Edward Arnold.

de Graaf, R. (2018) *Phantom Urbanism* [Lecture given at Harvard GSD] Available at https://www.youtube.com/watch?v=cLaY1to93Lw (Accessed on 30 July 2019).

de Soto, H. (1986) *El Otro Sendero: La Revolución Informal*. Lima: Editorial Ausonia.

Desmaison, B. (Ed.) (2019) *CASA [Ciudades Auto-Sostenibles Amazónicas] | HOME [Self-Sustainable Amazonian Cities]*. Lima: Fondo Editorial PUCP. Available at www.casapucp.com/publicaciones/libro/ (Accessed on 12 December 19).

Espinoza, A. and, Fort, R. (2017) *Inversión sin Planificación: La calidad de la inversión pública en los barrios vulnerables de Lima*. Lima: GRADE.

Fals Borda, O. and Rahman, M.A. (1991) *Action and Knowledge: Breaking the Monopoly with Participatory Action-Research*. New York: The Apex Press.

Fernández-Maldonado, A. M. (2014) Incremental housing in Peru and the role of the social housing sector, in Bredenoord, J., van Lindert, P. and Smets, P. (Eds.), *Affordable Housing in the Urban Global South: Seeking sustainable solutions* (pp. 271–285). New York: Routledge.

Freire, P. (1970) *Pedagogy of the Oppressed.* New York: Herder & Herder.

Gilbert, A. (2004) Love in the time of enhanced capital flows: Reflections on the links between liberalization and informality, in Roy, A. and AlSayyad, N. (Eds.), *Urban Informality: Transnational Perspectives from the Middle East, Latin America, and South Asia* [ebook version]. Retrieved from amazon.com. New York: Lexington Books.

Gilbert, A. (2014) 'Free housing for the poor: An effective way to address poverty?' *Habitat International*, 41: 253–261.

Graham, S. and McFarlane, C. (Eds.) (2014) *Infrastructural Lives: Urban Infrastructure in Context.* New York: Routledge.

Habitat, III (2016) *Habitat III Issues Papers: No. 11 – Public Spaces* [Online Document] Available at http://habitat3.org/wp-content/uploads/Habitat-III-Issue-Paper-11_Public-Space-2.0.compressed.pdf (Accessed on 29 July 2019).

Harvey, D. (2012) *Rebel Cities: From the Right to the City to the Urban Revolution.* London: Verso.

Hordijk, M. (2010) *Nuestra realidad es otra. Changing realities in Lima's peripheral settlement: A case study from San Juan de Miraflores.* Brussels: University of Amsterdam, NAERUS XI.

Klinenberg, E. (2018) *Palaces for the People: How Social Infrastructure Can Help Fight Inequality, Polarization, and the Decline of Civic Life.* New York: Crown.

Liba, A. (2020) 'El urbanismo puede ser una herramienta de redistribución y de cuidados'. *Pikasa: Online Magazine* [Online Article] Available at https://www.pikaramagazine.com/2020/01/urbanismo-puede-una-herramienta-redistribucion-cuidados/ (Accessed on 19 April 2020).

Madden, D. and Marcuse, P. (2016) *In Defense of Housing: The Politics of Crisis.* London: Verso [ebook version] Retrieved from verso.com.

McFarlane, C. and Silver, J. (2017) 'Navigating the city: Dialectics of everyday urbanism'. *Transactions of the Institute of British Geographers*, 42(3): 458–471. doi:10.1111/tran.12175

Monbiot, G. (2019) *The New Political Story That Could Change Everything*, in *TED Summit 2019.* [Online Video] Available at https://www.ted.com/talks/george_monbiot_the_new_political_story_that_could_change_everything/ (Accessed 1 August 2019).

Roy, A. and AlSayyad, N. (2004) Urban Informality: Crossing Borders, in Roy, A. and AlSayyad, N. (Eds.) *Urban Informality: Transnational Perspectives from the Middle East, Latin America, and South Asia* [ebook version]. Retrieved from amazon.com.

Sassen, S. (2014) Global cities, in Juergensmeyer, M. (Ed.) *Thinking Globally: A Global Studies Reader.* Berkeley: University of California Press, pp. 261–269.

Sendra, P. and Fitzpatrick, D. (2020) *Community-Led Regeneration: A Toolkit for Residents and Planners.* London: UCL Press. doi: 10.14324/111.9781787356061.

Turner, J. (1976). *Housing by People: Towards Autonomy in Building Environments.* Great Britain: Marion Boyars Publishers.

INDEX

Page numbers in '*Italic*' indicate Figures, and page numbers followed by 'n' indicate notes.

Ábramo, P. 334
Activist Graduate School 31
Adeniyi-Ogunyankin, Grace 116–120
Adey, P. 228
2020/2021 adjustment budget 179
advocacy 327, 328
Africa Is a Country 178
African economies 119
agnostic question 17
Ahlers, R. 259
Allen, J. 97
AlSayyad, N. 334, 341n1
Amarteifio, Nat Nuno 318
American Geographers' conference (AAG) 27
American University of Beirut (AUB) 327–329
Amin, Ash 6, 9, 17–24, 56, 163, 262, 338
Andueza, L. 263
Anglo-American hegemony 100
Anglo-North American theory 5
anglophone hegemony 66
Anthropocene 219, 220, 224, 225
anti-blackness 124
Anti Land Invasion Unit (ALIU) 178
anti-racist struggles 6
anti-slum campaigns 108
Arrevuoto 303
artistic residences 59
'assemblage of practices' 180
automated vehicle experimentation 244–245

Bab El-Falla market, Tunis: "*an addiction*"/"*a weekly sport*" 151; bale opening 148–152; buying bales blind 152; collective processes of valuation 152; consumption work 153; contingent global forms 149; cultural work of valuation 151; dense interactions 152; disclosure of bales 151; diverse processes of valuation 148; fripe materials 148, 150–152; fripe shopping 153; global form 153; global fripe circulations 152; modest dress 153; quotidian consumer politics 153; routinised probing and inspection 152; second-hand garments 148, 154n3, 154n4; second-hand Indian saris 152–153; second-hand objects 151; sheer heterogeneity 151; shock clients 153; socio-material density 151; tacit rules 150; thick layers of inscription 151; unparalleled creative energy 153; urban constellations 149; *Zahrouni* 150
Bădescu, Gruia 71–76
Baik, Y.-S. 67
Baloch Liberation Army 252, 256
Baloji, S. 174n8
barriadas 334–335
Barriera di Milano 56–60
Baudrillard, J. 167
bazaar, multifaceted commercial system 127
Becker, E. 167
The Bed 146
Bégout, B. 160
Beirut Madinati (BM) 328–329, 330n5
Benjamin, Solomon 284–289
Berlant, L. 126
beyond-the-network urbanism: heterogeneous infrastructural configurations 260; inequality and transformation 263; infrastructural power 263; infrastructural violence 264; non-state actors 262–263; political society 263; politics of anticipation 263; racialized/gendered bodies 263–264; scale and domain 263–264; services

259–260; state actors 262; urban fractures and fragments 262
Bhan, G. 176–178
Bhatt, Ela 294
Biocapital 231
bioeconomy 8
Biopolis 230, 231
black feminism 223
'black hole' cities 65
Black Lives Matter 30
Black Marxism 29
black outdoors 125–126
Black Reconstruction 29
Borisa, Dhiren 7, 49–54
boundary-spanning organisations 244, 246
Braun, B. 248
Brazil Country of the Future 269
Brazilian agro-industrial hinterlands 43
Brenner, Neil 8, 26, 34–46
Brexit outcome 81, 84
Brickell, Katherine 131–138
brick, global vertical urbanisms of: bipolar 133; blood bricks 132–135; Cambodia's brick kilns 132; economic liberalisation 132; embrace spectacular 134; financial-urban conjuncture 134; inspirational buildings 133; land titling rules 132; lurching upwards 132; trans-urban imaginaries 134
Brown, Gavin 7, 49–54
Brundtland, G.H. 219
Bulkeley, Harriet 203–206

Cambanis, T. 331n7
Cambodian People's Party (CPP) 132
Cambodia's brick kilns 132
Camorra 299
"Campaign for the protection Dalieh," 328
Cape Town water crisis 177–178
capitalism *versus* socialism 71
caricatures: artists 196–197; designer 197; flesh-and-blood actors 195; hacker 198–199; sentinel 195–196; subversive bureaucrat 199–200
Castán Broto, Vanesa 7, 206, 219–225
Castells, M. 26
CATference 93
Caust, M. 199
Cavalcanti, Mariana 267–274
centralized infrastructural systems 262
Chad Basin 127
Chari, S. 72
Charlesworth, S.J. 83
Chatterjee, P. 263
Chattopadhyay, Swati 6, 235–241
Chelcea, Liviu 71–76
Chhabria, S. 238
Chicago Anti-Eviction Campaign 31
child labour 138n1
"Chilling Prospects," 212
China's economic reform 65
Chinese livestock hinterlands 43
Chinese urbanism 64
Chiodelli, Francesco 8, 183–188
chronic unemployment 300
Ciliwung Merdeka 111, 113
Cité du Fleuve, construction project 168, 169, *169*, 171
Cité Joseph Kabila Kin-Oasis 174n1
cities and urban politics 198
'citizenship rights' 285
The City Is for All 312
city planning authority 323
City Possible 244
City Resilience Index 245
Civic Centre for Engagement and Community Service 327–328
civic crowdfunding 245
climate change: action planning 206; climate leadership group 204; climate-related urban disasters 204; climate storm 204; community-based adaptation 205; global environmental issues 203; greenhouse gas emissions 203; imaginary of ubiquitous crisis 205; IPCC 204; landmark Kyoto Protocol 204; political imaginary 203; purposive interventions 206; root causes of vulnerability 208; smart grids and digital technologies 206; socio-technical configurations 207; transformative agenda 207, 208; urban climate politics 206; urban ecological security 205; urbanised climate agendas 205
climate security 216
climate urbanism 7
coalfields 82
cohousing 185, 189n3
Cold War 71, 91
Coleman, G. 198
collective life 177
Common Front for Housing Rights 306
communal spaces 336, 337
community spaces 321
comparative urbanism: comparative imagination 97; "genetic" grounds 99; new "geography of theory," 97; practical challenges of building knowledge 96; reformatted comparative urbanism 99; reformatted urban comparativism 100
comparativism 97
concentrated urbanization 35
concrete totality 98
contemporary conceptualizations 34
continuous labour of 'normalisation' 179
coronavirus disease 2019 (COVID-19) pandemic 194, 228, 232, 303
Coronavirus Rapid Mobile Survey 181n1
corridor urbanism: Belt and Road Initiative 253; China–Pakistan Economic Corridor

252; epistemology 254–255; geographies and experiences 254; Global Infrastructure Connectivity Initiative 254; global trade and investment 254; large-scale opposition 251; Port of Piraeus 251, *252*; research agenda 255–257; shifting techno-territories 253
'cosmopolitan science' 230–231
counter-dominant global urbanism: Beirut Urban Lab 329–330; Lebanon's yearnings 327–329; mobilization 325; political and economic change 329; protestors 325–326; spatial production and city-making adapt 330; Utopias and imaginaries 326–327
Coward, N. 89, 93
Cox, E. 84
craft and craftiness 198
crisis management 178
critical social scientist 220–221
critical urbanism 23
critical urban theory 26

Davidson, K. 206
Dawn 253
'Day Zero' 177, 178
The Death and Life of Great American Cities 167
De Boeck, Filip 166–174
debt-bonded labour 131, 138n1
decentering global urbanism: absorb and depoliticise critique 27; dispossession 29; ethical commitment 32; expulsion and banishment 31; LA Tenants Union 31; North-South relationalities 26; political economy of 27; radical social movements 26; socio-spatial differentiation 27
decentre geographies of knowledge production 176
de-centring global (southern) urbanism: neo-Marxian political-economic perspective 65–67; politics of knowledge production and praxis 67–68; and re-centring urban studies 64–65; theorising 62–63
de Certeau, M. 269, 270
deep Sheng 146
Delaney, S.R. 51
Deleuze, G. 99
Deleuze's "general theory of the crack," 164n3
Delgado, M. 158, 164n3
Department of Urban Planning (DPU) 279
depeasantization 41
depression 121n13
Derickson, K. 27
Desmaison, Belen 333–341
de Sousa Santos, B. 12
developmental projects 317–318
dialectical orientation 177
dialectical urbanism: Cape Town water crisis 177–178; collective life 177; land occupation and eviction 178–179; Marxian dialectical thought 176; multi-scalar relations of indebtedness 179–181; 'situated' or 'global' urbanism 177
'Diamond Island' 134
digital auto-portraiture 7
digitality 23
digital life 140
diseases of affluence 229
displacement in urban theory 120n3
DIY sensibility 198
drugs market 301–302
Duckworth, Hilary 171
Duckworth, Russell 171

Easterling, K. 22, 280
Eastern Bloc 71
Eastern European (EE) cities: Belgrade Waterfront Project 309; capitalist investment 307; EU internal and external peripheries 308; financial relations 307; household debt 308; housing financialisation 309–310; housing movement 312–313; identity urbanism 310; international activist groups 312; international, national and local scales 309; issues 313; material and immaterial flows 306–307; neo-colonial dystopian West 314; nested orientalism 307; networks 312; opportunities and challenges 308; privatisation policies 308; radical politics 311; resistance movements 311; Roma communities 312; social movements 311, 313; societal property 314; Trojan horses 309
eco-habitat 18
'ecology of fear' 277
Eddo-Lodge, R. 223
EGU *see* everyday global urbanism
electricity blackouts 122n15
Elias, N. 83
Emery, J. 83
English coalfield communities 84
Englishpur ki Kothi 53
Environmental Impact Assessments (EIAs) 136, 328
epistemological chauvinism 67
epistemological vigilance 67
Espiner, S. 327
ethnic feminism 304
Euro-American epistemologies 66
Euro-American heartlands 39
Euro-American universities 57
European Action Coalition for the Right to Housing and the City (EAC) 306, 312
European city 56
European Union (EU) 58, 307–308, 313
everyday global urbanism (EGU) 156, 157, 159, 162–164
exposures of itineraries 129–130
ex-Soviet Bloc 71
extended urbanization 35

extrastatecraft 167, 280
Ezzati, M. 232

Farrugia, D. 80, 81
Fawaz, M. 328
Fehérváry, K. 74
female spatial experiences, public motor transport 157, 159–162
feminism 22
feminist perspective on global urbanism 116–119
Ferenèuhová, Slavomíra 71–76
Ferulano, Emma 9, 298–305
financialisation of home 121n11
Fitzgerald, D. 229
five public works program 168
Flint, John 8, 80–85
Florea, Ioana 9
Florea, Loana 306–314
Flyvbjerg, B. 21
footnote urbanism: image campaign 88; new geographies of theory 89; Vilnius 88
"Fora da Ordem," 270–271
foreign immigrants 59
Frederick, R. 142
Free Basic Water (FBW) 178
free education 120n2
Friedman, Yona 237
Fungible Life 228, 230
"Funiculì Funiculà" song 298

Garrone, Matteo 299
Gayatri Spivak 5, 18
Geddes, P. 36
gender discrimination 300, 304
"generative" grounds 99
"genetic" grounds 99
Genge hip-hop 145
gentrification 76, 300–301, 317, 318
geopolitical articulation 66
ghetto greenroom 144
Gilbert, A. 334, 341n1
Global Commission on Adaptation 219
Global East 62, 63
global environmental assessments 219–222, 224, 225
global financialisation 119
global 'gay community' 49
'global gay' lifestyles 50
global HIV prevention funding mechanisms 52
globalised lesbian, gay, bisexual, transgender (LGBT) 49
globalizing capitalism 91
globalizing postsocialist urbanism: debate about postsocialist city 75–76; ghosts and zombies of socialism 74; intransigent materialities 75; (post) Cold War urbanisms 72–74; project of global urbanism 75–76
Global North 62
global pandemic 302
global pharmaceutical firms 231–232
global philanthropies and charities 244
global social movements 246
Global South 63
Global South/East urbanism 65
global-urban imaginations 8, 108, 109
global urbanism 71; counter-political 3; decentering *see* decentering global urbanism; diverse range of interpretations of 4; epistemic and geographical 4; inside/out *see* kampungs; multiple political grounds 6–9; provincializing 5–6
"global urban studies," 100
Goldman, M. 280
Gomes, E. 157
Gomorra 299
Goozner, M. 231
Gorman-Murray, A. 50
Goveia, Anselmo 279, 281
Governa, Francesca 56–60
Government Reservation Areas (GRAs) 122n16
Graham, S. 163, 261
Gramsci, A. 240
Grazioli, Margherita 8, 183–188
green capital 245
Grüneisl, Katharina 148–154
Guattari, F. 9
Gupta, A. 53
Gurney, K. 196

Hae, L. 63, 66–68
Hage, G. 326–327
Haila, A. 68
Hall, P. 89
Harb, Mona 325–331
Hariri, Rafic 330n5
Hariri, Saad 330n5
Harris Ali, S. 228
Hart, G. 176
Harvey, D. 26, 167, 173, 333, 336
Hawkwood Fund Management 171
health innovation: air pollution 228; capitalist development histories 229; chronic and non-communicable disease 229; contagion and disease containment 228; genetic and physiological traits 229; global bioeconomy 230–232; mental health 228–229; Social Determinants of Health 229–230
Healy, D. 231
'Heat Relief Network' (HRN) 213
hijras 52
hinterlands of Capitalocene: Brazilian agro-industrial hinterlands 43; Chinese livestock hinterlands 43; cities without hinterlands 36–37; city-building processes 39–41; coloniality of 40; concentrated urbanization 34; context-transcendent universalism 39; cycles of creative

destruction 45; enclosure and distanciation 41–43; epistemological reorientation 35; extended urbanization 35; formal to real subsumption 43; hinterland enigma 38–39; infrastructuralization 41–43; metabolic rifts 45; metabolic urbanization 37; political-economic and ecological transformations 34; problematique 36, 39; specialized and export-oriented hinterlands 43
Hodson, M. 204
homeowners associations 188, 189n2
homosexuality 52, 53
hook-up apps 51
Housing Justice Conference and Network 30
Hsu, A. 205
hunger virus 146
Hypnose building 168, *168*

'idem' footnote 92
illegal kampungs 108
immense labour and tenacity 196
in-between space 58, 60
industrial capitalism 236
informal cemeteries 170
informal human settlements 294–297, 334
informal/illicit economy 300
Information Communication Technology (ICT) services 142, 253
infrastructuralization 41–43
'innovation districts' 231
Institute for Economic Justice 180
institutional archipelago 187–188
institutional hacking 198
institutional micro-geographies 187
Integration Syndicate 200
intentional voids 186
Inter-American Development Bank 271
Intergovernmental Panel on Climate Change (IPCC) 194, 219
Intergovernmental Science-Policy Platform on Biodiversity and Ecosystem Services (IPBES) 219
inter-kampung inequalities 111
international environmental policy 222
International Monetary Fund 180
international scientific assessment 220
'inter-referenced urbanism' 131
intersectional analysis 223
Isa, Dolkun 256
Italian city: areal identification 56; ordinary city 57, 59–60; Southern European cities 57–58; Turin 58–59; worlding practices 57
Itani, Jamal 330n5

Jacobs, J. 167
Jakarta Legal Aid Institute 113
Jakarta metropolitan 107
Jakarta's kampungs slums 108
Jakarta's urban minority 109
Jamieson, William 131–138
Jasanoff, S. 220
Jennings, W. 82
Jensen, C.B. 22
Jhabvala, Renana 9, 293–297

Kampung Akuarium 112–113
Kampung Improvement Program in Indonesia 108
kampungs: agentic spaces of alterity 111–113; Jakarta's kampung residents 107; lived realities 108–111
kampung street life 110
Kampung Tongkol 112
Katsikis, Nikos 8, 34–48
Katyal, A. 53
Katz, C. 27
Keil, R. 228
Kelly, F. 19
'Kenyan slums' 144
Kern, S. 237
"Killer Heat in the United States," 213
Kinnar Akhara 52
Kinshasa-style coronavirus disease 2019 (COVID-19) 174n6, 174n7
Kirmayer, L.J. 229
Klinenberg, E. 337
Koolfog's misting systems 212, 214
kothi 52

LA Community Action Network 31
Lagendijk, A. 59
Lancione, Michele 3–12, 18, 25–32, 293–306, 317–324
Land Acquisition Act 252–253
'land mafia politics' 288
landmark Kyoto Protocol 204
land rights 109
Landry, C. 199
land tenures 288
Larry Aminu, Yussif 9, 317–324
LA Tenants Union 31
Lawhon, M. 262
Lecavalier, Emma 203–208
Le Corbusier 169, 237
Lefebvre, H. 98–99, 119, 120n1, 164n4, 327
Lefebvrian sense 159
left-wing radicalism 186
Leitner, Helga 5, 7, 90, 107–115
lesbian, gay, bisexual, transgender (LGBT) subjectivities 52
Levi Strauss, C. 268–270
Lie, J. 228–229
Life as Surplus 231
liminal spaces and resistance, Mexico city: biographical events 164; biographies of city's inhabitants 164; degree of stability 156; dense fragments of urban life 156; EGU 156, 157, 159,

162–164; ephemeral temporality 163; female spatial experiences, public motor transport 157, 159–162; harassment of women 157; liminal male spaces 157–159; lines of fire 164; micro-territories 164; routinisation 156; small-scale urban situations 163; socio-cultural contexts 157; spatial behaviours 157; spatial-temporal simultaneity 156
Lindón, Alicia 156–165
Li, T.M. 284
'Little Paris' 134
Locke, J. 237

MacKaye, B. 36
Mah, A. 83
Making Massive Small Change 197
Making Social Science Matter 21
male liminal spaces 158–159
Manuel-Navarrete, D. 205
marginalisation 51
Maringanti, A. 5, 90
Marvin, Simon 204, 211–218, 261
Marx, Karl 68
Massey, D. 9, 27, 67, 97, 98, 141
Mastercard 244
Mayne, A. 108
McCann, E. 56, 219
McElroy, E. 73
McFarlane, Colin 3–12, 18, 262, 293–298, 317–324
McGuirk, Pauline 7, 243–249
McKay, L. 84
McKittrick, K. 125
McNeill, Donald 7, 227–233
"Medellin model," 271
Mediterranean city model 56, 57
megaport 135
metabolic urbanization 37
methodological cityism 137
methodological nationalism 66
methodological statism 66
The Metro 158
micro-practices 161
micro-urban stories 19
middle class, affective intensity 129
migrations 318
Millington, N. 177
MOBA Housing Network 312
The Monster at Our Door 228
monumental architecture: 'disciplined imagination' 278, 281; 'dis-individuated' abstraction 278; extrastatecraft 280; Maputo city centre 279, 280; physical buildings and monuments 279; physical markers 277–278; property investment strategies 277; speculative global capitalism 278; Waterfall City 279; 'world-city making' 280; 'worlding' process 281
Moosavi, L. 67
Morita, A. 22
Morton, E. 138n7
Moscow's hyperdiversity 76
movement-based scholarship 30
Muabilu, Pius 170
Muhammad, G. 111
Müller, Martin 73, 74, 88–93
multiple knowledge systems 224–225
multiple/plural assemblages 244
Mumford, L. 36, 276
Murray, M.J. 167

Nahr al-Bared Palestinian refugee camp 328
Nairobeez 144
Nairobi city: Along Juja Road framing 142; black vernacular culture 141; cultural forms 141; digital modes of self-fashioning and storytelling 141; DIY music videos 142; ethical and personal responsibility 140; ethnographic engagement 140; European hip-hop 142; *Genge* 141, 142; ghetto life 142; Ghetto University 143; global urbanism 141; Google 143; hip-hop soundtrack 140; hustle economies 142; hybrid cultures 141; ICT services 142; intercultural and transnational formation 141; *kuzurura* 143; leapfrog urbanism 143; low-income neighbourhoods 140; low-income youth 140; Mashanti 144–146; *matatus* 142, 143; material poverty 143; *muziki wa vijana* 141; pay-as-you-go phone credit 143; postcolonial scholarship 141; *Sheng* 141; 'slum' neighbourhoods 142; social innovation experiments 142; spatial openness 141; Straight Outta Dumpsite 143; 'telling young lives' 141; urban ethnography 140; youth entrepreneurs 142; youth music 141
Nash, C. J. 50
Natarajan, Nithya 131–138
National Consumption Institute of Tunisia 154n6
National Integration Strategy for RSC 299, 303
National Weather Service Heat Index 213
necrological notes, Kinshasa: capitalism 167; cemetery city 169–171; *The Death and Life of Great American Cities* 167; displaced urbanism 167–169, *168*, *169*; financial technologies 167; gift-giving 166; hedging 166; house ownership 166; landscapes of accumulation 167; Necropolis cemetery 171–173, *172*, *173*; thanatopolitics 167; urbanism of displacement 167–169, *168*, *169*; urbanisms of exception 167
Necropolis cemetery 171–173, *172*, *173*
neoliberal city 167
neoliberalisation 334
neoliberal managerialism 199
neoliberal urban projects 120
New Urban Agenda (2016) 3
The New Urban Crisis 231

New World cities: ethnographic motif 271–274; global south 269–271; monumental future 268–269
Ngobila, Gentiny 170
NGO Global Witness 136
Nicholls, W. 326
Nielsen, Morten 276–281
non-governmental organisation (NGO) 244, 323–324
non-Western single site 64
northern-centric politics 100
northern-centric vision of urbanisation 176
Northern England's 'left behind' urbanism: habitus of industrial ruination 82–84; historical complexity 81; reconfigured imaginaries 81, 84–85; revitalized concept of habitus 81; socio-spatial positioning 81
Northern Mind 84
northern urban studies 101
North-South binaries 62, 63

obsolescence 236–237
"official" urban life 125
Okoye, Victoria 317–324
Old World cities 268–269
Ollman, B. 176
Olympics-related infrastructures 272, 273
One Belt One Road 8
One Park: A New World within the City 138n6
Ong, A. 56, 134, 230, 277
'open-source urbanism' 235
Operation Burst 122n18
Operation Kin Bopeto 170
ordinary city 57, 59–60, 62
Orwell, G. 89
Osborne, G. 82
Os Sertões 269
Oswin, N. 5, 50
Otto, Frei 237
"out of body" experiences 127
"out there": anti-blackness 124; auto-construction 128; black outdoors 125–126; destination of abolition 124; exposure to time 126–127; extended urbanization 124; forward momentum 128; notion of exposure 127; self-constructed urbanization 128; unanticipated excesses and trajectories 125; urban modernity 125
over-emphasized homogeneity 71

Pakeha identity 53
Palermo Protocol 138n2
Paris Agreement 205
Parnell, S. 20
Parsons, Laurie 131–138
Peake, Linda 116–120
The Peak 134, 138n3
Pedersen, D. 229
The Penthouse Residence 134, 138n4
'perpetual hereditary tenancy' system 287
Peteet, J. 239
Petrova, S. 251
Pharmocracy 231
Pieterse, Edgar 193–200
Pieterse, H. 327
planetarity 5, 18
planetary urbanisation 50, 98
plantation life 125
policy maker 222–223
policy-making audience 222–224
policy tourism 246
political-ecological security 46
political-ecological vulnerabilities 46
political imaginations 194
political potentialities: caricatures 195–200; cities and urban politics 198; COVID-19 pandemic 194; cultural materialisations 194; infrastructural and land-use transitions 195; IPCC 194; localisation 194; overlapping and interdependent urban operating systems 194; political imaginations and experiments 194; regionalisation 194; sustainable development goals 194; transformative priorities 195
political sectarianism 327, 331n6
political wall writing 239
Popocivi, Veda 9, 307–315
Pósfai, Zsuzsi 9, 307–315
(post)Cold War urbanisms 72–74
postcolonial and decolonial theory 71
postcolonial feminism 28
postcolonialism 22
post-displacement strategies 120n3
post-industrial urbanism 82
postsocialism 72, 74, 311
postsocialist city 71, 75–76
post-structuralism 22
Poverty Capital 32
Powell, Ryan 8, 80–85
Prison Notebooks 240
problematique 35, 36
process-based urbanism 22
Prodromidou, A. 251
prolific and homogenizing formatting of residency 124
public mobilisation 238–239
public park 328
public-private partnership (PPP) 309
public-sector jobs 299
public space 301, 305
public transport 299–300
Pulido, L. 29

'queerer' global urbanism: global urban gay 51–52; *hijras* 52; homosexuality 52; *kothi* 52; LGBT subjectivities 52; planetary urbanisation 50; sexual and gender minorities 49; sexual politics 50; sexual tolerance 50; *takatapui* 52; urban gay spaces 49; urban queer life 52

racial anxieties 129
racial capitalism 28
radical transformative politics 119
Rákóczi Collective 306
real economy 271
Reckien, D. 206
reformatted comparative urbanism 99
reformatted urban comparativism 100
relational theorisation 255
Releasing the Commons 17
renowned and innovative antagonist movements 300
'revolution of modernity' program 170
'the Revolution of Modernity' 168
rhythmanalysis 120n1
1992 Rio Declaration 219
Robinson, C. 29
Robinson, Jennifer 8, 26, 90, 91, 96–102, 259
Roma communities 302–303
Rose, N. 228–229
Rousseff, Dilma 272
Roy, Ananya 9, 25–32, 56, 64, 66, 67, 91, 134, 259, 270, 334, 341n1

sacrifice zones 43
Saksouk-Sasso, A. 330n4
Salloukh, B.F. 331n6
sand, global vertical urbanisms of 135–137
sanitary operation 170
Santiago, Raull 273
SARS 228
Sassen, S. 340
Save the Bisri Valley campaign 328
Scheba, Suraya 176–181
Schmid, C. 8, 26, 34
Schwartz, K. 262
Sciascia, Leonardo 299
Seeing Like a City 17
Self-Employed Women's Association (SEWA): community-based organisations 296; cooperative bank 293; entitlements and legal issues 295; 'Governing the Plural City' 293; home-based workers 294, 295; housing finance 297; informal economy 295–296; informal settlements 294–295, 297; infrastructure 295; large-scale demolition 296; solid waste management 295; street vendors 296, 297; urban land 295
self-urbanism 8; definition 188; descriptive and normative differentiation 183; development and management of space and buildings 184; empirical manifestations 184; ground-conceptualisation 188; 'homeowners' associations' 185; and internal coordination 185; and market 185; neo-liberalisation 184; neo-localism 183; new *institutional archipelago* 187–188; political and normative implications 188; and public regulation 185–186; radical outcomes 188; radical urban autonomy 183; regulatory crisis 186–188; self-coordination 185; self-governance 185; subaltern urbanism 183; urban citizenship 187; urban enclavism 183; urban revanchism 183; urban sphere 184
sense of provisionality 129
SEWA *see* Self-Employed Women's Association
sexual and gender minorities 49
sexual politics 50
sexual tolerance 50
Sheng expression 144
Sheppard, Eric 5, 7, 90, 91, 107–115
Shin, Hyun Bang 62–68
shipbuilding communities 82
Shoki, W. 178
Silver, Jonathan 251–257, 262
Simone, AbdouMaliq 99, 111, 122n19, 124–130, 261, 327
Singapore's development and land reclamation 131
situated histories: bureaucracy and planning 289; East Delhi's Viswas Nagar (VN) 288–289; Hong Kong 287–288; Kowloon 288; land tenurial practices 285; philosophy of internal relations 284–285; political shift 285–286; Sham Shui Po (SSP) 288; time-space contexts 284; *tongbian* philosophy 286–287
Smith, J.P. 179
social and informal interventions 303
Social Determinants of Health 229–230
social infrastructure 337, 339–340
social investment bonds 245
socialism 74, 90
socialist cities 91
social media campaigns 246
social media platforms 331n10
social pressure 121n5
social stigma 51
social violence 193
Society and Space 176
socio-spatial configurations 83
socio-spatial differentiation 27
sociospatial inequality 120
Solnit, R. 180
solutions-oriented assessments 222
Song, J. 63, 66–68
South African capitalism 179
"southern" urban theory 99
South-to-South mobility 63
Soviet Bloc 71
spatial autonomy 188
'spatial dimensionality' 278
spatial imaginaries: crisscrossing economic circulations 10; cultural layering 10; everyday

life 11; multiple political grounds 6–9; new spatial experiments 10; resonance 11; societal structuring 10
spatial organisation and design 238
spatial orientation 160, 161
spatial production 326, 330
specialized and export-oriented hinterlands 43
'speculative urbanism' 280, 285
Spillers, H. 126
Spread-Out Initiative (SOI) 321–324
state 'beautification' schemes 121n12
state-sponsored events 238–239
static models of homogenous infrastructure 221
steel towns 82
Stiegler, B. 124
"Stop the Highway" campaign 328
Story Yetu 140, 144
Straight Outta Compton 143
subversive bureaucrat 199–200
supplychain capitalism 46
surplus labor 41
sustainable development goals (SDGs) 3, 20, 194
'Sustainable Energy for All' (SEA) 212
Sydney City Council 216

takatapui 52, 53
Tang, Wing-Shing 284–289
'technological zones' 255
Teleférico do Alemão 272–273
territorial trap 66
thanatopolitics 7, 167
thermal metabolics: climate modification 216–217, *217*; comfort and convenience 211; digital weather monitoring and control 211; life-threatening heatwaves 211–212; outside cooling and heating 213–215, *214*, *215*; overheating and urban thermal metabolism 211–212; perfect weather 211; 'ratcheting up' effect 212; thermal socio-technical infrastructures 211; urban thermal politics 212–213
Thieme, Tatiana 140–146
Thinking Global Urbanism 6
thinking urban grammars: on policy, practice, and publics 19–21; politics of emergent 22–23; styles of 23–24
Thrift, N. 56, 262, 338
Thunberg, Greta 220
time-limited concept-metaphor 74
Tiwa's morning 116–119
tongbian philosophy 286–287
Tôrres, A. 157
touristification 300–301
Tozer, Laura 203–208
Tristes Tropiques 268–270
The Triumph of the City 231
Truelove, Yaffa 259–265
Tshisekedi, Felix 170
Tucker, A. 52
Turin 58–59
Tuvikene, T. 73

ubiquitous construction materials and practices 136
Uistermark, J. 326
UNAR 303
UN-Habitat's World Cities Report 219
'Union of Concerned Scientists' (UCU) 213
unitary conceptualization 8
United Nations Environmental Programme's International Resource Panel 219
'United Nations Transitional Authority in Cambodia' (UNTAC) 132, 133
Universiadi 2019 298
university-based scholarship 30
UN Millennial Development Goals 108
urban activism 327–328
urban citizenship 187
urban code hacking: ephemeral architecture and 235; impermanence 238–239; permanence and durability 236–238; planning 238–239; politics of surprise 239–241
urban commons: co-construction *339*, *340*, 339–340; definition 333; economic-productive aspect 338–339; fragmentation and local scale 334–335, *335*; governance structures 337; land occupation 336; neoliberalisation 334; self-build housing 335–336; social and cultural spaces 337; socio-political structures 337, *338*; solidarity and communal work 336–337
urban gay spaces 49
urban governance innovation: dispositif thinking 245–246; fluid categories 247–248; management 248; multiplicity 247; urban unbound 246–247; world of cities 244–245
Urban Heat Island (UHI) 212
urban informality 270
urban infrastructure models 221
urbanization of anti-communism 74
'urban models' 277
Urban Poor Consortium 113
urban queer life 52
urban question 11
urban renewal policies 121n8
urban scholarship 80, 176
urban snowmelt infrastructure 215, *215*
urban South Korea 64
urban world 17

Vanolo, A. 58
Veloso, Caetano 270, 272
Verdery, K. 72
Vickers, E. 50
Vilenica, Ana 9, 306–315
von Thünen, J.H. 36
von Thünen's classic model 43

'vote bank' 288
vulture funds 171

Wacquant, L. 81
Wakefield, S. 205
Ward, K. 56
waste infrastructure 264
"wave of verticals" 269–270
West Bank 239
Western Cape province 178
Western-centric critical theory 12
Western normative approaches 3
Western-style 'gaybourhoods' 49
Weston, K. 49
Whiteis, D.G. 229
White, J. 142
Wiig, A. 253
Winter Olympic Games 2006, 58
the working poor 116–119
World Bank 63, 271, 308, 309
worlding practices 60
Wright, E.O. 327
Wright, M. 125
Wu, F. 64
Wyly's, E. 80
Wynter, S. 125

XXI century metropolis 57

Yang, Y.-H. 51
Yiftachel, O. 62
yin/yang diagram 286–287
youth urban 318–319, 323–324

Zweig, S. 269